ADVANCES IN CIVIL ENGINEERING AND ENVIRONMENTAL ENGINEERING, VOLUME 1

Advances in Civil Engineering and Environmental Engineering focuses on the research of civil engineering and environmental engineering. The proceedings feature the most cutting-edge research directions and achievements related to civil engineering and environmental. Subjects in the proceedings include:

- Civil engineering technology
- Civil engineering surveying
- Geological engineering
- Structural engineering
- Tunnel and bridge engineering
- Environmental protection materials
- Pollution control project
- Building environment and equipment engineering

The works of this proceedings can promote development of civil engineering and environmental engineering, resource sharing, flexibility and high efficiency. Thereby, promote scientific information interchange between scholars from the top universities, research centers and high-tech enterprises working all around the world.

PROCEEDINGS OF THE 4TH INTERNATIONAL CONFERENCE ON CIVIL ENGINEERING AND ENVIRONMENTAL ENGINEERING (CEEE 2022), SHANGHAI, CHINA, 26–28 AUGUST 2022

Advances in Civil Engineering and Environmental Engineering

Volume 1

Edited by

Wira Jazair bin Yahya
Universiti Teknologi Malaysia, Malaysia

Zawawi Bin Daud
University Tun Hussein Onn Malaysia, Malaysia

CRC Press
Taylor & Francis Group
Boca Raton London New York Leiden

CRC Press is an imprint of the
Taylor & Francis Group, an **informa** business
A BALKEMA BOOK

First published 2023
by CRC Press/Balkema
4 Park Square, Milton Park, Abingdon, Oxon, OX14 4RN
e-mail: enquiries@taylorandfrancis.com
www.routledge.com – www.taylorandfrancis.com

CRC Press/Balkema is an imprint of the Taylor & Francis Group, an informa business

SET
ISBN: 978-1-032-46734-4 (hbk)
ISBN: 978-1-032-46735-1 (pbk)

Volume 1
ISBN: 978-1-032-39397-1 (hbk)
ISBN: 978-1-032-39402-2 (pbk)
ISBN: 978-1-003-34956-3 (ebk)

DOI: 10.1201/9781003349563

Volume 2
ISBN: 978-1-032-46732-0 (hbk)
ISBN: 978-1-032-46733-7 (pbk)
ISBN: 978-1-003-38303-1 (ebk)

DOI: 10.1201/9781003383031

Typeset in Times New Roman
by MPS Limited, Chennai, India

Table of contents

 ISBN: 978-1-032-39397-1

Preface

The 2022 4th International Conference on Civil Engineering and Environmental Engineering (CEEE 2022) was successfully held in Shanghai, China, August 26–28, 2022. Due to the current COVID-19 pandemic, it was held as a virtual conference. CEEE 2022 provided a platform for all the participants to share scientific research achievements and cutting-edge technologies, understand academic development trends, strengthen academic research and discussion, and it promoted industrial cooperation of academic achievements.

In the conference, we had the honor of having invited both Prof. Chong Kok Keong from Universiti Tunku Abdul Rahman, Malaysia and Prof. Zhongliang Liu from Beijing University of Technology, China to serve as our Conference Chairs. This online conference was composed of three parts: keynote speeches, oral presentations and online Q&A discussion, attracting 200 experts, scholars and businessmen around the world. Firstly, keynote speakers were each allocated 30–45 minutes to hold their speeches. Then in the oral presentations, the excellent papers we had selected were presented by their authors one by one.

At the beginning of the conference, two distinguished keynote speakers were invited to address their speeches. For a start, A. Prof. Wira Jazair bin Yahya from Universiti Teknologi Malaysia, Malaysia performed a thought-provoking speech on the title Real-Time Non-Surfactant Emulsion Fuel Supply System: Bridge to a Sustainable Environmental Protection Solution. He presented to us that Real-Time Non-Surfactant Emulsion Fuel Supply System (RTES) is a device that can save diesel fuel consumption and reduce exhaust emissions of diesel engines and burners. It is mainly applied in road transport, marine vessels, locomotives, electrical generation, and industrial heating equipment. In addition, A. Prof. Zawawi Bin Daud from University Tun Hussein Onn Malaysia, Malaysia shared with us his study: Municipal Solid Waste Management in Malaysia. He introduced that sanitary landfills in developing countries are faced with the problem of technology as well as financial limitations. If not dealt with properly, it will become worse, affecting the country's development in various facets. Therefore, finding a sustainable method for leachate treatment has always been a priority. Their dramatic speeches had triggered heated discussion in the third part of the conference. Besides, every participant praised this conference for disseminating useful and insightful knowledge.

After months of hard work and preparation, we have selected and compiled a number of high-quality papers into the proceedings after rigorously reviewing them. These papers feature but are not limited to the following topics: Civil Engineering Technology, Geological Engineering, Environmental Chemistry and Biology, Geographic Information and Remote Sensing Science, etc. All the papers have been checked through rigorous review and processes to meet the requirements of publication.

On behalf of the conference organizing committee, we would like to extend our sincere acknowledgement to all the keynote speakers, peer reviewers, and all the participants. Especially, our thanks go to the CRC Press/Balkema–Taylor & Francis Group, for the endeavour of all its colleagues in publishing this paper volume. We truly believe that all the attendees have had fruitful discussions and gained valuable knowledge and will enjoy the opportunity for future collaborations.

Committee of CEEE 2022

© 2023 The Editor(s), ISBN: 978-1-032-39397-1

Committee members

Conference Chairs
Prof. Chong Kok Keong, *Universiti Tunku Abdul Rahman, Malaysia*
Prof. Zhongliang Liu, *Beijing University of Technology, China*

Technical Program Committee Chair
Prof. Zhongliang Liu, *Beijing University of Technology, China*

Organizing Committee Chair
Prof. Chong Kok Keong, *Universiti Tunku Abdul Rahman, Malaysia*

Publication Chairs
Prof. Chong Kok Keong, *Universiti Tunku Abdul Rahman, Malaysia*

Organizing Committee Member
Prof. Liborio Cavaler, *University of Palermo, Italy*
Prof. Guilherme Jorge Brigolini Silva, *Federal University of Ouro Preto, Brazil*
Prof. Lihui Wang, *Tianjin Modern Vocational Technology College, China*
Prof. Qiang Xu, *Dalian University of Technology, China*
Prof. Bo Wang, *Jilin Jianzhu University, China*
Prof. Der-Wen Chang, *Tamkang University, Taiwan, China*
Assis. Prof. Sin Jin Chung, *Universiti Tunku Abdul Rahman, Malaysia*
Dr. Chow Ming Fai, *National Energy University, Malaysia*
A. Prof. Denghu Jing, *Southeast University, China*
A. Prof. Ningshan Jiang, *Qinghai University, China*

Technical Program Committee
Prof. Sheng-Rui Jian, *I-Shou University, Taiwan, China*
Prof. Hong-wei Yen, *Tunghai University, Taiwan, China*
Assis. Prof. Wei Wu, *City University of Hong Kong, China*
A. Prof. Borliang Chen, *National United University, Taiwan, China*
Assoc. Prof. Stefano MARIANI, *The Polytechnic University of Milan, Italy*
Assoc. Prof. Belén González-Fonteboa, *University of A Coruña, Spain*
Dr. Jia Liang, *Southeast University, China*

Research on civil engineering materials and building's performance

Advances in Civil Engineering and Environmental Engineering – bin Yahya & Bin Daud (Eds)
© 2023 The Author(s), ISBN: 978-1-032-39397-1

Research on heating of light steel and wood structure house with biomass hot air furnace

Peibin Wang* & Zhuangzhuang Qu
Shandong Huayu University of technology, Dezhou City, Shandong Province, China

ABSTRACT: According to the building characteristics of modern wood structure and light steel structure, the heating application of hot air system of biomass hot air stove in the modern wood structure and light steel structure single house was studied, and the characteristics and adaptability of different heating methods of the rural single house were analyzed. This paper compares and analyzes two different heating methods of biomass fuel hot air and hot water, and points out the differences in initial system investment, annual operating fuel and electric energy cost, operation and maintenance, user comfort and energy saving adjustment. It also puts forward some suggestions for the research and development of single residential heating methods and corresponding hot air stoves in villages and towns.

1 INTRODUCTION

In recent years, modern wood structure and light steel structure single residential buildings in China have been gradually promoted and developed. "Steel, as a typical green material, has good ductility, strong plastic deformation ability and good seismic resistance to the wind; The light weight of the steel structure reduces the cost of building foundations."[1] When the insulation wall is used in the outer protective structure, the thermal insulation performance of light steel structure building is better than that of concrete and masonry structure, as well as in energy saving and environment. Therefore, "it conforms to the development principle of circular economy and is a kind of architectural structure system with the least impact on the natural environment of current urbanization construction (Li 2021)." "Town of the new modern timber-work building with Chinese traditional wooden construction has bigger difference, the approximate standard wooden combinations, in North America, 90% of the single residences, low-rise apartment buildings, high-rise apartment in internal partition, and some businesses and public buildings and a large number of low-level using timber-work building (Li 2005)." the commonly used building materials, only the wood products are renewable materials, light steel structure housing walls also can use heat preservation performance is good wood products, so the modern timber-work and light steel structure residential building is the national advocate of green energy-saving buildings. Corresponding to the requirement of ecological development, China government in recent years also introduced the new standard in the field of architecture. Because of the relatively high cost of modern wooden structure architecture, it is mainly used in the construction of high-standard single houses and characteristic towns in tourist scenic areas in China. With the development of the economy, the use of modern timber frame single houses will increase.

Due to the double carbon target requirements, as well as other kinds of energy prices and transport problems, biomass pellet fuel is a better choice in the villages and towns single residential heating. The hot air heating system with the traditional hot water heating method has certain characteristics

*Corresponding Author: 841738989@qq.com

DOI 10.1201/9781003349563-1

and advantages, so the biomass hot air furnace in the wood structure and light steel structure residential heating are to be studied independently.

2 HEATING METHOD AND BIOMASS HOT AIR FURNACE OF SINGLE HOUSE

For single residential villages and small towns, most families using the independent heating method, such as air source and ground source heat pump, can finish both winter heating and summer cooling, but the initial investment and operation cost is higher. When the indoor and outdoor temperature difference is bigger, its efficiency is reduced. Especially, it is not suitable for outdoor temperature lower in winter in northern China; Natural gas wall-mounted furnace heating has good emission reduction effect, are easy to use, but the price of natural gas is high. Most of the towns and villages do not have pipeline natural gas supply, so it is not easy to be widely used. Equipment of electric heater heating is simple and moveable, but it consumes a lot of electrical energy, resulting in high operating costs. It is generally used as an auxiliary temporary heating method but is not suitable for long-term heating of the whole building. Due to the requirements of environmental protection, the use of scattered coal is now restricted, and some regions have issued relevant regulations to limit the use of high-pollution scattered coal. Therefore, no matter the aspect of national policy, or the aspect of cleanliness and convenience, the traditional use of scattered coal heating and other ways are gradually eliminated. Due to the support of national policy and the continuous improvement of processing technology, biomass pellet fuel has a certain advantage in price, and has increased calorific value per unit volume. Although there is certain pollution in local areas, the use of abandoned crop straw, etc., is the national support to promote the use of environmental protection fuel. It is also a favorable guarantee for China to achieve the double carbon target within the specified time. Mo Shipeng investigated the heating situation of rural residential houses in Hebei Province, and pointed out that "biomass pellet heating can be used in areas with shortage electricity and natural gas supply, which is characterized by the lowest initial investment and easy access to raw materials (Rao 2017) of agroforestry waste biomass volatilization of generally higher than that of coal volatile, due to the increase of biomass combustion technology and the continuous improvement of biomass granule molding equipment, the structure and organization characteristics of biomass pellet fuel, combustion of volatile overflow rate reduced its combustion rate has always been able to make volatile put out heat transfer to heating surface in time, Reduced incomplete combustion loss and exhaust heat loss (Wang 2014)." The thermal efficiency of biomass boilers has also been greatly improved in recent years.

The combustion efficiency of the new biomass pellet hot air stove can reach more than 85%, the performance of the flue gas air heat exchanger is also improving, and the thermal efficiency of the biomass hot air stove can reach more than 70%. The heating method of small biomass pellet hot water boilers has been widely used. The hot air heating method of biomass pellet hot air stove is mainly used in crop drying, large space plant heating and facility agriculture heating, greenhouse planting, and breeding. It is rarely used in single residential buildings and other large residential and apartment buildings in China. Now there are small hot air stoves for a single room. The hot air stoves have no hot air ducts, and the heated hot air is directly sent into the room. The combustion system has smoke exhaust ducts, and the air for combustion comes from the room. The hot air stove is mainly used in small commercial rooms. The equipment is mobile and easy to use, but the air quality of the room is ordinary. Each room needs one hot air stove, so its use is subject to certain restrictions.

3 HOT AIR HEATING SYSTEM

Single-family homes in North America have used hot air heating for nearly 80 years, and "about 60 percent of households now use gas forced air furnace and another 9 percent use oil furnace."[5] Early hot air heating systems used oil-fired hot air furnaces. In the modern days, most of them are gas-fired hot air furnaces. "Hot air furnaces are arranged in the mechanical room or basement on the first floor. The indoor part of the return air and outdoor fresh air are mixed and heated in the heat exchanger in the hot air stove, and then sent to the heating room through the hot air duct (Wang 2014)." Because the interior walls and floors of modern wood and light steel structures are hollow, the air supply and return ducts can be conveniently installed in the wall, ceiling, or under the floor.

In general, according to the different structures of the building, the use of each room and the different heat load, the layout and size of the duct are different. "When the hot air furnace is arranged in the center of the building, the air supply duct is mostly arranged in radial mode, and the annular arrangement and the main duct and branch duct arrangement (Wang 2006). The arrangement of air supply ducts in hot air heating systems is similar to that in all air conditioning systems. The air supply port adopts adjustable grid type, manual or automatic adjustment mode, "by adjusting the relative opening of the baffle, adjust the air supply volume rate, the indoor temperature of a room can be controlled."] In most single residential buildings, the air supply outlet is mainly installed on the floor beside the wall or at the lower part of the inner wall of the room. By utilizing the density difference between hot air and indoor air, the hot air freely flows upward, and at the same time, the size of the duct is shortened. This arrangement also improves the comfort of the interior personnel.

The return air outlet is mostly fixed grid type, installed in the upper part of the room. Most of the return air into the circulating air mixing box, while a small part of is exhausted to the outdoors. The specific proportion depends on the requirements of indoor air quality and outdoor temperature.

Usually, the main temperature controller of the residence is installed in the main living room. The temperature can be set according to the indoor situation, and then the burning situation of the hot air stove can be adjusted to achieve the effect of economic operation. The room with high frequency can use the air supply outlet controlled by temperature, and other rooms can adjust the air supply volume through the baffle with an adjustable grille angle, and then adjust the temperature of the room. When there is no person in the room, the air supply outlet can be closed to achieve the purpose of energy saving.

4 ADVANTAGES AND DISADVANTAGES OF HOT AIR HEATING SYSTEM OF BIOMASS HOT AIR FURNACE

Compared with the biomass boiler hot water heating system, the hot air furnace is easy to operate, quick to start and stop, and the hot air generated by the heat exchanger in the furnace is clean and free of dust. In winter when the furnace stops and accidents occur, no furnace and duct are frozen and cracked. If the humidification device is installed in the outlet section of the heat exchanger, the temperature and humidity of the room can be flexibly adjusted according to the different needs of the heating space, which can solve the problem of low indoor air relative humidity and relatively dry in the northern heating space. The hot water heating system can only adjust the temperature of the heating space but cannot adjust the temperature and humidity of different rooms separately and cannot improve the indoor air quality. "The supply air in the hot air heating system is composed of fresh air and return air. By adjusting the proportion of the fresh air volume rate and return air volume rate, the ventilation times of the room can be guaranteed, and the indoor air quality can be improved (Yao 2018)." However, the hot water heating system can only provide outdoor air to the room when additional fresh air units are installed. In the season when there is no heating, the hot air system can also deliver fresh outdoor air into the room, which acts as a ventilation system to improve indoor air quality.

The hot air duct is larger than the hot water pipe of the hot water heating system. The open-installed hot air duct occupies the space of the room but also affects the layout and aesthetics of the room. Therefore, the radiant floor panel or radiator of the hot water system is mainly used in the single residence of concrete or brick structure in China. The total area of single residential buildings is small, and the size of the main air supply duct is not very large. Due to the characteristics of the interior walls and floors of modern wood structures and light steel structure buildings, the air supply and return duct of the hot air heating system can be arranged in the wall, ceiling or below the floor, occupying a small space of the room. The layout of air supply and return outlets are also more convenient, so the hot air heating system is more suitable for use in modern wood structure and light steel structure buildings.

Rao Yang conducted an economic analysis of heating with biomass hot air stoves in a large enterprise. "From the economic perspective, under the same heating load and installed capacity, the annual total investment of biomass hot air stove is 44% lower than that of natural gas hot air stove, and the operating cost is only about 60% of that of natural gas hot air stove. Therefore, biomass hot air stove heating method has advantages and feasibility in economic operation (Yao 2018)." Although biomass pellet fuel also has certain environmental pollution problems, its fuel source is part of the waste in agricultural and forestry production, and it is a renewable resource.

5 ECONOMIC COMPARISON OF THE TWO HEATING SYSTEMS

The design and economic analyses of a modern timber structure with a floor area of 130 m^2 and a light steel structure independent house with a floor area of 146 m^2 in Jinan city are made, respectively. Modern wood structure residential hot water heating system adopts horizontal abnormal pattern water supply mode and the radiator adopts mid-range steel radiator. The heat source is a biomass hot water boiler. The hot air heating system adopts the main air duct with branch ducts and a rectangular air duct made of galvanized sheets. The air supply port is manually adjustable grid type. The heat source is a biomass hot air furnace.

The price of biomass pellet fuel with a calorific value of 16700 kJ/kg is 10 RMB/kg. The operating cost includes the fuel cost and electricity cost of the fan in the hot air system and water pump in the hot water system. The heating period is 4 months. Table 1 shows the data of the wood structure house. In equipment and system investment respect, the price of residence of steel-structure and wooden structure is not quite different, but because of the difference in heat preservation performance, the running cost of a unit building area is slightly higher than that a wooden structure.

Table 1. Evaluation index and item of timber structure building.

Evaluation Index	Evaluation item	Biomass hot air system	Biomass hot water system
Economics	System investment	13,200 RMB	13,700 RMB
	Heating period operation cost	4,190yuan RMB /year	3,830yuan RMB /year
	Equivalent to the annual total cost	4,850yuan RMB /year	4,515yuan RMB /year
Technology and products	Hot air furnace /Hot water boiler	A few products	A series of products
	Heating system	Less domestic products	The product is perfect
Operation and maintenance	Maintenance and operation safety	Less maintenance, Safety	More Maintenance, Average safety
	Comfort	Available for fresh air	No fresh air
	Operation Energy saving	Room adjustment respectively	Adjustment is not convenient

From the above-mentioned data, the circumstance can be found that the biomass investment in the early phase is the same as the hot air furnace with the hot water. Hot air duct system initial investment is higher than hot water pipes, and in the terminal equipment of heating system, hot air heating system with normal manual regulation of grille in price is not high; however, if adopts the temperature automatic adjustment method, its price becomes higher. The price difference of different types of radiators in the hot water system is relatively different. The investment in a hot water heating system with a mid-standard hot water radiator is basically the same as that of a hot air heating system. Some users use radiation floor panels or high-standard steel aluminum or copper aluminum radiator, then the investment in hot water systems is higher than the investment in hot air systems. For small household heating systems, routine maintenance is less, but the maintenance and repair cost of hot water pipes and terminal equipment is higher than that of hot air systems. In terms of operating costs, because the thermal efficiency of a hot air furnace is lower than that of a hot water boiler and the hot air heating system has a fresh air heating load, the energy consumption of the hot air system is more than 15% higher than that of hot water system under the condition of full load room heating. Because the air supply for each room in the hot air heating system can be adjusted, in the actual operation condition, the operating cost of the hot water heating system and hot air heating system is almost the same if the appropriate energy-saving regulation operation mode is adopted. Due to the difference in enclosure structure, the insulation performance of modern wooden structure building envelope is better than that of light steel structure buildings, but the insulation performance is higher than that of concrete and masonry structure buildings.

6 CONCLUSIONS

Based on the analysis, we can conclude that the hot air heating system has certain advantages in energy-saving adjustment, user comfort, daily maintenance, and safe operation. Because biomass hot air furnaces for residential heating have high requirements for air cleanliness and air supply temperature, few biomasses hot air furnaces and hot air heating systems are applied in residential heating in China. Therefore, the development and research of biomass hot air furnaces should be increased to provide more choices for different users. When the equipment performance and operation safety of biomass hot air furnaces are improved, it is feasible to use the hot air heating method of biomass hot air furnaces in the modern wood structure and light steel structure single house.

ACKNOWLEDGMENT

This article is the research result of the 2021 science and technology research project of Shandong Huayu University of Technology “Research and Application of single Residential Hot air Heating System using biomass Hot air furnace” (Project Number: 2021KJ03).

REFERENCES

Earl laBonty. (2012). *Air Heating Systems* [M]. BCIT.35–36.
Li Xiaodi et al., (2021). Research and exploration of prefabricated steel structure residential design in China [J]. *Exploration and Discovery*. 101, 197–199.
Li Yin, Li Weili. (2005). Design and experiment of biomass brique-fired hot air stove [J], *Energy Saving Technology*. 4, 324–330.
Rao Yang et al. (2017). Technical and economic feasibility analysis of biomass hot air furnace heating method [J]. *Jiangxi Building Materials*. 12:268–269.
Wang Peibin et al. (2006). Application of hot air heating system in independent residential heating [J]. *Gas & Heat*. 12, 49–51.
Wang Peibin, (2014). Development of wood structure building and green building [J]. *Journal of Shenyang Institute of Engineering*. 01, 86–88.
Yao Shipeng. (2018). Clean heating scheme for rural residential buildings in Hebei province [J]. *Modern Rural Science and Technology*. 01,91.

Advances in Civil Engineering and Environmental Engineering – bin Yahya & Bin Daud (Eds)
© 2023 The Author(s), ISBN: 978-1-032-39397-1

Rain erosion characteristics of multi-stage loess high slope of highway[1]

Yan Liang* & Tian Zheng
School of Highway, Chang'an University, Xi'an, Shaanxi, China

Aidi Huo
School of Water and Environment, Chang'an University, Xi'an, Shaanxi, China

Guanghong He
Gezhouba Group Transportation Investment Co., Ltd, Wuhan, Hubei, China

ABSTRACT: To increase the stability of highway loess high slope, a multi-stage slope is generally adopted and platforms are set in the multi-stage slope. Soil loss is very serious in the loess region of China, but there are few studies on the erosion characteristics of multi-stage loess slopes, so this study is very important. Take the multi-stage loess slope of highway in the Qingyang area of Gansu Province as the research object, the evolution process of soil erosion of multi-stage loess slope is summarized and the factors affecting loess erosion are analyzed by means of field investigation and analysis. The soil erosion caused by platform catchment in the multi-stage loess slopes can't be ignored when the rain is heavy. The slope factor in the RUSLE model is modified according to the test results. The relationship of the quadratic polynomial between the slope factor and the slope is put forward based on the test results. It is suggested that soil and water conservation measures should be taken as soon as possible after cutting the loess high slope, so as to reduce the occurrence of soil erosion. This paper can provide a basis for preventing soil erosion on the highway loess high slope.

1 INTRODUCTION

Soil erosion and even landslide are serious environmental geological problems, which seriously threatens the sustainable development of the Loess Plateau in China (Shi, Shao, 2000; Liang, He, Liu, Chen, Huo, 2021; Zhao, Mu, Wen, Wen, Wang, Gao, 2013). There have been many studies on scour erosion (Huang, Wang, Song, Wang, 2015; Zou, Yan, Tan, Wang, 2019; Wu, 2010). Many scholars have modified the factors in the RUSLE model (Durigon, Carvalho, Antunes, Oliveira, Fernandes, 2014; Li, Lu, Zhang, Lu, Wang, 2019; Nikolaos, 2018; Chen, Takashi, Wu, 2017). The slope is an important factor affecting erosion. In the RUSLE model, the slope factor increases with the increase of slope angle, but this is inconsistent with the erosion law of loess slope according to indoor experiment and field observation, so taking the erosion of multi-stage loess high slope in Qingyang area of Gansu Province as an example in this paper, erosion evolution law and influencing factors of multi-stage loess slope are further studied based on field investigation, and the slope factor is improved according to the test results.

---

*Corresponding Author: 1638637767@qq.com
[1]Project supported by the National Natural Science Foundation of China (Grants No. 41790444 and 41877232); Key R & D projects of Shaanxi Province (Grants No. 2020SF-424); CAS (Grants No. SKLLQG1909).

DOI 10.1201/9781003349563-2

2 MATERIALS AND METHODS

In this paper, the evolution process and factors influencing soil erosion of multi-stage loess slope are summarized by the field investigation and analysis. The monthly mean temperature and rainfall in Qingyang were obtained from the data of field weather stations. The degree of soil erosion in the Qingyang area was calculated by the RUSLE equation and ArcGIS technology. The slope factor (S) in the RUSLE model was modified by analyzing the test data of rainfall erosion.

2.1 *Soil erosion in the Qingyang area*

2.1.1 *Overview of Qingyang area*

Qingyang area is located in the eastern part of Gansu Province, China (As shown in Figure 1), which belongs to the loess plateau area. Quaternary loess is widely distributed in this region. The climate of Qingyang is arid and semi-arid, belonging to the temperate continental climate. The annual precipitation in the Qingyang is between 400 and 600 mm, with more precipitation in mountainous areas and less precipitation in basins, and rainfall is concentrated in July, August, and September.

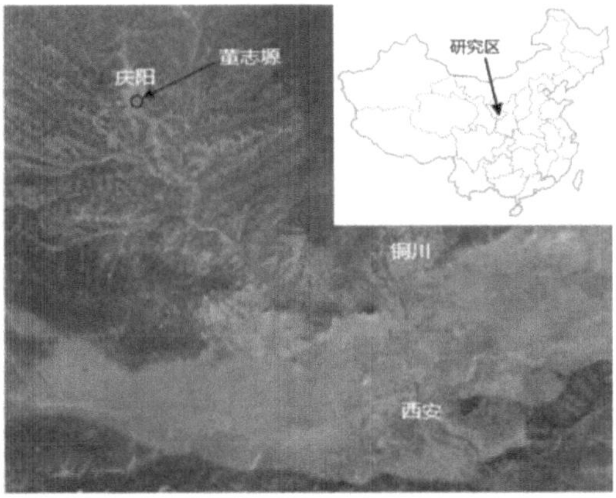

Figure 1. Location of the study area.

The annual average temperature in Qingyang is between 9.5°C and 10.7°C. It is drought, mild and rich light on a whole. The area of the study area is about $2.7 \times 10^4 km^2$ with altitudes between 868–2090m. The valleys are long, narrow, and deep in this area. There is serious soil erosion in this area for a long time.

2.1.2 *Degree of soil erosion in the Qingyang area*

The modified universal soil loss equation (RUSLE) is used to estimate the average annual soil erosion in the study area because the factors affecting soil erosion are comprehensively considered in this equation. The expression is as follows.

$$A = R \times K \times L \times S \times C \times P \tag{1}$$

In the formula, $A(t \cdot hm^{-2} \cdot a^{-1})$ is the annual average soil loss, that is, the soil erosion modulus; $R(MJ \cdot mm \cdot (hm^2 \cdot h \cdot a)^{-1})$ is the rainfall erosion factor; $K(t \cdot h \cdot (MJ \cdot mm)^{-1})$ is the soil erodibility factor; L is the slope length factor; S is the slope factor; C is the vegetation cover factor; P is the factor of the soil and water conservation measures. L factor, S factor, C factor, and P factor are all dimensionless.

The data used in this paper mainly include the data of DEM data, meteorological data, the data of soil texture spatial distribution, data on the normalized vegetation index spatial distribution, and data on land use. The data types and sources are shown in Table 1.

Table 1. Data type and sources.

Data type	*Format*	*The source of the data*
DEM data	Grid diagram	Geospatial Data Cloud Platform (http://www.gscloud.cn)
Vegetation Index	Grid diagram	
Weather data	Excel	China Meteorological Data Network (http://data.cma.cn)
Spatial distribution data of soil texture	Grid diagram	Resource Environment Data Cloud (http://data.cma.cn)
land utilization	Grid diagram	

The spatial distribution of the soil erosion modulus in the Qingyang region was obtained by RUSLE and ArcGIS grid calculator (as shown in Figure 2). It can be seen from Figure 2 that the soil erosion modulus was from 0 to $5.45 \times 10^4 t \cdot km^{-2} \cdot a^{-1}$ and the average soil erosion modulus was $4974.98 t \cdot km^{-2} \cdot a^{-1}$. Soil erosion was more serious in the central part of the study area. It can be found that soil erosion mainly occurs on both sides of the river (Figure 2b), in the steep slope area (Figure 2c), and at the head of the ditch (Figure 2d).

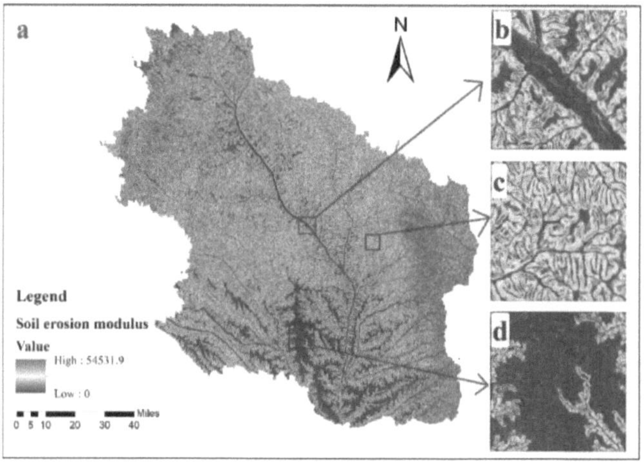

Figure 2. Map of soil erosion modules in the Qingyang area.

Based on ArcGIS software, the number of grids in different erosion intensity grades was calculated. The areas of different erosion grades and their area ratios were obtained. It can be found that the degree of soil erosion in the study area is relatively serious. The area of the soil erosion modulus greater than 5000 t/ (km2·a) occupies a considerable area ratio of 43.85%. The area of soil erosion modulus less than 5,000 t/ (km2·a) is 15026.66 km^2, accounting for 56.15% of the total area.

2.2 *The evolution law and influence factors of erosion of multi-stage high slope of loess*

The evolution law of loess slope erosion and the factors affecting slope erosion were studied by a field survey on the erosion of the loess multi-stage slope in Jing'ao village and Fengbuzi village in the Qingyang area.

2.2.1 *The evolution law of erosion of loess slope*

In August 2018, a survey of the multi-stage loess highway slope in the village of Fengbuzi found that the loess slope didn't take soil and water conservation measures after just cutting. Slope erosion developed continuously and relatively uniform and fine gullies were found on the slope as shown in Figure 3.

Figure 3. Fengbuzi Village slope photographed in Aug. 2018.

After more than a year, a field survey in December 2019 revealed that bryophytes grew on the slopes. The moss serves as a protective layer on the slope, so the erosion resistance of the slope is enhanced to a certain extent.

The November 2020 investigation found that the slope erosion is banded along the horizontal direction. The reason is that rainwater scour force is theoretically the same at the same height of the slope. During the rainy season from June to August 2020, as rainfall increases, it soaks and reduces slope strength. When the rain washes to a certain extent, small patches of moss layer are destroyed and washed down, gradually forming a certain scale of horizontal bands of erosion (Figure 4).

Figure 4. Fengbuzi Village slope photographed in Nov. 2020.

2.2.2 *Influence of catchment on slope scour*

The high slope of loess is cut into a multi-stage slope with platforms, which can not only improve the stability of the slope but also play a certain buffer role for rain erosion. However, if it's heavy rain and there is enough rainwater on the steps, the slope below it will begin to be eroded from the top of the slope. The larger the water collection, the larger the scour range is.

2.2.3 *Influence of joints and fissures on loess erosion*

Loess generally has large pores, more vertical joints, and fissures. When rainfall occurs, joints and fissures provide advantageous seepage channels for rainwater, which allows more water to seep into deeper parts of the soil, and easy to collect water to form a certain pressure head to promote infiltration. Meanwhile, the pore water pressure and moisture content around the vertical joints and fissures increase faster. The infiltration of rainwater softens the surrounding soil which is more prone to deformation and failure. Small cracks will develop into a gully, even slope collapse as time goes on.

2.2.4 *Loess slope erosion is related to slope*

Through the field investigation of the highway cutting slope in Jing'ao Village in November 2020 (Figure 5.), it is found that the slope is less eroded when it is steep, while the slope is small, and the degree of erosion is large. Wang et al. (2019) have proved through theory and experiments that when the slope is 41° to 43°, the erosion of the loess slope by rainfall is most severe, and then the amount of erosion decreases with the increase of slope. Huang et al. (2015) have tested to show that when the slope angle is 40° to 70°, the amount of erosion decreases with the increase of slope. Based on the analysis of field investigation and laboratory test results, it is not that the greater the slope is, the more serious the rainfall erosion is. USLE and RUSLE (Wischmeier & Smith 1978; Renard et al. 1991), widely used to calculate soil erosion, do not conform to the actual situation of loess slope erosion. Therefore, slope factors will be studied in this paper.

Figure 5. Jing'ao Village slope photographed in Nov. 2020.

Soil and water conservation measures, such as drainage ditch, slope protection walls, plant slope protection, retaining wall, etc., are effective means to reduce soil erosion.

2.3 *Revision of slope factor (S)*

Rainfall erosion factor, soil erodibility factors, and slope length factor refer to the formulas proposed by Brown and Foster (1987), Williams (1990), and Wischmeier and Smith (1978), respectively.

Some basic parameters were obtained from the indoor scour test results of Wang et al., Huang et al., Peng et al., and Du et al., respectively (Du et al. 2018; Huang et al. 2015; Peng 2020; Wang et al. 2019).

It is assumed that vegetation cover management factor C and factor of soil and water conservation measures P are 1.0 (Pradhan et al. 2012). Based on the results of various experiments, the slope factor S under different slope α is calculated according to Formula (1). The research results of Liu et al., Singer, and Blackar are also shown in Figure 6 (Liu et al. 1994; Singer & Blackard 1982).

By analyzing the data in Figure 6, the relationship between slope factor S and slope α can be obtained:

$$S = -0.0084\alpha^2 + 0.8753\alpha - 7.1688 \quad R^2 = 0.6062 \tag{2}$$

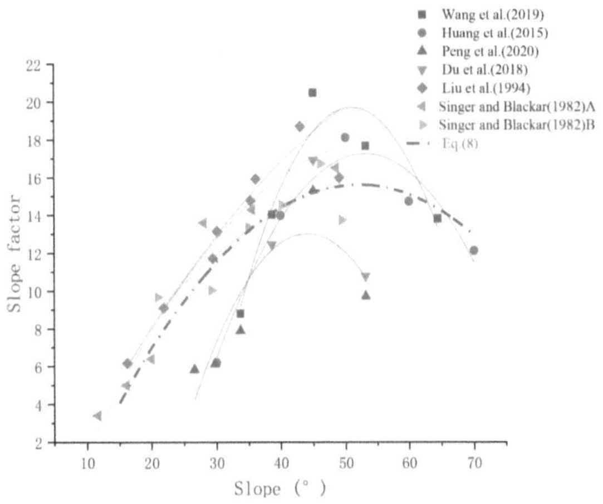

Figure 6. Relationship between slope factor and slope in different tests.

3 RESULTS AND DISCUSSION

The rainwater erosion process of the multi-stage slope is basically the same as the laboratory test results, but the difference is that bryophytes naturally grow on the site slope, which improves the erosion resistance of the slope to a certain extent.

Studies have shown that the slope angle corresponding to the maximum amount of rainwater erosion is between 40° to 43° (Huang et al. 2015; Wang et al. 2019). This study shows that slope factor S is the largest when slope α is in the range of 45° to 50° . Under the condition that other conditions remain the same, with the increase of α, S will increase first and then decrease. The maximum slope factors are between 15.32 and 20.49, corresponding to gravel soil and clayey silt soil respectively, which indicates that the rainfall erosion of clayey silt is the largest, while that of gravel soil is the smallest in these tests.

By observing Figure 6, the change rate of slope factor S with slope angle α is larger in gravelly soil and silty clay, and smaller in loam. The results show that the slope angle corresponding to the maximum slope factor is between 45° and 50°.

4 CONCLUSIONS

Through field investigation, it is concluded that the erosion evolution process of multi-stage loess slope is: splash erosion – relatively uniform small gully – slope with moss – horizontal bands of erosion – gully – collapse.

The factors affecting the erosion of the multi-stage slope of loess are analyzed. When setting up the platform in the multi-stage slope, corresponding engineering measures should be taken, such as building intercepting ditches and drainage ditches on the steps.

The relationship of the quadratic polynomial between the slope factor and the slope is put forward based on the test results, which provides a basis for accurately predicting the erosion of the loess slope.

ACKNOWLEDGMENTS

This work was supported by the National Natural Science Foundation of China (Grants No. 41790444 and 41877232). Key R & D projects of Shaanxi Province (Grants No. 2020SF-424);

Key Laboratory Open Project Fund of State Key Laboratory of Loess and Quaternary Geology, Institute of Earth Environment, CAS (Grants No. SKLLQG1909). We also would like to thank the reviewers and editors for their constructive comments, which helped to significantly improve the manuscript.

REFERENCES

Brown, LC, Foster GR. Storm erosivity using idealized intensity distributions [T]. *ASAE*. 1987; 30: 379–386.

Chen H, Takashi O, Wu P. Assessment for soil loss by using a scheme of alternative sub-models based on the RUSLE in a Karst Basin of Southwest China. *Journal of Integrative Agriculture*. 2017; 16(2): 377–388.

Du TT, Li ZQ, Wang XM, Bai YW, Qin DT, Chen We. Experimental study on rainfall scouring model of loess slope [J]. *Journal of Engineering Geology*. 2018; 26 (03):732–740.

Durigon VL, Carvalho DF, Antunes MAH, Oliveira PTS, Fernandes MM. NDVI time series for monitoring RUSLE cover management factor in a tropical watershed. *International Journal of Remote Sensing*. 2014; (35)2: 441–453.

Huang XH, Wang CM, Song PR, Wang TZ. Physical simulation test study on rainfall Erosion characteristics of loess slope [J]. *Journal of Engineering Geology*. 2015; 23(04): 725–730.

Li YP, Lu XP, Zhang H, Lu ZZ, Wang SY. Study on soil erosion in Huaihe river basin based on GIS and RUSLE – a case study of Shangcheng County, Xinyang City [J]. *Remote Sensing of Land and Resources*. 2019; 31 (04): 243–24.

Li J. Automatic extraction and analysis of slope length of Loess Plateau based on DEM [D]. Northwest University. 2007.

Liang Y, He GH, Liu SS, Chen C, Huo A. Influence range of loess landslide in the Southern Jingyang Plateau based on numerical simulation and mathematical statistics [J]. *Arabian Journal of Geosciences*, 2021, 14(12): 1–11.

Liu B Y, Nearing MA, Risse L M. Slope gradient effects on soil loss for steep slopes [J]. *Transactions of the ASAE*.1994; 37(6):1835–1840.

Nikolaos E. Evaluating the performance of different empirical rainfall erosivity (R) factor formulas using sediment yield measurements. *Catena*. 2018; (169): 195–208.

Peng SH. The simulation test and numerical analysis of rainfall scouring Characteristics of coarse-grained soil high embankment slope [D]. Changsha University of technology. 2020.

Pradhan B, Chaudhari A, Adinarayana J, Buchroithner, Manfred F. Soil erosion assessment and its correlation with landslide events using remote sensing. *Environ Monit Assess*. 2012; 184:715–727.

Renard KG, Foster GR, Weesies GA, Porter JP. RUSLE: Revised universal soil loss equation [J]. *Soil Water Conserv*. 1991; 46 (1): 30–33.

Shi H, Shao M. Soil and water loss from the loess plateau in China. *J. Arid Environ*. 2000; 45 (1): 9–20.

Singer MJ, Blackard J. Slope angle-interrail soil loss relationships for slopes up to 50% [J]. *Soil Sci. Sco. Am.* 1982; 46(6): 1270–1273.

Wang JY, Zhong C, Wang XC, Ding LT, Zhu JJ. Rainwater erosion characteristics of loess subgrade slope [J]. *Road Construction Machinery and Construction Mechanization*. 2019; 36(10): 46–51+57.

Wischmeier WH, Smith DD. Predicting rainfall erosion losses — A guide to conservation planning [J]. *The United States. Dept. of Agriculture. Agriculture Handbook* (USA).1978.

Williams JR. The erosion-productivity impact calculator (EPIC) model: A case history [J]. *Royal Society of London Philosophical Transactions*. 1990; 329(1225): 421–428.

Wu HJ. Study on the impact of climate change on highway slope erosion in the loess area of Gansu Province and its control technology [D]. Chang'an University. 2010.

Zhao G, Mu X, Wen Z, Wen Z, Wang F, Gao P. Soil erosion, conservation, and eco-environment changes in the loess plateau of China. *Land Degrad*. 2013; 24(5), 499–510.

Zou YJ, Yan QW, Tan XL, Wang J. Soil erosion assessment and driving factor analysis in Weibei mining area [J]. *Arid Area Geography*. 2019; 42 (06): 1387–1394.

Advances in Civil Engineering and Environmental Engineering – bin Yahya & Bin Daud (Eds)
© 2023 The Author(s), ISBN: 978-1-032-39397-1

Effect of organic LDHs on rheological properties of different bitumen undergoing UV aging

Song Xu*
College of Civil Engineering, Fuzhou University, Fuzhou, P.R. China

Lu Liu & Chuan Wang
School of Advanced Manufacturing, Fuzhou University, Fuzhou, P.R. China

Jitao Cao
College of Civil Engineering, Fuzhou University, Fuzhou, P.R. China

Chongyu Bian
School of Advanced Manufacturing, Fuzhou University, Fuzhou, P.R. China

Heyao Huang
College of Civil Engineering, Fuzhou University, Fuzhou, PR China

ABSTRACT: The resistances of two organic intercalated layered double hydroxides (PABA-LDHs and CA-LDHs) modified bitumens to aging were studied by physical and dynamic rheological properties tests. The results showed that the physical properties of bitumen modified by organic LDHs were improved to a certain extent, but not as remarkably as rheological properties. As for enhancing the resistance to UV aging, organic LDHs performed better than LDHs, and CA-LDHs showed the most outstanding improvement according to the experimental results of rheological parameters. And this effect was the same for two different types of bitumen.

1 INTRODUCTION

Bitumen has become one of the main types of high-grade pavement due to its advantages of driving comfort, low dust, and short construction cycle, which exhibits excellent road performance on account of its brilliant viscoelastic and plastic properties (Redelius & Soenen 2015). However, it can be affected easily by the environment and load during its service, resulting in its performance decline. The aging of bitumen during service can be mainly divided into thermal oxygen aging and UV aging (He et al. 2018). To resist UV aging of bitumen, it is more common to add inorganic materials or UV absorbents into bitumen to prepare modified bitumen (Li et al. 2021).

Recently, it was reported that LDHs could be used as UV screening materials, which was because their special layered structure could reflect and absorb UV light (Boccalon et al. 2020; Pillai et al. 2020). LDHs are a kind of inorganic material as well, the poor compatibility with bitumen, such as organic matter, make them difficult to form a uniform and dispersed composite system, resulting in the inconsistent performance of bitumen. However, unlike other inorganic materials, the organic anion can be intercalated into the interlayer of LDHs, thereby the basis of LDHs to greatly enhance the UV aging resistance of bitumen (Li et al. 2018).

To solve these problems and achieve the durability of bitumen, UV absorbents para-aminobenzoic acid (PABA) and cinnamic acid (CA) were intercalated into the interlayers of LDHs. Therefore, UV

*Corresponding Author: xusong@fzu.edu.cn

DOI 10.1201/9781003349563-3

light absorption materials PABA-LDHs and CA-LDHs with excellent UV absorption performance and stability were obtained. In this paper, the physical properties and rheological properties of organic LDHs modified bitumen were tested and characterized, which revealed the mechanism of organic intercalated LDHs on the properties of bitumen.

2 MATERIALS AND METHODS

2.1 *Materials*

Two original asphalts with penetration-grade of 70 and 90 were used as the matrixes to prepare organically modified bitumen, and the physical properties are listed in Table 1. LDH, a commercial product was used as the main antiaging agent of bitumen. Two kinds of organic compounds, para-aminobenzoic acid (PABA) and cinnamic acid (CA) were commercially available products and used as modifiers of LDHs.

Table 1. Physical properties of the original bitumen.

Items	70#	90#
Penetration (25°C, 0.1 mm)	78	92
Ductility (10°C, cm)	38.2	42.8
Softening point (°C)	47.5	46.6
Viscosity (135°C, mPa·s)	449	297
Viscosity (60°C, Pa·s)	178	126

2.2 *Preparation of organic LDHs*

LDHs were first dissolved in a solvent made of CO_2-deionized water and anhydrous ethanol at a ratio of 1:1, and the solution was poured into a three-neck flask with a condensing device and stirred at 65°C for 1 hour to prepare LDHs slurry. Then, PABA or CA solution was added into the three-neck flask. Thereafter, the mixture was adequately stirred with reflux in a nitrogen environment at the specified temperature (65°C for PABA, 20°C for CA). After stirring for 24 h, the precipitate was repeatedly filtered and cleaned with CO_2-deionized water, then the filter cake was dried to a constant weight. Finally, the organic LDHs (CABA-LDHs and CA-LDHs) were obtained after grinding and sieving.

2.3 *Preparation of modified bitumen*

The original bitumen was heated to 140°C, then a preweighted LDHs or organic LDHs (3 wt.% by weight of bitumen) was added into the bitumen in which the mixture was stirred at a speed of 4000 rpm and a temperature of about 140°C. The mixture was sheared for 45 min to ensure that it became adequately homogeneous. LDHs, PABA-LDHs and CA-LDHs modified bitumen were denoted as LMB, PABA-LMB and CA-LMB, respectively. The unmodified bitumen was treated with the same process for comparison.

2.4 *Aging procedures*

The thin film oven test (TFOT) was conducted in compliance with ASTM D1754. The UV aging was performed after TFOT in an oven equipped with a UV lamp under the conditions of a temperature of 60°C, UV intensity of 50 W/m^2, and an aging duration of 9 days.

2.5 *Measurements*

The physical characteristics of all binders, including viscosity, softening point, penetration, and ductility, were tested in accordance with ASTM D4402, ASTM D36, ASTM D5, and ASTM D113, respectively. The rheological tests were conducted using a dynamic shear rheometer (DSR) according to ASTM D7175.

3 RESULTS AND DISCUSSION

3.1 *Effect of organic LDHs on physical properties of bitumen*

The physical properties of two types of bitumen before and after modification were listed in Table 2. It could be noticed that the high-temperature of LMB, PABA-LMB and CA-LMB were improved evidently, such as softening point and viscosity compared with unmodified asphalt, while the effect on ductility was not obvious. As a whole, organic LDHs had a slight impact on the physical properties of bitumen.

Table 2. Physical properties of the modified bitumen.

Physical property	Original $70^{\#}/9^{\#}$	LMB $70^{\#}/9^{\#}$	PABA-LMB $70^{\#}/9^{\#}$	CA-LMB $70^{\#}/9^{\#}$
Penetration (25°C, 0.1 mm)	69.1/84.9	69.8/81.6	56.4/73.8	56.9/72.3
Softening point (° C)	47.9/46.6	48.9/48.9	50.4/48.3	50.7/49.1
Ductility (10°C, cm)	15.5/29.6	15.4/23.6	17.5/17.4	15.2/17.0
Viscosity (135°C, mPa·s)	463/276	468/323	444/372	475/396

3.2 *Effect of organic LDHs on rheological properties of bitumen*

3.2.1 *Complex Modulus*

In Figure 1, the influence of different organic LDHs can be reflected in the complex modulus of $70^{\#}$ bitumen and $90^{\#}$ bitumen. It could be seen in Figure 1 that the complex shear modulus(G^*) of both $70^{\#}$ bitumen and $90^{\#}$ bitumen after modification was larger than that of unmodified bitumen, which could effectively reflect the resistance to deformation, and CA-LMB showed better rheological properties compared with PABA-LMB. Compared with unaged bitumen, the G^* of bitumen after UV aging was significantly increased, and contrary to the G^* of unaged bitumen, the bitumen after UV aging was in the order of unmodified bitumen> LMB> PABA-LMB> CA-LMB, which represented that the addition of modifiers improved the UV aging resistance of bitumen, and the effect was consistent with the order of G^* after UV aging.

3.2.2 *Phase angle*

Figure 2 showed the effects of different organic LDHs on $70^{\#}$ and $90^{\#}$ bitumen respectively. As shown in Figure 2 (a) and (b), the phase angle (δ) of modified bitumen decreased evidently, while for $70^{\#}$ bitumen, the decreasing range of PABA-LMB was larger than that of CA-LMB, which was opposite to that of $90^{\#}$ bitumen. The δ of $70^{\#}$ and $90^{\#}$ bitumen modified by different organic LDHs after UV aging were shown in Figures 2 (c) and (d). After UV aging, the phase angle of all bitumen with modified materials was larger than those without, and in the order of CA-LMB> PABA-LMB> LMB> unmodified bitumen. The variation range of unmodified bitumen was the largest whether before aging or after aging, followed by LMB, indicating that unmodified bitumen was the most sensitive to UV aging.

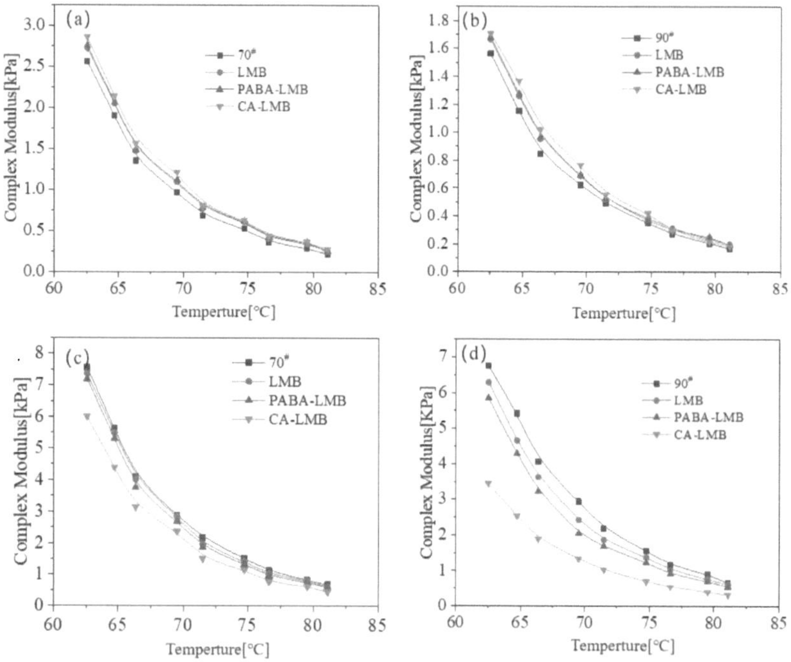

Figure 1. Complex modulus of original and modified bitumens before and after UV aging.

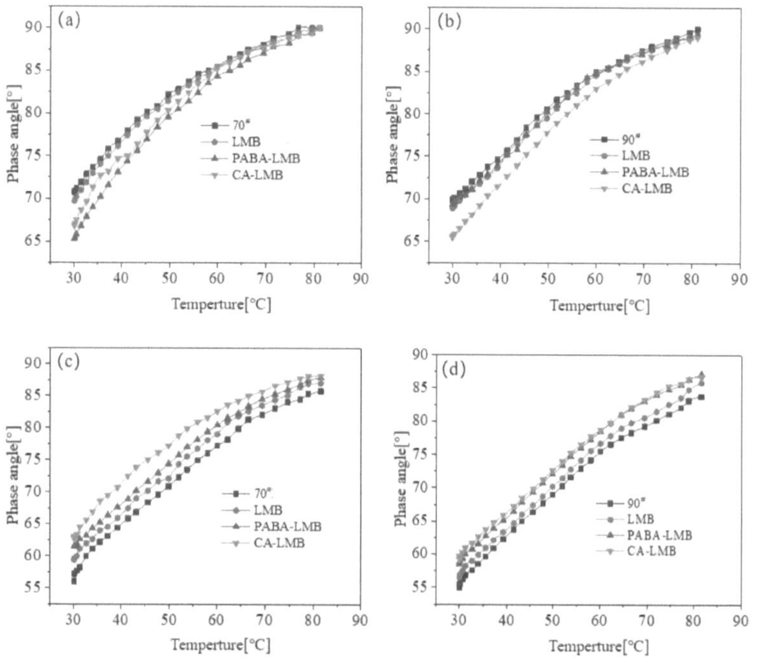

Figure 2. The phase angle of original and modified bitumens before and after UV aging.

3.2.3 *Rut factor*

Rut factor ($G^*/\sin\delta$) reflects the rutting resistance of bitumen at high temperatures. Figure 3 shows the influence curves of different organic LDHs on $G^*/\sin\delta$ of $70^{\#}$ and $90^{\#}$ bitumen. According to (a) and (b), the addition of PABA-LDHs and CA-LDHs would lead to the increase of $G^*/\sin\delta$ of bitumen at the same temperature, which indicated the improvement of rutting resistance of bitumen, and it was consistent with the phenomenon that organic LDHs could improve the physical properties of bitumen (softening point and viscosity). Besides, the changes of $G^*/\sin\delta$ caused by organic LDHs for the two kinds of original bitumen were the same. Compared to bitumen before aging, $G^*/\sin\delta$ increased significantly after UV aging, which confirmed that UV aging would increase the ability of bitumen to resist permanent deformation under high temperatures. Besides, the $G^*/\sin\delta$ of the original bitumen tended to become the largest after UV aging, which meant that it had a better effect to resist rut at high temperatures.

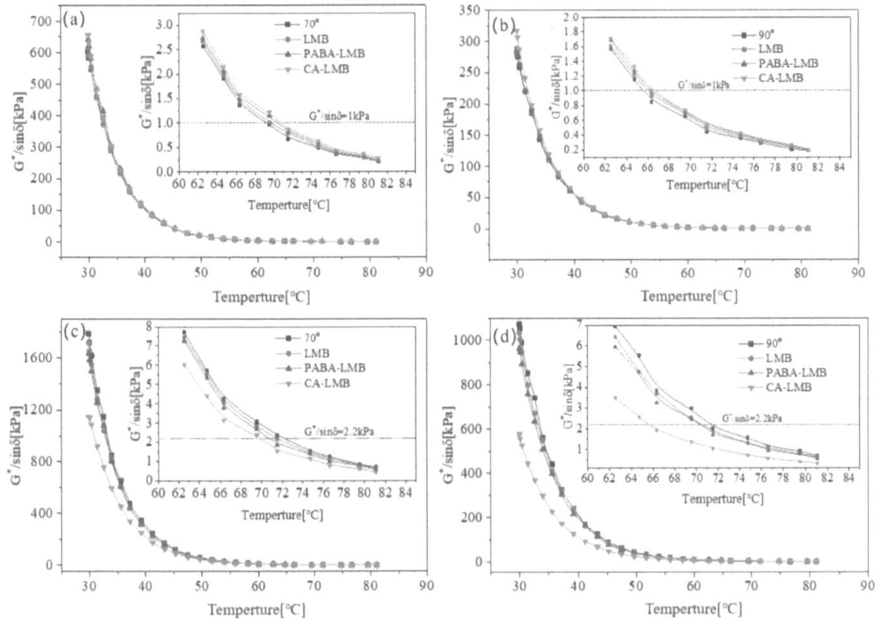

Figure 3. Rut factor of original and modified bitumens before and after UV aging.

3.2.4 *The aging index of complex shear modulus*

The aging index of complex shear modulus (G^*AI) referred to the ratio of G^* of bitumen before and after aging, which indicated the aging degree of bitumen, and its mathematical equation was shown as Equation (1).

$$G^*AI = G_2^*/G_1^* \quad (1)$$

Figure 4 (a) and (b) respectively showed the effects of different organic LDHs on G^*AI of $70^{\#}$ and $90^{\#}$ bitumen. For bitumen of different grades, the trends of G^*AI before and after modification were consistent, and they were in the following order: original bitumen> LMB> PABA-LMB> CA-LMB. It could be seen that the addition of LDHs reduced the aging degree of bitumen, while the reduction of organic LDHs was greater than that of LDHs, and CA-LDHs is greater than that of PABA-LDHs.

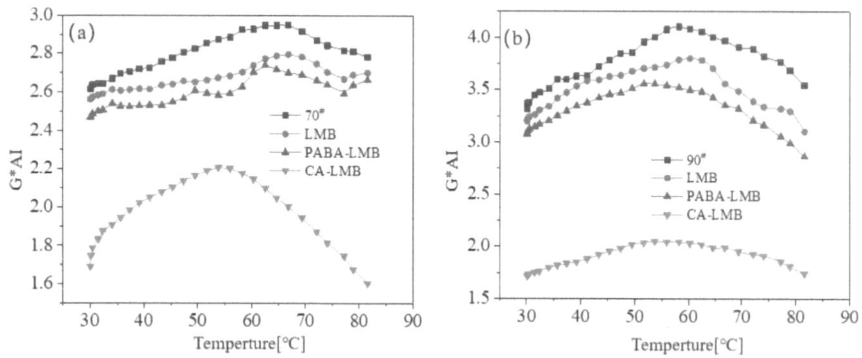

Figure 4. G^*AI of 70#bitumen after UV aging (a) and 90# bitumen after UV aging (b).

4 CONCLUSIONS

(1) The organic LDHs could enhance both high and low-temperature performance of bitumen to some extent even though the enhancement is not very remarkable.
(2) The type of original bitumen would affect the aging degree, both PABA-LDHs and CA-LDHs could improve the aging resistance of bitumen regardless of the type of bitumen.
(3) Compared with LDHs, Organic LDHs could further improve the UV aging resistance of bitumen, and the CA-LDHs showed the most outstanding improvement.

ACKNOWLEDGMENTS

This work is supported by the National Natural Science Foundation of China (No. 51708121) and the Project of Transportation Department of Fujian Province (No. 202014).

REFERENCES

Boccalon, E., Gorrasi, G., Nocchetti, M. (2020). Layered double hydroxides are still out in the bloom: Syntheses, applications and advantages of three-dimensional flower-like structures. *Adv. Colloid Interface Sci.*, 285: 102–284.

He, B.Y., Yu, J.Y., Gu, Y., Zhuang, R.H., Sun, Y.B. (2018). Rheological properties of lignosulfonate intercalated layered double hydroxides modified bitumen before and after ultraviolet aging. *Constr. Build. Mater.*, 180: 342–350.

Li, J., Yang, J., Liu, Y., Zhao, Z., Tang, X., Luo, J., Muhammad, Y. (2021). Fabrication of IPDI-LDHs/SBS modified asphalt with enhanced thermal aging and UV aging resistance. *Constr. Build. Mater.*, 302: 124–131.

Li, Y.Y., Wu, S.P., Dai, Y., Pang, L., Liu, Q.T., Xie, J., Kong, D.Z. (2018). Investigation of sodium stearate organically modified LDHs effect on the anti-aging properties of asphalt binder. *Constr. Build. Mater.*, 172: 509–518.

Pillai, S.K., Kleyi, P., De Beer, M., Mudaly, P. (2020). Layered double hydroxides: An advanced encapsulation and delivery system for cosmetic ingredients-an overview. *Appl. Clay Sci.*, 199: 105–868.

Redelius, P., Soenen, H. (2015) Relation between bitumen chemistry and performance. *Fuel*, 140: 34–43.

Advances in Civil Engineering and Environmental Engineering – bin Yahya & Bin Daud (Eds)

Variation law for aggregate physical stability of latosol particles in banana field

Zhang Xiumei*
Editorial Department of Journal, Hainan University, Haikou, Hainan, P.R. China

Wu Xinliang
School of Soil and Water Conservation, Beijing Forestry University, Beijing, P.R. China

Kou Lvheng, Fan Ben & Xie Zhongzhou
Mechanical and Electrical Engineering College, Hainan University, Haikou, Hainan, P.R. China

ABSTRACT: The latosol soils in the banana field of tropical Hainan were selected as the research object to explore the aggregate physical stability of latosol particles in the banana field by analyzing the tensile strength and penetration resistance of aggregates with different particle sizes (10-8mm, 8-5mm, 5-3mm and 2-1mm). The results show that: (1) the tensile strength of aggregates decreases with the increase in particle size, which shows that the tensile strength of aggregates is affected by soil bulk density and mechanical composition, and the tensile strength of aggregates increases with the increase of bulk density and sand content. (2) The long-term use of chemical fertilizer will harden the soil and increase the penetration resistance; the surface soil contains more organic matter than the surface and lower soil, and the water content is higher. On the one hand, the organic matter can improve the soil structure; on the other hand, because the water content of the surface is higher than that of the lower layer, it is convenient to form a thick water film on the particle surface, which makes the probe easier Penetration into the aggregate, which is why the penetration resistance of the topsoil is lower than that of the lower layer.

1 INTRODUCTION

Soil is the guarantee of human survival. In recent years, due to the use of chemical fertilizers and pesticides and the destruction of nature by human beings, the land has actually suffered a large degree of damage. And agricultural production activities are inseparable from the soil, without soil, there is no agriculture (Li et al. 2019). On the one hand, the soil is the carrier for the growth of various plants, providing nutrients for the growth of various plants including crops; on the other hand, it is the working object of farming machinery, and its mechanical properties directly affect the working effect of agricultural machinery. In some cases, soil acts as a bridge between crops and farm implements. China is a vast country. No matter from west to east or from north to south (She et al. 2017), when the terrain and topography change, the content of organic matter in the soil, the degree of dryness, and the composition of minerals will change regionally, which naturally makes the stability of soil aggregates different to varying degrees (Zhang et al. 2022), Among them, the different force stability will have a more obvious impact on the operation of agricultural implements (Wang et al. 2022). Since the working object of the machines and tools is soil, the mechanical performance parameters of the soil will put forward requirements and references for the design and manufacture of agricultural implements, which makes it very necessary to understand

*Corresponding Author: xiumei422@163.com

DOI 10.1201/9781003349563-4

the soil as much as possible in the early stage of the design of agricultural machinery. Soil scientists Utomo. et al. and Munkholm. et al. found that the higher the moisture content of the aggregate itself, the lower the tensile strength. In addition, some scholars found that the most critical factor affecting the tensile strength of aggregates is the content of dispersible clay, which is significantly positively correlated with the tensile strength. The effect of organic matter on the tensile strength of aggregates is uncertain because there are different research results: in a certain range, the content of organic matter increases, and the tensile strength increases. This is the same as that found by Yao. and Guimaraes., that the tensile strength of aggregates itself is positively correlated with the content of organic matter, while Zhang. et al. found that the tensile strength of aggregates in silty clay soil is negatively correlated with organic matter, and the performance of the negative correlation in clay is not obvious. Even some scholars such as Tormena and others found that there was no obvious relationship between the content of organic matter in soil and the tensile strength of soil aggregates in the study of cultivated land, that is to say, in this case, organic matter in soil is not the obvious influencing factor of the tensile strength of soil aggregates. Domestic scholars Hu Guocheng, Hua Zhongguang, and others found that the tensile strength of soil aggregates increased with the increase of free iron oxide content in the research process of the red soil in the middle and lower reaches of the Yangtze River. Moreover, in agricultural production activities, different methods of production activities management and land farming will greatly change the tensile strength of aggregates. Yang Wei of Huazhong Agricultural University found that different land use modes had a certain impact on the tensile strength of aggregates. Latosol is the zonal soil of tropical rainforest or seasonal rainforest in the southernmost part of China (Xing et al. 2022). Wu Peng study the characteristics of Lateritic soil, and a series of experiments were carried out to measure the soil cohesion force (Wu et al. 2020).

2 MATERIALS AND METHODS

The soils of tropical cultivated land such as Hainan are mostly latosol, which has large dispersion, weak flocculation and fixation, soft aggregates, and loose structure, as shown in Figure 1. This study mainly focuses on the lateritic soil of cultivated land in Hainan Province, analyzes the composition of soil particle size after dry sieving, then selects the aggregates with particle sizes of 10-8mm, 8-5mm, 5-3mm, and 2-1mm to test their tensile strength, and then tests the mechanical properties of penetration resistance for the aggregates with the particle size of 5-3mm to analyze the variation law of force stability and its influencing factors.

Banana field cultivated for many years is selected as cultivated land, in which the surface soil (0-10 cm) is collected according to the “s” shape, and the lower soil is collected with 30-40 cm soil layer samples. The collected samples include ring knife samples (120cm^3) and loose soil samples, in which four replicates are collected for each specification of ring knife samples. When collecting loose soil, in order to ensure that the undisturbed soil structure in the field will not be crushed and damaged during transportation, it shall be transported back to the laboratory in a special plastic box, and then spread out in the shading for air drying. When the humidity is appropriate, it shall be carefully broken along the natural fracture surface into about 10mm soil blocks, which shall be stored for standby after being completely air dried. According to the analysis requirements of soil physical and chemical properties, the samples shall be screened by 10mm and 2mm soil sieves respectively, bagged, and stored for standby.

The stability of aggregate strength is determined by compression test and penetration test. A self-designed combined tester for aggregate tensile strength and penetration resistance was used. The compressive test and penetration test are completed by replacing the test probe. The measured indexes are tensile strength (TS), fragility (f), and penetration resistance (APR).

Compression test: the probe of the aggregate joint tester is a flat plate probe, and the aggregates with different treatments are placed on the measuring platform of the joint tester. The movable platform moves upward slowly and controllably at the speed of 0.5 mm min-1, and the aggregate

a. Banana field b. Soil profile

Figure 1. Land use pattern and soil profile of Latosol.

contacts and compresses with the upper-pressure plate until it is completely broken. For the aggregate with high moisture content in the early stage, the aggregate is mainly plastic deformation during compression. Using the direct shear test for reference, the maximum compression displacement is set as 20% of the aggregate diameter. The tensile strength (TS) and fragility (f) are calculated. 30 capsules were determined for each treatment.

Penetration test: the determination process of penetration resistance (APR) of aggregates is similar to tensile strength. Replace the probe of the aggregate joint tester with a conical probe (with a diameter of 1 mm and a total cone angle of 30°C). 20 capsules were determined for each treatment.

3 RESULTS AND DISCUSSION

Figure 2 shows that after the dry screening, aggregates with different land use modes are easier to aggregate at particle sizes of 5-3 mm, 2-1 mm, 1-0.5 mm, and 0.5-0.25 mm. In the process of soil screening, the aggregates collide with each other on the screen wall and are subject to friction and tear action between each other and the bottom of the screen, resulting in mechanical crushing. The particle size distribution after dry screening shows that the stable state of aggregates subjected to mechanical crushing is at the particle sizes of 5-3 mm, 2-1 mm, 1-0.5 mm and 0.5-0.25 mm.

Figures (a) and (b) in Figure 3 show that 10-8 mm and 8-5 mm aggregates are crushed under pressure under different utilization modes, and are more likely to gather at the particle size of 5-3 mm after being crushed by mechanical impact; The two figures (c) and (d) show that under different utilization modes, 5-3 mm and 2-1 mm aggregates are crushed under pressure and are more likely to gather at particle sizes of 5-3 mm, 1-0.5 mm and 0.5-0.25 mm after being crushed by mechanical impact, which verifies the conclusion analyzed in Figure 3.

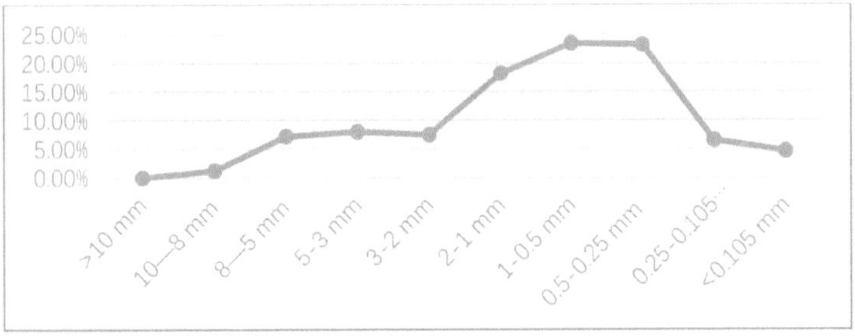

Figure 2. Particle size distribution of aggregates after soil dry screening.

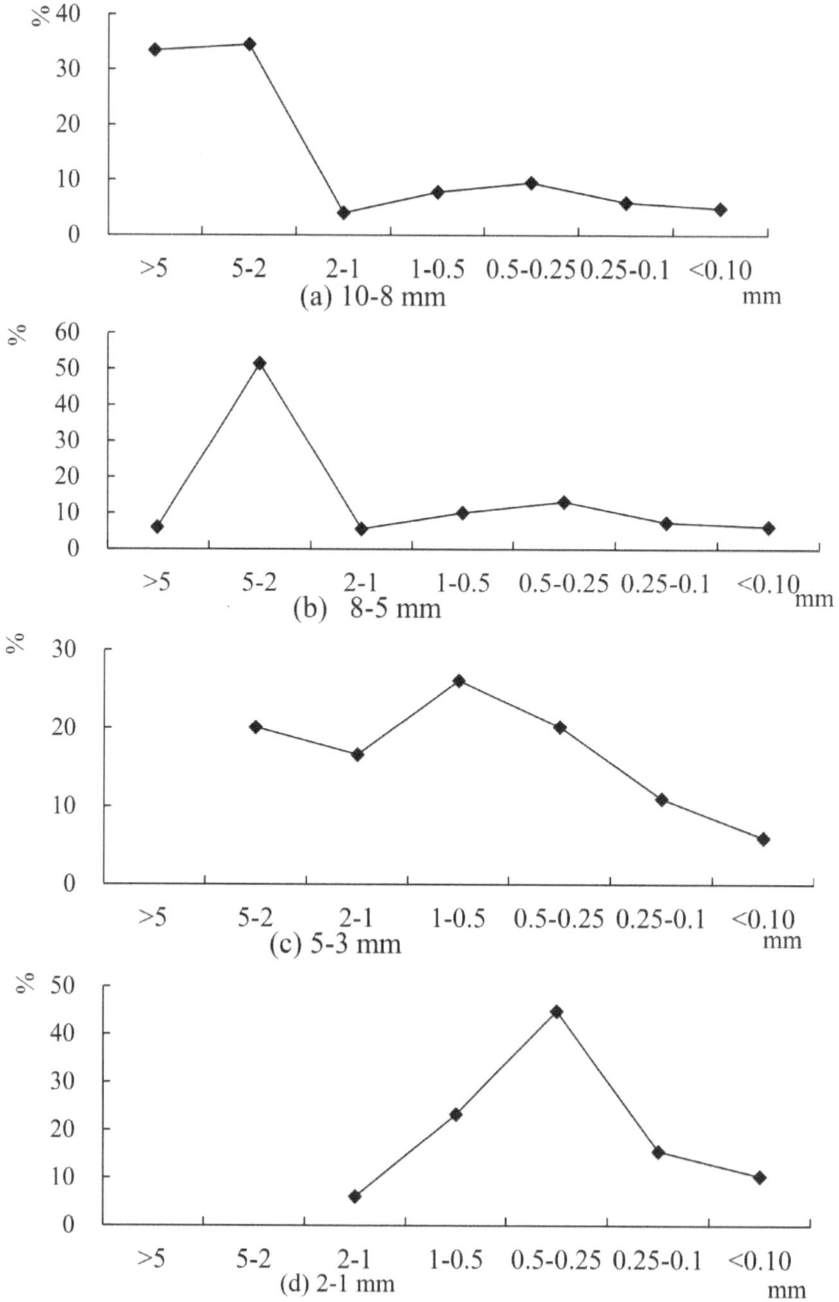

Figure 3. Tensile strength of soil aggregates with different particle sizes and particle size distribution of aggregates collected after crushing.

4 CONCLUSION

(1) The results show that the tensile strength of aggregates decreases with the increase of particle size; the bulk density of wasteland is the largest, the soil compactness is high, and the sand

content is the largest, while the content of silt and clay is low, which makes the tensile strength of aggregates larger. It is speculated that the soil bulk density and the content of sand and silt may be the influencing factors of the tensile strength of aggregates, and the larger the specific gravity is, the more the sand content is, the greater the tensile strength of aggregates is; the smaller the average value of tensile strength is the aggregate in the surface soil of forest land, and the pH value of the surface soil of forest land is the lowest, while the SOC content is the highest. It is speculated that the lower the pH value is, the more acidic the soil is, the higher the SOC content is, and the lower the tensile strength of the aggregate is; however, the main difference between forest topsoil and forest subsoil is SOC content. The SOC content of upper soil is high, but in this case, there is no obvious correlation between the average tensile strength of forest topsoil and forest subsoil with four particle sizes and SOC content. It is speculated that pH value and SOC content may be the influencing factors of aggregate tensile strength, but the influencing mechanism and the relationship are uncertain.

(2) There is no obvious relationship between fragility and land use mode under each particle size, and there is no obvious relationship between fragility and particle size under different land use modes.

(3) The long-term use of chemical fertilizer will harden the soil and increase the penetration resistance, which is not conducive to crop growth and farmland tillage by machines and tools. We should make rational use of chemical fertilizers and pesticides, return straw to the field, accelerate the promotion of conservation tillage, take good care of the earth and cherish our home; the topsoil contains more organic matter than the topsoil, and the moisture content is higher. On the one hand, organic matter can improve the soil structure; on the other hand, due to the high-water content in the surface layer and the thickening of water film on the particle surface for lubrication, the penetration resistance of the surface soil is lower than that of the lower layer.

(4) The particle size distribution after dry screening shows that the particle sizes of 5-3 mm, 2-1 mm, 1-0.5 mm, and 0.5-0.25 mm are the stable state of the aggregate after crushing and mechanical crushing.

ACKNOWLEDGMENTS

This work was financially supported by the Natural Science Foundation of Hainan Province (Grant No. 519QN192), and the Foundation of Hainan University (Grant No. KYQD(ZR)1807).

REFERENCES

Lijin, ZhengChao, Chen Bohao, et al. (2019) Effects of different land use modes on physical and chemical properties of latosols. *Chinese Journal of Tropical Agriculture*, 39(11):19–25.

She Li, CaiChongfa, WuXinliang, et al. (2017) Aggregate tensile strength and its influencing factors for typical [J]. *Journal of Soil and Water Conservation*, 31(5): 152–157.

Wang Mingen, Yan Limin, Chen Xuegeng, et al. (2022) Design and experiment of film picking and cleaning device for chain row type residual film recycling machine. *Journal of Agricultural Mechanization Research*, 214–218.

Wu Peng, Zhang Xirui, Li Yue, et al. (2020) Research on stacking characteristic parameters of lateritic soilmodel based on discrete element method, *Journal of Agricultural Mechanization Research*, 145–150.

Xing Jiejie, Zhang Rui, Wu Peng, et al. (2022) Parameter calibration of a discrete element simulation model for latosol particles in hot areas of Hainan Province. *Transactions of the Chinese Society of Agricultural Engineering*, 36(5): 158–166.

Zhang Xirui, Zeng Wangqiang, Liu Junxiao, et al. (2022) Design and experiment of lateritic soil inclined handle folding wing subsoiling shovel based on discrete element method, *Transactions of the Chinese Society for Agricultural Machinery*, 53(3): 40–49.

Advances in Civil Engineering and Environmental Engineering – bin Yahya & Bin Daud (Eds)
© 2023 The Author(s), ISBN: 978-1-032-39397-1

Theoretical investigation on mechanical properties of film-bag sand cofferdam

Feng Bo* & Chun-yuan Zhang
Department of Bridge Engineering, Southwest Jiaotong University, Chengdu, P.R. China

Cheng Fang
Chengdu Engineering Corporation Limited, Chengdu, P.R. China

Yong-jiu Qian
Department of Bridge Engineering, Southwest Jiaotong University, Chengdu, P.R. China

ABSTRACT: To study the mechanical properties of the film-bag sand cofferdam under the action of a geotextile hoop, the theoretical derivation of the film-bag sand enhancement mechanism is carried out. Via theoretical calculations, the relationship between the height of the membrane sand body bag and the tension of the geo-sand bag is proposed, and the relationship between the shape of the sandbag and its weight and the external load is further deduced, as well as the hoop effect of the geo-sand bag on the filled sand is verified on the influence of the overall shear strength of the internal sand body.

1 INTRODUCTION

Sand blasting cofferdam is widely used in the construction of temporary cofferdam projects (Kang et al. 2020), such as near-shore Bays and wharves of Rivers, Lakes, and Seas. Because of its advantages of simple construction, less pollution, high mechanization, and easy removal, traditional cofferdams can be divided into the earth-rock cofferdam, wooden cage cofferdam, steel sheet pile cofferdam, reinforced concrete cofferdam, wooden sheet pile cofferdam, pile-wall frame cofferdam (Peng et al. 2020; Xue et al. 2019), etc. Membrane bag sand cofferdam is a kind of cofferdam construction technology, which is made of artificial geomembrane bag material and filled with sand and stone by hydraulic power. Compared with the traditional cofferdam, the membrane bag sand cofferdam has a high degree of mechanization and fast construction speed advantages (Liu et al. 2020). Restalla et al., (2002) summarized the successful application of geomembrane bag sand technology in many projects, such as Green Island, Gold coast Broadwater and North Kirra Groyne in Australia. Saathoff et al., (2007) summarized the application of the geomembrane bag in scouring surface layer, coastal anti-corrosion, seawall, and sand reinforcement in Australia and Germany. In the NAJMAT project on Reem Island, Abu Dhabi, and United Arab Emirates (Yang & Zang 2009), a membrane bag sand cofferdam is used to construct a temporary cofferdam at the port, then the dry construction conditions for the project through continuous precipitation are provided.

The cofferdam technology is currently used in many offshore new construction projects, and many of them regard cofferdam as a cast-in-suit support foundation. However, few studies have used sand-blown cofferdams as the foundation of cast-in-suit bridges up to now, and there are few connections between engineering construction and theoretical research, so it is necessary to systematically study the mechanical properties of sand-blown cofferdams.

*Corresponding Author: fengboswjt@163.com

DOI 10.1201/9781003349563-5

2 GETTING STARTED STRENGTHENING MECHANISM OF FILM-BAG SAND

2.1 *Ultimate Strength of Clayey Soil Filled with Membrane Bag*

According to the strength theory of Mohr-Coulomb theory, when a certain point in the soil reaches the limit equilibrium state:

$$\frac{1}{2}(\sigma_1 - \sigma_3) = \frac{1}{2}(\sigma_1 + \sigma_3)\sin\varphi + c\cos\varphi \tag{1}$$

$$\sigma_1 = \frac{1+\sin\varphi}{1-\sin\varphi}\sigma_3 + 2c\frac{\cos\varphi}{1-\sin\varphi} \tag{2}$$

Make the $\frac{1+\sin\varphi}{1-\sin\varphi} = K_p$, which is expressed by the coefficient of passive earth pressure, according to the relationship of the trigonometric function $\frac{\cos\varphi}{1-\sin\varphi} = \sqrt{K_p}$. The formula (2) can be expressed as:

$$\sigma_1 = \sigma_3 K_p + 2c\sqrt{K_p} \tag{3}$$

For the film-bag sand, due to the hoop effect of the external film-bag, the limit equilibrium state of the sand body is:

$$\sigma_1' + \frac{2T}{B} = \left(\sigma_3' + \frac{2T}{H}\right)K_p + 2c\sqrt{K_p} \tag{4}$$

$$\sigma_1' = \sigma_3' K_p + 2\frac{T}{B}\left(\frac{B}{H}K_p - 1\right) + 2c\sqrt{K_p} \tag{5}$$

Bring Formula (5) into Formula (3), we can obtain:

$$\sigma_3 K_p + 2c\sqrt{K_p} = \sigma_3' K_p + 2\left[\frac{T}{B\sqrt{K_p}}\left(\frac{B}{H}K_p - 1\right) + c\right]\sqrt{K_p} \tag{6}$$

Among them, c is the cohesion of the sand body in the film-bag. φ is the internal friction angle of the sand body, and B, H, and L are the width, height, and length of the sand bag respectively.

According to Mohr-Coulomb theory, the shear strength of the reinforced film-bag is:

$$S = C + \sigma' \tan\varphi = c + c_f + \sigma' \tan\varphi = \left[\frac{\sigma_1' - \sigma_3'}{2} + \frac{T(B-H)}{HB}\right]\cos\varphi \tag{7}$$

For clayey soil, when a point in the soil reaches the limit equilibrium state:

$$\sigma_1 = \sigma_3 \tan^2\left(45^\circ + \frac{\varphi}{2}\right) + 2c\tan\left(45^\circ + \frac{\varphi}{2}\right) \tag{8}$$

$$\sigma_3 = \sigma_1 \tan^2\left(45^\circ - \frac{\varphi}{2}\right) - 2c\tan\left(45^\circ - \frac{\varphi}{2}\right) \tag{9}$$

Usually, the maximum principal stress σ_1 is the vertical compressive stress, which can be expressed as $\sigma_1 = \gamma H$. And γ is the weight of the soil, and H is the buried depth.

Therefore, the film-bag sand strengthens the ultimate shear strength of the thick soil:

$$S = \left\{\frac{1}{2}\left[\gamma H\left(1 - \tan^2\left(45^\circ - \frac{\varphi}{2}\right)\right) + 2c\tan\left(45^\circ - \frac{\varphi}{2}\right)\right] + \frac{T(B-H)}{HB}\right\}\cos\varphi \tag{10}$$

2.2 *Ultimate Strength of Non-cohesive Soil Filled with Membrane Bag*

For sandy soil, the cohesion is zero. According to Mohr-Coulomb theory, when a certain point in the sand body reaches the limit equilibrium state:

$$\sigma_1 = \frac{1+\sin\varphi}{1-\sin\varphi}\sigma_3 = \sigma_3 K_p \tag{11}$$

The internal limit equilibrium state of membrane bag sand under the action of the external membrane bag hoop can be expressed as follows:

$$\sigma_1' = \sigma_3' K_p + 2\frac{T}{B}\left(\frac{B}{H}K_p - 1\right) \tag{12}$$

The shear strength is:

$$S = \sigma' \tan\varphi = \sigma' \tan\varphi = \left[\frac{\sigma_1' - \sigma_3'}{2} + \frac{T(B-H)}{HB}\right]\cos\varphi \tag{13}$$

For non-cohesive soil, when a point in the soil reaches the limit equilibrium state:

$$\sigma_1 = \sigma_3 \tan^2\left(45^\circ + \frac{\varphi}{2}\right) \tag{14}$$

$$\sigma_3 = \sigma_1 \tan^2\left(45^\circ - \frac{\varphi}{2}\right) \tag{15}$$

Similarly, the first principal stress can also be expressed as: $\sigma_1 = \gamma H$, The formula (10) can also be written as follows:

$$S = \left[\frac{1}{2}\gamma H\left(1 - \tan^2\left(45^\circ - \frac{\varphi}{2}\right)\right) + \frac{T(B-H)}{HB}\right]\cos\varphi \tag{16}$$

3 THEORETICAL DERIVATION OF ULTIMATE BEARING CAPACITY OF FILM-BAG SAND

3.1 *The Relationship Between the Strength of Film-bag Sand and the Height of Tube Bag*

It is assumed that under the vertical external load, the shape of the pipe bag sand is shown in Figure 1, and the friction between the bag wall and the sand body is ignored.

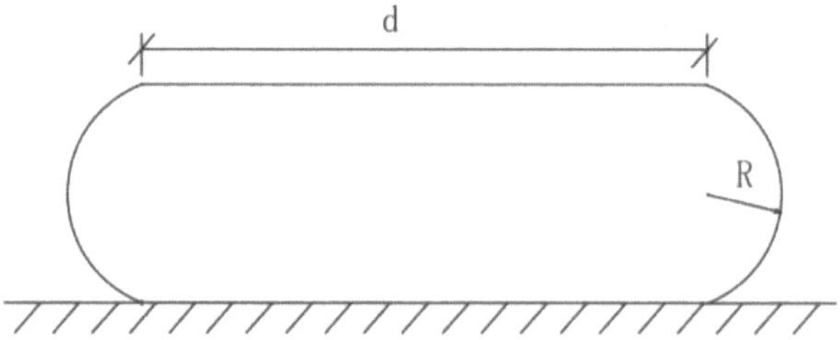

Figure 1. Approximate picture of film-bag sand.

If the total length of the bag is L, then:

$$L = 2d + 2\pi R; S = \pi R^2 + 2dR \tag{17}$$

The sand filling rate of the tube bag is:

$$w_0 = \frac{4\pi S}{L^2} = \frac{4\pi RL - 4\pi^2 R^2}{L^2}, 0 < w_0 < 1 \tag{18}$$

Assuming that the friction between the inner wall of the bag and the sand is not taken into account, according to the balance condition of the circumferential tension of the bag and the lateral extrusion pressure of the sand, the results are as follows:

$$\begin{aligned} 2T &= \int_0^{\pi} \rho g R(R + R\cos\theta)\sin\theta d\theta \\ &= \rho g R^2 \int_0^{\pi} \left(\sin\theta + \frac{\sin 2\theta}{2}\right) d\theta = 2\rho g R^2 \end{aligned} \tag{19}$$

Therefore, the circumferential tension of the bag can be obtained under the action of dead weight:

$$T = \rho g R^2 \tag{20}$$

When there is a vertical load on the bag, we can obtain:

$$\begin{aligned} 2T &= \int_0^{\pi} [\rho g(R + R\cos\theta) + pR] R d\theta \sin\theta \\ &= \rho g R^2 \int_0^{\pi} \left(\sin\theta + \frac{\sin 2\theta}{2}\right) d\theta + pR \int_0^{\pi} \sin\theta d\theta = 2\rho g R^2 + 2pR \end{aligned} \tag{21}$$

Therefore, the circumferential tension of the tube bag can be obtained under the external load:

$$T = \rho g R^2 + pR \tag{22}$$

$$2T = \gamma \int_0^H h dh = \gamma \frac{H^2}{2} \tag{23}$$

Under the action of internal pressure and external load:

$$2T = \int_0^H \left[(P_a + P_f) + \gamma h\right] dh = \gamma \frac{H^2}{2} + (P_a + P_f)H \tag{24}$$

$$T = \gamma \frac{H^2}{4} + \frac{1}{2}(P_a + P_f)H \tag{25}$$

Among them, P_a is the filling pressure and P_f is the external load, which can convert the vertical pressure.

3.2 *Relationship between shape of film-bag sand and internal and external load*

The micro-segment tube band is shown in Figure 2, and the pressure value at a certain point corresponding to the actual vertical position y of the filling process is as follows:

$$P = P_0 + \gamma y \tag{26}$$

According to the static equilibrium relationship between the two directions, the following can be obtained:

$$(T + dT)(\cos\theta\cos d\theta - \sin\theta\sin d\theta) - T\cos\theta + P\left(\sin\theta\cos\frac{d\theta}{2} + \cos\theta\sin\frac{d\theta}{2}\right) ds = 0 \tag{27}$$

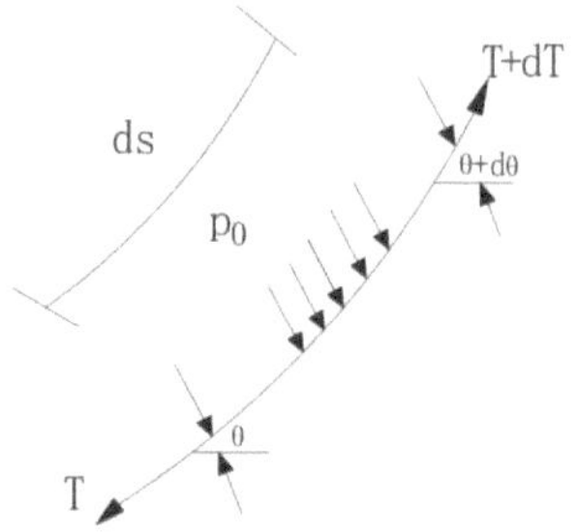

Figure 2. Calculation diagram of tube bag micro-segment.

According to the properties of the trigonometric function, it can be approximated as follows:

$$(T + dT)(\cos\theta - d\theta\sin\theta) - T\cos\theta + Pds\left(\sin\theta + \frac{d\theta}{2}\cos\theta\right) = 0 \tag{28}$$

Omitting the higher-order infinitesimal terms $dTd\theta$, $dsd\theta$ inverse:

$$dT\cos\theta - Td\theta\sin\theta + Pds\sin\theta = 0 \tag{29}$$

For the entire tube with:

$$\oint Tds = 0; \oint (T\theta_S - P)ds = 0 \tag{30}$$

When the friction between the filling sand and the bag is ignored, the bag carries the same tensile force in the extended circumferential direction, that is $dT = 0$:

$$Td\theta = Pds \tag{31}$$

For tube with micro-segments:

$$\begin{cases} ds = \sqrt{dx^2 + dy^2} = \sqrt{1 + y'^2}dx \\ \frac{dx}{ds} = \cos\theta \\ \frac{dy}{ds} = \sin\theta \\ y' = \frac{dy}{dx} = \tan\theta \\ t_S = 0 \\ t\theta_S - P = 0 \end{cases} \tag{32}$$

When the foundation is a straight rigid foundation, the contact length 2b between the bag and the foundation can be obtained according to the static equilibrium condition.

$$2b(P_0 + \gamma H) = W; b = \frac{W}{2(P_0 + \gamma H)} \tag{33}$$

W represents the bag filling weight.

$$S = \int_0^y \sqrt{1 + (x'(y))^2}dy \tag{34}$$

The tangent angle of a calculated position of the tube bag can be approximately written as a first-order function related to the cumulative arc length:

$$\theta = \frac{2(\pi - \theta_0)}{L - 2b}S + \theta_0 \tag{35}$$

Therefore, the ordinate and Abscissa of a calculated point on the pipe wall can be expressed as:

$$x(s)=\frac{L-2b}{2(\pi-\theta_0)}\times\left[\sin\left(\frac{2(\pi-\theta_0)}{L-2b}S+\theta_0\right)-\sin\theta_0\right]+b \quad x<-b, x>b \tag{36}$$

$$y(s)=\begin{cases} f(x) & -b\leq x\leq b \\ \frac{L-2b}{2(\pi-\theta_0)}\times\left[-\cos\left(\frac{2(\pi-\theta_0)}{L-2b}S+\theta_0\right)+\cos\theta_0\right]+f(b) & x<-b, x>b \end{cases} \tag{37}$$

Since the bag height H is unknown in the filling process, the filling mass W of the internal sand body can be written as a function of the bag area, which is a function of $x(s)$ and $y(s)$.

Therefore, the calculation of the shape of film-bag sand bag is a cyclic iterative process, so it is difficult to obtain a more accurate explicit expression. With the help of MATLAB software, the approximate figure of the tube bag can be obtained by iterative calculation, the results can be seen in Figure 3.

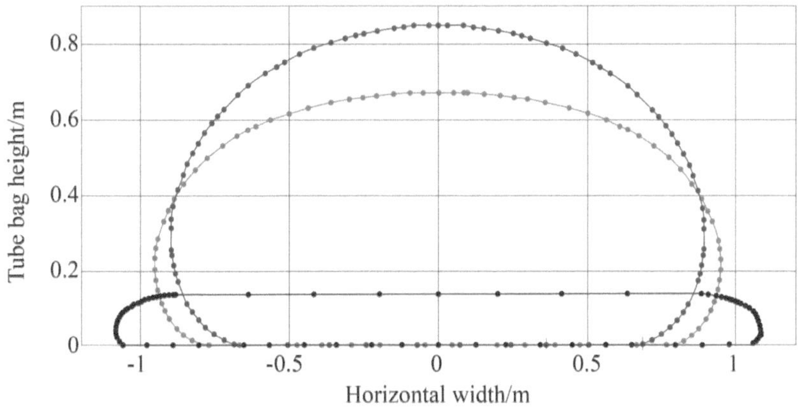

Figure 3. Shape diagram of tube bag under different filling pressure.

4 CONCLUSIONS

The main conclusions are as follows:

(1) The hoop effect of the geotechnical sand bag on internal filling sand is equivalent to providing additional cohesion to sand, which increases the overall shear strength. For non-cohesive soil, the main effect of the geotechnical sand bag is to increase the contact pressure between sand particles, increase the first and third principal stress of the sand body element, and improve the overall shear strength.

(2) After ignoring the friction between the geotechnical sand bag and the filled sand, the shape of the pipe bag can be approximately simplified to an expression related to the filling pressure, the dead weight of the sand body, and the boundary conditions of the foundation. Combined with the equal cohesion formula, the equivalent calculation formula of soil strength under different buried depths is obtained. Therefore, when calculating the cofferdam, the cofferdam can be equivalent to many kinds of materials according to the buried depth, avoiding the defect of the previous calculation of single material.

ACKNOWLEDGMENTS

The research reported in this paper is supported by China Natural Science Foundation (No. 51778532 and 52008047). Their support is gratefully acknowledged.

REFERENCES

Kang A, Zhu B, Lin P, et al. (2020). Experimental and numerical study of wave-current interactions with a dumbbell-shaped bridge cofferdam. *Ocean Engineering*. 210, 107433.

Liu A, Li B, Tong X. (2020). Research on Deformation Characteristics of New Type of Cofferdams Structure. *The 30th International Ocean and Polar Engineering Conference*. OnePetro, 1730–1733.

Peng Z, Wang T, Liang H. (2020). Numerical simulation on offshore artificial island cofferdam of Hong Kong-Zhuhai-Macao Bridge. *International Journal of Computer Applications in Technology*. 62(4), 285–295.

Restall S J, Jackson L A, Heerten G, et al. (2002). Case studies showing the growth and development of geotextile sand containers: An Australian perspective [J]. *Geotextiles & Geomembranes*. 20(5), 321–342.

Saathoff F, Oumeraci H, Restall S. (2007). Australian and German experiences on the use of geotextile containers. *Geotextiles & Geomembranes*. 25(4–5), 251–263.

Xue R, Bie S, Guo L, et al. Stability analysis for cofferdams of pile wall frame structures [J]. *KSCE Journal of Civil Engineering*. 2019, 23(9), 4010–4021.

Yang M, Zang D. (2009). Application of sand-bag cofferdam by hydraulic filling for NANJMAT project in UAE. *China Harbour Engineering*. 3, 45–48. (in Chinese)

Advances in Civil Engineering and Environmental Engineering – bin Yahya & Bin Daud (Eds)

Study on the application of permeable pavement materials in road rehabilitation

Yanjun Liu*
Zhengzhou Institute of Technology, Zhengzhou City, Henan Province, China

ABSTRACT: With the fast growth of Chinese cities in recent years, a considerable portion of the original land has been hardened. Most of them are hardened with impermeable building materials such as asphalt and cement, causing serious damage to the urban drainage function. In this paper, the writer introduces the road permeable pavement based on the analysis of two permeable paving materials, permeable asphalt mixture and permeable cement concrete, the application of road renovation, and their performance evaluation. After that, he combines these aspects with the real road renovation cases to analyze and summarize the full text, and make some recommendations.

1 INTRODUCTION

1.1 *Background and significance of the study*

The whole surface of modern cities are covered by reinforced concrete structures and impermeable pavements. Concrete paved roads improve people's mobility; nevertheless, when compared to natural soil, concrete pavements cannot absorb heat and penetrate precipitation, resulting in a slew of issues, including inadequate moisture in the soil, a lack of oxygen, and a rise in ground temperature.

Permeable pavement, which improves the permeability and breath-ability of the ground, might be considered to lessen the threats to cities by standard concrete paving. Permeable concrete pavements not only alleviate the drawbacks of standard concrete pavements, but also have their unique properties: Permeable paving enhances the ground's air and water permeability, keeps the soil wet, and provides moisture for urban plants.

1.2 *Domestic and international research status*

(1) Foreign research status
As early as the 1970s, some developed countries, such as the United States, Germany, and Japan, began to study drainage asphalt pavement technology, followed by the systematic development and application of various permeable engineering materials represented by permeable concrete. Kim et al. found that permeable asphalt pavements were more effective than permeable concrete, cement blocks. The role of ceramic blocks in decontaminating runoff was determined by comparing the evolution effects of different permeable materials on runoff pollutants, cement blocks, and ceramic blocks (Hyoungwon 2015; Hyunsuk 2015; Jaehun 2015). Christopher et al. found that the use of permeable asphalt pavements in cities can maximize the reduction in storm-water runoff rates while reducing the levels of pollutants, such as suspended solids and heavy metal polyaromatics in storm-water (Marion 2014; Murray 2014; Snyder 2014).

(2) Domestic research status
In China, the research on permeable pavement started at the end of 1990s, and it started late. In recent years, China has made great progress in the research of permeable bricks and permeable

*Corresponding Author: 11205628988@163.com

DOI 10.1201/9781003349563-6

concrete pavements, promulgating many standards one after another, and the application of permeable pavements has been increasing. Shen Zheng et al. explored the problem of engineering site selection and type selection for permeable pavement construction in China, and they put forward a practical method to comprehensively check the five conditions from five aspects of geology, climate, humanities, transportation, and engineering (Dong 2010; Shen 2010; Xiao 2010). Jiang Jianzhong investigated the application of asphalt pavement with permeable materials in the surface layer, subgrade as well as road base in sponge cities (Jiang 2021).

2 ANALYSIS OF THE APPLICATION OF PERMEABLE ASPHALT MIXES IN ROAD RECONSTRUCTION

2.1 *Brief description of permeable asphalt pavement*

The open-graded asphalt mixture with a large void on the anti-skid surface and occasional single size coarse aggregate is known as permeable asphalt pavement. This type of pavement is a skeleton void type structure, with a void rate of typically around 20% and a majority of connected voids. It uses a dense impermeable bottom layer to provide high anti-skid property, reduce water drift, water spray, water mist and glare, and has good noise reduction and high temperature stability. When it rains, rainwater can seep into the roadbed through material voids and enter the interior of the road surface, where it can then drain laterally down the cross slope of the road or continue to seep down into the interior of the roadbed, reducing noise from moving vehicles and relieving pressure on the urban drainage system.

2.2 *Performance evaluation of permeable asphalt mixtures applied in road rehabilitation*

2.2.1 *Evaluation of water penetration performance*

The permeability coefficient is a complete coefficient that measures how permeable porous materials are. As demonstrated by Darcy's law, for slower water velocities, the average infiltration velocity in the section is proportional to the hydraulic gradient under the influence of the head difference.

$$v = k \cdot i$$

Where v is the average water infiltration speed in cm/ s, k is the infiltration coefficient, and I is the hydraulic gradient in cm.

In this investigation, rutting slab specimens of PAC-13 and PAC-20 asphalt mixes were constructed in compliance with the specification for water penetration testing.

It's important to note that the percolation coefficient, or the total volume of water flowing out per unit time in flow units, is the test result obtained after the rutting slab percolation test according to our specifications. The units of the results must be converted from flow units to flow rate units for further analysis, which can be done by using the formula below. Both PAC-13 and PAC-20 show that water can be discharged in a short time, indicating that they have good permeability properties. The water permeability properties are shown in Table 1.

$$V = C_w \cdot S$$

Where Cw is the coefficient of water permeation within the pavement in ml/ min, and S is the percolator's bottom area in cm^2.

2.2.2 *Water stability evaluation*

Water stability is a highly essential characteristic of porous asphalt mixtures. In this study, water immersion Marshall tests and freeze-thaw splitting experiments were conducted on two porous asphalt mixtures, as shown in Tables 2 and 3. From the test results, it can be shown that the residual stability of PAC-13 and PAC-20 mixes are strong, and the splitting strength ratio also fulfills the

Table 1. Results of water penetration performance tests.

Mix type	Time spent	Seepage coefficient ml/min	Permeability coefficient cm/s
PAC-13	3.96	6044	0.56
PAC-20	3.73	6416	0.60

specification requirements. And from a complete perspective, the water stability of PAC-13 and PAC-20 mixes is good.

Table 2. Results of water immersion and Marshall tests.

Types of asphalt mixes	Stability after 48 h immersion in water, KN	Stability, KN	Residual stability, %	Residual Stability Technical Specification, %	Test method
PAC-13	5.08	5.22	97.3	≮ 85.0	T0709
PAC-20	5.83	5.94	98.1	≮ 85.0	T0709

Table 3. Freeze-thaw splitting test results.

Types of asphalt mixes	Splitting strength without freeze-thaw cycles, MPa	Splitting strength for freeze-thaw cycles, MPa	Splitting strength ratio, %	Residual Strength Technical Specification, %	Test method
PAC-13	0.72	0.81	90.0	≮ 80% of the total	T0729
PAC-20	0.47	0.54	87.1	≮ 80% of the total	T0729

2.2.3 *Evaluation of high-temperature stability*

The Rutting Test is suggested in China's Highway Engineering Asphalt and Asphalt Mixture Test Procedure (JTJE20-2011) for evaluating the high-temperature performance of mixes. As a result, the rutting test was utilized in this study to assess the mixture's high-temperature stability.

The rutting test is a method of stimulating the production of rutting grooves on a road surface caused by actual wheel loads. At 60°C, rutting plate specimens measuring 300 mm x 300 mm x 50 mm were tested for PAC-13 and PAC-20 with the specified gradation and the associated optimal oil to stone ratio, the results of which are shown in Table 4. The PAC-13 and PAC-20 mixes fulfill the standard dynamic stability values under heavy traffic situations. According to the data in the table, they show good high-temperature stability under the set gradation and optimal oil-to-stone ratio.

Table 4. Rutting test results.

Types of asphalt mixes	Dynamic stability, times/mm	Specification requirements	Test method
PAC-13	4645	1500 (general transport) 3000 (heavy traffic)	T0719
PAC-20	3031	1500 (general transport) 3000 (heavy traffic)	T0719

2.2.4 *Evaluation of low-temperature stability*

In this study, the low-temperature bending test was utilized to assess the low-temperature performance of two asphalt mixtures, PAC-13 and PAC-20. After that, the writer used the rutting test plates created by the wheel milling process, and cut each asphalt mix into prismatic beams of 250 mm 2.0 mm in length, 30 mm 2.0 mm in width, and 35 mm 2.0 mm in height (with a span diameter of 200 mm 0.5 mm).

For testing, the prepared beams were put in a $-10°C$ environment for over 10 hours of insulation. The test results are shown in Table 5. The maximum bending and tensile stresses of the PAC-13 and PAC-20 mixes satisfy the code required values, suggesting strong fracture resistance at low temperatures based on the test findings.

Table 5. Low-temperature bending test results.

Types of asphalt mixes	Bending and tensile strength, MPa	Max. bending and pulling strain. $\mu\varepsilon$	Modulus of stiffness, MPa	Technical specifications for maximum bending and tensile strain	Test method
PAC-13	3.70	3278	1130	$\not<$ 2500 pages	T0715
PAC-20	3.48	3983	878	$\not<$ 2500 pages	T0715

3 ANALYSIS OF THE APPLICATION OF PERMEABLE CEMENT CONCRETE IN ROAD RENOVATION

3.1 *Brief description of pervious cement concrete*

Permeable concrete is classified according to the different constituent materials, and it is commonly classed as cement permeable concrete pavement, polymer permeable concrete pavement, etc. Cement-based permeability concrete is a sand-free, porous concrete created with silicate cement as the cementitious material. It utilizes a single grade of coarse aggregate and no or little fine aggregate (Cao 2021; Liu 2021; Si 2021; Su 2021; Wang 2021; Yan 2021; Zhang 2021; Zhang 2021). This type of concrete is generally built with higher strength types of cement and demands a set-cement ratio of between 3.0 and 4.0 and a water-cement ratio of between 0.25 and 0.35. Polymeric permeable concrete is described as permeable concrete created by utilizing coarse particles of a single gradation and a binder agent based on asphalt or polymeric resin (Su 2020).

3.2 *Performance evaluation of pervious cement concrete for road rehabilitation applications*

3.2.1 *Evaluation of cooling performance*

The permeable cement concrete pavement belongs to test group one, the plain cement concrete pavement belongs to test group two, and the plain asphalt concrete pavement test group three. All three test groups utilized a pyrometer to directly measure the surface temperature. To decrease the test mistakes and base the tests on a single variable, this test program should be carried out concurrently to reduce errors related to time. The test began after 6 am and temperature changes were taken at different sites every three hours. And the temperature variations between 6 am and 6 pm were recorded and their measured pavement temperature T with each period are shown in Figure 1.

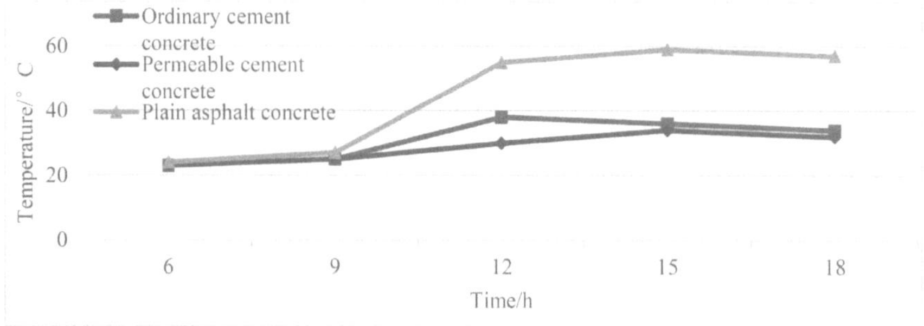

Figure 1. Temperature changing graph for pavement cooling test.

Based on the temperature profile, we can see that there are differences between the various types of concrete in terms of the temperature changing speed, but all three share a common changing line in that their surface temperatures all exhibit a positive correlation with the temperature profile. Ordinary concrete has an average temperature that is around 3°C higher than porous cement concrete, whereas asphalt concrete has an average temperature that is about 13°C higher. Permeable cement concrete may prevent the onset of high temperatures while it also increases the air humidity because of the pore structure of the material. As a result, permeable cement concrete is useful for enhancing urban environments. It matters when there is a "heat island effect".

4 EXAMPLE ANALYSIS OF ROAD IMPROVEMENTS

4.1 *Analysis of road improvement conditions*

The highway is located in the center of a city basin with a humid subtropical climatic zone, a temperate climate, and ample precipitation, with an average annual rainfall of 664.9 mm or more. The majority of the region's yearly precipitation, which ranges from 70% to 80% from June to September, and falls during the flood season. The average number of days with heavy-rain in July makes up 39.0% of all heavy-rain days; the average number of heavy-rain days in August makes up 29.9% of the total, and the two months from July to August make up 68.9% of all heavy-rain days. With maximum annual precipitation of 1,626.9 mm and low annual precipitation of just 316.3 mm, a difference of 5.1 times exists in yearly rainfall. With low inter-annual variance, the land surface evaporation is 520 mm and the multi-year average evaporation is 15.6 mm. With a level terrain and an average slope drop of 0.2% in the pilot area, the region is constructed with a high northwest and low southeast. Powdered clay, chalk, pebbles, conglomerate, mud-stone and other substances constitute the soil makeup from the surface below. With a soil permeability coefficient ranging from 1.8x10-6 to 3x10-6 m/ s, the soil permeability is good. Building low-impact development facilities are advantageous since the groundwater table is deeper than 10 meters.

4.2 *Determination of permeable pavement materials*

Permeable cement concrete and permeable asphalt mixtures are the two primary choices for permeable pavement materials.

When the design bending value index and asphalt layer bottom tensile stress calculation are done with the pavement accumulated equivalent axle Ne = 7656579 times, and the semi-rigid base layer bottom tensile stress calculation with the pavement accumulated equivalent axle Ne = 6169420 times, the project report survey data calculation can be obtained from the actual situation of the chosen road reconstruction project. The type of asphalt mix, the suggested thickness, the type of sub-grade, and the recommended thickness of the sub-grade structure are all mentioned in the pavement structure scheme. The pavement surface layer is built of asphalt concrete, with each layer's material having a 20°C compressive rebound modulus. Design bending and settlement values are utilized to determine the pavement thickness. All layers employ a 15°C compressive modulus of resilience and 15°C splitting strength to determine the tensile stress at the bottom of the layer.

Permeable asphalt mixtures are adopted in this study for this segment of the road as a result of this road restoration project.

5 CONCLUSION

By examining the use of permeable pavement materials in road reconstruction and combining them with particular road renovation instances, the writer clarifies the performance of permeable asphalt mixes and permeability cement concrete. The following findings are made from this study.

(1) Mixtures for permeable asphalt have strong water stability and permeability.
(2) Due to its porous nature, pervious cement concrete can postpone the onset of high temperatures while it also increases air humidity due to the increased internal moisture. The permeable road and ecological grass trench combination effectively removes a variety of contaminants and significantly contributes to the improvement of the surrounding ecosystem.
(3) After carefully analyzing a road example in the middle of a city basin, the writer compares the variations between permeable asphalt mixes and permeable cement concrete, assesses the performance of both, and offers reasonable recommendations for the selection of materials for this road improvement project.

REFERENCES

Christopher A. Murray, Kayla S. Snyder, Brooke A. Marion. Characterization of permeable pavement materials based on recycled rubber and chitosan [J]. *Construction and Building Materials*, 2014, 69.

Jiang Jianzhong. The application of permeable pavement on municipal roads in the construction of sponge city [J]. *Sichuan cement*, 2021 (12): 121–122.

Kim Hyoungwon, Ahn Jaehun, Shin Hyunsuk. Back-Calculation of Soil-Water Characteristics Curve for Permeable Pavement Material Based on Infiltration Tests [J]. *Journal of Korean Society of Hazard Mitigation*, 2015, 15(6).

Shen Zheng, Dong Xiang, Xiao Youyuan. Research on siting and type selection of permeable pavement projects for urban roads [J]. *Sichuan Construction Science Research*, 2010, 36(05): 156–159.

Si Zhengkai, Yan Liwei, Su Xiangdong, Zhang Meng, Zhang Yanli, Cao Lin Fang, Wang Wei, Liu Yanguang. Preparation and performance study of castor oil-based solvent-free polyurethane adhesives for permeable pavements [J]. *China Adhesives*, 2021, 30(02): 54–58.

Su Kai. Study on the effect of semi-permeable pavement runoff control under heavy rainfall conditions in Shanghai [J]. *Shanghai Highway*, 2020 (04): 102–106+151.

Advances in Civil Engineering and Environmental Engineering – bin Yahya & Bin Daud (Eds)
© 2023 The Author(s), ISBN: 978-1-032-39397-1

Fire protection design of external walls during the renovation of old residential areas

Yansheng Song* & Ke Wang
School of Civil Engineering, Shenyang Jianzhu University, Shenyang, China

ABSTRACT: In this paper, the fire simulation software PyroSim is utilized to simulate the exterior wall fire of a five-story old residential building in Beijing. Five working conditions shall be set according to the arrangement of the fire separation belt. The time of flame spreading to the top floor of residential buildings was studied under the different arrangements of fire barriers. The results show that the time of flame spreading to the top layer can be prolonged by 121.43% by centrally arranging a "fire isolation layer". During the reconstruction of the old residential area, if there is a residential building with corners, the use of fire-proof materials should be increased at the corners, and can be centrally arranged as a "fire isolation layer", which can not only ensure safety but also facilitate construction to the greatest extent.

1 INTRODUCTION

At present, the renovation of old residential areas is being carried out nationwide. As a northern city, the external wall insulation system is particularly important in the reconstruction of the old residential areas. The heating preservation materials are inflammable materials, so the number of casualties in a fire must be inestimable. (Dréan 2019) et al. studied the fire behavior of decorative materials. (Guillaume 2020) et al. conducted a numerical simulation of the vertical fire spreading process. (Bystrom 2012) et al. conducted building physical fire tests for different fire scenarios and obtained the differences between software simulation and physical fire through the comparative analysis of physical fire results and the fire simulation results. In this paper, by setting up a fire isolation belt on the exterior wall of the old building or laying a "fire isolation layer" with concentrated fire prevention materials, they can be used as "barriers" to block the longitudinal spread of fire, so as to avoid the formation of flame penetration on the external walls. Through the use of the software PyroSim for simulation analysis, the high-rise building exterior wall fire spread simulation is closer to the actual engineering.

2 THE NUMERICAL MODEL

2.1 *Building physical model*

In this paper, the writer takes a five-story L-shaped brick and the old residential building with concrete structure as the model. The first floor is 3.9 m high. The standard floor is 3 m, and the total height is 15.9 m. The location of the fire source is selected at the bottom near the corner; Wind speed is 1 m/ s, which is the direction of the right corner; The fire source area is set as 1 m^2, the fire source power 2 MW, the fire type t^2 type growth fire, the fire source growth coefficient

*Corresponding Author: ceyssong sjzu. edu. cn

DOI 10.1201/9781003349563-7

α 0.04689; The ambient temperature 20°C; The background environmental pressure 1.01325 × 105 Pa. The mesh size is 0.25 mm × 0.25 mm × 0.25 mm, and the range of simulation area is 7 m × 6 m × 16 m, which meets the requirements of simulation accuracy. The model is shown in Figure 1.

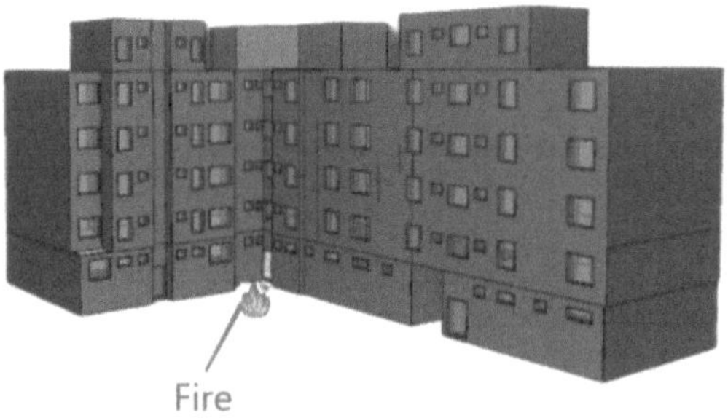

Figure 1. Architectural model.

2.2 *External wall insulation system*

The external wall insulation system of this model is set according to the existing "Code for Building Design Fire Protection" (GB50016-2014). The thermal insulation material extruded polystyrene board (XPS) and flame retardant material rock wool are used as the inner layer, and the outer layer is constructed with mortar material. The simulated fire isolation belt is set above the window at the height of 300 mm. The thickness of the first layer of the mortar protection layer is set to 15 mm. The thickness of the other layers is 10 mm, and the mortar protective layer wraps the whole insulation material. The thickness of the inner layer XPS and rock wool is set to 80 mm, and there is no gap among the protective layer, the wall, and the decorative layer.

In this simulation, the heat release rate (HRR) of XPS is 300 KW/ m^2, and the ignition point is 360°C. The physical parameters of the three materials are shown in Table 1.

Table 1. Thermal physical properties of building material.

Material name	Density kg/m^3	Thermal conductivity W/(m·k)	Specific heat capacity J/(kg·K)
XPS	25	0.03	5346.4
Rock wool	150	0.04	750
Mortar	1600	0.81	1050

3 FIRE SIMULATION AND RESULT ANALYSIS

3.1 *Working condition setting*

According to article 13 of "*The Standard for The Construction of Urban Fire Stations*", the layout of fire stations shall generally be determined on the principle that the fire brigade can reach the edge of the jurisdiction within 5 minutes after receiving the dispatch order. Considering the adverse conditions and the time required for rescue operations, in this paper, whether the internal and external wall flames can spread to the top within 15 minutes is regarded as the safety standard. In this simulation, the number and height of fire barriers are used as variables to analyze the impacts on the fire performance of external walls. Five working conditions are set, and the specific working conditions are shown in Table 2.

Table 2. Setting of fire prevention measures.

Scene number	Insulation material	Fireproof material	Protective layer	Fire prevention measures
Condition 1	XPS	Rock wool	mortar	No barriers
Condition 2	XPS	Rock wool	mortar	The isolation layer is equipped with a fire isolation belt
Condition 3	XPS	Rock wool	mortar	Set fire isolation belt on each floor
Condition 4	XPS	Rock wool	mortar	Set up half of the fire isolation layer
Condition 5	XPS	Rock wool	mortar	Set up a fire isolation layer

3.2 *Simulation results of each working condition*

3.2.1 *Condition 1*

In the case of no fire prevention measures, the flame spread speed of the external wall could not be controlled due to vertical penetration of flammable insulation material XPS. When the fire simulation reached 420 s, the flames spread to the top floor. The simulation results are shown in Figure 2.

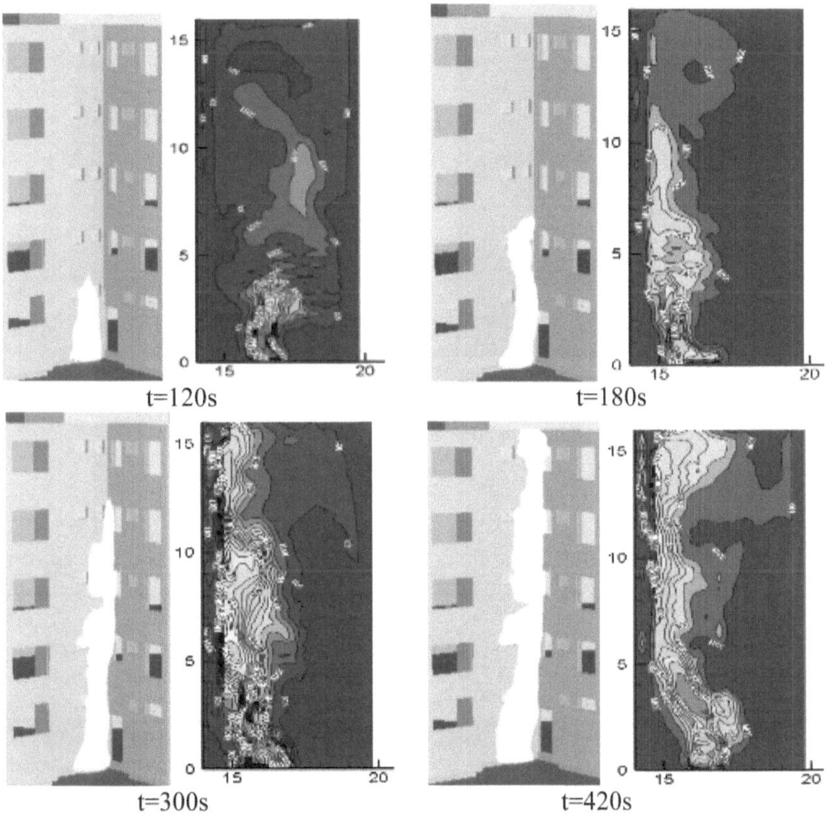

Figure 2. Simulation results of Working condition 1.

3.2.2 *Condition 2*

Although a layer of the fire isolation belt is laid above the windows on the second and fourth floors, the prevention and control effect is not obvious. In 600 s, the flame of the external wall spreads to the top floor, which is 42.86% longer than the time of the flame spreading to the top floor in working condition 1. The temperature at the top is about 550°C. It can be seen that at the corner of the chimney effect, only the fire isolation belt with the isolation layer can not effectively achieve the barrier effect of the external wall, and can not meet the safety requirements. The simulation results are shown in Figure 3.

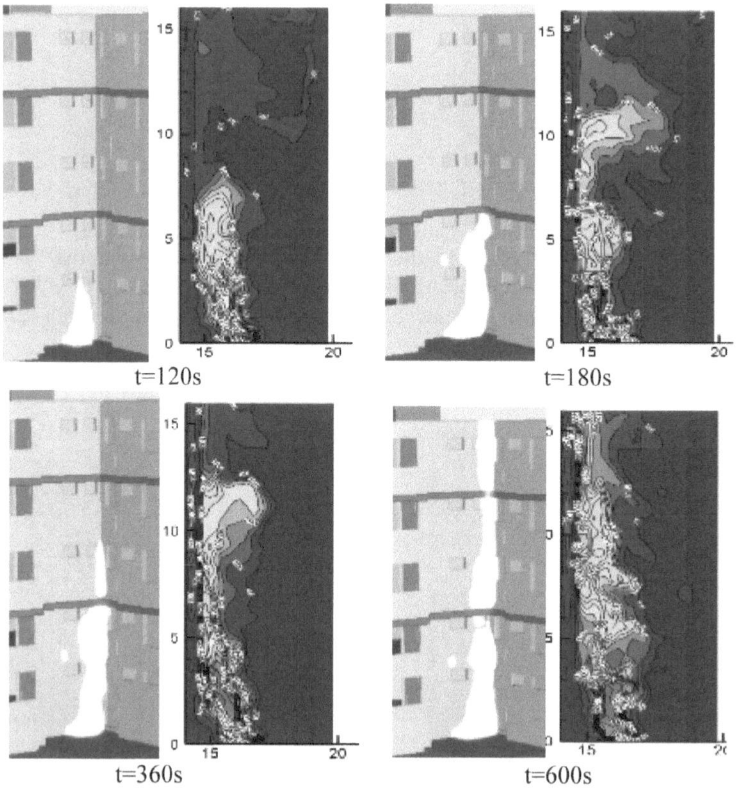

Figure 3. Simulation results of working condition 2.

3.2.3 *Condition 3*

Compared with working condition 2, the flame coverage area is significantly reduced compared with the laying of the fire isolation belt on the separation layer, and the flame still fails to spread to the top layer at 600 s. The temperature of the top layer is about 350°C, which is 200°C lower than that of working condition 2. However, due to the chimney effect, the flame continues to spread upward. When the fire simulation reaches 720 s, the flame spreads to the top floor, which is 71.43% longer than that in working condition 1. The top floor temperature is about 500°C. It can be seen that each layer is laid fire isolation belt, which can effectively delay the spread of external flame speed, and the time of spreading to the top is 120 seconds later than that of laying the fire isolation belt on the isolation layer. In addition, it can weaken the fire and reduce the temperature when the flame spreads to the top, but this condition does not meet the safety requirements. The simulation results are shown in Figure 4.

3.2.4 *Condition 4*

Concentrated rock wool material, laid into a 1.5 m high "fire isolation layer", can effectively block the the upward spread of flame and slow down the spread speed. However, the flame will break

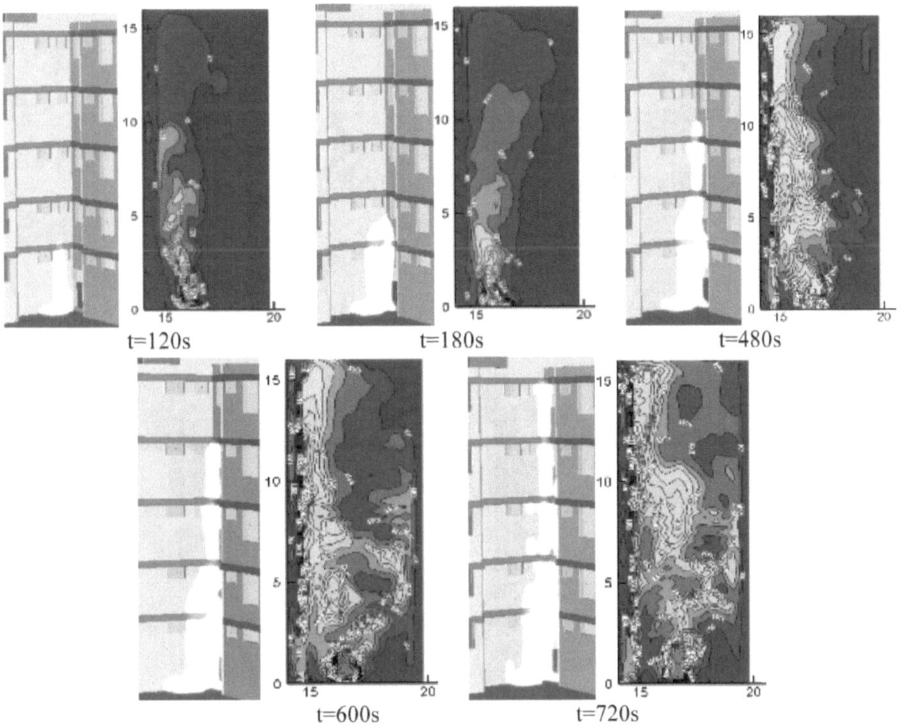

Figure 4. Simulation results of Working condition 3.

through the "fire isolation layer" after 600 s. Because of the smoke chamber effect, the fire does not decay and continues to spread upward. When the simulation time reaches 720 s, the flame spreads to the top floor. Compared with working condition 1, the time of flame spreading to the top layer is prolonged by 71.43%, and the top layer temperature is about 500°C. On the premise of no additional fireproof materials, centralized laying of fireproof materials can slow down the time when the flame of the external wall spreads to the top of the isolation layer, but there is no significant change in the temperature when it spreads to the top, which does not buy time for the fire rescue operation. The simulation results are shown in Figure 5.

3.2.5 *Condition 5*

Since the amount of rock wool material is doubled compared with working condition 4, the barrier effect on flame is obvious, and the speed of flame upward spread is effectively slowed down. At 720 s, the flame broke through the "fire isolation layer", and the fire continues to spread upward without attenuation. However, it does not spread to the top floor after 180 s, and the top floor temperature is about 300°C. When the fire is simulated to 930 s, the flame spreads to the top floor, which is increased by 121.43% compared with that in condition 1. Compared with working condition 4, working condition 5 greatly lengthens the time of the external wall flame spreading to the top floor. When the simulation time reaches 900 s, the top floor is still in a safe position, meeting the safety standard of 15 minutes. The simulation results are shown in Figure 6.

3.3 *Top window temperature comparison*

Through the comparative analysis, it can be seen that the temperature of the top window of the first 180 s is roughly the same. After 180 s, it can be seen that the temperature rise of the top window in

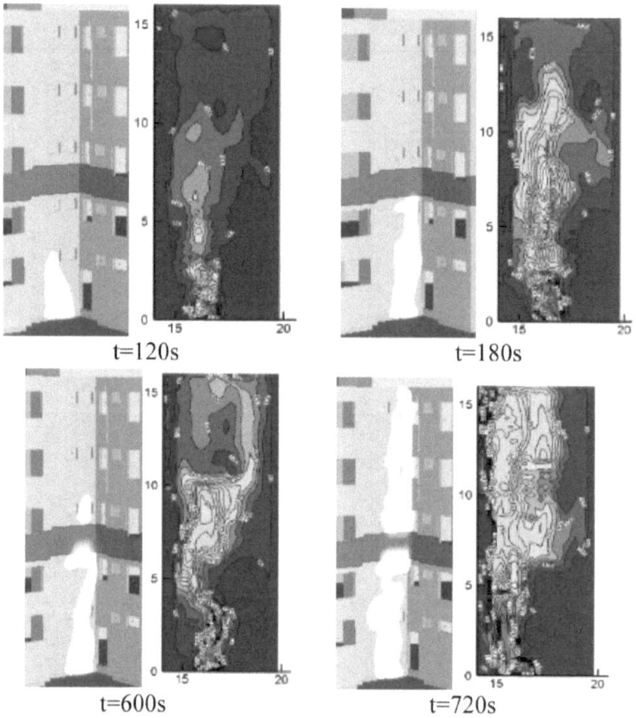

Figure 5. Simulation results of working condition 4.

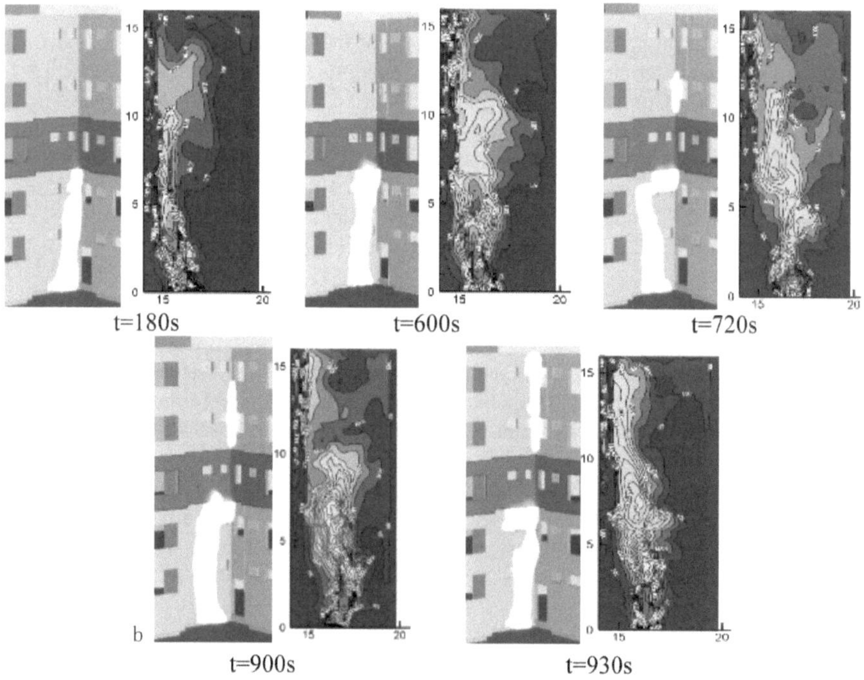

Figure 6. Simulation results of working condition 5.

working condition 3 is smaller than that of working condition 2. The reason is that the fire isolation belt is arranged on each floor. The temperature rise of the top window in the working condition with the "fire isolation layer" decreases more obviously. When the fire simulation reached 720 s, the outer wall flame could break through the "fire isolation layer" set in working condition 4, and the window temperature rises sharply. At 900 s, the top window temperature of working condition 3 is the same as that of working condition 4, which fails to meet the safety standards without additional fire prevention materials. In working condition 5, fireproof materials are added based on working condition 4 to delay the flame from spreading to the top floor. At 900 s, the top floor window temperature is reduced by 40% compared with that in working condition 4. The temperature measurement points and temperature curves are shown in Figure 7.

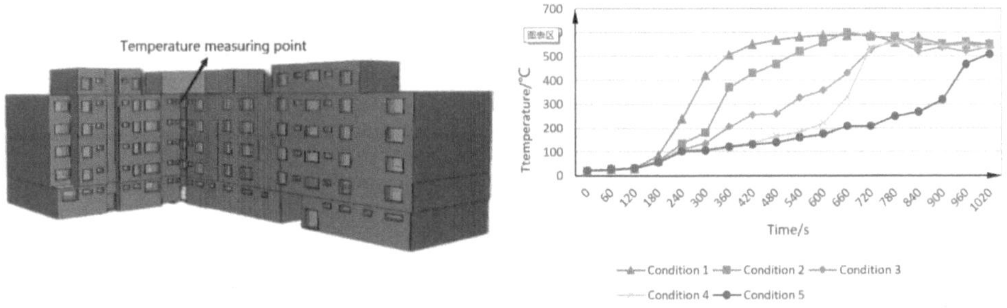

(a) Location of measuring point (b) Temperature time history curve

Figure 7. Measuring point position and temperature time history curve.

3.4 *Comparison and analysis of simulation results*

Based on the simulation results of the four working conditions, the following conclusions can be drawn: the fire separation belt can effectively block the flame spread speed of the external wall. The time of flame spreading to the top layer can be extended by 71.43% by setting fire separation belts between floors. However, at the corner with the chimney effect, the arrangement of a fire separation belt between floors can not meet the safety requirements of 15 minutes; Instead of adding incombustible materials, we can centrally arrange them as the "fire isolation layer" to delay the the spread of exterior wall flame above the "fire isolation layer height". However, the time of fire spreading to the top layer is basically the same as the temperature after the fire spreading to the upper layer; among the four working conditions, only the fourth working condition can meet the safety standard of 15 minutes. In terms of external wall fire protection of low-rise residential buildings, a "fire isolation layer" can more effectively delay the time when the flame spreads to the top floor, which is 121.43% longer than that without fire protection measures.

4 CONCLUSION

External thermal insulation of northern residential buildings is extremely important. Moreover, the thermal insulation material is combustible. Exterior fire caused by the exterior walls is common due to construction quality and the accumulation of debris in the corners. Moreover, there is a chimney effect at the corner, and the fire spread rapidly in the longitudinal direction. Only arranging a 300 mm high fire separation belt between floors can not meet the safety requirements. For the special structure at the corner, during the reconstruction of the old residential area, the number of fire-retardant materials should be added in the fire protection design and can be centrally set on one floor to form a "fire isolation layer", which can not only facilitate the construction but also meet the safety requirements.

REFERENCES

Alexandra Bystrom, Xudong Cheng, Uif Wickstrom & Milan Veljkovic (2012). Full-scale experiment and numerical studies on compartment fire under low ambient temperature *J. Building and Environment*, 51: 255–262.

Eric Guillaume, Virginie Dréan, Bertrand Girardin, Faiz Benameur, Maxime Koohkan & Talal Fateh (2020). Reconstruction of Grenfell Tower fire. Part 3— Numerical simulation of the Grenfell Tower disaster: Contribution to the understanding of the fire propagation and behaviour during the vertical fire spread *J. Fire and Materials*, 44 (1).

Rongxue Shang, Bin Zou, Peihong Zhang & Yujun Hao (2020). Fire characteristics of thermal insulation material XPS in ventilated environment, *J. Journal of Northeastern University* (Natural Science), 41 (02): 287–293.

Shoujiang Wang (2019). *Large-scale experimental study on fire behavior of building exterior wall organic thermal insulation materials and systems*, D. University of Science and Technology of China.

Virginie Dréan & Bertrand Girardin (2019). Numerical simulation of the fire behaviour of facade equipped with aluminium composite material-based cladding s-Model validation at large scale, *J. Fire and Materials*, 43(7): 839–856.

Yanan Hou, Xudong Cheng, Shenyou Liu, Changcheng Liu & Heping Zhang(2015). Experimental study on upward flame spread of exterior wall thermal insulation materials, *J. Energy Procedia*, 66.

Yuan Wang (2012), Simulation of fire spread in large buildings M. Beijing: *Dissertation of Beijing Institute of Civil Engineering and Architecture*, 2012: 32–36.

Yueqiang Wang & Baosheng Chen (2019). Study on fire characteristics of basic space of building *J. Fire Science and Technology*, 38(04): 508–511.

Yutao Zhang, Yujie Zhang & Yaqing Li (2020). Simulation study on fire spread characteristics of concave structure building façade *J. China Safety Science Journal*, 30 (08):70–78.

Zuo Q.L., Wang Y.J. & Li J.S. (2018). Safety performance of exterior wall insulation material based on large security concept, *J. IOP Conference Series: Materials Science and Engineering*, 359(1).

Advances in Civil Engineering and Environmental Engineering – bin Yahya & Bin Daud (Eds)

Discussion on problems in environmental management in the whole-process highway construction under new circumstances and the countermeasures

Qianjin Zhang, Xiaofeng Zhang & Long Liu
China Academy of Transportation Sciences, Beijing, China

Shoulin Li
Academy of Communications Science and Technology Group Co., Ltd., Beijing, China

Zening Xu*
Anhui Transportation Holding Group Co., Ltd., Anhui, China

ABSTRACT: At present, the construction of ecological civilization is paid more and more attention in our country. With the continuous improvement of environmental laws and regulations, the environmental management of highway construction projects is constantly optimized and adjusted, but there are still some problems in adapting to the requirements of environmental management under the new situation. Considering this, the status quo of environmental management in the whole process of highway construction is summarized, the main problems are discussed, and corresponding countermeasures and suggestions are put forward to provide support for the environmental management in the whole process of highway construction projects.

1 INTRODUCTION

With the continuous improvement of the national environmental protection legal system and the maturity of the environmental protection supervision and management system, a highway project-based system has been formed for environmental management in the whole-process highway construction projects. Generally speaking, with the assistance of professional environmental protection consulting agencies, the environmental management of highway construction projects includes environmental impact assessment (EIA), environmental supervision, environmental monitoring, and acceptance surveys for the completion of environmental protection. Among them, EIA is the foundation, for which laws and regulations such as the *Environmental Protection Law of the People's Republic of China*, *EIA Law of the People's Republic of China*, and *Regulations on Environmental Protection Management of Construction Projects* have provided clear provisions. According to the legal requirements, the construction unit shall submit the EIA documents to the authorized administrative department of ecological environment for approval before construction. Generally, environmental supervision and environmental monitoring are introduced in the construction stage, with the EIA documents and their approvals as the basis. The environmental protection acceptance upon completion shall be completed before the overall completion of the project. Strict regulations on the completion of environmental protection acceptance of construction projects are indicated in *the Regulations on Environmental Protection Management of Construction Projects*, and the Ministry of Ecology and Environment has formulated and issued the *Interim Measures for the Environmental Protection Acceptance of Construction Projects Upon Completion.* (National Environmental Regulation and Environmental Assessment [2017] No. 4).

*Corresponding Author: 18602552526@163.com

DOI 10.1201/9781003349563-8

Table 1. The environmental management of highway construction projects.

The environmental management of highway construction projects	Stage	Basis
EIA	before construction	*Environmental Protection Law of the People's Republic of China*, *EIA Law of the People's Republic of China*, and *Regulations on Environmental Protection Management of Construction Projects*
Environmental supervision	in the construction stage	EIA
Environmental monitoring	in the construction stage	EIA
The environmental protection acceptance upon completion	before the overall completion of the project	*The Regulations on Environmental Protection Management of Construction Projects*, *Interim Measures for the Environmental Protection Acceptance of Construction Projects Upon Completion.*

In the new era of ecological civilization construction, the focus should be on further improving the effectiveness of environmental protection management of highway construction projects. It is necessary to sort out and summarize the problems existing in the environmental management of highway construction projects, thereby putting forward countermeasures.

2 CONCEPT OF ENVIRONMENTAL MANAGEMENT IN THE WHOLE-PROCESS HIGHWAY CONSTRUCTION

At the conference on environmental management of construction projects held in 2002, in view of the problems existing in the environmental management of construction projects, the concept of environmental management in the whole-process construction projects was put forward, that is, starting from the whole life cycle of construction projects, environmental management is carried out in the whole process from project proposal to project disposal residual value (Dong 2008). The whole-process environmental management of highway projects refers to starting from the proposal of the project, to the feasibility study report, preliminary design, construction drawing design, construction period, and trial operation period, which have to carried out in accordance with national environmental policies, laws, regulations, standards, etc. Following the principles of sustainable development and environmental friendliness and a combination of macro-decision-making with micro-supervision, commitment is made to coordinate the construction unit to regulate various behaviors in the design, construction, supervision, consultation, and operation management. By doing so, we can reduce activities damaging the environmental quality and protect the ecological environment along the highway by use of scientific management methods (Cao 2006). A lot of practices have been carried out since the concept of the whole-process environmental management of highway projects was put forward. The purpose is to continuously optimize and improve the content and operating methods of the whole-process environmental management of highway projects. The working mode of environmental management of highway construction projects has been continuously innovated and developed, forming two relatively stable modes at present. One is to have the individual environmental protection management work independently entrusted, and the other is to implement package management in the "environmental protection stewardship" mode.

3 ANALYSIS OF PROBLEMS OF ENVIRONMENTAL MANAGEMENT IN WHOLE-PROCESS HIGHWAY CONSTRUCTION PROJECTS UNDER NEW CIRCUMSTANCES

3.1 *Insufficient legal basis for environmental management in the whole-process highway construction*

At present, the environmental management of highway construction mainly comprises EIA, environmental supervision, environmental monitoring during construction, and environmental protection acceptance upon completion. Among them, EIA and environmental protection acceptance on completion are procedures stipulated by laws and regulations, so they have more unified and standardized relevant standards and specifications, work modes, and work results. However, environmental supervision and environmental monitoring during construction have no laws and regulations to follow, so they have quite different work modes and work results. For example, some provinces follow the *Notice of the Ministry of Communications on Carrying out Environmental Supervision of Traffic Engineering* (Jiaohuanfa [2004] No. 314) to include environmental supervision in the project supervision. But the project supervision unit does not have such professionals, resulting in a certain influence on the implementation of environmental supervision. Besides, many construction units entrust environmental supervision to professional third parties. They lack effective restraints on construction units in actual work, but it difficult to punish them for violations. In addition, construction only cares about the construction progress and quality and often offers fines for environmental protection violations (CUI 2008). The greatest environmental impact of highway construction projects occurs during the construction period, and the ecological destruction has been difficult to recover by the time of completion acceptance (Zhu 2019).

3.2 *Too early or lagged EIA*

EIA serves as the foundation and key link of environmental management in the whole-process highway construction. The "Regulations on the Management of Environmental Protection of Construction Projects" was revised in 2017, adjusting the approval time of the environmental impact report (form) from "project feasibility study stage" to "before the start of construction". Before the start of highway construction, there are generally three stages: project proposal, engineering feasibility study, and survey and design. Different stages of research and design have different priorities and depths. The research goes from rough to detail, qualitative to quantitative, optimizing, adjusting, and improving layer by layer from the project proposal to the construction drawing design. Due to the continuous adjustment or optimization of the highway engineering design (route) scheme, the environmental sensitive points often change greatly, which directly leads to the corresponding environmental management problems such as environmental protection engineering changes (Gao 2014). However, careful consideration needs to be given to which stage prior to commencement of work is most appropriate for the EIA.

It is defined as significant changes when the EIA is completed in the feasibility study stage, while optimization and adjustment are made to the route plan of the feasibility study stage, causing major changes (especially aggravated adverse environmental impacts) to one or more of the five factors including the nature, scale, location, production process, and environmental protection measures of the project. In this case, according to the *EIA Law of the People's Republic of China* and *Environmental Protection Management Regulations for Construction Project*, the EIA documents should be re-approved. Two problems may appear when preparing the EIA documents in the project feasibility study stage. First, major changes are discovered in time, in this case, the re-approval of the EIA documents will increase the time and expense cost, and affect the progress of the project. Second, if a major change occurs without being discovered in time, the continued construction of the project constitutes illegal acts of "constructing without approval" and the violation of the "three simultaneous" system of environmental protection.

When carrying out EIA in the construction drawing stage, with a determined construction plan and sufficient project data, an accurate analysis of the impact of highway construction on the environment along the route can be made, therefore, we can propose targeted ecological protection and pollution prevention measures. If the intervention is too late, problems may also arise. If the EIA is started late in the route selection stage, attention will fail to be paid to avoid the core areas and buffer areas of nature reserves or the first-class water source protection areas and other areas prohibited by law during the route selection stage, and it will be too late on the discovery of legal obstacles. In this case, previous line selection, design, and other thematic work may be canceled, costing huge amounts of manpower, time, and resources. In addition, EIA in the construction drawing stage fails to include the demonstrated measures to prevent and control environmental pollution and ecological damage and the investment in environmental protection facilities in the investment budget, affecting the environmental protection measures in the construction drawing or the design of facilities, and violating the *Environmental Protection Management Regulations of Construction Project*.

Therefore, the selection of time to carry out the EIA must be carefully studied and judged.

3.3 *Coordination between works at different stages*

At each stage of highway construction, the construction unit needs to select the appropriate engineering consulting unit, design unit, construction unit, and other participating units by bidding. Different units are responsible for the tasks in different stages, and the work results of the previous stage are the basis for the work of the next stage. According to the regulations, environmental protection must be carried out in each work stage. Through the handover of work results at different stages, environmental protection measures are continuously refined. However, in actual work, due to insufficient attention to environmental protection measures, there are cases of omissions in the coordination of works at different stages. The writer has participated in dozens of highway environmental protection acceptance upon completion and finds that some of the measures proposed in the EIA stage of highway projects fail to be effectively implemented. By searching in the reverse order of construction drawing approval, the writer finds that these measures are omitted in the coordination process at different stages. In addition, there may be an incomplete or revised description of environmental protection measures. For example, the environmental impact report of a project proposed "installation of sound barrier measures for a village where the noise is expected to exceed the standard in the mid-term operation", but in the construction drawing design document, it became a necessity of "taking noise reduction shelter forest measures for a village where the noise is expected to exceed the standard in the mid-term operation." The noise reduction shelter forest needs land which is difficult to acquire, so it is hard to realize. Therefore, the environmental protection management at different stages has gone wrong in the coordination process which may cause significant changes to the project. Therefore, it is required that these environmental protection measures, especially the measures proposed in the approved environmental impact report (table), can be perfected and optimized in the coordination of different stages, but cannot be omitted.

In addition, EIA, environmental supervision and environmental protection acceptance upon completion are three important components of environmental protection work of highway construction projects, so each unit should maintain information exchanges (Gao 2014). At present, for highway construction projects, many units concerning EIA, environmental supervision, and environmental protection acceptance check have insufficient information exchange and communication, so it is difficult to timely and effectively solve the environmental protection problems existing in the highway construction and the acceptance survey (Wang 2008).

3.4 *Emphasis on construction while ignoring design*

The highway construction project has such characteristics as linear, wide area, and long time for the preliminary investigation and demonstration. However, due to the short investment plan cycles, some highway projects reduce the preliminary work time, failing to fully demonstrate the impact of highway construction on the environment in the feasibility study and design stages based on the actual situation of the area where the highway is located. It leads to the fact that even complete environmental protection measures are taken in construction, the construction still has a great impact on environment. In addition, designers sometimes lack professional judgments and cannot accurately predict and analyze the changes in environmental impact caused by project changes, and the environmental protection design of construction drawings they work out fails to meet the corresponding environmental protection goals (Qu 2016). There are a lot of cases like that. A national highway project in a province crosses the edge of a nature reserve and wild animals often appear in some sections of the road. Due to limited conditions, insufficient environmental investigation, and the lack of due importance attached to wildlife protection, the wild animal passage is not reserved, resulting in the phenomenon that some wild animals will be killed by passing vehicles after the road are made passable. To solve this problem, the highway administration department has set up fences in places where wild animals often appear. Although it has reduced the probability of wild animals being crushed to death, the quality of the habitat suffers, a complete habitat has been divided into two separate ones. The species are also separated, resulting in inbreeding of wild animals and the failure to gene exchange, which has reduced the adaptability of species.

3.5 *Insufficient funds for environmental protection*

When participating in the completion environmental protection acceptance survey, the writer found that some highway projects failed to implement the environmental protection measures proposed in the environmental impact report, especially those implemented in the later stage of the project such as sound barriers, sound insulation windows, and emergency accident pools. One important reason is the lack of funds. Specifically speaking, there are two reasons. First, because the upfront cost exceeds the budget, the fund for environmental protection facilities is misappropriated. Second, the environmental protection measures proposed during the EIA into the preliminary design hasn't been considered, and the full investment in environmental protection measures into the design budget hasn't been included. In the construction drawing design stage, the investment cost of environmental protection measures hasn't been included in the project budget due to a lack of accounting basis. For example, a highway in a province passes through a densely populated area, and the environmental impact report and its approval requires the installation of sound barriers. In the preliminary design, the sound barrier measures were not considered, the environmental protection investment was not included in the budget estimate and the construction drawing stage. In the end, the project couldn't pass the environmental protection acceptance upon completion, and they could do nothing but re-apply for special funds to carry out special design, re-construction, and other work. It not only seriously affects the overall progress of the project, causing unnecessary investment waste, but it is also suspected of violating the "three simultaneous" system of environmental protection.

3.6 *Insufficient environmental protection professionals*

For a long time, the environmental management of construction projects mainly relies on the EIA in the early stage and the environmental protection acceptance upon completion in the later stage. In the construction stage, the development organization, the engineering supervision unit, and the construction unit are at the center, while the environmental supervision unit is marginalized. Since most environmental management personnel of development organizations and construction units are part-timers with little professional background in environmental protection, they focus on project progress, quality, cost, etc., failing to attach due importance to environmental protection (Xu 2021). At present, the environmental management of highway projects is mainly entrusted to third party professional institutions, and there are very few practitioners in the highway industry with environmental expertise. For example, development organizations and construction units often incorporate environmental management into the safety management department, and there are few designers with environmental backgrounds, especially in the highway design field.

The author has visited many central enterprises and local design institutes and found that few designers majoring in the environment have been engaged in highway design. In addition, environmental protection management in the whole-process highway construction involves a wide range of professions and many units and personnel. With the continuous strengthening of ecological civilization construction in the new era, higher requirements have been proposed for highway environmental protection. Practitioners are required to not only have comprehensive knowledge in environmental protection, gardening, and aesthetics, but also master relevant knowledge in the field of engineering. Besides, they are expected to be able to adapt to local conditions, using flexible and diverse methods to maximize the application of the concept of environmental protection to each stage of highway construction, from proposal to highway operation, to minimize the negative impact of highways on the environment. Under the current situation, the shortage of environmental protection professionals limits the implementation of the concept of environmental protection management in whole-process highway construction.

4 STRATEGIES AND SUGGESTIONS FOR IMPROVING THE EFFECTIVENESS OF WHOLE-PROCESS ENVIRONMENTAL MANAGEMENT

4.1 *Establishment and improvement of an effective environmental protection management system*

The highway project construction unit shall establish and improve the environmental protection management system and assign qualified management personnel according to the actual work needs; efforts should be made to formulate enterprise-level and project-level environmental management systems that meet the needs of work, form an internal system for environmental protection management, and dynamically update according to changes in internal and external conditions of the enterprise. It is necessary to improve the construction of highway environmental management institutions, set up environmental management institutions and assign full-time environmental management personnel, defining the responsibilities of environmental management institutions through systems. Besides, it can regularly organize experts to revise and improve the environmental management system to ensure effective operation. According to environmental protection regulations and policies, it is required that the construction project should be organized, planned, coordinated and controlled throughout the entire process, including the proposal, decision-making, implementation, use and maintenance (Zhu 2006).

4.2 *Early intervention of the whole-process environmental management and timely EIA work*

The earlier the whole-process environmental management is introduced to the construction project, the earlier the environmental constraints faced may be discovered, and the easier it is to take effective measures, giving full play to the advantage of "prevention first" in the whole-process environmental management. If we intervene in the route selection stage to select the route with the least impact on the environment without reducing the road function, the inspection of environmentally sensitive areas and the environmental comparison of the corridor can be carried out together with the engineering consulting unit. After on-site surveys to determine accurate routes and major structures while ensuring that no major changes will occur, we need to carry out EIA timely and obtain the approval of the competent environmental department as soon as possible. By doing that, we can include various environmental protection measures proposed during EIA and the approvals into the preliminary design, and include the environmental protection investment into the investment budget.

4.3 *Coordination of environmental protection work at different stages*

For effective coordination of environmental management at different stages, it is required that the construction unit should take the lead to work out effective coordination measures. It is possible to innovate on the existing environmental management model to implement the "environmental protection stewardship" model for highway construction projects. There are two ways to achieve it. One is to take the construction project as the main body to implement this model, which not only provides integrated environmental protection services and solutions such as EIA, environmental supervision, and environmental monitoring during the construction, environmental protection acceptance, environmental protection consulting, and other professional work, but also provides "environmental protection plus" value-added services to improve the environmental protection level of highway construction projects, intensively save various input resources, and help highway construction achieve green and low-carbon goals. The other is to take the construction unit as the main body, and the "environmental protection stewardship" provides professional environmental management services for the construction unit, which can effectively coordinate the environmental management work at each stage and solve the problem of the lack of professional environmental management personnel in the construction unit. The first "environmental protection stewardship" model has been gradually promoted and applied in practice in recent years.

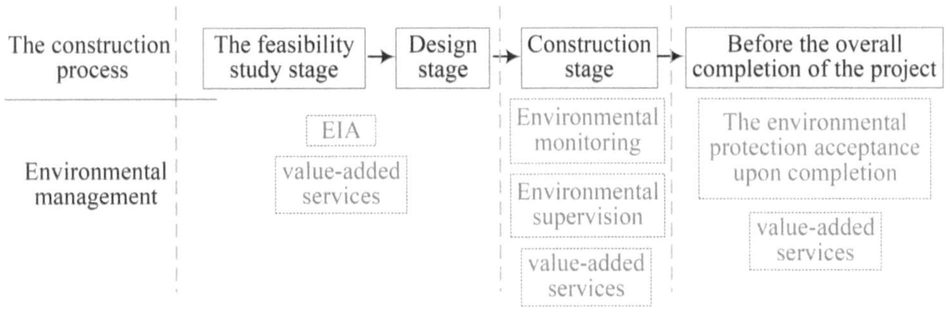

Figure 1. The environmental management of highway construction projects.

4.4 *Strengthening the professional design of environmental protection for highway construction projects*

It is stipulated in *Construction Project Environmental Protection Management Regulations*: "In the preliminary design of a construction project, it shall, in accordance with the requirements of environmental protection design specifications, compile an environmental protection section to introduce measures to prevent and control environmental pollution and ecological damage, and estimate the investment in environmental protection facilities." Therefore, it is necessary to link the preliminary design and construction drawing design of highway engineering with EIA to implement the EIA document and its approvals in the preliminary design and construction drawing design. In the preliminary design and construction drawing design review stage, the construction unit should also organize a special review of environmental protection to ensure that the design documents fully implement various environmental protection measures.

4.5 *Strengthening the role of environmental supervision*

Although there is no clear legal provision on environmental supervision at present, it is an important environmental management means for the implementation of environmental protection measures during construction. Therefore, construction units need to raise awareness, ensure the proper use of environmental supervision funds, improve environmental supervision and assessment mechanisms, grant environmental supervisors assessment authority, and incorporate environmental protection into the assessment indicators of the construction unit.

4.6 *Enhancing the involvement of environmental protection professionals*

To improve the involvement of environmental protection professionals in the highway construction industry, environmental supervision had been introduced into the highway system as early as 2002 to supervise and guide the environmental protection management and measures implementation during construction. In recent years, as the concept of whole-process environmental management has been recognized by the industry, the scope of highway environmental supervision has been expanded in some highway investment projects. They have adopted EPC, PPP, and general contracting of design and construction with the whole-process environmental supervision intervening from the preliminary design stage of the project. And these aspects have been widely adopted by highway construction units all over the country. In addition, to strengthen the environmental management in the whole-process highway construction, some construction units have tried to adopt a new management model of environmental protection stewardship. In short, to improve the environmental management capacity in the whole-process highway construction, it is necessary to adopt a reasonable management model, introduce a professional environmental protection team, and utilize the professional advantages to improve the quality of project environmental protection management.

4.7 *Fund guarantee*

Insufficient funding is a salient restrictive factor for carrying out the whole-process environmental management of highway projects. Without enough funds, not only the environmental protection measures of the project are difficult to be implemented, and the whole-process environmental management is hard to be put into place or produce concrete results. Therefore, from the time of project approval, sufficient funds must be reserved according to national regulations and the actual situation, and the cost of environmental protection measures should be included in the investment estimate, budget estimate, and budget for review when the project feasibility study report and design documents are reviewed. In addition, in the whole-process environmental protection management, it is important to strengthen funds management, use special funds for special purposes, learn from the successful experience of highway safety management, and ensure that environmental protection investment is not misappropriated. By doing so, we can ensure the normal operation of environmental protection.

5 CONCLUSION

In this paper, the writer analyzes the typical problems existing in the current environmental management in the whole-process highway construction and proposes countermeasures to improve its effectiveness from such aspects as the management system, working mechanism, professional design, environmental supervision, personnel, and funds. The writer hopes to provide a reference for the in-depth implementation of the environmental management in the whole-process highway construction.

ACKNOWLEDGMENTS

This work is supported by the Science and Technology Program of Anhui Transportation Holding Group Co., Ltd. (Grant No. jkkj-2019-16).

REFERENCES

Cao G. H. Research on technology and method system of whole-process environmental management of highway construction projects [D]. Xi'an: Chang'an University, 2006.

Cui C. X., Wu R. H., Shi L.A. The application status and trend of environmental supervision in China's highway construction [J]. *Shanxi Architecture*, 2008, 34(26): 226–228.

Dong X. L. Research on the whole-process environmental management system of highway construction projects [J]. *Journal of China Highway*, 2008, 21(1):100–105.

Gao S. H., Wang X. J., Li C. J., et al. Status quo of environmental management of highway construction and the countermeasures [J]. *Transportation Standardization*, 2014, 42 (9):16–22.
Qu Y. Analysis of the whole-process environmental management system of highway construction projects [J]. *Heilongjiang Transportation Science and Technology*, 2016, 39 (9): 191–192.
Wang F. Discussion on the problems in the environmental protection acceptance survey of highway construction projects upon completion and the countermeasures [J]. *Arid Environment Monitoring*, 2008, 22 (2): 126–128.
Xu H. H. Measures for environmental protection management of expressway construction projects under new circumstances [J]. *Western Transportation Science and Technology*, 2021 (9): 3.
Zhu H. M. Analysis of main problems in environmental management during expressway construction and countermeasures [J]. *Western Transportation Science and Technology*, 2019 (7): 187–189.
Zhu J. Study on the environmental management system of highway construction projects [D]. Xi'an: Chang'an University, 2006.

Advances in Civil Engineering and Environmental Engineering – bin Yahya & Bin Daud (Eds)
© 2023 The Author(s), ISBN: 978-1-032-39397-1

Analysis of the advance of flood control project in Jingjiang with satellite earth observation and remote sensing

Jiuyi Pan*
Suzhou Foreign Language School, Suzhou, Jiangsu Province, China

ABSTRACT: Jingjiang basin is rich in natural resources, fertile soil, and a pleasant climate. In the middle reaches of Changjiang River, it twists and turns with frequent floods since antiquity, affecting the living standards of citizens and the development of economics. Various projects of flood control have been carried out, including levee construction, flood-diversion project, water conservancy project, and cutting and straightening project. Satellite earth observation and remote sensing techniques are used to view alternations of watercourses in the most essential project of cutting and straightening directly. Principles of projects are introduced, as well as contributions to the flood control project. Though all the projects have played a great role in flood control work, several existing problems are proposed with appropriate solutions that can be referred to the further development of flood monitoring and control. In this article, the developing satellite earth observation and remote sensing technology are introduced and the advance of projects that are done to the flood control work is analyzed.

1 INTRODUCTION

1.1 *Geographical location*

The middle reaches of the Changjiang River begin at Yichang in Hubei province in the west and end at Chenglingji in Hunan province in the east. The section from Zhicheng in Hubei province to Chenglingji in Hunan province is called Jingjiang, with a total length of 337km. It is divided by Ouchiko into upper and lower Jingjiang, around 167km and 170km respectively (Xiong 2007). The Dongting Lake area on its south bank and the Han River Plain on its north bank are important granaries of China that are densely populated and flourishing in industrial and agricultural production (Wang 1992).

1.2 *Essential flood control section*

The terrain on the sides of Jingjiang is high in the south and low in the north for around 5m. During flood season, the flood level is higher than the surface of Jingbei by 10~14m. Han River Plain would be dangerous if the levee burst. And the twists and turns of Jingjiang make the current extremely impeded with frequent floods in past dynasties. In 1870, Yichang station of Changjiang River investigated that the flood peak has reached 110000m^3/s, while the flood discharge capacity of Jingjiang Wuhan section was only 60000~70000m^3/s, which is called a catastrophic flood in the annals of history (Xiong 2007). In the 450 years from 1499 to 1949, the levee broke 186 times, with an average of once every 2.5 years (Wang 1992). Flood control projects need to be carried out in Jingjiang.

*Corresponding Author: 3180200051@caa.edu.cn

 DOI 10.1201/9781003349563-9

1.3 *The flood control project*

Jingjiang levee water conservancy project was built in 1954 to give direct protection of flood control safety in Jingbei Plain. In 1967 and 1969, artificial cutting and straightening work was carried out for the two bending sections of Zhongzhouzi and Shangchewan in the lower reaches of Jingjiang. Research about the flood control capacity of Jingjiang and the blossom of the industrial and agricultural economy around Oxbow lake should be further enhanced. Changes in the value of water transport and water conservancy also need to be taken into consideration.

1.4 *Aim of this paper*

Satellite images of Landsat 8 and 7 are used to introduce the project of straightening and cutting. The purpose is to show the principle and significance of projects with the extent to which the project promotes flood control work.

2 METHODOLOGY

2.1 *Satellite earth observation and remote sensing*

Remote sensing refers to detecting and monitoring the physical characteristics of an area by measuring the reflected and emitted radiation from a distance. It has the advantage of being able to acquire information on a wide area at once which is useful for disaster situation monitoring, weather forecasting, and global environment understanding. In addition, once a satellite is launched into space, it will continue to do observation constantly, contributing significantly to some long-term and continuous projects.

Satellite images could give a clear view of changes in watercourses from space due to the cutting and straightening project. For human beings, light receptive cells in the retina in the eyes are sensitive to visible light with a wavelength from 400 nm to 700 nm and color formation could be done by the addition of primary colors (RGB). Landsat 8 and 7 bands carry two push-broom instruments: the Operational Land Imager (OLI) and the Thermal Infrared Sensor (TIRS), forming color images from spectral band measurements. Various spectral bands are used:

- Red (band 3), 600...630-680...690 nm, is used for imaging man-made objects, in water up to 30 feet (9 m) deep, soil and vegetation.
- Green (band 2), 515...520-590...600 nm, is used for imaging vegetation and deep water structures, up to 90 feet (30 m) in clear water.
- Blue (band 1), 450-515...520 nm, is used for atmosphere and deep water imaging and can reach depths up to 150 feet (50 m) in clear water.
- Near-infrared (NIR) (band 4), 750-900 nm, is used primarily for imaging vegetation.

2.2 *Images*

Selecting part of Jingjiang in 1986 (the earliest database) that the project had carried out through Landsat 8 in space (Figure 1). With the help of QGIS, images could be made by myself to show the change in watercourses.

Figure 2 shows the true color image with RGB: 3-2-1 (band 3- red input of imaging device; band 2- green input of imaging device; band 1- blue input of imaging device). Yellow stripes in the image represent Jingjiang with green plants surrounded. It is easy to recognize things on the ground because they have similar colors to what we see in daily life. However, it is hard to see the discarded river channels to verify the project.

To highlight the alternation of watercourses in an image, the false-color image is made with RGB:4-3-2 (band 4- red input of imaging device; band 3- green input of imaging device; band 2- blue input of imaging device). In the enlarged image of Figure 3, both blue-green and blue

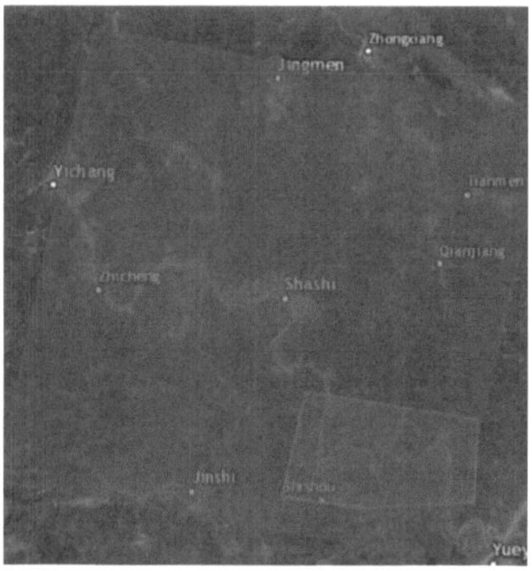

Figure 1. Region selection.

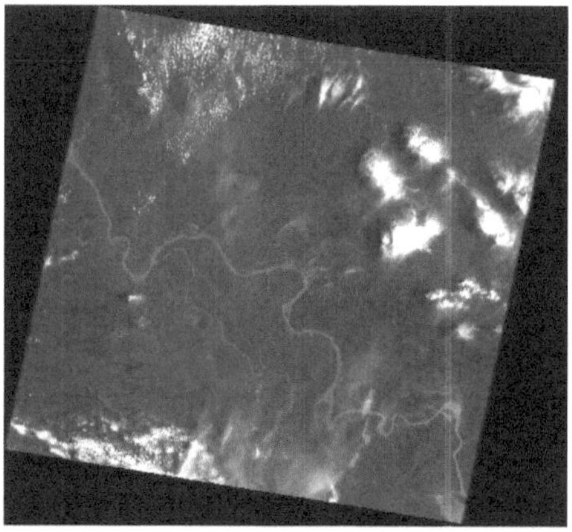

Figure 2. True-colour satellite image.

shows the watercourses of Jingjiang, while dark-blue stripes above the existing green-blue one are abandoned river channels. Governments cut some bending watercourses to promote the flood control work of Jingjiang.

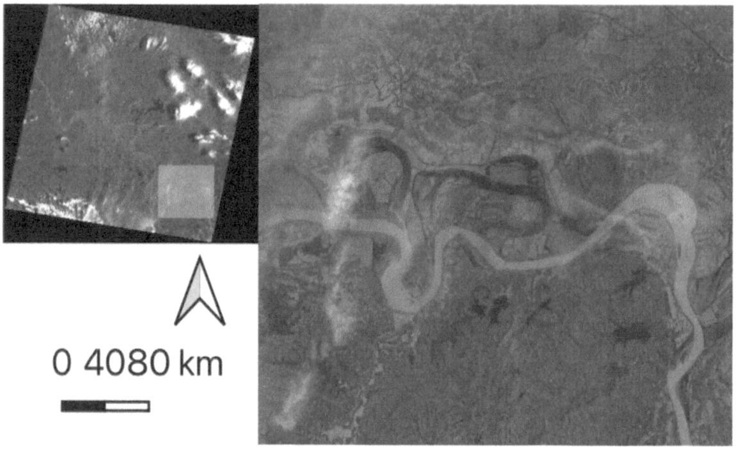

Figure 3. False-colour satellite image.

3 OVERVIEW OF JINGJIANG FLOOD CONTROL PROJECT

3.1 *Levees*

The construction of Jingjiang levee began in the Yonghe period of the Eastern Jin Dynasty (AD 345). In 1542, Haoxue, the last diversion port on the north bank, was blocked. And the levee was connected into a line, with a total length of 124 km, which was called the Wancheng levee. In 1951 and 1954, the dike of 8.35 kilometers above the Dujin platform and the original main dike 50 km downstream were included in Jingjiang levee. So far, the total length of the levee is 182.4 km, which is an essential flood control barrier in Jiang Han Plain (Li & Chen 1992).

The protection scope of Jingjiang levee is around 18000 square kilometers, with 11 million cultivated areas and more than 10 million population, including important industrial cities, oil fields, and trunk lines. It is a primary embankment with a crest elevation of 2.00 m above the flood level. Jingjiang levee is the most significant levee in Changjiang River basin and is a national guaranteed levee section (Zhang & Xu 2007).

3.2 *Flood-diversion project*

Jingjiang flood-diversion project is located in Gongan county, Hubei province, and was founded in 1952. It is 13.6 km wide and 68 km long, covering an area of $921km^2$. The designed water storage level is 42 m, and the flood storage capacity is 5.4 billion m^3. More than 300,000 soldiers and civilians who participated in the construction of the project completed the first phase of the main work at an amazing speed of 75 days (Guo etc. 2005). When the great flood occurred in 1954, after the flood-diversion project was used, the water level in Shashi was reduced by about 1m (Li & Chen 1992). The project played a role in ensuring the safety of Jingjiang levee and Jingbei Plain.

The flood discharge capacity of Jingjiang has been significantly improved, which alleviates the contradiction that it is incompatible with the huge and frequent flood volume upstream.

3.3 *Water conservancy project*

There is water from Qingjing River and Juzhang River above Jingjiang. The discharge of Qingjiang River $5000m^3/s$ can raise the water level of Shangshi by 0.1m. Although the inflow of Juzhang River is not as large as Qingjing River, its closeness makes it non-negligible.

In August, 1998, a catastrophic flood occurred. The fortified water level in Jingjiang section lasted for more than 20 days, and the warning water level lasted for more than 10 days. The flood water level continued to rise after breaking 45 m. The Geheyan water control project of Qingjiang River has brought obvious engineering benefits. The flood of Qingjiang River was blocked by the flood limit water level exceeding the flood limit so that the flood of the Three Gorges flowed out first. After the peak staggering regulation, the water level in Shashi stabilized at 45.22m, avoiding the losses caused by the start of the flood diversion project of Jingjiang and ensuring the safety of the section (Wang 2006). (When the flood level reaches 45.30m, the Jingjiang flood diversion project would be started to ensure the safety of the embankment.)

3.4 *Cutting and straightening project*

The lower Jingjiang is winding with a small water surface ratio during the flood period. The flood level was high endangering the safety of the embankments on both banks, especially the Jingjiang levee. At the same time, banks collapsed seriously and shoaled hindering navigation, which affect urban safety, industrial and agricultural production, as well as transportation. Therefore, the artificial cutting and straightening project of the lower Jingjiang system has been implemented since the 1960s (Li & Chen 1992).

The cutoff works refer that at the narrow neck of the excessively curved river section, the engineering measures of excavating the river diversion and accelerating the river channel from bending to straightening. Figure 4 shows the whole process of the project. The watercourses twisted and turned initially. The river was constantly scouring and eroding the bank, and the meander developed accordingly. With the process of the project, the original meanders were abandoned. Eventually, the watercourses were straightened and the previous watercourses became Oxbow lake which provides habitats for wildlife or bases of aquatic products.

From 1966 to 1971, the river channel cutting works of Zhongzhouzi and Shangchewan were completed in the lower Jingjiang section. Shatanziwan upstream was also naturally cut in July 1972 (Li, Xiong & Xiao 2003).

- Zhongzhouzi cutoff work: the curve is 37 km long, and the water surface on both sides of the narrow neck is about 4km away. The length of the approaching river is 4.3km, and the mean radius of curvature is 2.3km with a cut-off ratio of 8.5. The upper clayey soil of the river diversion zone is 6m thick, and the following is medium-fine sand. The bottom of the river diversion is

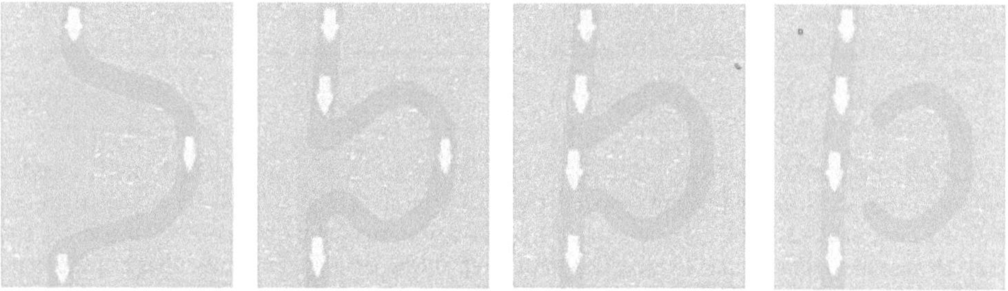

Figure 4. Principle of cutting and straightening project.

30m wide. The excavation depth is subject to the removal of clayey soil from river scouring. The cross-section area is 1/30 of the original river.

- Shangchewan cutoff work: the original length of the river was 33 km, and the water surface on both sides of the narrow neck is about 3 km away. The length of the approaching river is 3.5km, the mean radius of curvature is 2 km and the cut-off ratio is 9.3. The upper clayey soil covering layer of the river diversion zone is 20~30 m, and the lower part is medium-fine sand. The bottom width of the approaching river is 30 m, the excavation depth is 3 m below the designed dry water level, and the excavation section is 1/25~1/17 of the original river.

After the cutting and straightening projects, the length of lower Jingjiang decreased. Also, the flood level in the 200km upstream reach was reduced, alleviating the threat of flood to both banks. The voyage was shortened by 78km, and four shoals that hindered navigation were cut off. It also plays a great role in stabilizing the river regime and developing industry and agriculture.

4 IMPROVEMENTS

4.1 *Increase of diversion*

When the water level in Shashi is 43m, the total diversion discharge is only around $10000m^3/s$. Therefore, when the flood volume in Yichang is greater than 70000 m^3/s, the water level in Shashi will exceed the fortification water level and reach the warning water level. So, it is necessary to increase the diversion flow of Songzikou, Taipingkou and Ouchikou. At the same time, Tiaoxiankou should play its diversion role to reduce the pressure of upstream flood in Jingjiang section (Wang 2006).

4.2 *Further dredging*

To change the situation that the south bank of Jingjiang is higher than the north bank (5~7 m), ensure the safety of Jingjiang levee, and expand the discharge of Jingjiang, the Changjiang River basin planning office proposed the reinforcement scheme of Jingjiang levee with the theme of Jingbei desilting project in the early 1960s. However, the project was not implemented, only some field tests were carried out and some experience was gained.

At present, the combination use of flood storage and flood diversion works enable the relatively large flood peak flow through the upper Jingjiang. However, due to the relatively curved channel of the lower Jingjiang, the water flow is slow, and the sediment deposition makes the riverbed silted up. Below Jianli, the flood level often exceeds the warning line, and the high water level lasts for a long time, which seriously threatens the safety of people's lives and properties in Jiang Han Plain. Therefore, the Jingjiang section, especially the lower Jingjiang should be artificially widened, and the corresponding bends should be further cut and straightened to make it easier for flood discharge (Wang 2006).

4.3 *Renovation*

The construction standard of Jingjiang levee is not that strict. For example, in terms of the vertical height of embankments, other embankments are generally less than six meters. Although the Jingjiang levee has been reinforced many times with both internal and external platforms being built, its vertical height of more than eight meters reaches 150.65 km long, including the vertical height of more than ten meters reaching 6.97 km long, which brings great pressure to the flood control project.

Jingjiang levee was gradually built through many times of heightening and thickening in the long historical process. Limited to the historical conditions and construction level at that time, its building materials could only be taken from nearby soil and filled on site. As materials could not be selected, many impurities made the quality poor. The composition and compactness of the embankment-filling soil are quite different, and the permeability is extremely uneven. The existence of sandy soil, miscellaneous fill in the embankment, the irregular distribution of biological caves, and the interference of human activities make the flood resistance of embankments low. Among the 91 dangerous cases that occurred during the flood season in 1998, 62 belonged to the embankment danger. And 22 among 30 dangerous cases in 1999 (Zhang & Xu 2007).

4.4 *Afforestation*

Vigorous afforestation is needed in the upper and middle reaches of Changjiang River, Hubei province and four river basins in Hunan province. The forest per capita in Hubei province is $0.081hm^2$, which is lower than the national one of $0.113hm^2$. The forest coverage rate in Hunan province is 55% at present. Increasing vegetation is a must to achieve the target of 57% at the end of the eleventh five-year plan. Due to a large amount of cultivated land in the upper reaches of Changjiang River, the forest is seriously damaged with soil erosion intensified, and the concentration of sediment increases (Wang 2006). The government needs to do tree planting to keep a favorable water environment.

4.5 *Forecast*

Government departments should take responsibility for hydrological forecasting. The accuracy of the forecast directly affects the flood control measures and disaster relief actions of flood control headquarters and is related to the safety of people and national properties. Therefore, rainstorm and flood forecasting should be accurate and timely.

The application of remote sensing technology may provide the natural environment information of the whole hydrological underlying surface for flood forecasting and the background information for runoff estimation. Multi-temporal remote sensing records periodic hydrological information and the process of flood occurrence, which can be used as dynamic monitoring. The meteorological satellite data upstream is an important basis for river diversion and arranging the release of reservoirs. Landsat information can be used to estimate the boundary of the flood, the area of the inundated region, and economic losses. The disaster information system based on remote sensing can be collected, stored, and used to establish a database for comprehensive disaster analysis, prediction and forecast, and provide a scientific basis for macro decision-making and relief measures (Wang 2006).

5 CONCLUSION

After years of construction including levee construction, flood-diversion project, water conservancy project, cutting and straightening project, etc., Jingjiang initially could safely prevent the general flood of Changjiang River. However, there are still some problems affecting the flood control project. The constructed levee and completed projects need to be properly adjusted and further optimized. Long-term water environment monitoring and flood control work should be carried out, as well as the forecast. Satellite earth observation and remote sensing are newly-developing technology that is helpful for the development of flood control work. The basin of Jingjiang is flourishing in industrial and agricultural production, safety of lives and properties should also be guaranteed by the flood control project which needs to be further inspected and developed.

REFERENCES

Duan G., Peng Y. & Wang Q. (2007). Influence of channel evolution on flood control in Jingjiang reach of Yangtze River in recent years. *Yangtze River.* Vol. 38 No. 2 100–102.

Guo H. etc. (2005). Design and Realization in Information Demonstration System of Jingjiang Flood Diversion Area. *Resources and Environment in the Yangtze Basin.* Vol.14 No. 5 655–659.

Li M. & Chen B. (1992). Study and Application of Ellod Control Database of the Jingjiang River. *Remote Sensing for Land & Resources*. No. 1 27–33.

Li Z., Xiong F. & Xiao H. (2003). The change of water distribution relationship between Jingjiang River and Dongting Lake and its influence on flood control. *Symposium of Water and Sediment Committee of China Hydropower Engineering Society*. 59–62.

Lin B. & Zhou J. (2003). Improvement of flood control over dongting lake & Jingjiang strtch by clean water from TGP. *China Three Gorges Construction*. 4–6.

Wan Z. etc. (2019). Evolution of Jingjiang section of the Yangtze River based on historical maps and remote sensing over 100 years. *Scientia Geographica Sinica*. Vol. 39 No. 4 696–704.

Wang, M. (1992). River-lake relationship and regulation strategy for the area of Jingjiang River and Dongting Lake. *J. of Wuhan Univ of Hydr. & Elec Eng*. Vol. 25 No. 2 1–8.

Wang, Z. (2006). Flood management countermeasures in Jingjiang section of Yangtze River and downstream of Hanshui River. *China Water Resources*. 40–55.

Xiong H. (2007). Analysis on the influence of Changjiang River and Qingjiang River flood on Jingjiang River flood control. *China Hydropower Engineering Society Informatization Special Committee 2007 academic exchange meeting*. 372–383.

Zhang C. & Xu M. (2007). *Comprehensive improvement will be carried out to ensure the flood control safety of major dykes along the Jingjiang River*. The impact of the three gorges project on the middle reaches of the Yangtze river after completion of the special forum – the 10th symposium of the 2007 China Association for Science and Technology Annual Conference. 76–78.

Advances in Civil Engineering and Environmental Engineering – bin Yahya & Bin Daud (Eds)
© 2023 The Author(s), ISBN: 978-1-032-39397-1

Model tests on temperature and displacement characteristics of freezing underground engineering in water-rich gravel stratum

Wenxuan Ji*
College of Civil and Architectural Engineering, Xi'an University of Science and Technology, Xi'an, China

ABSTRACT: Taking a subway tunnel connecting passage in a water-rich gravel stratum as the research background, indoor strength tests were used to determine the mixing ratio of similar materials of gravel. According to the similarity criterion of freezing temperature field, moisture field, and stress field, a large-scale model test box with functions of loading, refrigeration, and monitoring was established to measure the temperature field, frost heave, and thaw settlement in real-time during the freezing process. The model test showed that when the soil temperature was above zero, the temperature drop gradient is greater. The soil around the freezing pipe became negative after an average of 150 minutes of freezing.The temperature drop rate decreased as the freezing process progressed. When the freezing rate was large, the drop rate of soil temperature was greater. During the frost heave stage, the displacement of the frozen wall and the ground surface increased gradually. The displacement of the frozen wall was greater than the displacement of the ground surface.The initial soil displacement in the thawing stage changed little, but with the continuous increase in soil temperature, the change in displacement increases. When the soil temperature reached the same as room temperature, the thawing displacement of the soil also tended to be stable. During the process of frost heave and thaw settlement, the displacement change of the frozen wall was larger than that of the ground surface.

1 INTRODUCTION

In the development of underground space, rock and soil formations with water content, weak crushing, and poor stability are often encountered. Common construction methods are difficult. Unpredictable accidents will occur in the process of tunnel excavation, so the freezing construction method is extremely suitable for such strata. The freezing construction method is mature, but this construction method will cause changes in the temperature field of the rock and soil layers, which will cause frost to heave and thaw and sink in the surrounding strata, which will then affect the ground buildings. Therefore, for rock and soil layers with different characteristics, a targeted design is required. The model test is an auxiliary means of freezing method construction design, which has been favored by many scholars at present. In 1980, Alfreds first proposed to use the method of model testing to solve the related problems of freezing engineering. Cai et al.conducted a model test study on the freezing method of a subway tunnel by using a self-developed model test box using similar criteria. Lei summarized the similar theories used in different model tests and their different derivation methods. Li used the method of model test to study the freezing method of a subway connection channel in Wuhan, and combined it with theoretical analysis and numerical simulation to obtain the variation law of formation frost heave and thaw subsidence, and obtained a good verification effect. Hu et al. used a large-scale freezing method model test to simulate the construction of a tunnel of the Hong Kong-Zhuhai-Macao Bridge by freezing method, analyze and

*Corresponding Author: 522012268@qq.com

DOI 10.1201/9781003349563-10

verify the rationality of the freezing scheme proposed for this project, and study its effectiveness and controllability. Cheng et al. deduced the similarity criterion of the coupling of the temperature and humidity fields and the mechanical field based on the similarity principle and finally designed a large-scale model test system to study the freezing of the rock and soil mass in a Guangzhou subway section as the background. The formation law of the wall and the problem of frost heave and thaw settlement. Huang et al. studied the influence of ground fissures on the changes in surrounding rock structure and lining structure caused by the subway connecting channel through the method of model test. Xiao et al.used the self-developed heat conduction model of the surrounding rock of subway tunnel to study the temperature field distribution of surrounding rock and obtained the soil thermal conductivity through inverse calculation.

The research in this paper is based on the engineering background of a subway tunnel connection channel in the water-rich pebble-gravel stratum, and the temperature field, frost heave, thaw-sink displacement field changes during the freezing process of the water-rich pebble-gravel stratum are studied through model tests.It provides a basis for the construction of urban rail transit engineering under the condition of water-rich pebble and gravel strata.

2 DETERMINATION OF MODEL PARAMETERS

2.1 *Model size design and determination of similarity constants*

Considering the limitations of existing test conditions, the convenience of machining models, and the spacing of frozen holes in the field, the geometric similarity ratio obtained from the similarity ratio criterion is $C_l = 20$. According to the frozen construction drawing of the tunnel, the length of the communication channel is 13.5m. According to the similarity ratio of 1:20, the length is 0.675m, and the model box is 1m. The height of thecommunication channel is 4.02m. According to the similarity ratio of 1:20, the length is 0.201m, the upper influence range is 3 times the hole diameter and 0.603m, the lower influence range is 2 times the hole diameter and 0.402m, and the height of the model box is 1.2m. The width of the communication channel is 3.5m. According to the similar ratio of 1:20, the width is taken as 0.175m, and the two sides of the communication channel are respectively 3 times the diameter of the hole and 0.525m, and the width is taken as 1.225m. Taking into account the main properties of the laboratory space and excavation problems, the size of the model test box was finally determined to be 1m × 1.225m × 1.206m. Then, there are similar ratio constants in the model test, as shown in Table 1.

Table 1. Similarity ratio constant table.

Parameter	Geometry	Displacement	Temperature	Time	Stress	Load	Gravity	Velocity
Similarity ratio	20	20	1	400	1	1	1/20	1/20

2.2 *Soil simulation design*

Because the subway communication channel is in the water-rich pebble gravel ground, the most representative soil layer of this type is selected, as shown in Table 2.

Table 2. Physical and mechanical properties of soil.

Name	Unit weight $kN{\cdot}m^{-3}$	Cohesion $c{\cdot}kPa^{-1}$	Internal friction angle ϕ	Moisture content %
Miscellaneous fill	19.2	11.0	8.0	27.5
Round gravel	20.6	0.0	33.0	6

Table 3 shows the particle size composition of the prototype soil sample and the model test soil sample composition. Figure 1 shows the particle gradation curve of the field soil layer and the soil in the model. It can be seen from the figure that the fine particle content of the model material is higher than that of the undisturbed soil.

Table 3. Soil particle grading.

Undisturbed soil sample	Particle size (mm)	40~20	20~10	10~5	5~2	2~1	1~0.5	< 0.5
	Mass percentage (%)	37.3	35.4	13.0	5.9	1.3	1.5	5.6
Test soil samples	Particle size (mm)	20~10	10~5	5~2	2~1	1~0.5	0.5~0.25	< 0.25
	Mass percentage (%)	15.4	30.8	20.9	14.3	3.2	4.1	11.3

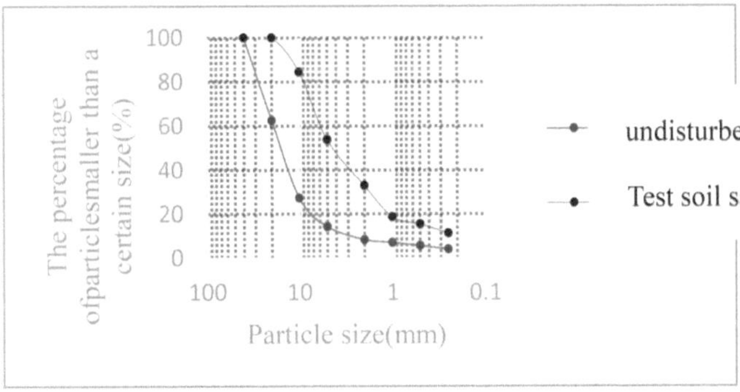

Figure 1. curves of particle grading.

3 DESIGN OF THE MODEL TEST SYSTEM

As shown in Figure 2, the model is mainly composed of loading, freezing, and measuring systems. The loading system is used to simulate the in-situ stress of the stratum where the connection channel is located; the freezing system runs through the entire test process to meet the needs of freezing temperature; the measurement system is based on the change data of the displacement field and temperature field during the test process. It provides a reference for the research on the stability of the frozen wall of the stratum during the freezing process, to guide the on-site construction.

(1) Model loading system
It can be seen from the similarity criterion $\sigma/\gamma H$ that the stress similarity ratio $C_\sigma = 1$ and the severity similarity ratio $C_\gamma = 1/C_l = 1/20$ are similar. Therefore, the test fill density is 20 times that of the field soil density. In this test, the vertical load is applied to simulate the pressure of the overlying soil layer of the connecting channel.

(2) Calculation of overlying load
The load value γH applied by the model depends on the weight and thickness of the overlying soil layer of the connecting channel, so the calculation process and results of the overlying load: the top of the connecting channel is buried at a depth of 20.35m, and the gravitational acceleration g is 10N/kg. The soil layer information is plain fill, silty soil, silty clay ②2-2, silty clay ②3-2, silty clay ②4-2, silt, silty (fine) sand, and pebble. Self-weight stress of overlying soil layer is: $\sigma_z = \sum_{i=1}^{n} \gamma_i h_i$ =435.89kN.

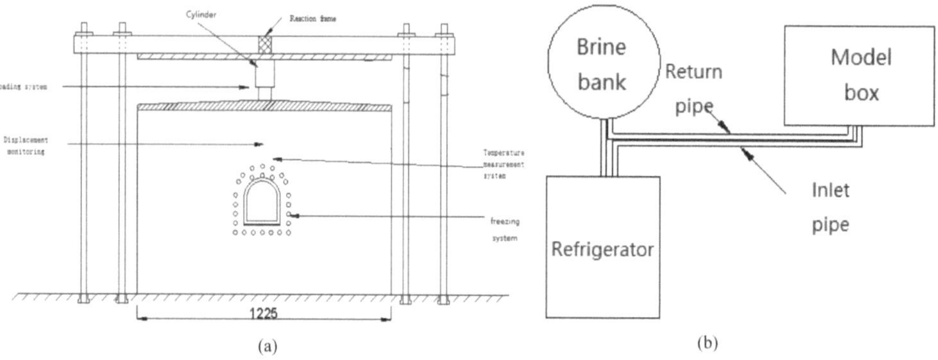

Figure 2. Test bench structure diagram.

(3) Layout scheme of freezing pipes
The arrangement of the frozen tubes in the field is significantly different from the arrangement of the frozen tubes in the model test, which is due to the effects of the size. Thus, we should freeze the tube for redistribution.

According to the on-site design plan, it is known that there are 56 frozen tubes with a specification of Φ89×8. Using similar quasi-sides to deduce, it is concluded that the indoor model test should use frozen tubes with an outer diameter of 4.5mm. At the same time, to ensure that the freezing wall can be formed stably, the freezing tubes were rearranged in this experiment according to the principle of total heat dissipation conservation.

The total cooling capacity of the freezing tube is:

$$Q = ndHk \tag{1}$$

In the formula, n is the number of freezing tubes,d is the outer diameter of the freezing tubes,H is the freezing depth, andk is the heat dissipation coefficient of the freezing tubes. Then $Q_1 = n_1 d_1 Hk = 89 \times 0.0045 \times H \times k = 0.4005Hk$. Since the frozen tube selected in this model is Φ16, there are $Q_2 = n_2 d_2 Hk = n_2 \times 0.016 \times H \times k = 0.016 n_2 Hk$. $Q_1 = Q_2$, and 25 frozen tubes of Φ16 are selected. Therefore, the specific arrangement of the freezing holes in this test is shown in Figure 3.

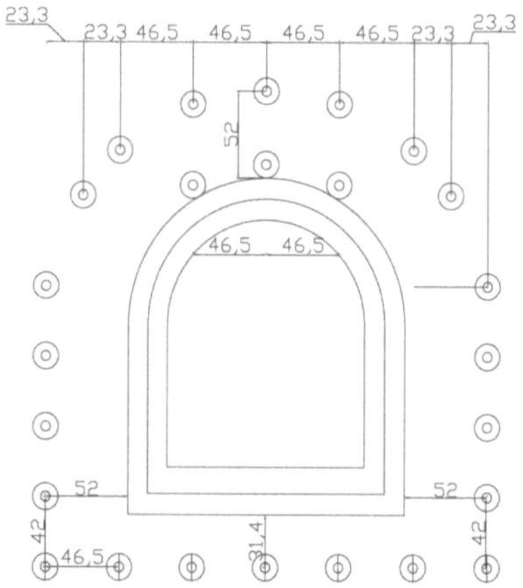

Figure 3. Freeze hole layout.

To ensure that the arrangement circle diameter of the frozen pipes in the test is consistent with the field, the frozen pipes of the model test are respectively arranged around the communication channel. The excavation diameter of the arch bottom is set to 52mm, and the rest is set to 31.4mm.

At the same time, the measurement results of the model test are closely related to the selection of the frozen pipe material. To ensure the accuracy of the test, this test uses two layers of casing test copper pipes. The piping returns to the refrigeration unit. The size of the copper tube is the inner layer Φ6, the outer layer Φ16, and the length is 1000mm.

(4) Refrigeration cycle system

The refrigeration cycle system is mainly composed of a brine tank, refrigeration unit, water pump, liquid collecting pipe, refrigeration pipe, etc. In the test preparation stage, the refrigeration pump switch is turned on, and the brine is pumped to the refrigeration unit. At this time, the compressor is turned on to cool the brine with ethylene glycol. Then, the cooled brine is passed into the model box to form a whole refrigeration cycle. When it is determined that the freezing wall is formed and reaches the design temperature, the refrigeration process is stopped and the compressor switch is turned off. It maintained a constant supply of cooling water in the brine tank to the model box to keep the freezing wall at the design value.

(5) Temperature measurement system

The main function of the temperature measurement system in this experiment is to observe the temperature changes of the freezing system and the freezing wall in realtime. The refrigeration system is mainly responsible for monitoring the temperature of the cooling brine flowing in the freezing tube and the working temperature of the refrigeration unit in realtime during the test and displaying it on the display screen of the refrigerator. The copper-constantan T-type thermocouple embedded in the soil is used to connect to the data acquisition instrument to measure the temperature change of the soil around the communication channel in realtime.

The temperature measurement points of this test arrangement are shown in Figure 4. Three sections are selected, and 8 thermocouple measurement points are arranged on each section, which is located at the arch corner and the arch bottom of the tunnel. The formation process and distribution law of the freezing wall temperature field is obtained through the data fed back by the actual measurement. With the progress of the whole experiment, the change in the temperature field can be observed through the data acquisition equipment.

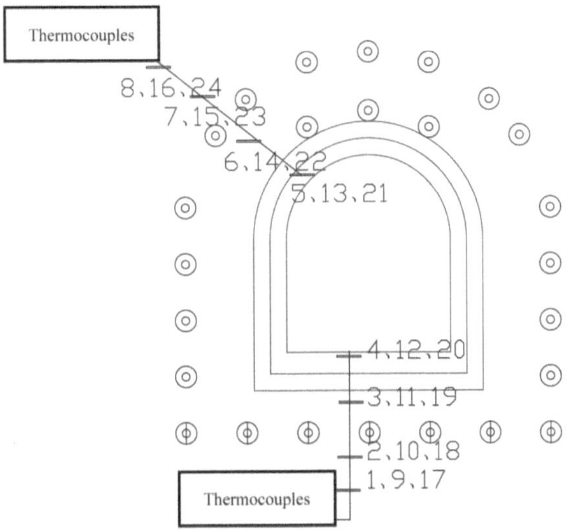

Figure 4. Schematic diagram of test temperature measurement point layout.

(6) Displacement measurement system
Three sections are selected radially along the connecting channel, and two displacement sensors are arranged on each section at 850mm and 150mm from the top of the model box. In this way, the displacement changes of the frozen wall and the ground surface during the freezing, excavation, and thawing process of the communication channel can be approximated.

4 MODEL TEST RESULTS ANALYSIS

4.1 *Monitoring and analysis of freezing temperature field*

In the model test process, the real-time data of 24 temperature monitoring points in three sections were monitored for nearly 6 hours, and the sections at the entrance and 1/2 of the distance from the entrance were selected to obtain the relationship between the temperature change with time of each measuring point. The relationship between temperature and freezing time is shown in Figure 5.

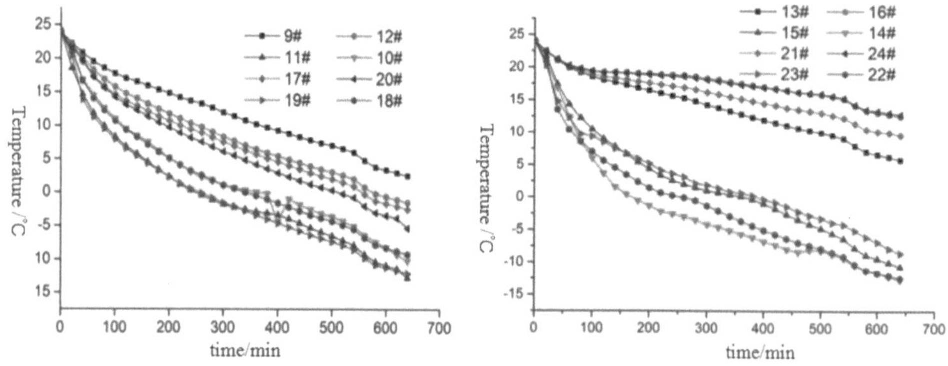

(a)Temperature change graph of arch bottom (b)Arch angle temture change graph

Figure 5. The temperature measurement point changes with time.

It can be seen form Fig5. that the temperature of the monitoring points arranged decreases continuously with time. When the soil temperature at the measuring point is above 0°C, the temperature drop rate is relatively large, the maximum can reach $5.7°C{\cdot}h^{-1}$, and the average gradient temperature is $4.98°C{\cdot}h^{-1}$; the maximum gradient is $2.3°C{\cdot}h^{-1}$.

At the same time, the temperature of the measuring points located on both sides of the freezing tube decreases faster, and the temperature of the temperature measuring points on the outside is relatively slow. The main reason for this phenomenon is that the outer monitoring point is less affected by freezing and continuously exchanges heat with the external environment, while the inner temperature measurement point has an adiabatic boundary, which also leads to the inability to supply heat in the external environment.

Through the analysis of the whole freezing process, it can be seen that each temperature measurement point is above 0°C in the initial stage of freezing, with a linear downward trend. After 150 minutes of freezing, the temperature reached negative temperature, and the temperature drop rate appeared to be a process of gradually increasing to leveling off.

4.2 *Displacement field monitoring results and analysis*

(1) Distribution law of frost heave displacement field
After about 640 minutes, a 100mm thick frozen wall formed around the communication channel, and the average temperature of the frozen wall was –10°C. During the freezing process, we measured the frost heave of the soil every 20 minutes using vertical displacement meters

arranged in the soil at a distance of 150mm and 850mm from the top of the model box. To measure the displacement change more accurately, three different sections are selected for measurement. The buried displacement gauges are numbered along the excavation direction of the communication channel, and the data measured by the three measuring points close to the excavation surface are approximately considered to be the displacement changes of the frozen wall, numbered 1, 2, 3 respectively. The displacement of the measuring points near the roof of the model box is approximately considered to be the frost heave displacement of the ground surface, numbered 4, 5, 6, respectively. Finally, according to the similarity theory, the frozen wall and the frost heave displacement of the ground surface under the actual working conditions can be deduced. The specific displacement measurement results are shown in Figure 6.

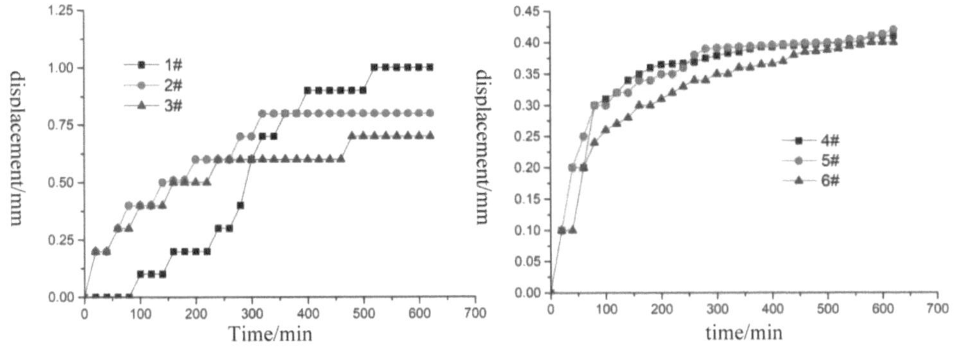

(a) The frost heave displacement of the frozen wall Changes with time

(b)The ground surface frost heave displacement changes with time

Figure 6. The frost heave displacement of each displacement measuring point changes with time.

According to the measurement results of frost heave displacement, during the freezing process, with the continuous decrease of the freezing temperature, the soil body undergoes different degrees of frost heave displacement. In the process before the formation of the frozen wall, the frost heave displacement is large and has a certain linear trend, because this stage is the first stage of freezing in the process of the transformation of the soil temperature at the measurement point from the positive temperature to 0°C; the second stage is a process of switching between positive and negative temperatures. At this stage, the frozen wall gradually begins to form, and the frost heave displacement continues to increase but has a slowing trend; in the third stage, the formation of the frozen wall is completed, and the frost heave displacement tends to be stable. The temperatures are all negative. At the end of the freezing process, the displacement change of the frozen wall was measured to be 0.834mm, and the frost heave displacement of the ground surface was measured to be 0.4mm. The corresponding field displacement changes of the two are 16.67mm and 8mm, respectively.

(2) Distribution law of melting and sedimentation displacement field

The thawing method of this experiment adopts the method of natural thawing. When the freezing, excavation, and other processes are completed, the freezing system is closed and the excavation surface of the communication channel is sealed and insulated. At this point, the analysis stage of the melting and settlement displacement of the frozen wall is entered. The change of the melting displacement with time is shown in Figure 7.

According to the actual measurement results, the following conclusions can be drawn: the frozen wall and the surface will not change the displacement immediately in the initial stage of soil melting, but with the continuous increase of the soil temperature and the continuous progress of the melting process, the displacement changes. The curve shows a linear increasing trend. When the soil temperature reaches the same as the room temperature, the melting displacement of the soil also tends to be stable. During the whole thawing process, the maximum

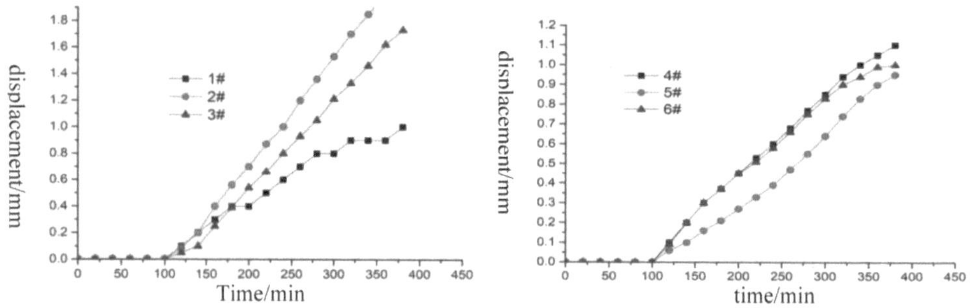

(a)The law of thawing displacement of frozen wall with time

(b)The law of thawing displacement of ground surface with time

Figure 7. The melting and settlement displacement of each displacement measuring point changes with time.

thawing amount of the frozen wall and the ground surface is 1.2mm and 0.9mm respectively, that is, the corresponding displacement of the construction site is 24mm and 18mm.

5 CONCLUSION

This paper analyzes and summarizes the test results of the freezing wall temperature field and displacement field of the connecting channel model test at each monitoring point, analyzes the changes of the temperature field and displacement field under different freezing temperature conditions, and draws the following conclusions:

(1) Heat transfer is also affected by the external environmental heat source at the freezing temperature, especially the freezing temperature. The temperature drop rate of the measuring points closer to the freezing tube is greater than that of the remote measuring points.
(2) The displacement between the frozen wall and the surface gradually increases during the frost heave stage, in which the displacement of the frozen wall is greater than that of the surface, and the stress at the frozen wall is greater than that of the surface, and the external load perpendicular to the frozen wall will inhibit the freezing of the soil. The amount of frost heave is closely related to the burial depth of the soil layer, that is, the smaller the burial depth, the greater the amount of frost heave. In the initial stage of the subsidence stage, there is no sudden change in the displacement of the soil, but with the continuous increase of the soil temperature and the continuous progress of the subsidence process, the change curve of the displacement shows a linear increasing trend. When the soil temperature reaches the same as the room temperature, the melting displacement of the soil also tends to be stable. The displacement change of the frozen wall is larger than that of the surface.
(3) For the first time, the model test was carried out for the subway connection channel in the water-rich pebble-gravel stratum, and a large number of measurement data of the displacement field and temperature field of the freezing method of the water-rich pebble-gravel stratum were obtained through the test, which provided information for the design and construction of the stratum freezing method.

REFERENCES

Alfreds Juikis R. Thermal modeling of freezing soil systems [A]. *The 2nd International Symposium on Ground Freezing* [C]. Trondheim, Norway, 1980.

Cai Haibing, Cheng Hua, Peng Limin, Yao Zhishu, Rong Chuanxin. Model test on displacement field of double-route metro constructed with horizontal freezing method [J]. *Chinese Journal of Rock Mechanics and Engineering*, 2009, 28(10): 2088–2095.

Chen Xueying, Tan Zhongsheng, Yuan Jie, Wang Guan, Li Junwei. Case study on soil reinforcement method for the cross-passage constructionof shield tunnel in water-rich round gravel stratum [J]. *China Civil Engineering Journal*, 2017, 50(S1):105–110.

Cheng Hua, Yao Zhishu, zhang Jingshuang, Rong Chuanxin. A model test study on the effect of freeze heaving and thaw subsidence for tunnel construction with artificial horizontal ground freezing [J]. *China Civil Engineering Journal*, 2007(10): 80–85.

Huang Qiangbing, Gao Huan, Liu Nina, et al. Shaking table model test on seismic response of metro tunnel crossing ground fissure site [J]. *Journal of Geomechanics*, 2018, 24(06): 785–794.

Hu Xiangdong, Ren Hui, Chen Jin, Cheng Yong, Zhang Jun. Model test research on active freezing scheme of tube-screen freezing method [J]. *Modern Tunnelling Technology*, 2014, 51(05): 92–98.

Li Liguo. Study on deformation law of upper tunnel caused by freezing and thawing [D]. *China University of Mining and Technology*, 2015.

Lei Min. Model Test similarity theory research [J]. *Technological Development of Enterprise*, 2014, 33(27):154+156.

Xiong Tianfang, Shao Shengjun, Wang Tianming, et al. Research on physical model experiment of xi'an metro tunnel orthogonally crossing ground fissure [J]. *Rock and Soil Mechanics*, 2010, 31(01):179–186.

Xie Xiongyao, Wang Qiang, Liu Huan, Li Jun, Qi Yong. Settlement control study of shield tunnelling crossing railway station in round gravel strata [J]. *Chinese Journal of Rock Mechanics and Engineering*, 2016, 35(S2):3960–3970.

Yang Yang, Tang Zhongsheng, Peng Bin, Li Jiangpeng, Wang Guan. Study on optimization boring parameters of earth pressure balance shield in water-soaked round gravel strata [J]. *China CivilEngineeringJournal*, 2017, 50(S1):94–98.

Yao Yao, LuoTingwen, Xiao Lin, et al. Urban land space evaluation and early warning control system: Design and implementation [J]. *Bulletin of Surveying and Mapping*, 2017(08):125–128.

Zhang Xiaofeng. Reinforcement type selection and application of connecting channel in water-rich pebble formation [J]. *Journal of Shijiazhuang Tiedao University (Natural Science)*, 2016, 29(02):50–55.

Advances in Civil Engineering and Environmental Engineering – bin Yahya & Bin Daud (Eds)
© 2023 The Author(s), ISBN: 978-1-032-39397-1

Research on the evaluation method of the treatment effect of subway shield tunnel crossing karst cave

Bin Zhu, Xiaogang Luo & Jian Wang
Shenzhen Metro Construction Group Co., Ltd., Shenzhen, Guangdong, China

Fengting Li* & Zhihao Xing
School of Civil Engineering and Water Conservancy, Shandong University, Jinan, Shandong, China

ABSTRACT: The shield method is currently the construction method chosen for the construction of most urban rail transit tunnels. When the shield tunnel passes through the karst area, it is necessary to use grouting reinforcement to improve the karst geological disaster and ensure the safety of the shield tunnel construction. Relying on the karst characteristics and disaster grouting control project of Shenzhen Metro Line 16, the whole process is controlled from the beginning of karst geological disaster grouting control, and the grouting control of karst geological disaster characteristics is studied by using grouting field records and drilling core methods. The main conclusions are found to be: (1) The grouting field records can quickly and economically judge the grouting effect of karst caves or soil caves in the effective depth of the formation; (2) Drilling and coring can most intuitively reveal the compactness of the karst cave filled with slurry, and reflect whether the degree of cement slurry and rock bonding meets the requirements.

1 INTRODUCTION

At present, the construction of urban subways mostly adopts the shield construction method, because the shield construction has the advantages of safety, high efficiency, high degree of mechanization, and little interference with urban ground traffic. However, when a shield tunnel passes through a karst area, the karst caves are mostly highly concealed, which easily leads to an increase in the permeability of the rock mass structure and a decrease in the mechanical parameters of the rock mass. Engineering disasters such as insufficient power, water leakage, surface karst collapse, and shield tunneling are frequently occurring (Chen Wende & Zhou Wei 2012). Grouting reinforcement is a common method for karst geological disaster control (Liang Shurong 2009). However, whether the grouting reinforcement effect meets the requirements of engineering construction requires a reasonable and scientific detection method to evaluate.

Sun Jiapo (Sun Jiapo et al. 2008) carried out grouting reinforcement for the foundation composed of artificial spoil and filled soil, and conducted nondestructive tests before and after grouting using geological radar. Chen Wende (Chen Wende & Zhou Wei 2012) used high-density electrical method supplemented by coring and ultrasonic CT to detect the effect of karst grouting, to make the detection results more accurate. Yu Jingcun (Yu Jingcun & Li Zhidan 1998) used the high-density electrical method to compare the detection results before and after foundation grouting, indicating that the method is effective and the results are ideal. Zhao Jinlong (Zhao Jinlong et al. 1999) used the transient surface wave method to analyze and evaluate the grouting effect of the foundation. Liang Shurong (2009) proposed that after karst grouting, drilling cores, water injection tests, and high-density electrical methods should be used to detect the grouting effect. Liu Yi (2010) compared two construction techniques of the casing and buried pipe grouting, and pointed out that

*Corresponding Author: lifengting_123@163.com

 DOI 10.1201/9781003349563-11

the buried pipe grouting method is effective and has many advantages. Wang Yongzheng (2010) believed that the grouting effect can be detected by water injection tests, drilling core sampling, and geophysical methods. Silas C. Nichols (Silas C. Nichols & Deborah J. Goodings 2000) proposed to design effective and economical grouting procedures and main control parameters, including geotechnical conditions, overlying rock, and soil pressure, grout combination ratio, grouting speed, and grouting pressure.

From the above analysis, it can be seen that the commonly used control measures for karst geological disasters are grouting reinforcement methods, Whether the grouting effect meets the expected requirements, and the detection of the grouting effect is an important means to evaluate whether the grouting effect meets the relevant requirements. For this reason, according to the karst characteristics of Shenzhen Metro Line 16 and the purpose of disaster grouting control, the whole process is controlled from the beginning of karst geological disaster grouting control. Pulp treatment effect evaluation.

2 PROJECT OVERVIEW

2.1 *Engineering situation*

The section between Longping Station and Shuanglong Station of Shenzhen Metro Line 16 is located on Longping East Road, Longgang District, Shenzhen. It is laid along Longping East Road from west to east, 952.947m. Shield + open-cut method is adopted for construction, the structure bottom plate is buried at a depth of 11.6–27.4m, and the elevation is about 5.3–20.9m. The soluble rock is generally distributed in the proposed construction interval, and the bedrock burial depth varies greatly. Therefore, the interval structure is mostly located in limestone, and the large mileage is located in sand layer. Through special karst and detailed exploration drilling, it can be seen that 52 of the 169 soluble rock exploration holes have exposed karst caves, the porosity is 30.76%, the line karst rate is 1.40-45.70%, and the bus karst rate is 5.39%. Among them, the geophysical exploration revealed that the cave height is 0.3–15.9m, the average is 3.3m, generally 2.0-5.0m, and the cave span is 0.3-38.3m, with an average of 4.1m. The top of the cave is buried at a depth of 8.42–42.85m, corresponding to an elevation of -10.70-24.30m, and the bottom of the cave is buried at a depth of 9.32–46.95m, corresponding to an elevation of 14.80–23.20m, and the roof thickness is 0–33.2m. The geological profile of the right line from Longping Station to Shuanglong Station is shown in Figure 1 below.

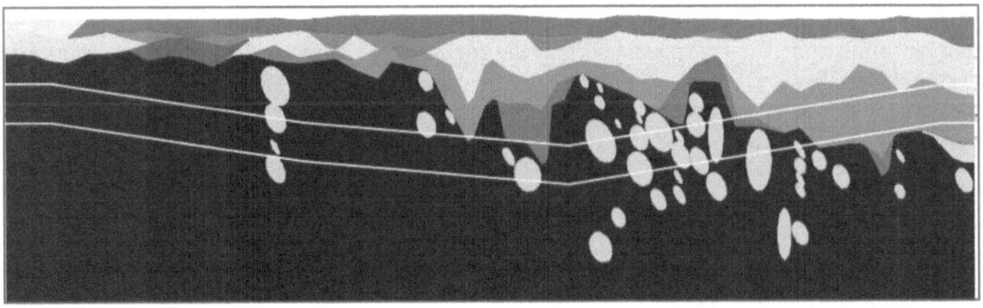

Figure 1. Geological longitudinal section of the right line between Longping Station and Shuanglong Station.

2.2 *Karst treatment method*

The influence of karst on the tunnel structure is related to the location of its distribution and the size of the karst form. According to the relationship between the depth of karst distribution and the structural floor, the evaluations are as follows:

(1) Above the tunnel roof: Most of them are unfilled karst caves, and some are filled karst caves, filled with cohesive soil, medium-coarse sand, and fine sand. According to the geophysical results, the connectivity is good. It is recommended to fill the karst cave with sand and gravel, cement mortar, or plain concrete for the cavity with a thin roof and large hole diameter, and then carry out high-pressure grouting treatment. High-pressure grouting filling is performed again, and for filled soft plastic clay and small-scale karst caves, high-pressure grouting is recommended.
(2) Within the tunnel body: Most of them are unfilled karst caves; some are filled karst caves, filled with cohesive soil, medium-coarse sand, and fine sand. According to the geophysical results, the connectivity is good. It is recommended to fill the cavities with sand and gravel, cement mortar, or plain concrete, and then perform high-pressure grouting treatment. For semi-filled karst caves, it is recommended to perform high-pressure flushing treatment first, and then high-pressure grouting filling. Clay and small-scale karst caves are recommended to be treated by high-pressure grouting.
(3) Within the range of 0–5m below the tunnel floor: most are unfilled karst caves; some are filled karst caves, filled with silty clay, medium-coarse sand, and fine sand. According to the geophysical results, the connectivity is good. It is recommended to fill the cavities with sand and gravel, cement mortar, or plain concrete, and then perform high-pressure grouting treatment. For semi-filled karst caves, it is recommended to perform high-pressure flushing treatment first, and then high-pressure grouting filling. Clay and small-scale karst caves are recommended to be treated by high-pressure grouting. For the karst cavity at the base surface, it is recommended to fill the base with concrete about 30 cm below the base.
(4) Within the range of 5–10m below the tunnel floor: all karst caves without filling. According to the geophysical results, the connectivity is general. It is recommended to fill the karst cave with sand and gravel, cement mortar, or plain concrete for the cavity with a larger diameter, and then perform high-pressure grouting. For filled soft plastic clay and small-scale karst caves, high-pressure grouting is recommended.
(5) Beyond 10m below the tunnel floor: Because it is within 10-15m below the tunnel floor, it has little impact on the operation of the tunnel after completion, but for areas with strong karst development, there may be karst caves connected. Since the karst development has the characteristics of dynamic change, the construction and operation of the project will have side effects on the karst development. Considering the long-term safety of the project, the karst caves found in this depth are dealt with in a targeted manner according to the stability check results.

Based on the 100-year operation cycle of the subway project design, from the perspective of safety, all the karst adverse geological phenomena found within 10m below the floor of the site should be reinforced. The soft plastic soil on the rock surface of the trough should be reinforced to ensure the quality and safety of the project. For some small-scale karst caves within the range of 15m below the bottom plate, because the upper part has a thick stable rock and soil layer, the impact on the structure is small, so no special treatment is required.

3 ANALYSIS OF THE KARST GROUTING EFFECT

3.1 *On-site record analysis of grouting*

Longping Station-Shuanglong Station interval LSR041 has a top-buried depth of 16.84 m and a bottom-buried depth of 22.54 m. The estimated karst cave volume is 488.9 m^3, of which the estimated amount of plain concrete is 391.1 m^3 and the estimated amount of cement slurry is 97.8m^3. The cave is an unfilled karst cave, the design treatment method is filling + grouting treatment. The layout of the LSR041 planar holes is shown in Figure 2.

The design treatment method is filling + grouting treatment, and the on-site analysis is carried out according to the drilling data to select the appropriate treatment method. When the filling

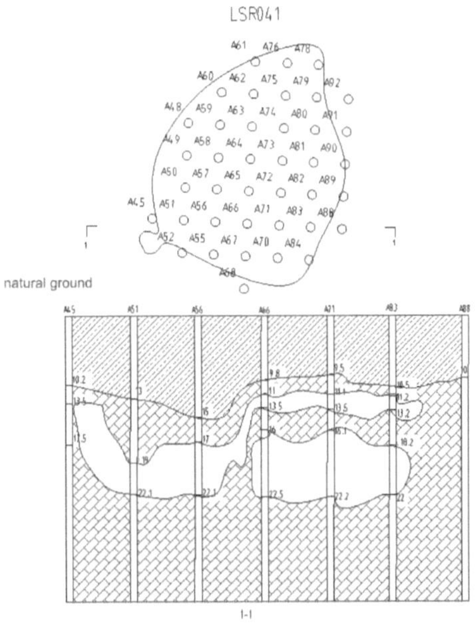

Figure 2. LSR041 Plane Hole Layout.

treatment method is adopted, the diameter of the feeding hole is 200 mm, and the feeding pipe is a 150 mm steel casing. The casing is lowered to the cavity below the roof of the cave for not less than 1.0 m, and the ground pump is used to put in C15 fine stone concrete (to ensure the treatment effect), at least one vent hole should be set in each hole, and the distance should not be less than 4m, and then grouting is used for reinforcement. If the karst caves found on the site are filled, grouting is used directly for treatment. The distance between the grouting holes is 2m, the hole diameter of the drilling is 110 mm, the grouting pipe is grouted with a sleeve valve pipe, and the pipe diameter is 50mm. The principle of grouting sequence is to inject the outer ring holes first, then the inner ring holes; grouting holes in the same sequence at intervals, after filling one hole, skip one hole or grouting with multiple holes; for a large amount of slurry loss during the grouting process Or when the grouting pressure cannot meet the design requirements, the slurry concentration should be adjusted or the area around the cave should be sealed with double-liquid slurry; When the design requires, the peripheral hole and the central hole shall be grouted alternately.

Figures 3 and 4, respectively, show the grouting pressure and grouting amount statistics of different grouting holes of LSR041. It can be seen from the figure: (1) The final grouting pressure of the karst formation is less than the designed 3MPa, and the grouting pressure is relatively stable. (2) The total amount of grouting is much higher than the estimated cave volume of 488.9 m^3.

3.2 *Drilling and coring analysis*

After the completion of grouting in the LSR041 hole between Longping Station and Shuanglong Station, the drilling core test analysis was carried out. The drilling depth was 22.23m. The core map after grouting is shown in Figure 5.

(1) 0~2 m is miscellaneous fill, the silty clay is medium dense, slightly wet, and has poor stability, and the recovery rate is 80%;
(2) 2~8 m is silty clay, brownish yellow, the soil is soft and slightly wet, and the recovery rate is 85%;

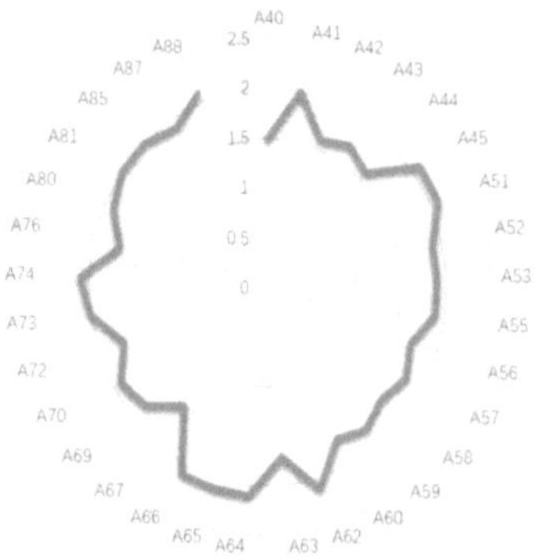

Figure 3. Statistics of grouting pressure of different grouting holes.

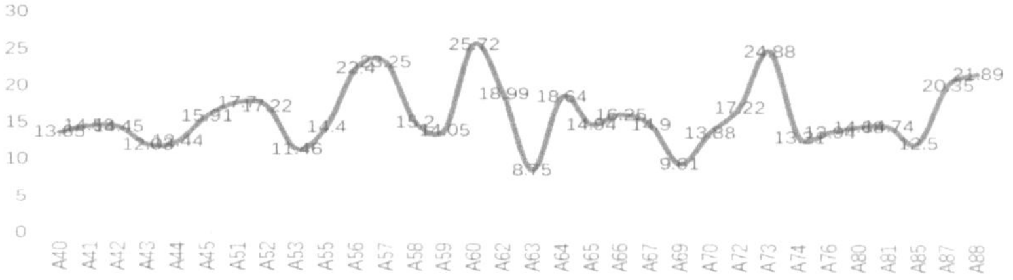

Figure 4. Statistics of grouting amount of different grouting holes.

(3) 8~22 m is lightly weathered limestone, which can be seen to be filled with cement crystals, and the recovery rate is 92%.

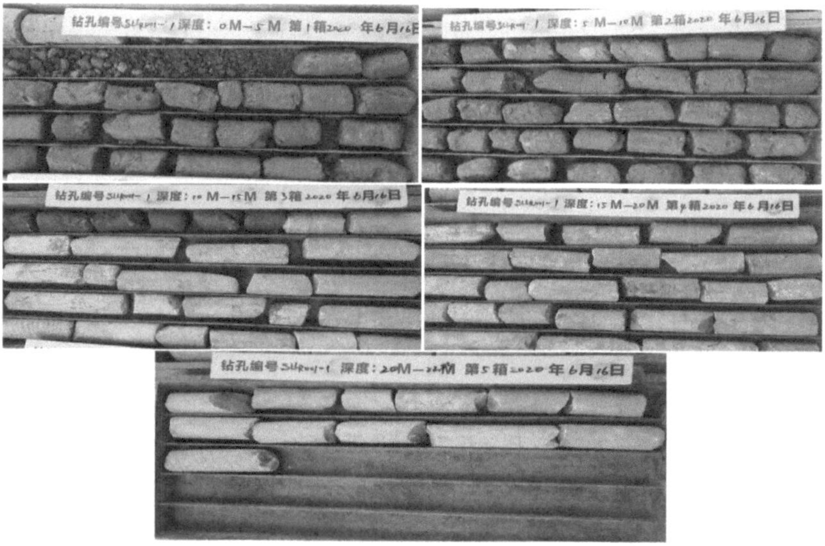

Figure 5. Drilling and coring after grouting.

Core samples were taken from the borehole, and it was found that the performance of the core samples before and after grouting was significantly different. The core samples were broken before grouting, and there were few complete core samples. After grouting, the borehole core sample is complete, and it can be found that there are grout veins filled with grout in the core, indicating that the grouting filling is sufficient and that a good grouting effect has been achieved. There are cemented stones exposed in inspection holes, and all have a greater than 70% recovery rate. This indicates that the cement slurry injected has filled karst caves and fissures.

4 CONCLUSION

Relying on the karst characteristics and disaster grouting control project of Shenzhen Metro Line 16, the whole process is controlled from the beginning of karst geological disaster grouting control, and the grouting control of karst geological disaster characteristics is studied by using grouting field records and drilling core methods. The main conclusions are as follows:

1) The grouting field record can quickly and economically judge the grouting effect of karst caves or soil caves in the effective depth of the formation.
2) Drilling and coring can most intuitively reveal the compactness of the karst cave filled with slurry, reflecting whether the degree of cementation between the cement slurry and the rock meets the requirements. The grouting vein expands in a band-like manner and extends by extrusion. This kind of cementation effect is better. In this state, the soil or broken rock mass can be squeezed and compacted, and the grouting reinforcement effect is good.

REFERENCES

Chen Wende, Zhou Wei. Analysis of the application of high-density electrical method in karst grouting detection [J]. *Enterprise Herald*, 2012, (4):255–256.

Liang Shurong. Analysis of subgrade karst grouting [J]. *Inner Mongolia Science and Technology and Economy*, 2009, (13):60–61.

Sun Jiapo, Chen Yuanyuan, Wang Feiyue. The application of ground grouting effect monitoring by geological radar [J]. *Construction Science and Technology*, 2008, (15):114–115.

Yu Jingcun, Li Zhidan. Detection of foundation grouting effect by high-density resistivity method [J]. *Geology and Exploration*, 1998, 34(5):48–51.

Zhao Jinlong, Du Yongqing, Shi Guangyu, Guo Wensheng. Transient Rayleigh wave detection of grouting foundation and evaluation of grouting effect [J]. *Survey Science and Technology*, 1999, (2):61–63.

Liu Yi. Application of grouting method in karst control of Baishazhou Avenue [J]. *Railway Survey and Design*, 2010, (4):74–77.

Wang Yongzheng. Application of karst grouting in Wuhan-Guangzhou Guest College [J]. *Friends of Science*, 2010, (6):49–50.

Silas C. Nichols, Deborah J. Goodings. Effects of grout composition, depth and injection rate on compaction grouting [A]. Advances in grouting and ground modification [C]. Denver, Colorado: *American Society of Civil Engineers*, 2000, 18–31.

Advances in Civil Engineering and Environmental Engineering – bin Yahya & Bin Daud (Eds)

Design and application of ventilation and cooling systems for high geotemperature tunnels

Haibin Xu, Shaoqiang Zhang, Chenchen Jiang & Weiyu Tang
Power China Road Bridge Group Co., Ltd, Beijing, China

Ping Zhang*
College of Resources and Safety Engineering, Chongqing University, Chongqing, China

ABSTRACT: Thermal damage in high geotemperature tunnels has become a hot issue in underground engineering. To study the high geotemperature tunnel cooling systems, the Nige tunnel was taken as the engineering background. The cooling capacity and the insulation thickness of the high-temperature tunnel refrigeration system were analyzed using theoretical analysis and field tests. The tunnel refrigeration system is established, the cooling capacity per unit time is 1209 kJ, and the thickness of the rubber and plastic insulation material of the tunnel ventilation and cooling system is not less than 12mm. The application results show that the cooling effect of the tunnel refrigeration system is remarkable.

1 INTRODUCTION

With the development of economic construction and progress in road construction technology, tunnels are developing in the direction of more complex geological conditions. A large number of high-geotemperature tunnels have emerged in recent years. High-temperature tunnels not only affect the surrounding rock structure of the tunnel but also produce many heat damage problems, which can affect tunnel construction. Therefore, the study of high-geotemperature tunnels has more important practical significance.

In recent years, many scholars have carried out relevant research on the comprehensive temperature control of high-geotemperature tunnels. Gu et al. (Go 2007) took the Gaoligongshan railway tunnel as the engineering background and analyzed the causes and hazards of high-geotemperature tunnel, and used the method of increasing the air supply volume of the ventilation system to control the temperature of the tunnel at 30°C. Yang et al. (Yang 2010) analyzed the causes of high temperature in the Luquan tunnel and put forward the numerical simulation calculation method of ventilation and cooling in high-temperature tunnel, which can provide engineering reference for the numerical simulation of high geotemperature tunnels. Chen et al. (Chen et al. 2019) used fluent numerical simulation software to study the temperature distribution of high-geotemperature tunnels under different parameters of a ventilation system and concluded that the ventilation temperature has a very obvious impact on the temperature distribution of the tunnel. Li et al. (Li et al. 2014/Miao et al. 2019) analyzed the advantages and disadvantages of hybrid ventilation, side wall press in ventilation, and vault press in ventilation in solving high-geotemperature tunneling disasters, and compared the tunnel cooling effects of the three ventilation modes. Zhu et al. (2019) summarized the application of the existing tunnel temperature comprehensive control technology in the construction period and put forward a new composite lining technology of waterproof and

*Corresponding Author: zhangpgut@163.com

 DOI 10.1201/9781003349563-12

thermal insulation for high-temperature tunnels. Wu et al. (Nie et al. 2020) carried out experimental research on the properties of thermal insulation materials and designed a new type of thermal insulation material for high-geotemperature tunnels. Wu et al. (Wu & Wang 2017) optimized the thermal insulation materials of high-geotemperature tunnels through fuzzy evaluation methods and compared and analyzed the advantages and disadvantages of different laying methods of the thermal insulation layer. It can be seen from the above research, there are many studies on ventilation and insulation cooling of high-geotemperature tunnels, and there are no relevant reports on the research on comprehensive cooling technology of high-geotemperature tunnels.

This paper studied the cooling system of high-geotemperature tunnels under different conditions by using the method of theoretical analysis and field tests, to provide a reference for the construction of similar tunnels.

2 PROJECT OVERVIEW

Nige Tunnel is located in Kele Village, Honghe Prefecture, with a total tunnel length of 3366m and a maximum burial depth of 643m. The two-way double tunnel construction model is adopted. The tunnel crosses the Triassic Middle Tectonics Jielao Group tuff and Yanshan Stage intrusion granite, the maximum temperature of the limestone section is 65.1°C. The maximum temperature of the granite section is 88.8°C. The Yashadi, Nige, and Laohutan hot springs are distributed around the tunnel site and the hot spring water temperature is 51.8~86.6°C. Based on the results of the field temperature test of the tunnel, the temperature curves of the tunnel exit at different distances from the tunnel face are shown in Figure 1, and the temperature change of the inner and outer second lining and inner surrounding rock of the left tunnel exit LZ5K47+045 right side is shown in Figure 2.

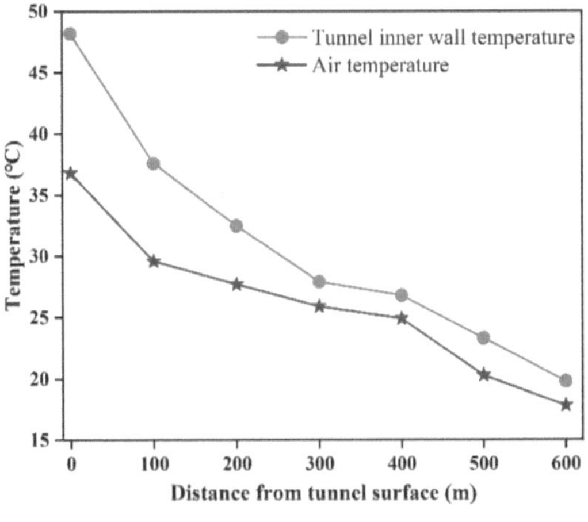

Figure 1. Temperature profile of the tunnel at different distances from the tunnel face.

3 CLIMATIC ANALYSIS OF THE NIGE TUNNEL AREA

Nig tunnel is located in Jianshui County, Honghe Prefecture, Yunnan Province. The annual temperature is shown in Figure 3. According to the actual temperature measurement on site, the local maximum temperature is 35°C, which is in good agreement with the temperature measurement results on site.

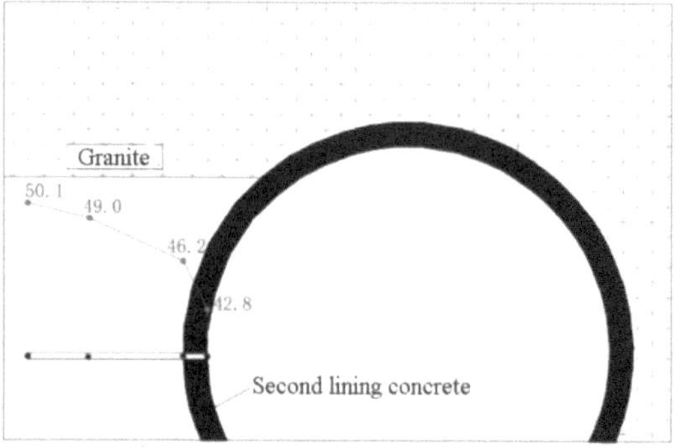

Figure 2. Temperature variation of tunnel transverse borehole.

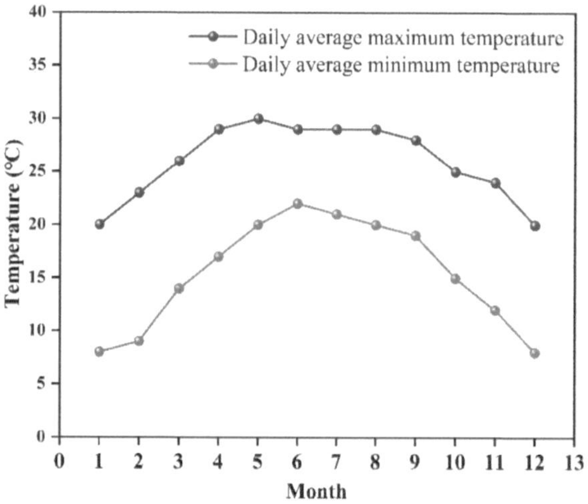

Figure 3. The temperature profile of Jianshui County throughout the year.

4 STUDY ON NIGE TUNNEL REFRIGERATION SYSTEM

4.1 *Cooling capacity calculation of cooling system*

According to the field measurement, the maximum temperature at the outlet of the Nige tunnel was 35°C. According to the calculation, the maximum inlet air temperature of the ventilation system shall not be higher than 20°C. Dry bulb temperature of fresh air inhaled in summer t_1=35°C, relative humidity φ=50% RH, according to the air enthalpy and humidity diagram, the enthalpy value of fresh air h=80.89 kJ/Kg. After cooling treatment, the air supply drying bulb temperature t_2=20°C, relative humidity φ=90% RH, according to the air enthalpy and humidity diagram, the air supply enthalpy value h=53.66 kJ/kg. Then, the enthalpy difference between inlet air and supply air isΔH=80.89-53.66=27.23 kJ/Kg.

The cooling capacity of the refrigeration system can be expressed as:

$$Q = m \times \Delta H \times t$$

where the Q is the cooling capacity of the ventilation refrigeration system, m is mass of air required by the refrigeration system, ΔH is the enthalpy difference between the inlet and outlet air of the refrigeration system, and t is the time.

According to the air demand of Nige tunnel, the cooling capacity of Nige tunnel ventilation is calculated as:

$$Q = 1.2 \times 37 \times 27.23 = 1209 \text{ Kj/s} = 1209\text{kW}$$

4.2 *Insulation thickness of air duct*

According to the actual temperature test results of Nige tunnel, the temperature at the inlet of the air duct was 33.1°C and the temperature at the outlet of the air duct was 46.2°C. It showed that the temperature of the airflow coming in from the entrance of the tunnel increased by 13.1°C when it reached the air outlet of the air duct. When the wind flow reached the tunnel driving face, the heating wind marked the air supply and cooling effect of the tunnel ventilation system as unsatisfactory.

The air duct material commonly used in tunnels is high-density plastic, and the thermal conductivity was 0.32 W/m·k. Due to the continuous high temperature in the tunnel, the wind at the inlet of the air duct goes through a heating process when it reached the outlet of the air duct. To cooperate with the use of the refrigeration system, the air duct needed to be insulated and wrapped with a layer of rubber and plastic insulation material.

The thickness of thermal insulation material for thermal insulation of the cold storage tank and the connecting pipeline is calculated as follows:

$$(t_a - t_f)/(t_a - t_s) = 1 + \delta \times \alpha/\lambda$$

where δ is the required thickness of thermal insulation material, λ is the thermal conductivity of thermal insulation material, and the thermal conductivity of rubber plastic insulation material is 0.035 w/m·K, α is the heat release coefficient of the outer surface of the thermal insulation material, t_α is the temperature of the outer surface of the thermal insulation material, t_f is the temperature of the inner surface of the thermal insulation material, and t_s is the temperature of the surface of the insulation material

The thickness of rubber and plastic insulation material can be obtained:

$$(50 - 20)/(50 - 40) = 1 + \delta \times 5.8/0.035$$

The thickness of rubber and plastic insulation materials for tunnel ventilation and cooling system shall not be less than 12 mm.

4.3 *Design of refrigeration system for Nige tunnel*

As the air outside the tunnel was directly input into the ventilation system, the tunnel can not be cooled effectively, the Nige tunnel must be cooled by artificial refrigeration. The artificial cooling method was to use cooling equipment outside the tunnel to cool the air, and then through the tunnel ventilation system, the cooled air was fed into the tunnel. Therefore, the outside cave cooling system was the key to the success of its artificial cooling method.

The refrigeration system consisted of an air conditioning cabinet machine, a water-cooled (screw) chiller, a cooling water tower, a cooling water pump, and a chilled water pump. The tunnel adopted air-conditioning fan cabinet for ventilation and cooling and was internally equipped with an air-water heat exchanger to exchange heat with low-temperature cold water, to reduce the tunnel temperature.

The schematic diagram of the refrigeration system outside the tunnel is shown in Figure 4. The blue arrow in the figure referred to the air delivery pipe. The air conditioner was set at the outlet of the air supply fan and as close as possible to the end of the air supply pipe. The blue pipe connects the air conditioner cabinet unit, chilled water pump, and chiller. Chilled water was delivered to the air conditioner cabinet unit through the chilled water pump to absorb heat from the air. The air was then returned to the chiller to cool down the temperature. The orange pipeline connects the cooling water pump, the chiller, and the cooling water tower through the pipeline. The chiller can finally transfer the heat from the chilled water side (cold production) to the cooling water side of the unit (heat release). The cooling water pumped and cooling water pipes carry the heat generated by the system out to the outdoor cooling water tower, which was then dissipated into the atmosphere by fans.

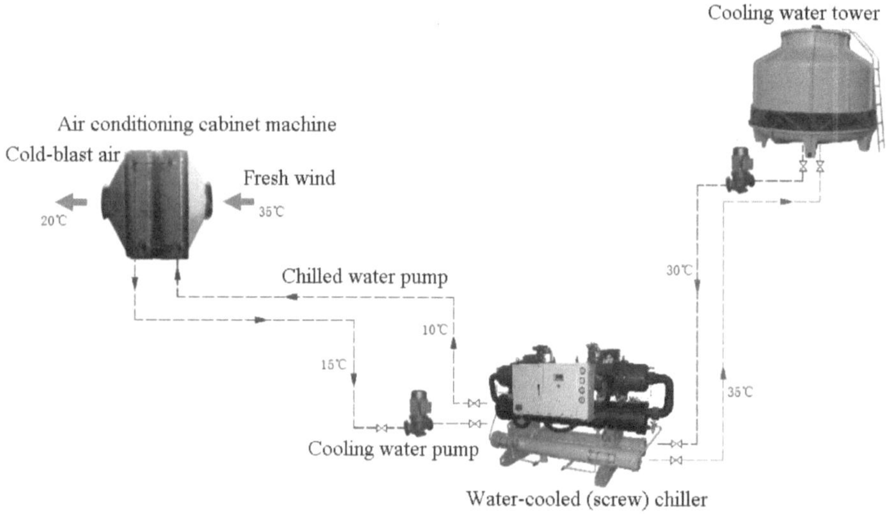

Figure 4. The schematic diagram of the layout of the refrigeration system outside the tunnel.

According to the site conditions of Nige high-temperature tunnel, the heat exchanger, chiller, and cooling water tower were arranged outside the tunnel, as shown in Figure 5.

This arrangement has the following advantages:

① The system is safe and reliable;
② Convenient heat removal;
③ Easy to install, easy to manage, easy to operate;
④ Natural cold source on the ground can be used in winter;
⑤ No need to excavate large-section electromechanical chambers in the cave.

5 APPLICATION EFFECT ANALYSIS

The ventilation and cooling system of high-geotemperature tunnels were applied to Nige tunnel, and the cooling experiment was carried out at the left outlet lz5k47 + 045 of Nige tunnel. The results are shown in Figure 6.

Figure 6 showed the temperature distribution of the tunnel with or without a cooling system. It was found that the temperature after using the cooling system was significantly lower than that without the cooling system. The temperature of the tunnel face decreased by 6.8°C, the temperature 100 meters away from the face decreased by 2.9°C, and the temperature 200 meters away from the face decreased by 2.5°C. At the same time, it can be seen that the greater the distance from the face, the smaller the temperature change. The cooling system can be placed close to the tunnel face.

6 CONCLUSIONS

In this study, the tunnel refrigeration system is established, and the effect of the refrigeration system is analyzed. The following conclusions are obtained:

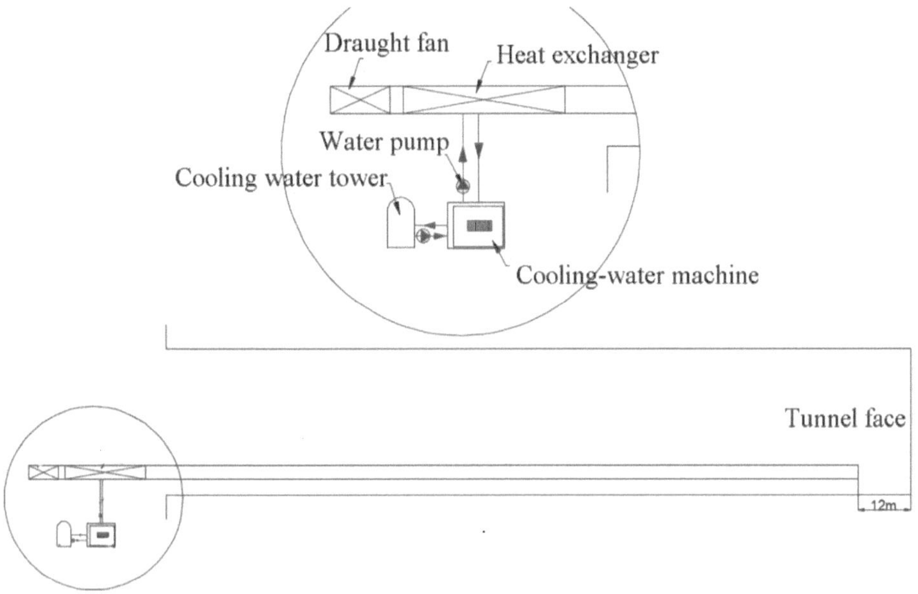

Figure 5. The schematic diagram of high geotemperature tunnel ventilation and cooling system.

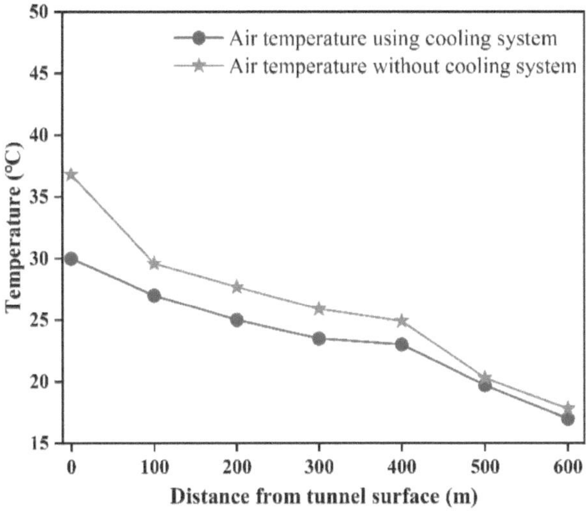

Figure 6. Temperature distribution at different distances.

(1) The cooling capacity of the Nige tunnel cooling system was calculated by combining the site temperature of the Nige tunnel. It is concluded that the refrigeration system requires that the cooling capacity per unit time is 1209 kJ, and the thickness of the rubber and plastic insulation material of the tunnel ventilation and cooling system is not less than 12 mm.
(2) A tunnel refrigeration system was established, which is composed of air-conditioning wind cabinets, chillers, cooling water towers, cooling water pumps, and chilled water pumps.
(3) The tunnel refrigeration system is applied to Nige tunnel and a good refrigeration effect is obtained, the average temperature decreased by 4.6°C after using the cooling system.

(4) The research results can provide theoretical guidance and technical support for the prevention and control of thermal hazards in the construction of high-geotemperature tunnels.

ACKNOWLEDGMENTS

This work was financially supported by the National Natural Science Foundation of China (No. 52074048).

REFERENCES

Chen QY, Chen SG, Chen Q. Study on the effect of different ventilation parameters on the temperature drop of high temperature tunnel face [J]. *Sichuan Architecture*, 2019, 39 (02): 141–143.

Go BS. Tunnel high ground temperature response measures and ventilation design – super long tunnel of Gaoligongshan railway and feasibility study [J]. *Modern Tunnel Technology*, 2007 (2): 66–70.

Li JS, Zong RD, Ao XB. Study on the effect of different ventilation modes on the temperature drop of high temperature tunnel face [J]. *Journal of Hunan University of Technology*, 2014, 28 (05): 32–37.

Miao YW, Chen ZL, Peng XJ. Discussion on blasting and cooling measures for tunnel construction in Plateau and high temperature [J]. *High Speed Railway Technology*, 2019, 10 (04): 95–98.

Nie Xx, Wang ty, Sun FG. Influence of hot and humid environment of high temperature mine on human function [J]. *Metal Mine*, 2020 (04): 186–193.

Wu GQ, Wang ZJ. Study on thermal insulation layer scheme of high temperature railway tunnel [J]. *Journal of Railway Science and Engineering*, 2017, 14 (08): 1715–1726.

Yang CS. Study on comprehensive construction technology of high temperature tunnel [J]. *Railway Construction Technology*, 2010 (10): 39–46.

Zhu u, Zhou JM, Zhao DQ. Study on cooling and thermal insulation technology of high temperature tunnel during construction [J]. *Modern Tunnel Technology*, 2019, 56 (S2): 563–571.

Advances in Civil Engineering and Environmental Engineering – bin Yahya & Bin Daud (Eds)

Design and application of prefabricated top air duct in subway station

Jian Xiang
CCCC Third Harbor Consultants Co. Ltd, Shanghai, China

Xinyu Yan* & Wang Yuan
CCCC Second Harbor Engineering Co. Ltd, Wuhan, China

ABSTRACT: In order to solve the long-standing construction problem of the cast-in-place concrete top air duct in the subway station, in the construction of Dameisha Station of Shenzhen Metro Line 8 Phase II, a prefabricated top air duct at three layers of underground island platform was designed. Based on the finite element simulation calculation, the deformation and stress difference of the integral prefabricated top air duct under the positive and negative wind pressure load, overhead contact system (OCS) load, and dead weight are analyzed, and the applicability of the designed components is verified. The results show that the prefabricated top air duct meets the requirements of deformation and stress, and the reinforcement of the bottom slab can effectively enhance the structure.

1 INTRODUCTION

1.1 *Top air duct instructions*

With the increasing urbanization of the country and the increasingly rapid development of underground space, the subway and rail transit has become the first choice for people's daily travel due to their fast and affordable characteristics, but there are still many technical problems to be optimized and solved in the construction process. The top air duct is located above the subway station, as shown in Figure 1, which can play an important role in ventilation and heat dissipation, maintaining low temperature, bearing and sealing shielding doors, and fixed contact systems.

Based on the design and construction process of the top air duct, it is found that the top air duct mostly adopts cast-in-place concrete construction. This often brings to small construction space, large material area, more construction personnel, slow overall construction progress, and other problems that are not conducive to the construction, and the site construction is shown in Figure 2, resulting in the increase in construction difficulty and cost investment, in order to solve

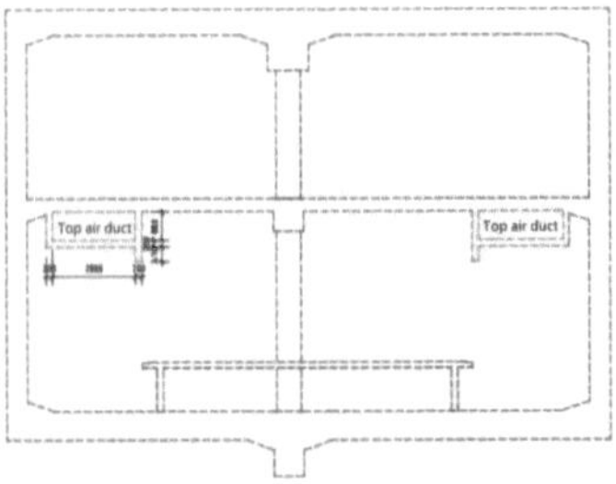

Figure 1. Top air duct structure.

Figure 2. Cast-in-place concrete construction.

*Corresponding Author: 347832087@qq.com

DOI 10.1201/9781003349563-13

this problem, and combined with the needs of promoting the new prefabricated policy, the use of prefabricated top air duct for construction is a better choice.

1.2 *Cast-in-situ and prefabricated type analysis*

The prefabricated top air duct has obvious advantages in time, space, personnel, and other aspects. The specific comparison is shown in Table 1.

Table 1. Comparison table of cast-in-place and prefabricated rail track.

Influencing factor	Cast-In-Situ top air duct	Prefabricated top air duct
Construction time	Installation of steel pipe bracket, formwork fixing, concrete pouring and maintenance and formwork removing is very long	Precast components can be directly transported and hoisted, greatly shortening the construction period
Construction space	Construction materials are piled up together, and scaffolding construction makes the construction space narrow	Precast concrete blocks are installed in different sections to operate efficiently in a small area of space and reduce the construction space
constructor	Large numbers of workers are needed at all stages	A small number of workers are arranged during the installation of prefabricated blocks
Construction cost	Low construction cost	High construction cost

Therefore, the development of prefabricated top air ducts has broad application prospects. In the Dameisha subway station of Shenzhen metro line 8, engineers designed a new type of prefabricated top air duct structure. Through the finite element analysis, the rationality and applicability of the design are verified, which provides a useful reference for the application of prefabricated assembly technology in the construction of subway stations.

2 ENGINEERING SITUATION

2.1 *Top air duct structure*

Dameisha metro station is about 240 meters long, and the station is three floors underground. It is constructed by the open excavation method. The length of the top air duct is about 382.8 meters. The layout is shown in Figure 3.

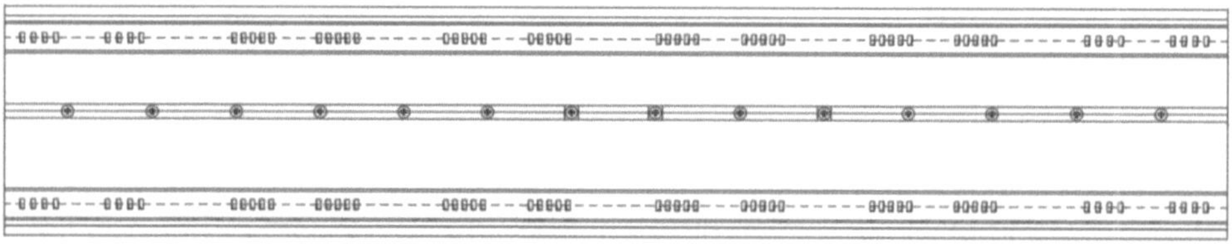

Figure 3. Layout of the top air duct.

The top air duct is designed and constructed by prefabricated assembly technology, and the prefabricated reinforced concrete components are assembled by hoisting and lifting methods. Workers secured the entire prefabricated block with high-strength bolts and hung it directly above the station track. The top air duct mainly includes the bottom plate and side wall. The design size is shown in Figure 4. The spacing of the embedded suspension steel bar is 300mm, the standard section length is 2.1m, the height is 1.45m, the width is 2.55m, and the thickness of the bottom plate and side wall is 150mm. The prefabricated members are C35 concrete, the steel protective layer is 30mm thick, and the steel dimensions are shown in Table 2.

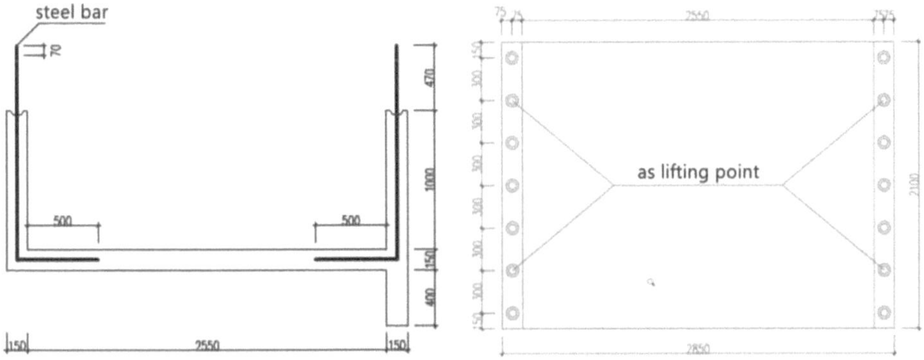

Figure 4. Design drawing of top air duct structure.

Table 2. Steel bar size design of precast blocks.

Steel bar	Type	Size and spacing (mm)
Longitudinal reinforcement	HRB400	14@150
Distribution bar	HPB300	8@150
Pulling steel bar	HPB300	8@300*300
Suspension steel bar	HRB400	25@300

3 FINITE ELEMENT ANALYSIS

3.1 *Model foundation*

As a secondary structure of the subway station, the load test of the top air duct is separate from the overall test of the main structure. The main structure is calculated by considering its load. The load of the top air duct mainly considers its dead weight, contact system tension, and positive and negative wind pressure. Midas Gts, whose finite element model is shown in Figure 5. The

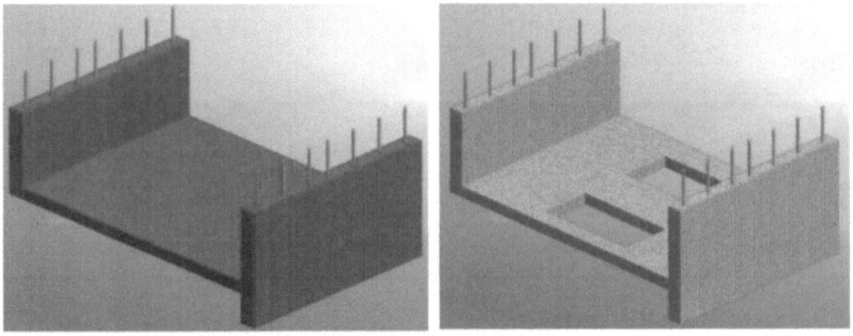

(a) Standard prefabricated block (b) Preformed blocks with holes

Figure 5. 3 D computational model.

precast concrete block adopts a 3D unit, and the steel bar adopts a 1D beam unit. The dead weight load is automatically calculated according to the software, the catenary load is 5kN, and the wind pressure load is ± 1 kN/m^2. The four models are formed according to the combination of positive and negative air pressure and hole opening.

3.2 *Results and analysis of numerical simulation*

The stress distribution and deformation of the top air duct are shown in Figure 6. The maximum central stress of the bottom plate is 0.99 MPa. The peripheral stress was symmetrically distributed and gradually decreased. There is stress concentration around the suspension steel bar. The structural model is symmetric under the action of load, the maximum deformation is 0.46mm, and the deformation between the bottom plate and the high and low wall is the minimum.

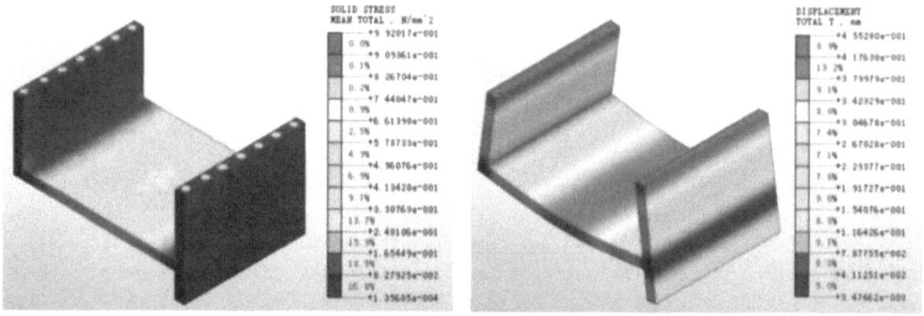

(a) Stress distribution (b) Overall deformation

Figure 6. Stress and strain diagram.

The deformation and stress data were extracted from the X and Y directions of the model base plate, as shown in Figure 7. In Figure 7 (a), the X-direction vertical displacement shows a

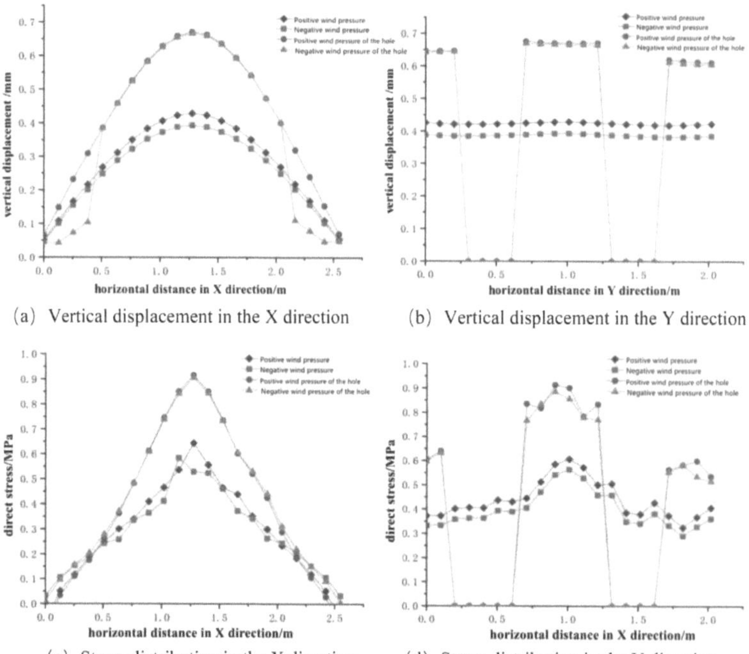

(a) Vertical displacement in the X direction (b) Vertical displacement in the Y direction

(a) Stress distribution in the X direction (d) Stress distribution in the Y direction

Figure 7. Deformation and stress distribution of prefabricated top air duct.

symmetric parabolic distribution. The existence of the holes weakens the integrity of the structure. The maximum displacement of the open block is 0. 67 mm, significantly greater than the 0.46mm of the standard block. The suspended rebar area of the standard block has a 0.2~0.4mm displacement difference under the influence of positive and negative air pressure, and the central position is the largest. The opening block without a suspension bar is large and not affected by wind pressure, but under the influence of negative air pressure, the vertical displacement of the suspension bar area of the opening block is significantly reduced, reflecting the significant effect of steel reinforcement. Figure 7 (b) shows that the displacement of the hole in the opening block is zero. The Y-direction vertical displacement is stable and tends to a straight line. The vertical displacement of the area outside the opening hole is significantly increased, and the maximum value of 0.67mm. In Figure 7 (c), the stress distribution in the X direction shows a symmetrical distribution, and the wind pressure influence is small. In the suspension steel reinforcement area, the stress of the standard block and the opening block are similar. The stress growth rate of the unsuspended reinforcement area with holes is larger, the maximum value is 0.92MPa, and the stress difference reaches the maximum. In Figure 7 (d), in the Y direction, the center stress of the bottom plate is the largest, both sides are gradually smaller, the stress of the opening block is large, and the maximum is 0.91mm, so it is necessary to strengthen the distribution of reinforcement at the opening.

4 CONCLUSION

Based on the results and discussions presented above, the conclusions are obtained as below:

(1) Under the load action of dead weight, contact system, and wind pressure, the deformation and stress of the prefabricated top air duct shall not be greater than 0.91mm and 0.92MPa, which can meet the strength requirements.
(2) The opening of the bottom plate will deform the structure and increase the stress and should be arranged to strengthen the reinforcement.
(3) When the structure of the prefabricated air duct can meet the requirements of subway traffic operation, the thickness of the bottom plate can be adjusted and used on a large scale.

REFERENCES

Hui Shiqian, Diao Muguo (2021). Application of door-type mobile support system in construction of rail roof air duct of metro station. *Yunnan Water Power.* 37(12), 130–133.

Huo Li (2020). Experimental study on mechanical properties of prefabricated components of an air duct on the top of the rail. *Railway Construction Technology.* (11), 42–44.

Liu Bocheng, Ding Jian (2019). Study on mechanical properties of the assembled air duct in metro station considering the effect of sleeve grouting defect. *Journal of Railway Science and Engineering.* 16(11), 2809–2814.

Pan Zhentao, Liu Xiaojun (2019). Study on steel structure rail head air duct pre-assembly technique. *Building Technique Development.* 46(22), 39–40.

Su Liyong, Lu Qingquan (2021). Assembled technology of top air duct of a metro station: a case study on phase 1 project of Beijing metro line No. 19. *Tunnel Construction.* 41(5), 814–821.

Advances in Civil Engineering and Environmental Engineering – bin Yahya & Bin Daud (Eds)
© 2023 The Author(s), ISBN: 978-1-032-39397-1

Challenges faced by landscape architecture in creating information models and reasons behind

Xilun Cao* & Zhe Li
Scientific and Technological Commission, China Urban Construction Design & Research Institute Co., Ltd., Beijing, China

ABSTRACT: In recent years, Building Information Modeling (BIM) technology has developed rapidly in the engineering architecture industry in China because of its great influence in enhancing efficiency and quality of design results. Under the trend of digitalization and informatization, landscape designers have also attempted to introduce landscape information modeling into their practical works. However, building BIM models in the landscape is relatively time-consuming and increases the designer's workload, preventing further application promotion. This paper collects problems landscape designers commonly encounter in using BIM software and then discovers, analyzes the reason, and provides plausible suggestions. It discovers that the problems are mainly caused by a combination of the drawing process, information format, and level of detail between professions. It summarises that standardization in the above aspects is necessary for the subsequent development of BIM in landscape architecture.

1 INTRODUCTION

1.1 *Background information*

In order to solve practical problems, including significant inefficiency and cost wastage in the field of engineering construction, countries all over the world have carried out research on digital expressions of engineering during the past decades. In 2002, the famous US design software company Autodesk proposed the concept of BIM, and after nearly 20 years of continuous development, BIM has been gradually and deeply applied to all kinds of engineering construction projects because of its significant improvement of work efficiency and quality.

In China, BIM is also widely adopted with the participation and promotion of the government, industry associations, and various construction enterprises. It especially plays a great role in the construction of complexly designed large-scale projects, such as the Winter Olympics venues (China Construction Industry Association 2019). In all kinds of construction projects, BIM has now developed into an important tool for information integration, presentation, and collaboration between professions throughout the construction lifecycle.

As an essential part of the engineering construction industry, with the rise of BIM applications in the fields of architecture and municipalities, landscaping architects have also made relevant explorations in conjunction with their situation, thus gradually forming the research direction of landscaping information modeling (LIM) (Landscape Institute 2016). In China, several researchers and architects have been exploring BIM practice in landscaping since 2008. For example, from 2011 to 2012, Beijing Forestry University Professor Cao Likun led his students Xu Tianxin and Liu Wen to apply BIM technologies in several landscaping design projects. The application scenarios varied from terrain modeling to earthwork calculation, planning comparisons, etc. According to

*Corresponding Author: caoxilun@cucd.cn

 DOI 10.1201/9781003349563-14

the scale and complexity of landscaping projects, different BIM and GIS software such as ArcGIS, Map 3D, and Civil 3D was mixed and used for problem-solving and decision making. Since 2013, the Landscape Architecture Centre of Tsinghua Tongheng Institute of Planning and Design has been developing a LIM platform, implementing functions such as terrain modeling and planting design based on a number of practical projects.

1.2 *Research problem*

However, generally speaking, the Chinese landscape industry has not yet fully stepped into the age of informatization, but is still in the transition stage from traditional CAD-aided drawing to BIM-assisted design. BIM technologies are still relatively uncommon in practical landscape projects, and the applications are limited in basic functions such as site analysis and result display. Even in those projects where BIM is applied, a great number of landscape architects believe that BIM software reduces their work efficiency due to its inconvenience in operation and would rather return to traditional CAD workflow, which is completely opposite to the original intention of BIM application. Therefore, without solving the confusion landscape architects have in practical work, there will be little possibility to promote further BIM application in the landscape field and subsequent research.

1.3 *Research objective*

Based on the above situation, in order to explore the specific challenges faced by landscape designers in using BIM technologies, this study takes the perspective of front-line designers to find current problems that hinder the BIM work process. By transferring specific drawings into BIM models, this study summarises problems that occurred in both software applications and working habits and then analyzes the reasons behind them, finally providing countermeasures and suggestions for enhancing the efficiencies of using BIM in landscape architecture. It also gives ideas for promoting BIM applications within the industry and further improving the establishment of BIM standard systems and workflow.

2 RESEARCH METHOD

2.1 *Workflow*

This study is conducted by building BIM models based on CAD mapping and construction drawings of a particular landscape design engineering project. BIM software Civil3D is used to create the original terrain, design terrain, and road models of the planning site, which is the fundamental work for a vertical design. It is worth noting that this process is not a BIM-based forward design process, as the design plan has already been determined, but rather a transition between the traditional CAD drawing process and the BIM design process. By sorting out the difficulties and problems faced in translating data between design platforms in different dimensions, designers are able to recognize the huge distinctions between BIM and traditional design in terms of the logical framework, which is not just 2D versus 3D.

2.2 *Software selection*

Autodesk Civil 3D is a powerful BIM solution for civil engineering design and documentation, which has been used in a wide range of applications, including transportation and hydropower engineering projects. It could accurately create terrain and quickly complete design plans for the site, road, water supply, and drainage work, as well as conduct earthwork calculations and other data statistics, allowing engineers to quickly compare the feasibility of the plans and make adjustments at any time. In addition, it shows good compatibility in data transition with Revit, Infraworks,

Dynamo, and other commonly-used BIM software, which indicates a smooth workflow in BIM cooperation (Zhang 2005). Civil 3D has been used in different types of landscape projects and performed excellently, especially in creating topography models of complex terrains in large-scale landscape design.

3 SUMMARY OF PROBLEMS

Based on the drawing process, the main problems in creating original terrain models are listed below:

Firstly, in the process of creating original terrain models, since the requirement for data accuracy of elevation points in traditional mapping drawings is lower than that in BIM models, it is quite common that designers find the models created by using existing elevation data contain blank or other abnormal areas, which are usually caused by missing information of elevation data (Figure 1). As a result, designers have to spend extra time checking the data completeness for preparation before building BIM models. On the one hand, inaccurate data reduces the efficiency of the model creation process; on the other hand, it also affects the following works, such as slope design and earthwork calculation results.

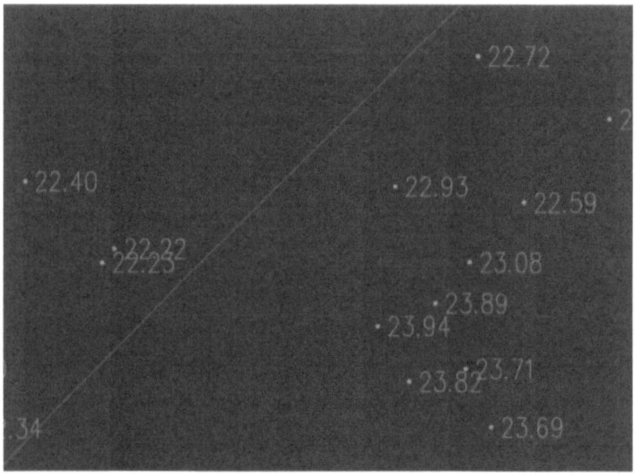

Figure 1. Elevation points with insufficient information.

Secondly, the main problems of creating design terrain models could be divided into the following two types:

(1) Data within the same class in CAD drawings often have inconsistent formats and contain confusing parameters. For example, the design terrain contours are in several formats, such as 3D polylines, points, and so on, which contain discontinuous or blank elevation values. In traditional CAD workflow, the above problems also exist in 2D drawings, whereas they seldom lead to serious consequences, because data format is separated from data value. It is possible that contours with different format clash with each other in 3D space, but still look normal from the plane view (Figure 2). However, using these data to create BIM models usually causes obvious modeling errors such as broken surfaces (Figure 3) or abnormal connections between two blocks. Therefore, to ensure the accuracy of input data, designers have to unify data formats first and then assign correct elevation values to contours before creating topography information models.

(2) The junction areas between the original topography and the designed ones are not clearly expressed. For example, in CAD drawings, some contours of landscape design sites extend to the boundary line of roads but remain unclosed, which could result in wrong connections of topographic surfaces or gaps between design and road surfaces when establishing BIM models (Figure 4). In

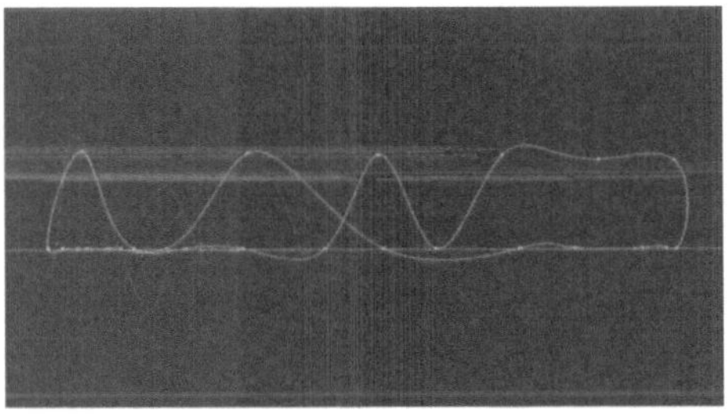

Figure 2. Different data formats of contours.

Figure 3. Broken surfaces in the BIM model.

other cases, even if the contours are closed, because of different requirements in information level of detail, the junction areas between landscape design and other professions in models also contain various abnormal data and sometimes need designers to check and rework CAD drawings.

Figure 4. Unclosed contours at the junction area between design and road surfaces.

4 RESULT ANALYSIS

The above problems in building landscape BIM models are not the result of using different software, but due to interactions of multiple factors. In a narrow sense, it shows the contradictions in drawing habits and workflow processes in different dimensions. From a higher perspective, it is also related to barriers among different professions as well as standardization level of landscape architecture.

Firstly, the characteristics of BIM applications lead to contradictions with traditional drawing processes. According to the MacLeamy curve (Figure 5), along with the project progresses over time, the degree of human intervention decreases, and the cost of changing design solutions increases. Therefore, the sooner problems are identified, the sooner solutions and adjustments can be made, and the more cost could be saved (Serginson et al. 2013). In a traditional workflow, many problems are not discovered until the construction drawing stage, in which complete data has been integrated into one drawing. The low accuracy and fragmentation of data in early-stage drawings result in significant rework and associated costs. In contrast, in a BIM-aided forward design workflow, since specific information must be integrated at the beginning of the design phase to establish the foundation of models, most of the modification works in drawings have already been finished in the early stages of design, in which changes could be made with less cost than in the construction drawing stage. However, although many landscape designers are familiar with detailed operations for BIM model creation, their thinking set in design has not transferred to BIM workflow, and they have not experienced significant modification cost reduction in earlier stages. As a result, these designers believe that BIM only increased their workload in collecting and processing data in the early stages of design, which could be addressed later in the traditional CAD drafting process.

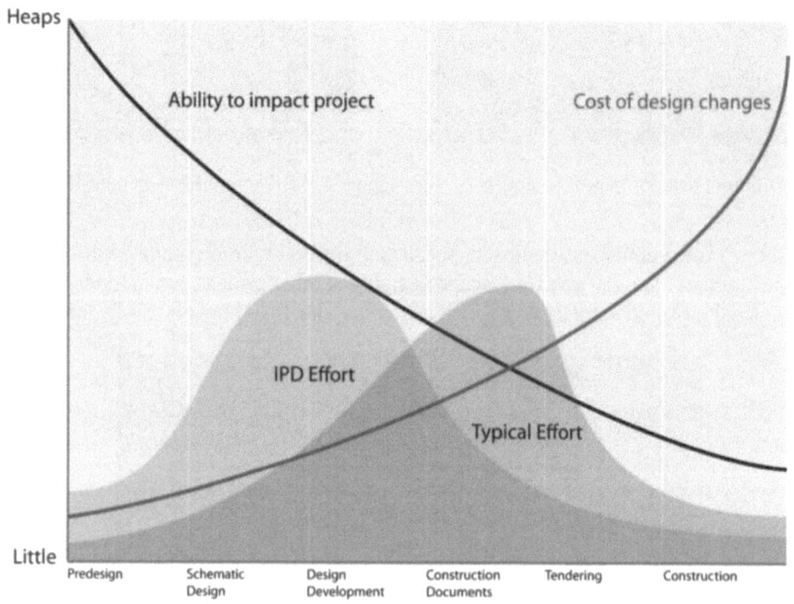

Figure 5. MacLeamy curve.

Secondly, the level of detail in design data varies between different professions. Due to professional characteristics, different professions have their requirement for depth of information expressed in drawings at the same design stage. For example, in the municipal profession, designers need a series of data, including original and design elevation at each mileage point, to draw specific road profile tables and build road models. However, in the landscape field, designers only need to mark approximate distances and slope values in road drawings. In addition, landscape designers usually extend design contours to the boundary line when drawing the junction areas between the design site and the original terrain, without clearly showing the connections between different blocks, as these issues are generally left to the construction stage, in which adjustments are made according to specific construction situation as well as subsequent communication. However, when landscape designers use BIM software which has already been commonly applied in other professions, due to the consistent information depth requirements of BIM for the same type of element, once the landscape profession could not provide as much information as other professions do, it is difficult to establish an integrated BIM model at the same level of information detail.

5 SUGGESTIONS AND CONCLUSIONS

In the trend of engineering construction informatization, in order to promote the application of BIM in the whole process of engineering and participate in multi-professional collaborations, The first step for landscape architecture is to focus on solving the current coordination problem with BIM software and other professions, which requires unity of all aspects. Only by establishing a standardized information format and consistent workflow, information transition between different professions and different stages could be accurate enough to ensure the progress of the project runs smoothly.

Therefore, on the one hand, it is necessary to classify various elements contained in different design phases, supplement information not collected in the traditional process, and determine the corresponding information depth, data format, and workflow. On the other hand, for integrating with models in other professions, if there are clear information model standards or guidelines, the model in landscape design should also be constructed with reference to those standards to ensure consistent depth of information, which is helpful in promoting data sharing and enhancing the level of collaboration among various professions, improving the efficiency of BIM applications in the landscape architecture, and providing a practical basis for subsequent BIM software development, BIM application standard formulation as well as whole life cycle management exploration.

ACKNOWLEDGEMENTS

This research was supported by the Scientific and Technological Commission of China Urban Construction Design & Research Institute Co. Ltd. We also would like to express our sincere gratitude to Professor Wang Panyan for her valuable comments on the technical aspects of landscaping standardization.

REFERENCES

China Construction Industry Association. (2019). *China Construction Industry Association's BIM Application Analysis Report for Chinese Construction Enterprises*. Beijing: CAB Press.

Landscape Institute. (2016). *BIM for Landscape*. Abingdon: Routledge.

Serginson, M., Mokhtar, G. & Kelly, G. (2013). A Theoretical comparison of traditional and integrated project delivery design processes on international BIM competitions. *J. International Journal of 3-D Information Modeling*. 2(4), 52–64.

Zhang, J.P. (2005). *Digital Civil Engineering Design: Autodesk Civil 3D*. Beijing: CCI Press.

Advances in Civil Engineering and Environmental Engineering – bin Yahya & Bin Daud (Eds)
© 2023 The Author(s), ISBN: 978-1-032-39397-1

Analysis of mechanical properties of cold-formed thin-walled steel wood composite columns

Ying Wang*, Chuanwei Zhong, Futing Pan & Yuanchao Hu
Key Laboratory of Building Structure of Anhui Higher Education Institutes (Anhui Xinhua University), Hefei, Anhui, China

ABSTRACT: The cold-formed thin-walled steel members are simple to process and convenient to connect, but their plates are thin in the wall and large in width and thickness. For the uniaxial symmetric open section, the torsion resistance is poor, and the plates are prone to local buckling when pressed, resulting in the loss of stability of the members. In order to popularize the use of cold-formed thin-walled steel members, it is necessary to try to limit the local buckling of such members and combine wood with cold-formed thin-walled steel to form cold-formed thin-walled steel wood composite columns. High-strength steel is used to bear the external load, and wood is used as a surface material to provide sufficient lateral rigidity for the core steel plate. Under the action of axial compression, wood can restrain the local and global instability of steel plate, so that the composite member has a high bearing capacity and seismic performance.

1 INTRODUCTION

In recent years, the state has strongly encouraged the development of green buildings. The energy consumption of traditional cement and concrete building materials is relatively high, and the processing process is easy to pollute the environment and damages the ecology. With the continuous improvement of people's requirements for living environment, low-carbon, green, energy-saving and high recycling rate building materials have become the new favorite of building structures. Under this trend, steel structures, wood structures, and combined structures have ushered in new development opportunities (Zhong 2002). Wood has a long history in architecture as a recycling environmental protection material. At present, some modern new buildings have begun to adopt the structural type of wood structure or steel wood combination, especially the multi-storey buildings in urban villas and villages. The steel wood composite components are used as the stress-bearing components, which can enhance the seismic performance of the buildings. At the same time, it is unified with the wood architectural style of indoor roofs and columns, and the force and aesthetics are perfectly coordinated.

2 COLD-FORMED THIN-WALLED STEEL WOOD COMPOSITE COLUMN

With the innovation of light steel structure and new technology of cold-formed thin-walled steel members, the single cold-formed thin-walled steel member has been difficult to meet the design and use requirements in terms of bearing capacity and stability performance. Therefore, the multi-limb composite section member came into being, with excellent mechanical performance and a wide application range (Zhou 2012). The combined section is usually based on cold-formed thin-walled C-section steel and U-section steel. If wood plates are added, steel wood composite columns can be

*Corresponding Author: 153342707@qq.com

 DOI 10.1201/9781003349563-15

formed by bolt or screw combination, for example, double limb and four limb composite columns shown in Figure 1. Wood can be used to provide lateral rigidity for cold-formed thin-walled steel, restrain local buckling of steel plate, and at the same time, it can effectively use wood properties to enhance the warmth of the building environment and build a wood like building through the texture of the wood itself (Chu 2017).

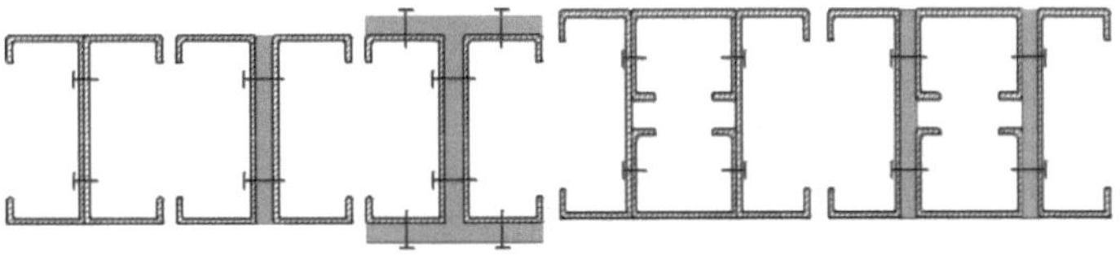

Figure 1. Example of cold-formed thin-walled steel wood composite column.

3 STRESS PERFORMANCE ANALYSIS

3.1 *Calculation of effective width and effective area of the plate*

According to Chinese codes, the effective width method is used to check the bearing capacity of compression members. For C-shaped steel, the web is a stiffened plate, the flange is a partially stiffened plate, and the flange is a non-stiffened plate (Nie 2019). The effective width ratio is calculated according to formula (1) to formula (3):
when $\frac{b}{t} \leq 18\alpha\rho$,

$$\frac{b_e}{t} = \frac{b_c}{t} \tag{1}$$

when $18\alpha\rho < \frac{b}{t} < 38\alpha\rho$

$$\frac{b_e}{t} = \left(\sqrt{\frac{21.8\alpha\rho}{\frac{b}{t}}} - 0.1\right)\frac{b_c}{t} \tag{2}$$

when $\frac{b}{t} \geq 38\alpha\rho$

$$\frac{b_e}{t} = \frac{25\alpha\rho}{\frac{b}{t}} \cdot \frac{b_c}{t} \tag{3}$$

Where: b—the width of cold-formed thin-walled steel sheet; t—the thickness of cold-formed thin-walled steel sheet; b_e—the effective width of cold-formed thin-walled steel sheet; α—the calculation coefficient, $\alpha = 1.15 - 0.15\phi$, ϕ—unevenness coefficient of compression stress, $\phi = \frac{\sigma_{\min}}{\sigma_{\max}}$, b_c—the height of compression zone of cold-formed thin-walled steel sheet, ρ—calculation coefficient. They shall be determined according to article 5.6 of the technical code for cold-formed thin-walled steel structures (ZNSJY2002).

Therefore, after the type of C-shaped steel is determined, the finite width of the web, flange, and bead can be calculated, and then the effective cross-sectional area A_e of C-shaped steel can be calculated by combining the thickness of C-shaped steel.

3.2 *Calculation of ultimate bearing capacity and stable bearing capacity of members under compression*

Under the action of axial pressure, the compressive ultimate bearing capacity of single leg C-shaped steel is calculated as: $N_u \leq f_y A_e$, and the compressive stable bearing capacity is calculated as: $N \leq \varphi f_y A_e$. According to the conditions for checking and calculating the ultimate bearing capacity and stability of split section columns in Chinese codes, the ultimate bearing capacity of double leg C-shaped cold-formed thin-walled steel columns can be calculated as follows: $N_u \leq 2 f_y A_e$, and its compressive stability bearing capacity is calculated as: $N \leq 2\varphi f_y A_e$.

3.3 *Approximate calculation of compression stability bearing capacity of steel wood composite members*

The inertia moment and sectional area of the combined section are calculated by the approximate conversion method. In the conversion, wood is generally converted into steel. It is considered that the composite column conforms to the assumption of a flat section, and the relative sliding between steel and wood and the initial defects of wood are not considered. The wood is converted into the cross-sectional area and cross-sectional moment of inertia of steel according to formulas (4) and (5).

$$E_W I_W = E_s I_s' \tag{4}$$

$$E_W A_W = E_s A_s' \tag{5}$$

Where: E_s—elastic modulus of steel; E_W—elastic modulus of wood; A_s'—converted sectional area; A_W—wood sectional area; I_s'—converted section moment of inertia; I_W— the moment of inertia of wood section.

Theoretically, in the process of compression, the wood plate will restrain the buckling deformation of cold-formed thin-walled steel. Meanwhile, the radius of gyration of the section in the y-weak axis direction increases by about 10% ~ 20%, the slenderness ratio of the member will decrease, and the stability coefficient of the member will increase accordingly. Therefore, the compression stability bearing capacity of steel wood composite columns will increase in theory, and the longer the member length, the greater the increase of the stability bearing capacity.

4 EXAMPLE CHECKING CALCULATION

4.1 *Axial compression performance of steel wood composite members*

The bearing capacity of double limb cold-formed thin-walled C-shaped steel columns and double limb cold-formed thin-walled C-shaped steel wood composite columns under different slenderness ratios are checked and calculated, as shown in Figure 2. The size of single leg C-shaped steel is C200×70×20×2.5. The web height is 200mm, the flange width is 70mm, the crimping height is 20mm, and the section steel thickness is 2.5mm. It is assumed that the calculated heights of the columns are 1000mm, 1500mm, 2000mm, and 3000 mm, respectively.

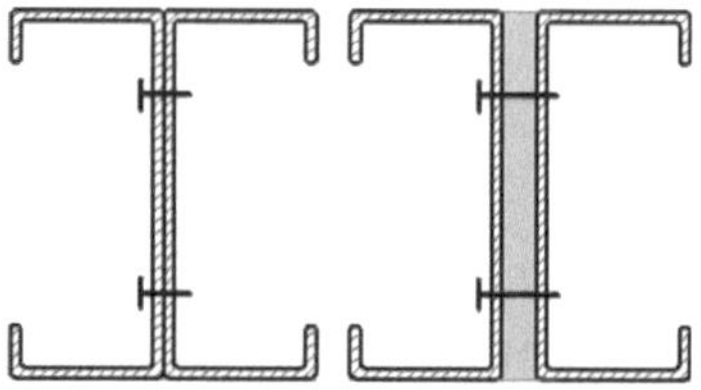

Figure 2. Double limb cold-formed thin-walled C-shaped steel wood composite column.

The steel model is Q235, and the section parameters of single-leg C-shaped steel are shown in Table 1. According to the calculation formula of wood plate thickness and bolt spacing (Pan 2021). The thickness of the wood plate is selected as 20mm. The longitudinal spacing, edge spacing, and lateral edge spacing of bolts are shown in Table 2.

The theoretical calculation results of the ultimate bearing capacity and stable bearing capacity of single limb C, double limb C, and double limb C + wood composite columns are shown in Table 3 and Figure 3. The ABAQUS finite element software is used for analysis in which shell element is adopted for C-shaped steel, the solid element is adopted for end plate, coupling connection is adopted for bolts, the surface to surface contact is adopted, and binding is adopted between end plate and components. The calculation results are shown in Figures 4, 5, and 6.

Table 1. Section parameters of single leg C-shaped steel.

$C200 \times 70 \times 20 \times 2.5$	$fy/N/mm^2$	As/mm^2	Isx/mm^4	Isy/mm^4
	235	898	5.38×10^6	5.63×10^5

Table 2. Screw longitudinal spacing table (mm).

Member length	1000	1500	2000	3000
Margin	100	150	100	150
spacing	200	200	300	300

Table 3. Pressure performance analysis table.

			Ultimate bearing capacity/kN		Stable bearing capacity/kN		
Member length/mm	Slenderness ratio	Effective area/mm^2	Single limb C	Double limb C	Single limb C	Double limb C	Double limb C + wood
1000	26	648	152	304	141.5	283	290
1500	39	658	154.5	309	137	274	287
2000	52	667	156.5	313	132.5	265	283
3000	78	695	163	326	119.5	239	275

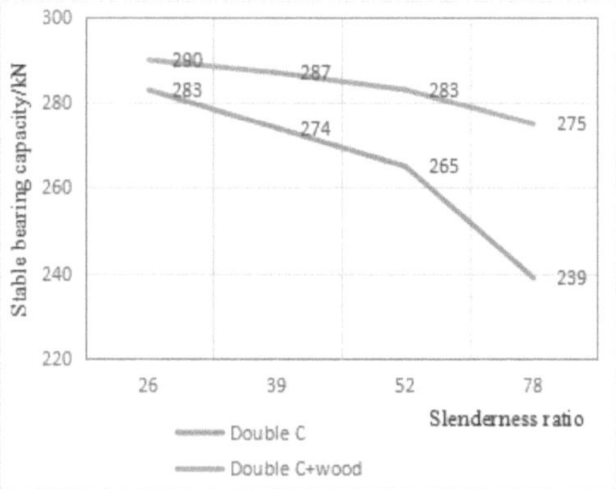

Figure 3. Analysis of stable bearing capacity.

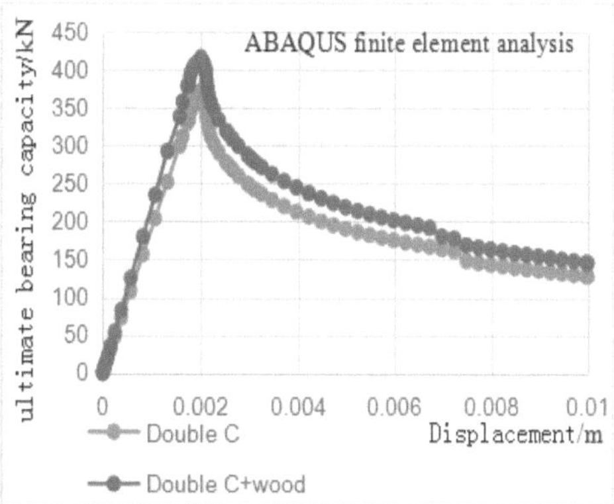

Figure 4. Analysis of ultimate bearing capacity.

From the calculation results, the effective width of the plate under axial compression decreases, and the maximum decrease of the effective area is 27.8% compared with the gross cross-sectional area; With the increase in slenderness ratio, the effective area of the plate increases, but the effect is not significant. With the increase of slenderness ratio by 52, the effective area of the member increases only by 6.8%; At the same time, with the increase of slenderness ratio, the ultimate bearing capacity of single limb C increases, while the stable bearing capacity decreases, with a maximum decrease of 15.5%. It shows that the slenderness ratio has a great influence on the stability and bearing capacity of members.

According to the specification, the compression ultimate bearing capacity and stable bearing capacity of double limb C are not less than 2 times that of single limb C. Since the wood board

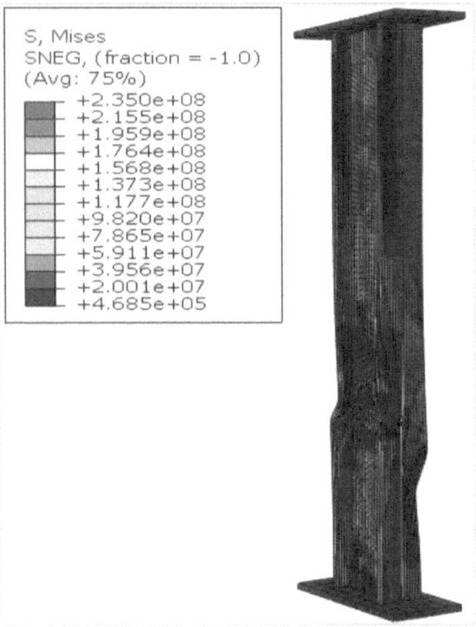

Figure 5. Axial pressure of double limb C.

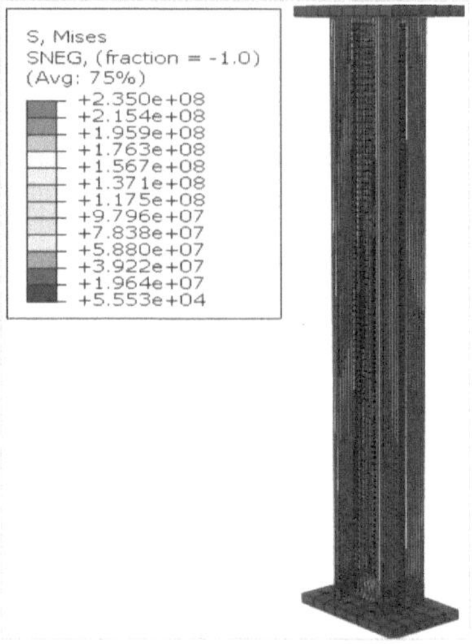

Figure 6. Axial pressure of double limb C + wood.

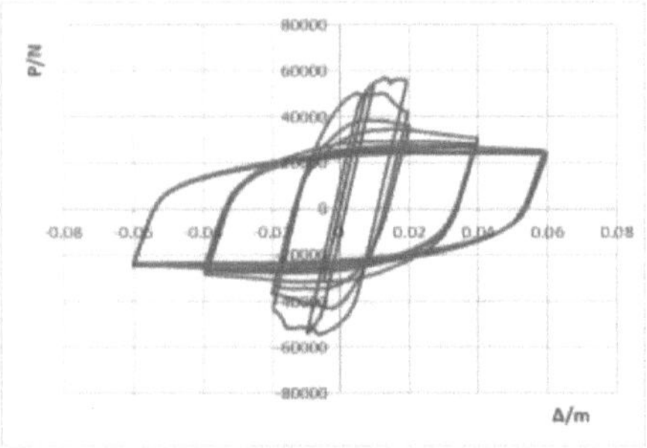

Figure 7. Load-displacement hysteretic loop double limb C.

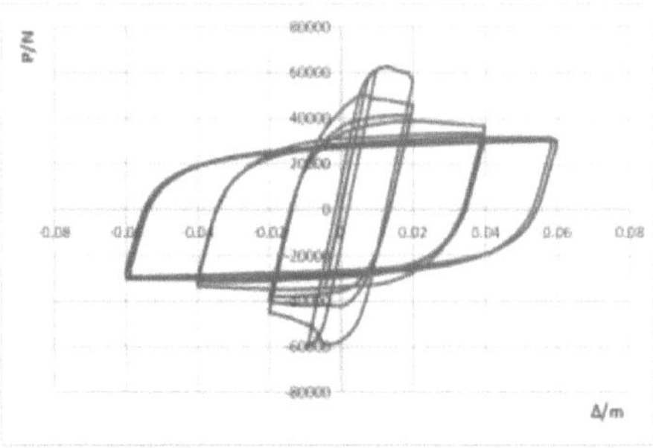

Figure 8. Load-displacement hysteretic loop of double limb C + wood.

can restrain the lateral deformation of C-shaped steel, the stable bearing capacity of double limb C + wood is theoretically larger than that of double limb C, and it increases with the increase of slenderness ratio, with a maximum range of about 13%. According to ABAQUS finite element analysis, the ultimate bearing capacity of double limb C + wood is about 9% higher than that of double limb C. meanwhile, and the finite element result is slightly higher than the theoretical result, which indicates that the approximate calculation modified by the effective width method is conservative. It can be seen from Figure 5 and Figure 6 that the local buckling of the plate can be effectively suppressed by adding wood plates at the web of the double limbs C.

4.2 *Seismic performance of steel wood composite members*

The seismic performance of steel wood composite members is studied by using finite element software, taking double C and double C + wood composite columns as examples. C250×75×20×3 is selected. The thickness of wood is 50mm. The analysis results are shown in Figure 7 and Figure 8:

It can be seen from the figure that the horizontal peak load of the double limb cold-formed thin-walled steel wood composite column is nearly 10% larger than that of the double limb cold-formed thin-walled steel. The plastic deformation and ductility of the composite column are more fully developed, and the hysteretic curve is fuller.

5 CONCLUSION

For the wider application of cold-formed thin-walled steel, the research group is committed to the research of cold-formed thin-walled steel wood composite columns. Through theoretical calculation and finite element analysis, it can be concluded that:

1) The effective width method is used to analyze the axial compression of cold-formed thin-walled steel columns. The effective width of the plate decreases, and the bearing capacity decreases accordingly; the bearing capacity of the member can be effectively improved by adopting the double or multi-limb composite section.
2) By adding a certain thickness of wood, the local buckling deformation of the section steel can be suppressed, the overall stability and bearing capacity of the member can be improved, and the seismic performance can also be improved.
3) The mechanical performance of cold-formed thin-walled steel wood composite columns needs further experimental research and simulation analysis to provide a reference for the engineering application and design of such members.

ACKNOWLEDGMENTS

The research described in this paper was sponsored by the "Natural Science Research Funds for universities of Anhui Province" (KJ2020A0795 and KJ2019A0885) and the key projects of the Key Laboratory of Building Structure of Anhui Higher Education Institutes (Anhui Xinhua University) (KLBSZD202004 and KLBSZD202007), and the national innovation project (202112216038). The support are gratefully acknowledged.

REFERENCES

Chu, Y.P. & Yao, Y. (2017). Experimental study on axial compression performance of cold-formed thin-walled square steel tube wood composite long columns. *J. Building Structure*. 47, 75–80.

Nie, S.F. & Zhou, T.H. (2019). Experimental study and numerical analysis of compression performance of three leg cold-formed thin-walled steel split double cavity box columns. *J. Journal of Southeast University*. 495, 25–33.

Pan, F.T. & Wang, Y. (2021). Design and bearing capacity analysis of cold-formed thin-walled steel-timber composite members. *3rd International Conference on Civil Architecture and Energy Science* China: Hangzhou.

Zhong, G.H. (2002). Research and development of cold-formed thin-walled steel in building construction. *J. Progress of Building Steel Structure*. 4, 31–38.

ZNSJY. Technical specification for cold-formed thin wall steel structure. Beijing: *China Planning Publishing House*.

Zhou, T.H. & Yang, D.H. (2012). Experimental study and numerical analysis on axial compression performance of cold-formed thin-walled steel column with four limbs. *J. Journal of Civil Engineering*. 45, 77–85.

Advances in Civil Engineering and Environmental Engineering – bin Yahya & Bin Daud (Eds)
© 2023 The Author(s), ISBN: 978-1-032-39397-1

Spatial analysis of commercial land prices in urban agglomeration using the gravity model

Yawei Xu*
School of Economics and Management, Beijing Jiaotong University, Beijing, China

ABSTRACT: This paper takes eight major cities in the Beijing-Tianjin-Hebei urban agglomeration as the research objects, and builds a gravity model based on the commercial land price and economic development-related data in each city from 2011 to 2019, so as to quantify the inter-city land price gravity. On this basis, the spatial structure analysis and regression analysis of commercial land prices are carried out using software such as ArcGIS and SPSS. The result indicates that the regional land price gravity between Beijing and its southeastern cities is stronger than that in its northwestern region. Furthermore, we find that as an important factor affecting commercial land prices, the degree of influence of land price gravity has been increasing over time. The findings will provide a reference for the development and utilization of land resources, population, and industrial structure optimization in the process of coordinated development of Beijing-Tianjin-Hebei.

1 INTRODUCTION

Land is an important material resource supporting economic and social development. Industrial restructuring, population mobility, and urban infrastructure construction will all affect the reallocation of land resources to a certain extent. The Beijing-Tianjin-Hebei urban agglomeration is the largest and most dynamic region of northern China. However, the urban hierarchy within the urban agglomeration is clearly divided due to the differences in geographic location and economic development level among cities. Therefore, in order to effectively relieve Beijing's noncapital functions and promote the coordinated development of Beijing-Tianjin-Hebei, it is necessary to focus on solving problems such as industrial layout and industrial land conflicts. As a direct manifestation of the supply and demand of land resources, land prices can effectively reflect the development status and existing problems of different levels of cities in terms of land use efficiency and urban carrying capacity. Therefore, analyzing the spatial distribution characteristics of commercial land prices within urban agglomerations will help to promote the rationalization of industrial land in the region.

In fact, many scholars have applied the extended gravity model to the calculation of the interaction between cities. Lowry (1979) constructed a classic urban spatial interaction gravitational model. Tsekeris (2006) applied the gravity model to the study of the dynamic change of urban traffic interactions. Zhang et al. (2021) analyzed the spatial interaction between cities in the Chengdu-Chongqing urban agglomeration by adding factors such as traffic and education into the urban gravity model. In terms of land price research, Goffette-Nagot (2011) studied the spatial heterogeneity of residential land prices in Belgium, arguing that accessibility and environmental suitability are the main reasons for this difference. Kheir & Portnov (2016) used time trend analysis and multiple regression analysis to explore the impact of geographic location and population size on urban land prices in the Arab sector of Israel. Gao et al. (2020) used national industrial land

*Corresponding Author: xyw49376217@163.com

 DOI 10.1201/9781003349563-16

price sales data to confirm that factors such as government intervention capacity, distance from urban centers, and population size significantly affect industrial land prices. Garang & Wu (2021) analyzed the temporal and spatial evolution of commercial land prices in the main urban area of Hangzhou by using multiple regression analysis.

In referring to the literature introduced above, the gravity model is widely used in the study of spatial interactions between regions. It has been confirmed that commercial land prices are more significantly affected by transportation, economy, and social services. Therefore, by constructing a gravity model, this paper discusses the spatial distribution characteristics and regional connection strength of commercial land prices in urban agglomerations to reflect the current situation of land resource utilization within urban agglomerations through land prices.

2 AN OVERVIEW OF THE STUDY AREA AND DATA SOURCES

The Beijing-Tianjin-Hebei urban agglomeration includes the two municipalities of Beijing, Tianjin and Baoding, Tangshan, Langfang, Shijiazhuang, Qinhuangdao, Zhangjiakou, Chengde, Cangzhou, Hengshui, Xingtai, Handan in Hebei Province and Anyang in Henan Province. According to the annual data on land price monitoring in key areas and major cities in the country released by the China Land Price Information Service Platform, in 2019, the average commercial land price in the Beijing-Tianjin-Hebei urban agglomeration was 8,422 yuan/m^2, of which the Beijing commercial land price was as high as 25,669 yuan/m^2, the price in Tianjin is 8707 yuan/m^2, while the commercial land price in various cities in Hebei Province is around 2000 yuan/m^2. In addition, from the perspective of the annual growth rate of the overall commercial land price in the Beijing-Tianjin-Hebei urban agglomeration, after the review and approval of the "Beijing-Tianjin-Hebei Coordinated Development Plan Outline" in 2015, the overall commercial land price has increased significantly. The data shows that in 2016, the annual growth rate of commercial land prices in the urban agglomeration jumped directly from 3.1% in 2015 to 5.54%. It can be seen that with the increasing frequency and degree of policy linkages, population mobility, and industrial interaction between cities within the urban agglomeration, the spatial correlation of land prices between cities has become more and more obvious.

Figure 1. Location and commercial land price of the study area in 2019.

Based on the principle of data availability, this paper further determines the research objects as eight key cities in the Beijing-Tianjin-Hebei urban agglomeration, which are included in the statistical scope of the National Urban Land Price Dynamic Monitoring System. They are Beijing, Tianjin, Baoding, Tangshan, Langfang, Shijiazhuang, Qinhuangdao, and Zhangjiakou. They all have good regional representation in the development and utilization of urban land at different levels. In addition, due to the impact of COVID-19 on the financial, catering, and other business service industries, this paper is based on relatively stable data from 2011 to 2019. The urban commercial land prices data over the years comes from the land price dynamic monitoring system of the China Land Price Information Service Platform, and the social and economic data of each city comes from the "China Urban Statistical Yearbook," "Statistical Communique of the People's Republic of China National Economic and Social Development," etc.

3 THE CONSTRUCTION OF THE URBAN LAND PRICE GRAVITY MODEL

3.1 *The modification of the traditional gravity model*

The gravity model mainly emphasizes the spatial correlation between various elements in regions. In order to specifically study the land price attraction between cities, this paper constructs a comprehensive variable M_i, M_j that can reflect the quality of urban commercial land prices and uses it to replace the regional development quality in the original model. Further, this paper also re-measures the geographical distance variables between regions and refers to most scholars taking the gravitational adjustment coefficient as 1; that is, the urban land price gravity model constructed in this paper is:

$$F_{ij} = \frac{M_i M_j}{D_{ij}^2} \tag{1}$$

where F_{ij} = the land price gravity between city i and city j, M_i, M_j = land price quality in city i and city j and D_{ij} = distance between city i and city j.

3.2 *The construction and calculation method of the index system*

3.2.1 *The construction of an urban commercial land price-quality index system*

Referring to the factors that have been proved by relevant scholars to have an important impact on the price of commercial land prices, this paper selects a total of 11 indicators from the three levels of agglomeration factor, input-output level, and comprehensive service level. The index system that reflects the quality of urban commercial land prices is comprehensively constructed, as shown in Table 1.

3.2.2 *The calculation method of urban commercial land price quality*

In order to eliminate the influence of dimensions, this paper first uses the extreme value standardization method to standardize all indicator data from 2011 to 2019. The calculation formula is:

$$Z_{ij} = 100 * \frac{C_{ij} - C_{min}}{C_{max} - C_{min}} \tag{2}$$

where Z_{ij} = the normalized value of the factor C_{ij} = the raw value, $C_{\max}$ = the maximum value of the raw value (C_{ij}), and $C_{\min}$ = the minimum value of the raw value (C_{ij}).

Second, the weight of each index in the entire comprehensive index system is determined by the objective weighting method of the coefficient of variation. This method can give a higher weight to the index that has a greater degree of variation and can distinguish differences between cities. The calculation formula is:

$$W_j = \frac{V_j}{\sum_{j=1}^{n} V_j} \tag{3}$$

Table 1. The index system of urban land price quality.

Factor Type	Influencing Factors	Impact Factor
Agglomeration factor	Population agglomeration	Total population
	Business agglomeration	Value added to the tertiary industry
		The tertiary industry's share of GDP
Input-output level	Input level	Investment in fixed assets of unit land
	Output level	General public budget revenue of unit land
		Total retail sales of social consumer goods per unit of land
Comprehensive service level	Traffic accessibility	Road freight
		Road passenger traffic
	Educational level	Primary and secondary school numbers
	Medical level	Number of hospital beds
	Environmental quality	Green coverage rate of built-up area

where W_j = the weight of the factor j, V_j = coefficient of variation, and it is equal to the ratio of the standard deviation (A_j) to the mean (U_j) for each factor.

Finally, the comprehensive index method is used to calculate the comprehensive quality score of land price (M_j). The calculation formula is:

$$M_j = \sum_{j=1}^{n} (Z_{ij} * W_j) \tag{4}$$

3.2.3 *The distance between cities*

For urban agglomerations, the factors affecting interregional accessibility are mainly reflected in the perfection of the intraregional highways, railways, and other traffic road networks. But considering the large differences in the daily departure frequency of railway transportation between different cities, this paper regards the accessibility of highways as an important distance factor affecting the spatial connection between cities. The shortest distance traveled by highways between cities was obtained through Baidu Map, which will be used as the distance variable for constructing the urban land price gravity model. And the shortest travel routes between cities' connectivity are shown in Figure 2.

4 THE ANALYSIS OF SPATIAL INTERACTION BETWEEN CITIES BASED ON THE LAND PRICE GRAVITY MODEL

4.1 *The calculation of intercity commercial land price gravity and comprehensive urban influence*

By substituting the measured urban land price quality and the distance between cities into the gravity model, the commercial land price gravity between cities over the years can be calculated. Taking 2019 as an example, the calculated gravity matrix and the regional urban interaction strength of commercial land prices are shown in Figure 3 and Table 2.

In urban agglomerations, the influence of regional factors on a city can be regarded as the sum of the gravity among the city and all other cities in the region, and it is usually called the regional comprehensive influence of the city. The summary table of the regional comprehensive influence of commercial land prices in each city from 2011 to 2019 is shown in Table 3.

Figure 2. The shortest intercity highway route.

Figure 3. Regional urban interaction strength in 2019.

4.2 *The analysis of empirical results*

Combining the urban commercial land price gravity and the regional comprehensive influence obtained through model building and calculation, we can see the following facts:

First, the urban commercial land price gravity of Beijing compared to the other seven cities in the Beijing-Tianjin-Hebei urban agglomeration is the highest, and they are far higher than the gravity of other cities. The results fully prove that Beijing has an absolute centrality in the Beijing-Tianjin-Hebei region.

Table 2. The urban commercial land price gravity between cities in 2019.

City	Beijing	Tianjin	Baoding	Tangshan	Langfang	Shijia zhang	Qinhuang dao	Zhangjia kou
Beijing	——	226.31	58.96	54.08	284.83	27.89	3.72	9.22
Tianjin	226.31	——	24.97	50.37	48.67	12.43	2.43	1.74
Baoding	58.96	24.97	——	3.70	7.00	22.00	0.32	0.77
Tangshan	54.08	50.37	3.70	——	5.92	2.86	3.01	0.56
Langfang	284.83	48.67	7.00	5.92	——	3.23	0.43	0.64
Shijiazhuang	27.89	12.43	22.00	2.86	3.23	——	0.31	0.72
Qinhuangdao	3.72	2.43	0.32	3.01	0.43	0.31	——	0.06
Zhangjiakou	9.22	1.74	0.77	0.56	0.64	0.72	0.06	——

Table 3. Regional comprehensive land price gravity from 2011 to 2019.

City	2011	2012	2013	2014	2015	2016	2017	2018	2019
Beijing	612.82	622.71	679.54	651.90	677.88	669.10	669.75	671.23	665.01
Tianjin	385.62	401.98	470.40	427.42	433.55	423.53	399.06	367.67	366.93
Baoding	109.48	109.44	129.07	108.64	121.75	107.79	110.99	121.22	117.73
Tangshan	124.24	127.91	152.24	130.69	129.73	126.10	126.38	117.27	120.49
Langfang	266.80	269.66	290.35	297.82	325.09	322.65	338.73	364.19	350.72
Shijiazhuang	55.60	56.31	68.76	58.43	69.80	58.63	65.51	62.47	69.43
Qinhuangdao	13.95	14.39	16.43	21.66	11.33	13.43	11.68	11.01	10.28
Zhangjiakou	8.79	9.73	11.15	11.40	12.22	17.76	14.82	12.44	13.70

Second, still from the perspective of Beijing's gravity on other cities, from 2011 to 2019, Tianjin and Langfang always ranked in the top two, followed by Baoding and Tangshan, while Beijing's gravity to Zhangjiakou has always been at a low level. It can also be clearly seen from the regional urban interaction strength in 2019, Beijing has a stronger urban interaction strength in cities located in its southeastern region than in cities in its northwestern region.

Third, for Tianjin, its interaction strength with Tangshan and Langfang, which are adjacent to its geographical location, is relatively high. But it is not very attractive to other cities in the urban agglomeration. Within Hebei Province, only the gravity between Baoding and Shijiazhuang is strong, and the urban interaction strength between other cities is weak.

Fourth, according to Table 3, it can be seen that the regional comprehensive influence of commercial land prices in different cities shows different time trends, such as Langfang and Tianjin. After 2018, the regional comprehensive influence in Langfang has been nearly equal to Tianjin. But overall, all cities in the Beijing-Tianjin-Hebei urban agglomeration except Tianjin and Qinhuangdao have a clear upward trend in the regional comprehensive influence of commercial land prices over time.

4.3 *The coordination analysis of regional comprehensive land price gravity and commercial land price*

In order to further quantitatively analyze the relationship between the regional comprehensive land price gravity and commercia land price between the major cities in the Beijing-Tianjin-Hebei urban agglomeration, this paper uses SPSS 20 to take linear regression analysis, which takes the commercial land prices of eight major cities over the years as the dependent variable and the regional comprehensive gravity as an independent variable. The results are shown in Table 4.

Table 4. Linear regression analysis results.

	2011	2012	2013	2014	2015	2016	2017	2018	2019
R^2	0.9207	0.9190	0.9175	0.7059	0.7031	0.7026	0.6926	0.7341	0.7662
Regression coefficient	17.66**	17.48**	16.92**	53.67**	53.26**	59.07**	63.47*	63.47*	31.74**

** represents significant at 0.01 level; * represents significant at 0.05 level.

It can be seen that the regression coefficients between the two variables from 2011 to 2019 are significant at the level of 0.05 and the R^2 is large. At the same time, the regression coefficients are all positive numbers, which indicates the comprehensive influence of urban commercial service land prices and regional land prices. There is a significant positive correlation between them; in addition, except for 2019, the regression coefficient increases year by year from 2011 to 2018. On the one hand, the above regression results confirm the important influence of the spatial interaction of regional land prices on the price of commercial service land, and on the other hand, it also reflects that with the advancement of the Beijing-Tianjin-Hebei coordinated development strategy, the influence of regional land price gravity on the price of commercial service land is also constantly improving.

5 CONCLUSION AND DISCUSSION

By analyzing the calculation results of the model, it can be seen that although there is still a large gap in the commercial land prices among different grades of cities in the Beijing-Tianjin-Hebei urban agglomeration, the spatial correlation of the overall commercial land prices in the region is becoming more and more obvious. As the core city, Beijing has established strong spatial connections with Langfang, Baoding, Tangshan, and other places. From a regional perspective, Beijing shows that the commercial land price gravity with cities in its southeastern region is obviously higher than that of its northwestern cities. But in any case, the spatial interaction of commercial prices in the region has been proven to be an important factor affecting commercial land prices, and the influence on the prices has been increasing over time.

Based on the gravity model, this paper measures the urban commercial land price gravity and its comprehensive influence on regional land prices. At the same time, using ArcGIS to visualize the land price gravity between cities can more intuitively reflect the spatial radiation intensity and direction of Beijing and Tianjin. Therefore, the conclusion will have reference significance for thinking about rationally planning industrial land and realizing coordinated regional development from the perspective of spatial interaction and regional connection. However, because only partial socio-economic factors and highway-based traffic accessibility factors are included in the model, especially there are less considered in the index model of traffic accessibility, the extent of the impact of commercial land prices and the size of the scope of action still needs to be further studied.

REFERENCES

Anderson, J E. (1979). A theoretical foundation for the gravity equation. *The American Economic Review* 69(1), 106–116.

Dobkins L H. & Ioannides Y M. (2001). Spatial interactions among US cities: 1900–1990. *Regional Science and Urban Economics* 31(6), 701–731.

Gao B., Luo H. & Huang Z. (2020). The spatial pattern and influencing factors of China's industrial land transfer prices. *Journal of Geoinformation Science* 22(06), 1189–1201.

Garang Z., Wu C. & Li G. (2021). *Spatio-temporal non-stationarity and its influencing factors of commercial land price: A case study of Hangzhou*, China. Land 10(3), 317.

Goffette-Nagot F., Reginster I. & Thomas I. (2011). Spatial analysis of residential land prices in Belgium: accessibility, linguistic border, and environmental amenities. *Regional Studies* 45(9), 1253–1268.

Kheir N. & Portnov B A. (2016). *Economic, demographic, and environmental factors affecting urban land prices in the Arab sector in Israel*. Land use policy 50, 518–527.

Liu L. (2021). A comparative study on the degree of economic connection between the Chengdu-Chongqing economic circle and its adjacent areas based on the gravity model. *Economic Research Journal* 22(06), 1189–1201.

Liu Y. & Nath N. (2022). Murayama A, et al. *Transit-oriented development with urban sprawl? Four phases of urban growth and policy intervention in Tokyo*. Land Use Policy 112, 105854.

Song J., Jin X. & Tang J. (2011). Analysis of influencing factors for urban land price and its changing trend in China in recent years. *Acta Geogr. Sin* 66, 1045–1054.

Tsekeris T. & Stathopoulos A. (2006). Gravity models for dynamic transport planning: Development and implementation in urban networks. *Journal of Transport Geography* 14(2), 152–160.

Xie H. & Wang W. (2015). Exploring the spatial-temporal disparities of urban land use economic efficiency in China and its influencing factors under environmental constraints based on a sequential slacks-based model. *Sustainability* 7(8), 10171–10190.

Yuan S. & Yang L. (2013). A spatial econometrics analysis on the relationship between construction land and socio-economic development: an empirical study of 69 countries in Zhejiang province. *Chinese Land Science* 27(6), 79–85.

Zhang X. & Liu G. & Huang Z. (2020). Spatial interaction analysis of urban agglomeration oriented to urban gravitational network model—taking the Chengdu-Chongqing urban agglomeration as an example. *Journal of Geomatics*, 1–5.

Zhao S. & Zhao K. & Yan Y. (2021). *Spatio-Temporal Evolution Characteristics and Influencing Factors of Urban Service-Industry Land in China*. Land 11(1), 13.

Zheng H., Akita N. & Araki S. (2022). *Provision of Allotment Gardens and Its Influencing Factors: A Case Study of Tokyo*, Japan. Land 11(3), 333.

Zhuang X. & Zhao S. (2014). Effects of land and building usage on population, land price and passengers in station areas: A case study in Fukuoka, Japan. *Frontiers of Architectural Research* 3(2), 199–212.

Advances in Civil Engineering and Environmental Engineering – bin Yahya & Bin Daud (Eds)
© 2023 The Author(s), ISBN: 978-1-032-39397-1

Experimental analysis on corrosion resistance of extruded anchor type recoverable prestressed anchor cable

Xiaofang Zhou
Gansu Province Product Quality Supervision and Inspection Research Institute, Lanzhou, Gansu, China

Shuaihua Ye* & Liting Luo
Key Laboratory of Disaster Mitigation in Civil Engineering of Gansu Province, Lanzhou University of Technology, Lanzhou, Gansu, China
Western Center of Disaster Mitigation in Civil Engineering, Ministry of Education, Lanzhou University of Technology, Lanzhou, Gansu, China

ABSTRACT: In order to analyze the corrosion resistance of extruded anchor type recoverable prestressed anchor cable and cement paste grouting body, seven groups of extruded anchor type recoverable prestressed anchor cable electrochemical specimens and four groups of cement paste specimens were made, and the coupled salt solution containing Cl^- and SO_4^{2-} ions was prepared. The mechanical properties of cement paste grouting body specimens with different water-binder ratios in the corrosion solution were analyzed. The non-destructive testing electrochemical workstation was used to test the corrosion law of extruded anchor type recoverable prestressed anchor cable specimens with different types of protective layers in the corrosion solution. The test results show that the corrosion resistance of 0.50 water-binder ratio cement paste test specimens is the best, and the epoxy resin coating composite cement paste protective layer with the water-binder ratio of 0.50 has the best protection effect on the extruded anchor recoverable prestressed anchor cable.

1 INTRODUCTION

The pull-anchor retaining structure has the characteristics of low cost, short construction period, and ease to control the deformation of the foundation pit. However, the unrecoverable anchor cable left in the foundation will cause waste of resources, environmental pollution, high-cost treatment, and infringement of other people's rights. Therefore, it is necessary to develop recoverable anchor cable technology. On the basis of summarizing the previous research on recoverable anchor cable, the author developed extruded anchor type recoverable prestressed anchor cable. The main component of extruded anchor type recoverable prestressed anchor cable structure is made of steel. Whether it is used in temporary slope protection engineering or foundation pit support engineering with extended service, if there are corrosive salt ions in the geotechnical and groundwater environment where the recoverable anchor cable is located, such corrosive salt ions will undoubtedly cause a certain degree of corrosion damage to the recoverable anchor cable. Furthermore, it will affect the integrity of the recoverable anchor cable structure, and eventually affect the safety and stability of the extended service foundation pit engineering or temporary slope engineering. Therefore, it is necessary to study the corrosion law of the main structure of the recoverable anchor cable immersed in the salt solution for a long time.

*Corresponding Author: yeshuaihua@163.com

 DOI 10.1201/9781003349563-17

The main salt ions for corrosion of reinforced concrete are Cl^- and SO_4^{2-} (Jin et al. 2007; José et al. 2019; Neville 2004, Wang et al. 2020; Zhang et al. 2020), Cl^- will cause the corrosion of steel bars, and SO_4^{2-} will invade into the concrete to occur salt crystallization reaction so that the concrete expansion damage (Hekal et al. 2002). In this paper, the corrosion resistance of cement paste grouting body is analyzed by testing the changes of mechanical properties of cement paste test specimens in a coupling solution containing Cl^- and SO_4^{2-}, and the corrosion law of recoverable anchor cable in the coupling solution is analyzed by electrochemical method, so as to provide some reference value for the design and construction of recoverable anchor cable.

2 WORKING AND RECYCLING PRINCIPLES OF EXTRUDED ANCHOR TYPE RECOVERABLE PRESTRESSED ANCHOR CABLE

2.1 *Working Principle*

The extruded anchor type recoverable prestressed anchor cable belongs to the pressure type recoverable anchor cable, and its actual working state is shown in Figure 1. Its working principle is: rotate the steel strand counterclockwise, make the extrusion anchor contact the bearing plate, stretch the steel strand, make the bearing plate under pressure, transfer the pressure to the surrounding grouting body, and then transfer the pressure to the surrounding rock and soil in the form of shear stress through the grouting body.

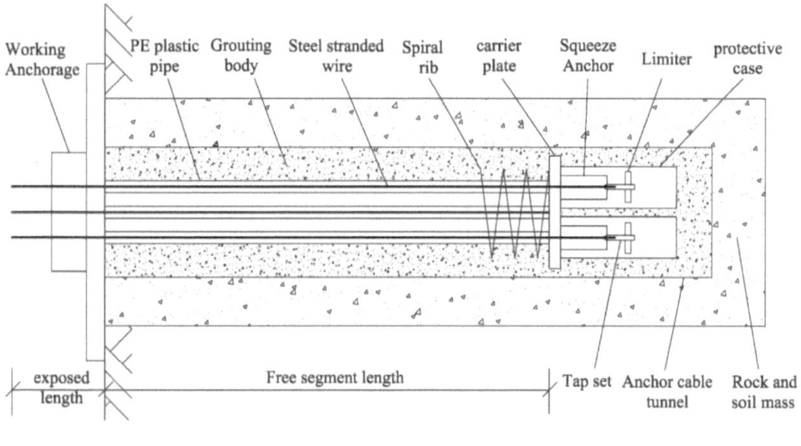

Figure 1. Working diagram of extruded anchor type recoverable prestressed anchor cable.

2.2 *Recycling principle*

The specific recycling process of extruded anchor type recoverable prestressed anchor cable is to recover the working anchor and fixture first, and then there are two methods to choose from. The first method is to rotate the steel strand clockwise, so that the extrusion anchor moves inward until it is in contact with the limiter, leaving the extrusion anchor and the bearing plate at a distance of 2–3 cm. Then, the tap sleeve is pulled out of the steel strand by a single tension jack, and then the steel strand is continued to be tensioned to make it out of the extrusion anchor. Finally, the steel strand is pulled out from the PE plastic pipe by manpower to complete the recovery of the steel strand. The second method is to pull out the middle straight steel wire of a steel strand from the extrusion anchor with a small jack, and then continue to tension the steel strand so that the remaining six steel wires of the steel strand are pulled out from the extrusion anchor. Finally, the steel strand is pulled out from the plastic pipe manually, and the recovery of the steel strand is completed. In the recovery operation, the first method is usually used. If the first method fails, the second method is used.

3 RAW MATERIALS AND TEST SCHEME

3.1 *Raw materials*

The main raw materials used in the test are cement, epoxy resin, water, and extruded anchor recoverable prestressed cable main components.

3.2 *Test scheme*

According to the specification (Ministry of Housing and Urban and Rural Construction of the People's Republic of China JGJ120-2012), the water-cement ratio should be 0.50–0.55 when ordinary anchor grouting slurry is used. Therefore, four kinds of cement paste water-cement ratios are used in this experiment, which are 0.45, 0.50, 0.55, and 0.60, respectively. The electrochemical test pieces of recoverable anchor cables selected in this paper are shown in Table 1. According to the research of Joe (2017), the main corrosive salt ions in groundwater and rock, and soil in Lanzhou are Cl- and SO_4^{2-}, in which the highest concentration of Cl- is 1613 mg/L and the highest concentration of SO_4^{2-} is 4250.7 mg/L. Therefore, the coupling immersion solution of 2000 mg/L NaCl and 5000 mg/L Na_2SO_4 is used in this experiment.

Table 1. Electrochemical specimen of recoverable anchor cable.

Serial number	Abbreviation	Explanation
1	LG specimen	Recoverable anchor cable exposed specimen
2	HY specimen	Recoverable anchor cable epoxy-coated specimen
3	C1 specimen	Recoverable anchor cable specimen with 15 mm thickness 0.45 water-binder ratio cement paste protective layer
4	C2 specimen	Recoverable anchor cable specimen with 15 mm thickness 0.50 water-binder ratio cement paste protective layer
5	C3 specimen	Recoverable anchor cable specimen with 15 mm thickness 0.55 water-binder ratio cement paste protective layer
6	C4 specimen	Recoverable anchor cable specimen with 15 mm thickness 0.60 water-binder ratio cement paste protective layer
7	C5 specimen	Recoverable Anchor Cable specimen with Epoxy coating + 15mm thickness 0.50 water-binder ratio cement paste protective layer

During the standard curing period, the flexural and compressive strength of cement paste test blocks were tested at 3 d, 7 d, and 28 d. After the standard curing of the cement paste test block was completed, four kinds of water-binder ratio test blocks were immersed in the coupled salt solution. The flexural and compressive strength of a group of test blocks were tested every 60 days, and the coupled salt solution was prepared again every 30 days. The bending test and automatic pressure testing machine of the cement paste test block are shown in Figures 2 and 3.

After the standard curing of C1-C5 electrochemical test pieces for 28 d, the first electrochemical test was carried out. Firstly, the C1-C5 specimens were soaked in the laboratory tap water for 24 h to ensure the good conductivity of the specimens during the electrochemical test, and then the first electrochemical test was started. After the first test, the seven groups of electrochemical specimens were immersed in the configured coupled salt solution, and an electrochemical test was carried out with 60 d as a cycle, as shown in Figure 4 and Figure 5. A total of six electrochemical data were collected in this test.

Figure 2. Bending test of cement paste test block.

Figure 3. Fully automatic pressure testing machine.

Figure 4. Immersion specimen.

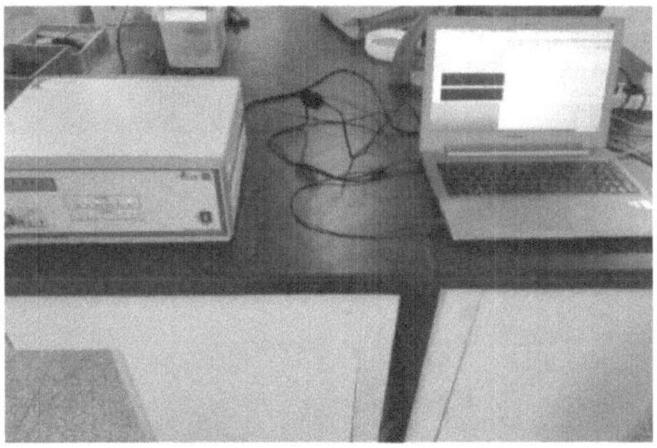

Figure 5. Electrochemical test.

4 ANALYSIS OF THE CHANGE OF MECHANICAL PROPERTIES OF CEMENT PASTE TEST BLOCK IN COUPLED SALT SOLUTION

4.1 *Analysis of the change in flexural strength*

It can be seen from Figure 6 that during the curing period, the flexural strength of four kinds of cement paste specimens with water-binder ratios of 0.45, 0.50, 0.55, and 0.60 decreases with the increase of the water-binder ratio. The smaller the water-binder ratio is, the faster the flexural strength increases. It can be seen from Figure 7 that with the increase of corrosion time in a coupled salt solution, the changing trend of flexural strength of the four kinds of water-binder ratio cement paste is the same, and the overall trend is to increase at first and then decrease and then increase, but the range of change is different. The decrease of flexural strength of 0.45 water-binder ratio cement paste is the largest, and that of 0.60 water-binder ratio cement paste is the smallest. Therefore, the flexural strength properties of the four kinds of water-binder ratio test pieces are in the following order: 0.50 water-binder ratio > 0.60 water-binder ratio > 0.55 water-binder ratio > 0.45 water-binder ratio.

4.2 *Analysis of the change in compressive strength*

It can be seen from Figure 8 that during the curing period, the compressive strength growth rates of the four water-binder ratios of cement paste test blocks all show a trend of first fast and then slow. In addition, the smaller the water-binder ratio is, the faster the growth rate of the compressive strength is. The reason is that the smaller the water-binder ratio is, the greater the cement dosage is, and the greater the compactness of the test piece is, which leads to the higher its compressive strength. It can be seen from Figure 9 that in the coupled salt solution, with the increase of immersion time,

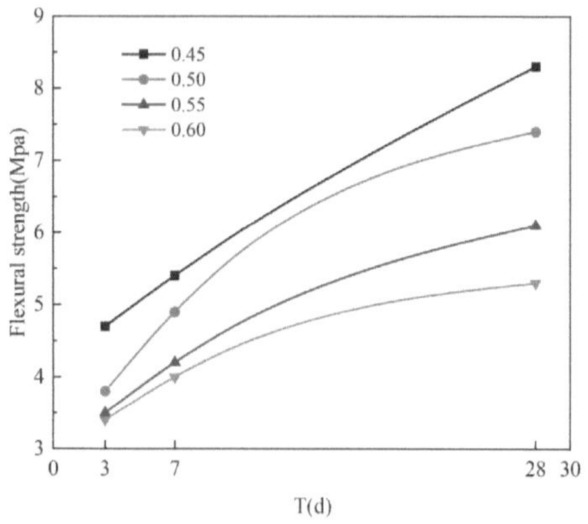

Figure 6. Change of flexural strength during the curing period.

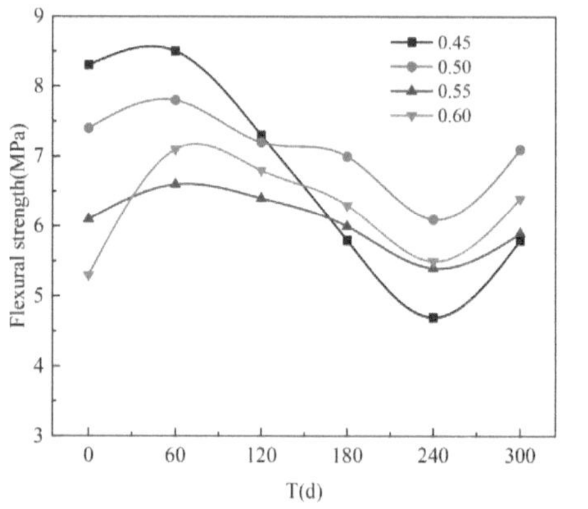

Figure 7. Change of flexural strength during the corrosion period.

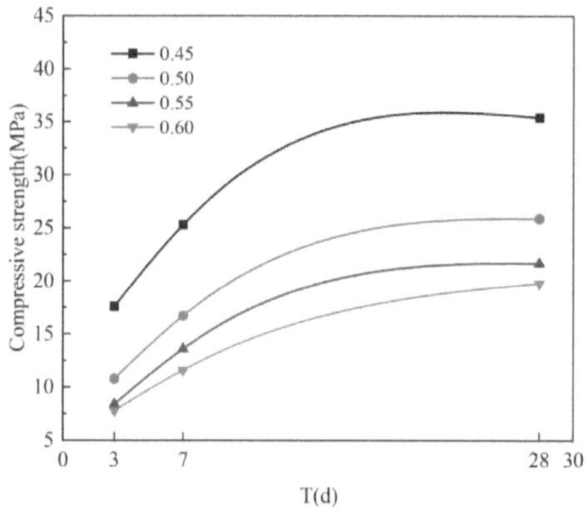

Figure 8. Change of compressive strength during the curing period.

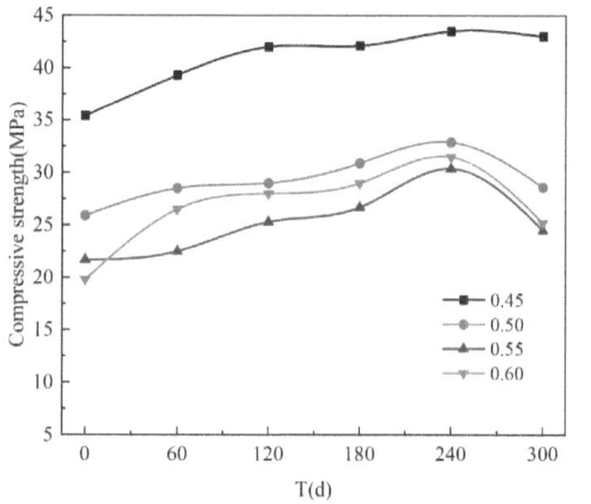

Figure 9. Change of compressive strength during the corrosion period.

the compressive strength of four kinds of water-binder ratio cement paste increases at first and then decreases. Therefore, the corrosion resistance of four kinds of cement paste test pieces is in the following order: 0.50 water-binder ratio > 0.60 water-binder ratio > 0.55 water-binder ratio > 0.45 water-binder ratio.

5 CORROSION RESISTANCE ANALYSIS OF RECOVERABLE ANCHOR CABLE SPECIMEN IN COUPLED SALT SOLUTION

In order to better analyze the best corrosion resistance of the recoverable anchor cable under the action of which protective layer, the corrosion current density i_{corr}, is used to evaluate the corrosion degree of the recoverable anchor cable and analyze the protective effect of different protective layers. The relationship between corrosion current density and corrosion degree is shown in Table 2 (Joe et al. 2017), and the variation of the corrosion current density of different specimens is shown in Figure 10.

It can be seen from Figure 10 that the corrosion current density of seven groups of recoverable anchor cable specimens increases at first and then decreases and then increases with the increase of immersion time in a coupled salt solution. Among them, the variation of LG specimen and HY

Table 2. Relationship between corrosion current density and corrosion degree.

Corrosion current density	i_{corr} <0.1	0.1< i_{corr} <0.5	0.5< i_{corr} <1.0	i_{corr} >1.0
Corrosion degree	Not corroded	Low corrosion	Moderate corrosion	Severe corrosion

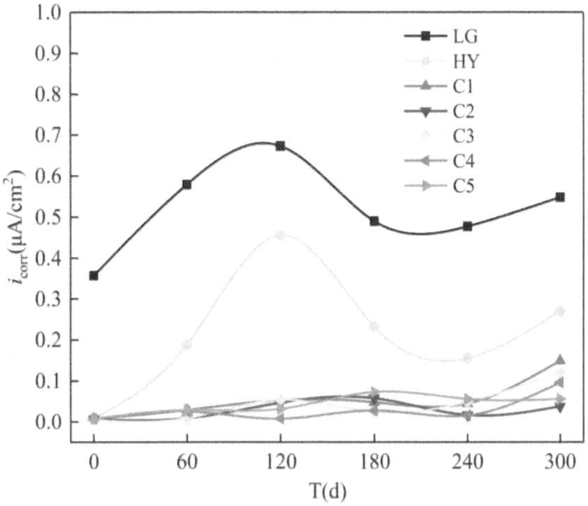

Figure 10. Variation diagram of corrosion current density of different specimens during 0–300d.

specimen is the largest, and the variation of C1-C5 specimen is the smallest, indicating that the 15 mm thick cement paste protective layer can reduce the chloride ion penetration in the coupled salt solution. As a result, the corrosion resistance of recoverable anchor cable members is improved.

Within 0–120 d, the corrosion current density of LG and HY specimens showed a rapid upward trend. At 120 d, the corrosion current density reached a peak. And i_{corr} of the LG specimen exceeded 0.50, indicating that the LG specimen was in a medium corrosion state. And i_{corr} of the HY specimen was 0.456, indicating that the HY specimen was corroded and in a low corrosion state. Within 120–240 d, the corrosion current density of LG and HY specimens showed a downward trend, indicating that the corrosion process of recoverable anchor cable components was hindered, and its corrosion resistance was improved. After 240 d, the corrosion current density values of LG and HY specimens increased, the rust could not continue to hinder the corrosion process, and its corrosion resistance decreased.

For C1-C5 specimens, the corrosion current density fluctuates up and down within 0–240 d, but the values are not more than 0.1, indicating that the passive protective layer on the surface of the recoverable anchor cable component is not damaged, and its corrosion resistance is good. After 240 d, with the increase of the number of cracks on the surface of the specimen, the channels of chloride ions in the coupled salt solution to the surface of the component will also increase, thereby increasing the amount of chloride ions, resulting in an increase in the corrosion current density of C1-C5 specimens and a decrease in the corrosion resistance. At 300 d, only the i_{corr} of C1 and C3 specimens exceeded 0.1, and the i_{corr} of C2 specimens was the smallest, followed by C4 and C5 specimens, indicating that the recoverable anchor cables of C2, C4, and C5 specimens were not corroded and the passive protective layer was still intact when immersed in a coupled salt solution for 300 d, while the passive protective layer of C1 and C3 specimens was just destroyed in a low degree of corrosion.

In summary, within 300 days of immersion in a coupled salt solution, the corrosion resistance of seven groups of recoverable anchor cable specimens is in the following order: C5 specimen > C2 specimen > C4 specimen > C3 specimen > C1 specimen > HY specimen > LG specimen.

6 CONCLUSION

(1) The flexural strength of the cement paste test block with a water-binder ratio of 0.50 was the best, and the compressive strength of the cement paste test block with a water-binder ratio of 0.45 was the best.

(2) The corrosion resistance of the C5 specimen is the best in the coupling salt solution used in the test, and the cement paste protective layer with epoxy resin coating composite water-binder ratio of 0.50 has the strongest protection ability for extruded anchor recoverable prestressed anchor cable.

(3) LG specimen has the highest corrosion current density in all electrochemical specimens of recoverable anchor cable, and the C2 specimen has the strongest resistance to salt ion corrosion in the C1-C4 specimen, which again shows that the cement paste with a water-binder ratio of 0.50 has the best corrosion resistance in this coupled salt solution.

REFERENCES

Hekal E E, Kishar E, Mostafa H (2002) Magnesium sulfate attack on hardened blended cement pastes under different circumstances. *Cement and Concrete Research*. 32(9), 1421–1427.

Jin ZQ, Sun W, Zhang YS (2007) Interaction between sulfate and chloride solution attack of concretes with and without fly ash. *Cement and Concrete Research*. 37(8), 1223–1232.

Joe HX, Zhang ZW, Gao S (2017) Electrochemical corrosion behavior of steel bar in sulfate – chloride coupling environment. *Journal of Lanzhou University of Technology*. 43(4), 132–136. (in Chinese)

José Marcos Ortega, Victoria E García-Vera, Afonso Miguel Solak, Antonio José Tenza-Abril (2019) Pore structure degradation of different cement mortars exposed to sulphuric acid. *Applied Sciences*. 9(24), 5297.

Ministry of housing and urban and rural construction of the people's republic of China (2012) JGJ 120-2012, *Technical Specification for Retaining and Protection of Building Foundation Excavations*. Beijing: China Architecture and Building Press. (in Chinese).

Neville A. (2004) The confused world of sulfate attack on concrete. *Cement and Concrete Research*. 34(8), 1275–1296.

Wang XM, Yuan J, Wei P, Zhu MW (2020) Effects of fly ash microspheres on sulfate erosion resistance and chlorion penetration resistance in concrete. *Journal of Thermal Analysis and Calorimetry: An International Forum for Thermal Studies*. 139(10), 3395–3403.

Zhang RL, Liu P, Ma LN, Yang ZJ, Chen HS, Zhu HX, Xiao HG, Li J (2020) Research on the corrosion/permeability/frost resistance of concrete by experimental and microscopic mechanisms under different water–binder ratios. *International Journal of Concrete Structures and Materials*. 14(8), 2988–2996.

Advances in Civil Engineering and Environmental Engineering – bin Yahya & Bin Daud (Eds)
© 2023 The Author(s), ISBN: 978-1-032-39397-1

Study on the spatial characteristics of Lingtang Hui nationality township, Gaoyou City, Jiangsu Province

Zimo Li*
The High School Affiliated to Renmin University of China, HaiDian, China

ABSTRACT: Hui nationality is a minority nationality with a large population in China. The Hui people believe in Islam and have a unique national culture, which greatly enriches the cultural diversity of our country; in the development process of our country, the Hui people have also made positive contributions. In the long process of development and ethnic integration, the southern Jiangsu culture and Muslim culture have merged to form a unique culture that is different from other cultural forms. Among them, the village layout is closely related to local customs and has distinctive local characteristics, which is one of the important ways to show the urban culture and the living habits and folk customs of local residents, so the study of Hui villages in Jiangsu province can help discover the Muslim culture in Jiangsu region. This study takes Lingtang Hui Nationality Township in Gaoyou City, Jiangsu Province as the research object, and summarizes the spatial characteristics of Lintang Hui Township in three aspects, namely, the overview, village and street patterns, and residential architecture, using the documentary research method and the graphic background method, to provide a theoretical basis for the development of Lintang Hui Township and other Hui villages, to promote and protect the cultural heritage of Hui people living in Jiangsu Province in terms of architectural characteristics and living patterns, and to promote the culture of minority groups in China.

1 INTRODUCTION

Ethnic minority culture is an integral part of Chinese culture. Since the reform and opening up, with the development of the country, China has become more and more conscious of the protection of minority cultures, and ethnic villages have been further strengthened and developed as an important way to show ethnic cultural customs. In 2012, the State People's Committee issued the Outline of the Plan for the Protection and Development of Ethnic Minority Villages (2011–2015), emphasizing the promotion of infrastructure construction and cultivation of special industries in ethnic villages to protect and inherit the excellent traditional culture of ethnic minorities and generate income for the masses. (The National People's Committee issued the outline of the plan for the protection and development of minority characteristic villages, www.gov.cn, 2012)The distribution of the Hui people is mainly characterized as "large mixed and small clusters", where the mixed population is unevenly distributed in 31 provinces, autonomous regions, and municipalities. However, due to the cohesion of religion and national culture, the Hui people are also in small clusters. Jiangsu Province is one of the major areas where the Hui people live, and they are widely distributed in 11 cities in the province (Wen & Zheng 1990). With the support of policies, eight Hui villages in Jiangsu Province have flourished, among which Lingtang Hui Township in Gaoyou City, as the largest Hui village in the province and the provincial key center town, has always adhered to the concept of "common unity and struggle, common prosperity and development", and has achieved remarkable results in

*Corresponding Author: dwxmxgbxhjpz@126.com

DOI 10.1201/9781003349563-18

economic and social development (Hui 2011). At the same time, in the process of development, Lingtang Hui Township also faces some problems, such as the weakening of Hui characteristics, the deterioration of the ecological environment, and the degradation of tourism resources. Therefore, it is necessary for Lingtang Hui Township to explore a development method that is more friendly to the environment and can better protect the traditional ethnic culture. Among them, optimizing the spatial layout and creating a spatial feature with vivid cultural characteristics is an important part of sustainable development (Xiong et al. 2011). This study will draw conclusions from the analysis of three aspects: the general situation of Lingtang Hui Township, village street patterns, and residential architecture, and sort out design inspirations to provide reference ideas for the subsequent development of Lingtang Hui Township.

2 OVERVIEW OF LINGTANG HUI TOWNSHIP

2.1 *Historical background of Lingtang Hui nationality township*

On May 12, 1988, the people's Government of Jiangsu Province approved the establishment of the Lingtang Hui nationality township. It covers an area of about 53 square kilometers and has a population of more than 23000, including more than 7800 Hui people, accounting for more than 30% of the population of the township (Liu 2016). Although Lingtang Hui nationality township was formally established in 1988, it has a very long history. According to the Yangs' genealogy in Lingtang halal village, more than 300 Muslims took root in Lingtang and continued to thrive as early as 700 years ago when the Arab puhardin came to Yangzhou to preach. After hundreds of years of development, Lingtang Hui Township has become the largest Hui Township in Jiangsu Province (Ma & Cang 2019). By June 2020, Lingtang Hui Township has two communities, Wenchang and Shuguang, and six administrative villages, Lingtang, Qilong, Gaomiao, Gongjia, Halal, and Wangyao. There are two temples in the village: the market town mosque and the ancient mosque. The Hui people live around the ancient mosque and in the Hui Community (Ma & Cang 2017). Since the Hui nationality township was approved to be established in 1988, the government has focused on promoting the economic development of the village through township enterprises. As the factories have high requirements for traffic convenience, most of the factories are concentrated in the market towns. At the same time, the Muslim primary school in the halal village was moved to the market town, and the infrastructure such as shops and health centers in the market town was gradually improved. The market town became the main activity space for the residents of Lingtang Hui Nationality Township.

2.2 *Natural environment of Lintang Hui nationality township*

Lingtang Hui Nationality Township is located on the West Bank of Gaoyou Lake, Gaoyou City, Jiangsu Province, adjacent to Tianchang City, Anhui Province in the west, Shenju mountain in the south, Songqiao town in the East, Gaoyou Lake in the north, surrounded by lakes on three sides and mountains on one side, with a unique ecological environment (Qiao 2011). Hui Township is located at the northern foot of Shugang, the township presents a topography of high in the middle and low around, high in the south and low in the north, shaped like a tortoise, with many hills, depressions, and mountainous areas. Gaoyou City is located in the mid-latitude region, so Lingtang Hui Township belongs to the subtropical humid monsoon climate, with a mild climate, cold winter and warm summer, and abundant light and heat, which provides unique climatic conditions for agricultural development. The precipitation in Lingtang is concentrated and has a remarkable plum rain climate. Meanwhile, the east, north, and west face Gaoyou Lake, and there are Xiangyang River in the east and Shuangqing River in the west, and the two rivers converge into Gaoyou Lake, which shows that Lingtang Hui Township has abundant water resources. As of 2011, the administrative area of Lingtang Hui Township is about 53.92 square kilometers, and the arable land area accounts for about 28,200 mu. Since there are many rivers and developed fisheries in Lintang Hui Township, there is an aquaculture area of about 13,000 mu in the township (Qiao 2011) (see Figure 1).

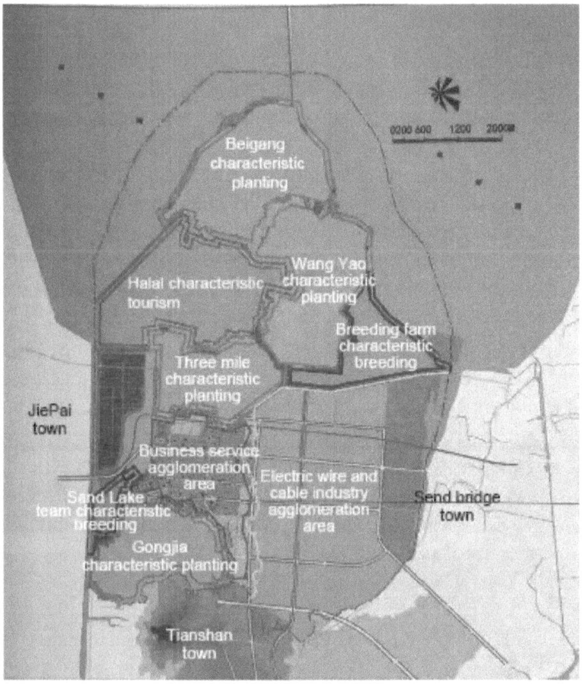

Figure 1. 2010–2030 Lingtang township overall planning map.

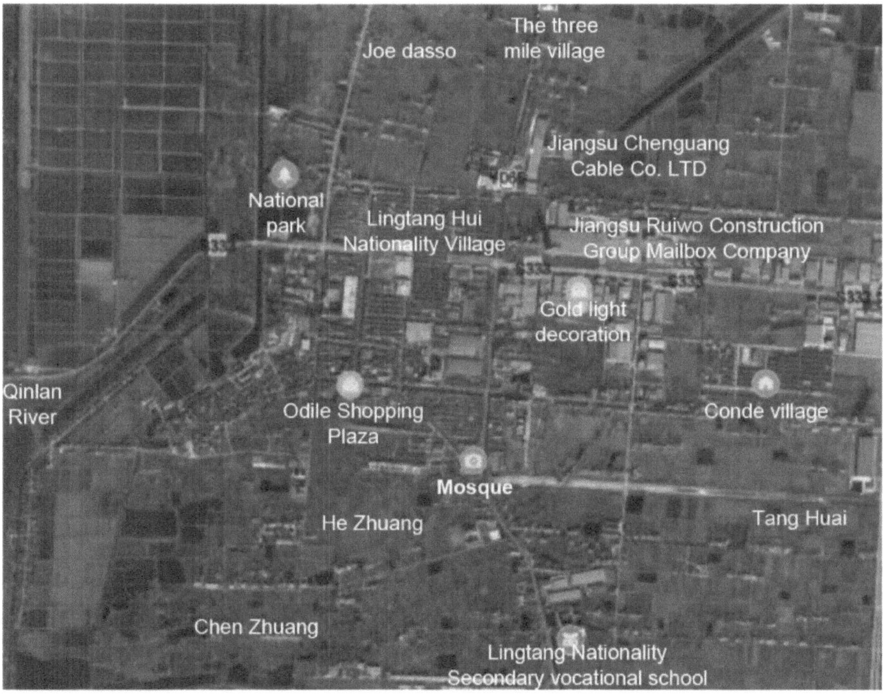

Figure 2. Satellite map of Lingtang Hui nationality township.

2.3 *Folk customs of Lingtang Hui nationality township*

Lingtang Hui Township has a long history, and many of the habits of Hui residents have gradually integrated with those of the Han Chinese, as they get along well with each other and intermarry commonly. However, at the same time, the Hui people still have their unique folk customs (Ma & Cang 2017).

Due to the unique Mosque of the Hui nationality, that is, the residential form takes the mosque as the center and is surrounded by layers, the residential community structure in the Hui Township is also this form, which strongly maintains the cohesion of the Hui people and is also an important part of the Hui religious culture (Pan 2012). Secondly, the Hui people's religion has the dietary habit of "eating clean", so they do not eat pork, fish without scales, or animal blood. Therefore, the diet in the Hui township fully respects the customs of the Hui people, and even during the Hui-Chinese marriage, the daily diet is mainly the Hui-hua diet (Wang 2018). There are many unique festivals in Islam that the Muslims believe in, and the tradition of celebrating these festivals has been preserved in Lingtang Hui Township. Every Friday at 1:30 p.m., the Muslims gather at the ancient mosque to perform the rituals of the Lord's Day, and on the three major Islamic festivals of Muharram, Eid al-Fitr, and Eid al-Adha Corban Festival, the Hui people in the township also gather at the ancient mosque to celebrate the festivals. The wedding and funeral ceremonies of the Hui people in the township are held in the ancient mosque, in addition to birthdays and New Year's events are held in the ancient mosque. (Ma & Cang 2017).

3 ANALYSIS OF THE VILLAGE AND STREET PATTERN OF LINGTANG HUI VILLAGE

3.1 *Analysis of the morphology of Lingtang Hui village*

Lingtang Hui nationality township is mainly in the form of cluster and linear settlements and is a water network enclosed village. In the Hui nationality township, where the terrain is relatively flat and the land is relatively fertile, the settlements are in a cluster shape and develop around the mosque, while the settlements at the intersection of residential areas and farmland are linear. Among them, the cluster settlements are in the majority, while the linear settlements are very rare (as shown in Figure 2). There are two decisive factors in the formation of the rural settlement form of Lingtang Hui Nationality: the natural environment and human activities.

3.1.1 *Natural environmental factors*

Lingtang Hui nationality township is mainly composed of flat depressions and hills, thus forming a cluster and linear settlement form. Cluster settlements are generally formed in areas with relatively flat terrain and superior land resources, and develop around one or more centers. Linear settlements generally develop along rivers or roads and grow axially (Peng 2004). The advantage of cluster settlement lies in a higher land utilization rate, a more flexible building layout, and a better combination of man-made buildings and natural elements. Therefore, Lingtang Hui Township has a wide area of cultivated land, which is suitable for agricultural activities. The advantage of linear settlements lies in the convenient connection between buildings and farmland in the township. Therefore, linear settlements in Lingtang Hui nationality township are mostly located at the junction with farmland. However, its disadvantage lies in the difficult layout of public facilities, so there are only a few linear settlements in Lingtang Hui Township.

The rainy season in Lingtang Hui Township is obvious and precipitation is concentrated, so Lingtang Hui Township actively develops aquaculture, and there are several fish ponds in the township. Since the village is close to Gaoyou Lake, Lingtang Hui Township is a water-net enclosed village. The characteristics of this type of village are that there are both natural water systems and artificial fish ponds around the village. The advantage of the water network enclosed village is that it can provide abundant water resources for the village, which can help the development of agriculture and aquaculture industries and create economic value for the village. Therefore, Lingtang Hui Township makes full use of its advantages, and the existing aquaculture area is about

13,000 mu. However, the disadvantage of the water network enclosed village is that it is prone to flooding, and once flooding occurs, aquaculture and fisheries industries are vulnerable to the impact. In response to this problem, the Lingtang Hui Township government has set up water relief funds and other funds to reduce the losses suffered by the people. (Yang 2017).

3.1.2 *Human activity factors*

Due to the special religious beliefs of the Hui people in Lingtang Hui Township, the main gathering place of the villagers is the mosque, which directly leads to the development of the regimented settlement with the mosque as the center. However, due to the development of the market town and the special industries in the township, not all the buildings in the township strictly follow the layout of the temple square. Such a layout can meet the custom of the Hui people to live around the temple, but also to respond to the development of the times, through the special industry to improve the economic level. The concentration of housing in Lingtang Hui Township, in addition to the custom of living around the temple, also facilitates subsequent planning and creates space for the development of the township. At the same time, the public facilities in the township, such as schools and health stations, are also concentrated, making it convenient for residents to use. In addition, the stores in Hui Township are concentrated on both sides of the main roads in the township, making it convenient for residents to shop and for tourists to shop, which can contribute to the development of tourism.

3.2 *Analysis of street and lane morphology in Lingtang Hui nationality township*

The streets in Lingtang Hui Township attach great importance to external communication, and the roads are well connected, with main traffic roads such as Xingling Road, Guoling Road, and Zhongda Road running through the village as a backbone and a channel for external traffic. Other secondary streets are extended from the main road, and the main road is used as the axis to form a road network perpendicular or parallel to the axis and is connected to the residential area through the alleyway. Lingtang Hui Township, as a provincial key center town, has undergone continuous planning by the government, such as the relocation of industrial area households. The network of mutually perpendicular or parallel roads can preserve more complete land and facilitate the government's subsequent planning. Public facilities such as health centers are generally located along both sides of the secondary streets, which are also linked to other public buildings such as mosques and ethnic squares. Lingtang Hui Township focuses on the cultivation of special industries, especially tourism and industrial development this year, so it needs a well-connected traffic main road to transport products and facilitate tourists to enter the Hui Township. Secondary streets connect mosques, public facilities, residential areas, and main roads, and can facilitate religious activities such as pilgrimage for the Hui people, as well as provide convenience for residents. The early Muslims who entered China were mainly merchants, and due to the war, they stayed in China to continue developing their commercial economy, forming a tradition of business for the Hui people (Wen & Zheng 1990). Most Hui neighborhoods have an Islamic commercial street, whose business scope includes food, clothing, handicrafts, etc., and is one of the main places for residents' activities. The commercial streets in Lingtang Hui Township are distributed in a ribbon shape, connecting to the main traffic roads, and there are almost no traditional forms of stores along the streets in the neighborhood, but some stores selling Hui specialty products still retain the traditional characteristics of Muslim architecture. In addition to buying goods, residents of the Hui township can also carry out ethnic activities in the streets, such as the "Halal Food Festival", and the commercial function of the streets is continued to this day.

The form of residence of the Hui people follows the Jiaofang system, also known as the "the wooden Jiaofang system of the pavilion", which is a religious organization system with the Jiaofang as the basic unit. This system relies on the common religious culture of the Hui people and is an important form of social organization to maintain the cultural heritage of the nation. The living space in Lingtang Hui Township is centered on the mosque, which is integrated through the Jiaofang system. The mosque usually serves as the center of the Jiaofang. Lingtang Hui Township has two mosques, the Jizhen Mosque and the Ancient Mosque. The Jizhen Mosque is located in the market town, one of the most important activity sites for Hui Township residents. The ancient mosque is located in the halal village and also serves as the center of the village. The road networks around the two mosques are radial. These centripetal and directional roads can reflect the importance of mosques to the Hui people and maximize the religious activities of local residents.

The part of Lintang Hui Township facing farmland forms a belt-like, single distribution feature with few and mostly vertical roads linking residential areas and farmland. Some of the streets in the residential areas facing the mountains and the river change according to the topography, forming streets of different widths and varying heights. The single road in the part facing the farmland can facilitate the residents to reach the farmland quickly and facilitate agricultural activities. The streets facing the mountains and rivers are flexible and changeable, winding along the mountains and rivers, reflecting the harmony between man and nature, and adding interest to the streets.

4 ANALYSIS OF RESIDENTIAL ARCHITECTURE IN LINGTANG HUI TOWNSHIP

4.1 *Layout analysis of Lingtang Hui township*

When studying the layout of Lingtang Hui Township, this study focuses on mosques and residential courtyards. As a building group, the mosque is composed of different buildings, and the main axis is in the east-west direction. Residential courtyards are generally divided into living, sleeping, kitchen, bath, and other different spaces, with relatively flexible locations and layouts.

4.1.1 *Mosque layout*

The mosque, as the spiritual bond of the Hui, is undoubtedly in the most important position in Islamic architecture. There are two mosques, in Lintang Hui Township, the Jizhen Mosque and the Ancient Mosque. Two mosques, one in the south and one in the north, are intended to facilitate the religious activities of all residents, which also shows the importance of the mosque. The Ancient Mosque is located in the mosque village, which is the main settlement of the Muslim people. The Ancient Mosque is located in the center of the village and is still the main location for large religious events for the Hui people. The Jizhen Mosque is located in the town, which is also in a relatively central location.

Unlike other religious buildings, a mosque is a group of buildings composed of different functional monolithic buildings. The mosque in Lingtang Hui Township is a combination of different buildings such as the worship hall, the back kiln hall, and the water room. The mosque in Lingtang Hui Township is located from the west to the east, facing the direction of Kerbai, the west-oriented courtyard system is formed. The main axis is east-west, which is different from the traditional north-south axis in China. The courtyard of the mosque is not necessarily surrounded by buildings, but only in the form of a compound or triple compound to form a more closed courtyard (Pan 2012).

4.1.2 *Residential courtyard layout*

The main settlement pattern of the Hui people follows the temple-place system. Therefore, within the halal village, the people have formed a community of the Zhematizu community living around the temple. In addition, the government is also gradually incorporating the village of Beihuang into the mosque village because of its proximity to the mosque village, which also has the form of a temple enclosure. The Jizhen Mosque is surrounded by a 60-acre community of Muslims, which is also intended to meet the needs of the Muslim people in the form of a temple. It can be seen that the distribution of residential houses in Lingtang Hui Township is basically determined by the change in the location of the mosque.

In traditional Hui communities, dwellings generally have living space, sleeping space, kitchen space, bathing space, etc. The living space is mainly located on the north side of the house. In the courtyard, the kitchen space is usually located in the west wing, while the layout of other spaces is relatively flexible. (Li 2016).

4.2 *Analysis of architectural details in Lingtang Hui township*

The residential compounds and mosques in Lingtang Hui Township have been integrated with Chinese culture in the process of development, forming a unique local culture that blends Chinese and Islamic cultures. Therefore, the architecture of Lingtang Hui Township also has both traditional Chinese architectural features and Islamic architectural features.

Figure 3. Ancient mosque in Lingtang Hui nationality township.

Figure 4. Mosque proclamation platform.

4.2.1 *The mosque*

The mosques in Lingtang Hui Township follow the architectural form of traditional Chinese architecture on the premise that they have the necessary conditions for religious activities. The styles of Ancient Mosques are different from those of Jizhen Mosques. The ancient mosques are courtyard style and adopt a wooden frame structure. There are also Taoist cultural symbols such as "Fu Lu Shou" and Tai Chi eight trigrams on the tiles, which have distinctive characteristics of the Jiangnan garden style. They are a typical example of the integration of Chinese culture and Islamic culture (Figure 3). The mosques in the market town are magnificent and mostly decorated with flowers and plants in Islamic style, so the exotic customs are more intense.

Inside the mosque, because Islam opposes idol worship, there are no statues, photos, etc. in the hall, so the main hall must set up a shrine as the focus of worship for believers. Most of the shrines are concave wall shaped and richly decorated. Most of the shrines are located on the west wall of the central axis of the main hall. In front of the shrine on the left is a pulpit for preaching. The proclamation platform is equipped with wooden walking sticks, stairs, etc., which are also exquisitely carved, highlighting its important position with its strong decoration (Figure 4).

4.2.2 *Residential courtyard*

Due to the long-term intermingling of Hui people and Han Chinese in Lingtang Hui Township, many of the dwellings in Hui Township have been Sinicized and do not differ much from the traditional dwellings in China. The Hui people in the market towns live next to the Han people, and the communities are often mixed with Hui and Chinese, thus the Islamic characteristics are not prominent.

However, in the traditional Hui community, there are usually living spaces, sleeping spaces, kitchen spaces, and bathing spaces. The living space is the main activity space of the Hui people and has basically the same function as the living room. The living space is also used for small cleanings, washing hands, and faces, etc. to keep the body clean. The sleeping space is used as a resting place for the Hui people and also for entertainment, and some families also place the stove in the sleeping space. Due to the Islamic concept of cleanliness, the Hui people's kitchens are generally very clean and tidy, and the ingredients are mostly halal. As the conditions of the Hui people have improved, the kitchen space is mostly brick. In addition, the Muslim idea of cleanliness is also reflected in the bathing space. The bathing space is usually not very large, and the Muslims use hanging jars for large purifications and soup bottles for small purifications.

5 CONCLUSION

Chinese Islam is a culture with great Chinese characteristics, which can greatly enrich the diversity of our culture and is an integral part of Chinese culture. And Jiangsu province, as one of the settlements of the Hui people, contains a blend of Chinese and Muslim cultures that gives the Hui villages in Jiangsu a unique aesthetic character. This study hopes to enhance the understanding of Hui culture and religious culture and development process, inherit and protect minority cultures, promote traditional Chinese culture, and promote and publicize the precious cultural heritage of Hui people living in Jiangsu province, such as food, clothing, housing, lifestyle, and architectural features.

REFERENCES

Li Wei. Comparative study on the spatial form of hui dwellings in henan and Jilin [D]. *Jilin University of Architecture*, 2016.

Lingtang Hui. Township of Gaoyou City—the only ethnic township in the province, the pearl on Gaoyou Lake [J]. *Zhongshanfengyu*, 2011(02):65.

Liu Xing. Study on the construction of ecological civilization in ethnic villages in Jiangsu Province [D]. *Jiangsu University*, 2016.

Ma Haiyan, Cang Ming. The "compatible coexistence" and "harmony but difference": A survey of the Hui Culture in Lintang, Jiangsu Province under the Habitat of Diaspora [J]. *Journal of Central South University for Nationalities (Humanities and Social Sciences Edition)*, 2017, 37(01):43–47.

Ma Haiyan, Cang Ming. The image of the other and tourists' perceptions: a qualitative study of the rooting theory of Hui villages in Jiangsu[J]. *Journal of Yunnan Agricultural University (Social Science)*, 2019, 13(02):37–44.

Pan Ruya. Study on traditional Islamic architecture in Jiangsu [D]. *Nanjing University of Technology*, 2012.

Peng Song. From architecture to village form [D]. *Southeast University*, 2004.

Qiao Yuanjiang. National sustainable development of the top 100 townships–Lingtang Hui Township [J]. *China Old Area Construction*, 2011(12):2.

Qiao Yuanjiang. Top 100 towns for sustainable development in China–Lingtang Hui Township [J]. *China Old Area Construction*, 2011(12):2.

The National People's Committee issued the outline of the plan for the protection and development of minority characteristic villages (www.gov.cn), 2012

Wang Jinyang. On Hui dietary taboos [J]. *Food Science*, 2018(14):96+98.

Wen Qichang, Zheng Mianzhi. The Hui of Jiangsu [J]. *Ningxia Social Science*, 1990(02):96–100.

Xiong Guoping, Rong Qingwen, Tian Hui. Exploration of Hui township planning–Jiangsu Lintang Hui township planning as an example [J]. *Journal of Architecture*, 2011(04):35–38.

Yang Dawei. The research report on the development of modern fishery in Lintang Hui Township, Yangzhou, Jiangsu [J]. *A Fishery to Get a Rich Guide*, 2017(04):21–23.

Advances in Civil Engineering and Environmental Engineering – bin Yahya & Bin Daud (Eds)
 ISBN: 978-1-032-39397-1

Development and engineering application of environment-friendly pavement markings

Xiaoyong Zou & Wenxiu Wu*
Jinhua Highway and Transportation Administration Center, Jinhua, China

Jing Liu
Dongyang Highway and Transportation Administration Center, Dongyang, China

Yonglin Hu
Highway and Transportation Administration Center of Jindong District, Jinhua, China

Dapeng Li
Zhejiang Institute of Communications Co., LTD, Hangzhou, China

ABSTRACT: The expansion of highway and urban road repair and reconstruction has put forward higher requirements on the performance of traffic line clearing. To solve the problems of visual residue and road surface damage caused by line marking removal in previous studies, a kind of microwave quickly removing multifunctional marking line for road engineering is designed by adding a functional bonding layer and microwave efficiently removing layer between asphalt pavement and two-component epoxy marking line. Then the structure, formula composition, clearance performance, and road performance of the two-component epoxy multifunctional marking lines are studied. Thus, it can reduce the difficulty of construction, which has been marked in road reconstruction and expansion projects to avoid damage to the road surface due to traffic guidance, extend the service life of two-component epoxy multifunctional marking, provide additional guarantee to improve the highway driving safety, and meet the requirements of the construction of an "environmentally friendly" society in China.

1 INTRODUCTION

With the rapid growth of China's highway construction mileage, traffic line markings of the market demand are growing (Burghardt et al. 2021; Solla et al. 2021), the visibility of road signs, durability, resistance to sliding and easy in addition to performance requirements also more and more big (Chen et al. 2022; Li et al. 2021; Liu et al. 2022; Wang et al. 2022), need to study environmental protection line of development and carry out the corresponding engineering application, in order to build the highway asphalt pavement multi-function environmental marking method of research and development and layout (Liu et al. 2021, 2021, 2022, 2022, 2022, 2021; Xu et al. 2021).

In existing research (Scovill 1976) (Koo et al. 2009; Pastor & Lemonnier 1994; Shi et al. 2014), there are many different kinds of road marking paint (Wang et al. 2022), and one of the two-component markings relies on unsaturated resin and curing chemical crosslinking curing reaction taking place between crosslinking agent to generate a space interpenetrating network (IPN), not only has short table dry time, and its strength has been effectively improved, compared with traditional solvent type marking and hot melt marking (Wang et al. 2020, 2019). The two-component marking line has obvious superiority in road performance such as structural strength, adhesion and wear resistance (Liu et al. 2022, 2022, 2022). However, violent methods are used to destroy the cohesion of the marking line or the cohesion between the marking line and the road surface. This kind of

*Corresponding Author: 277182929@qq.com

DOI 10.1201/9781003349563-19

method not only consumes a lot of energy, but also has low efficiency, and will cause different degrees of damage to the original road surface (Liu et al. 2017). The marking microwave clearing technology proposed in this project is simple and efficient, which can effectively avoid the above problems (Chen et al. 2022; Liu et al. 2019, 2021; Ren et al. 2021; Yang et al. 2004).

In this paper, by adding a functional bonding layer and microwave easily removing layer between asphalt pavement and two-component epoxy marking line, a kind of microwave easily removing multifunctional marking line for road engineering is designed, and then the structure, formula composition, clearance performance and road performance of two-component epoxy multifunctional marking line are studied. Thus, it can reduce the difficulty of construction which has been marked in road reconstruction and expansion projects to avoid damage to the road surface due to traffic guidance, extend the service life of two-component epoxy multifunctional marking, provide further guarantee to improve the highway driving safety and meet the requirements of the construction of an "environmentally friendly" society in China.

2 LABORATORY PREPARATION OF MARKINGS

In this study, the functional marking specimen (FMS) of waterborne epoxy marking (WEM) was designed successively for the purpose of improving clearance performance based on the requirements for basic pavement performance. The structures of WEM are shown in Figure 1.

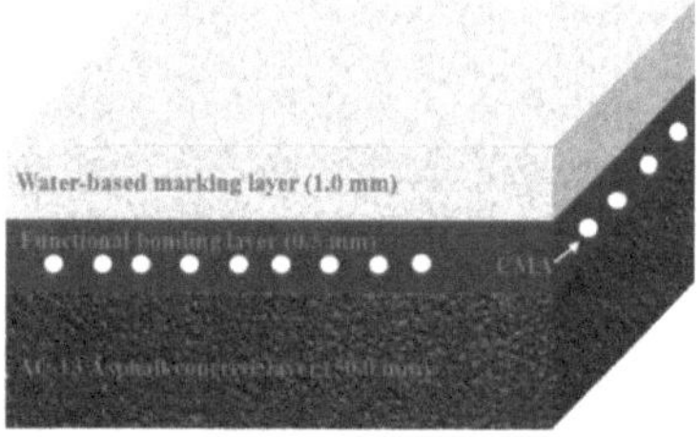

Figure 1. Schematic diagram of functional marking specimens.

Then, Figure 2 shows the laboratory preparation of the marking materials of WEM as follows:

1) The water-based epoxy emulsified asphalt was placed in an empty beaker, and corresponding mass fractions of composite microwave absorbent, dispersant, defoamer, and silane coupling agent were added into the mixture in turn. The functional emulsion (FM) was obtained by stirring the mixture with a stirring speed of 1500 rpm for 15 min.
2) After mixing components A and B in FM at A mass ratio of 2:1, it is needed to put them in A beaker containing deionized water, and stir them with A stirrer at A speed of 1000 rpm for 10 min to obtain WEM.

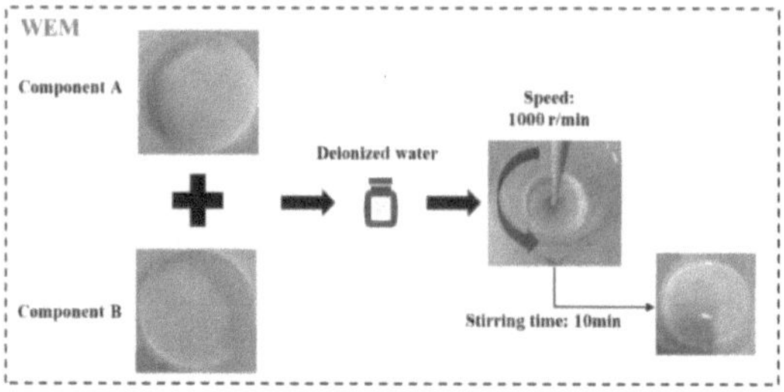

Figure 2. Schematic for the preparation of waterborne epoxy marking.

3 FILED CONSTRUCTION OF PAVEMENT MARKINGS

3.1 *The template file*

Relying on the highway maintenance plan of Guizhou Province in 2021, the newly completed AC-10 cover surface of G245 (K961+000 – K963+000 and the middle line and sideline of nearby sections) in Kaili City, Guizhou Province was constructed by the easy-to-clear marking process. The total construction area of the marking line was about 1000 square meters, and the construction marking layer structure was as follows: easy to clear functional layer + spontaneous light two-component marking layer.

As described in Figure 3, the construction process of waterborne epoxy marking was performed as follows:

1) Prepare the materials in Figure 2: water-based epoxy emulsified asphalt, hydroxyl iron powder, graphite powder, silane coupling agent, and defoaming agent. Then add the four materials in turn to the water-based epoxy emulsified asphalt, and use a blender to mix well (continue stirring for about 30 seconds, until all materials are evenly mixed).
2) Road cleaning: clean the original road surface, remove debris, and put safety cones according to the requirements of construction specifications.
3) Pre-laying out of line marking: Lay out the line with a long rope and pre-mark the water line to distinguish the positions of dotted lines, solid lines, and double lines for later construction of line marking.
4) Easy to clear functional layer construction: use a roller brush to evenly coat the easy-to-clear material at the marking position (to ensure that the thickness is at least 0.4 mm, and the roller coating is uniform, if the thickness is not enough, it should be coated).
5) Two-component line marking construction: After the functional layer for easy cleaning is put into practice, the operator drives the construction vehicle for line marking, and the construction of the line marking layer is carried out according to the pre-laying position and the position of the line marking layer for easy cleaning, and the self-lighting microbeads are spread synchronously (pay attention to covering the functional layer for easy cleaning, and try to avoid exposing the functional layer for easy cleaning).
6) Completion of construction: After the two-component marking layer is dried (the drying time is determined by the actual site temperature), the cone bucket will be recovered and traffic will be opened.

Figure 3. Schematic for the construction process of waterborne epoxy marking.

4 ANALYSIS OF ENGINEERING APPLICATION EFFECTS

4.1 *Functional results*

As shown in Figure 4, the marking samples heated in the microwave oven for 3 min were removed and placed on the surface of a flat table. Then, the surface line was gently removed from the upper surface of the samples using heated forceps. It can be seen that there was almost no visual residue on the surface of the marked specimens, and the anti-slip performance is greatly improved after microwave heating.

(a) (b)

Figure 4. The removal function: (a) before microwave heating; (b) after microwave heating.

As can be seen from Figure 5, by adding long afterglow powder to the WEM, the luminous marking line with high luminous performance was prepared. When used on the road, the luminous marking line has both luminous indicator and warning function, which is of great significance to road guidance at night.

(a) (b)

Figure 5. Self-lighting function: (a) day; (b) night.

4.2 *Economic benefits*

The economic benefits of the prepared markings can be calculated using Eq. (1) as follows:

$$\begin{gathered} C_M = C_1 + C_2 + C_3 \\ C_1 = \frac{1}{0.15m \times 0.15m} \times (22.73g \times 0.00077\$/g + 2.84g \times 0.0077\$/g + 2.84g \times 0.0046\$/g) = 2.339\$/m^2 \\ C_2 = \frac{1}{13311.7875 \times 10^{-6} m^2} \times 0.7kw \times \frac{3}{60}h \times 0.1037\&/kw \cdot h = 0.2727\$/m^2 \quad , \\ C_3 = 0.772\$/m^2 \end{gathered} \tag{1}$$

Where the total cost C_M included three components: the cost of removing materials C_1, the cost of removing energy consumption C_2, and the labor cost C_3. In this study, the selling prices of WEA, CIP, and GP were 0.772, 7.72, and 4.632 \$/kg, respectively. The cost comparison between different cleaning methods for waterborne markings of asphalt pavement is shown in Figure 6.

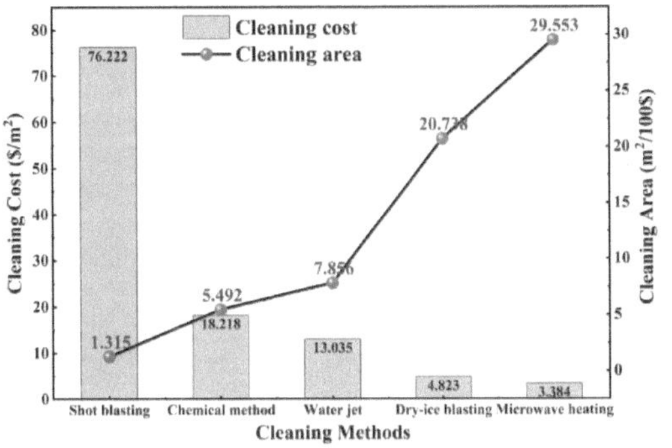

Figure 6. Schematic for the construction process of waterborne epoxy marking.

Compared with the other asphalt pavement marking-cleaning methods, the microwave heating method had the lowest cleaning cost; it was 4.44, 18.57, 36.59, and 40% of those of the shot blasting, chemical solvent, water jet, and dry-ice blasting methods, respectively. It also covered the largest cleaning area (per 100 $) which was 22.53, 5.38, 3.76, and 1.43 times those of the above four methods, respectively. This indicated that the microwave heating method proposed in this study had good economic benefits and could significantly reduce the cost and increase the efficiency of marking removals.

5 CONCLUSIONS

This paper presented a new method of removing traffic line marking on asphalt pavement by microwave heating. The main conclusions are as follows.

1) There was almost no visual residue on the surface of the marked specimens, and the anti-slip performance was significantly improved after microwave heating.
2) The developed pavement markings have both luminous indicator and warning functions, which are significant to road guidance at night.
3) The microwave heating method proposed in this study had good economic benefits and could significantly reduce the cost and increase the efficiency of marking removals.

ACKNOWLEDGMENTS

This study was sponsored by the Key Science and Technology Research Project of Jinhua, China under Grant 2021-3-177.

REFERENCES

Burghardt, T. E., Popp, R., Helmreich, B., Reiter, T., Bohm, G., Pitterle, G., and Artmann, M. (2021). "Visibility of various road markings for machine vision." *Case Stud Constr Mat*, 15.

Chen, Y. H., Gu, X. Y., Liu, Z., and Liang, J. (2022). "A fast inference vision transformer for automatic pavement image classification and its visual interpretation method." *Remote Sens-Basel*, 14(8).

Koo, J. K., Moon, D. S., Bernold, L. E., and Lee, T. S. (2009). "Study on the application of dry-ice blaster for development of automatic stripe-line removal." *Journal of the Korean Society of Civil Engineers*, 29(2D).

Li, S. W., Gu, X. Y., Xu, X. R., Xu, D. W., Zhang, T. J., Liu, Z., and Dong, Q. (2021). "Detection of concealed cracks from ground penetrating radar images based on a deep learning algorithm." *Constr Build Mater*, 273.

Liu, W., Miao, P. H., and Wang, S. Y. (2017). "Increasing microwave heating efficiency of asphalt-coated aggregates mixed with modified steel slag particles." *J Mater Civil Eng*, 29(10).
Liu, Z., Chen, Y. H., Gu, X. Y., Yeoh, J. K. W., and Zhang, Q. P. (2022). "Visibility classification and influencing-factors analysis of airport: A deep learning approach." *Atmospheric Environment*, 278.
Liu, Z., Gu, X. Y., and Dong, Q. (2019). "*Multi-Scale 3D Display of the Internal Quality of the Pavement Based on BIM.*" Proc., 19th COTA International Conference of Transportation Professionals (CICTP) – Transportation in China 2025, 4265–4273.
Liu, Z., Gu, X. Y., Dong, Q., Tu, S. S., and Li, S. W. (2021). "3D visualization of airport pavement quality based on BIM and WEBGL integration." *J Transp Eng B-Pave*, 147(3).
Liu, Z., Gu, X. Y., Ren, H., Wang, X., and Dong, Q. (2022). "Three-dimensional finite element analysis for structural parameters of asphalt pavement: A combined laboratory and field accelerated testing approach." *Case Stud Constr Mat*, 17.
Liu, Z., Gu, X. Y., Ren, H., Zhou, Z., Wang, X., and Tang, S. (2022). "Analysis of the dynamic responses of asphalt pavement based on full-scale accelerated testing and finite element simulation." *Constr Build Mater*, 325.
Liu, Z., Gu, X. Y., Wu, C. Y., Ren, H., Zhou, Z., and Tang, S. (2022). "Studies on the validity of strain sensors for pavement monitoring: A case study for a fiber Bragg grating sensor and resistive sensor." *Constr Build Mater*, 321.
Liu, Z., Gu, X. Y., Wu, W. X., Zou, X. Y., Dong, Q., and Wang, L. T. (2022). "GPR-based detection of internal cracks in asphalt pavement: A combination method of DeepAugment data and object detection." *Measurement*, 197.
Liu, Z., Gu, X. Y., Yang, H. L., Wang, L. T., Chen, Y. H., and Wang, D. Y. (2022). "Novel YOLOv3 model with structure and hyperparameter optimization for detection of pavement concealed cracks in GPR images." *Ieee T Intell Transp*.
Liu, Z., Gu, X., and Wang, L. (2021). "*Research on information management of Airport pavement quality based on BIM and GIS integration.*" Green and Intelligent Technologies for Sustainable and Smart Asphalt Pavements, CRC Press, 432–436.
Liu, Z., Gu, X., Chen, Y., and Chen, Y. "System architecture and key technologies for the whole life cycle of smart road." Proc., *Journal of Physics: Conference Series*, IOP Publishing, 012105.
Liu, Z., Wang, S. Y., Gu, X. Y., Li, Z. G., Dong, Q., and Cui, B. Y. (2022). "Application of a novel EWMA-phi chart on quality control in asphalt mixtures production." *Constr Build Mater*, 323.
Liu, Z., Wu, W. X., Gu, X. Y., Li, S. W., Wang, L. T., and Zhang, T. J. (2021). "Application of combining YOLO models and 3D GPR images in road detection and maintenance." *Remote Sens-Basel*, 13(6).
Pastor, M., and Lemonnier, G. (1994). "Shotblasting, a technique in the service of safety." *Revue Generale Des Routes Et Des Aerodromes*.
Ren, H., Gu, X. Y., and Liu, Z. (2021). "*Analysis of Mechanical Responses for Semi-Rigid Base Asphalt Pavement Based on MLS66 Accelerated Loading Test.*" Proc., 20th and 21st Joint COTA International Conference of Transportation Professionals – Advanced Transportation, Enhanced Connection, 732–742.
Scovill, D. A. (1976). "*Hydraulic Sandblasting of Pavement Striping.*" Highway Focus, 8.
Shi, H., Gu, J., and Bao, Y. (2014). "Study on nozzle angle of ultra high pressure water jet for removing road marking." *Chinese Hydraulics & Pneumatics*(6), 51–54.
Solla, M., Perez-Gracia, V., and Fontul, S. (2021). "A review of GPR application on transport infrastructures: Troubleshooting and best practices." *Remote Sens-Basel*, 13(4).
Wang, D., Liu, Z., Gu, X., Wu, W., Chen, Y., and Wang, L. (2022). "Automatic detection of pothole distress in asphalt pavement using improved convolutional neural networks." *Remote Sens-Basel*, 14(16).
Wang, H. P., Yang, J., Lu, G. Y., and Liu, X. Y. (2020). "Accelerated healing in asphalt concrete via laboratory microwave heating." *J Test Eval*, 48(2), 739–757.
Wang, H. P., Zhang, Y., Zhang, Y., Feng, S. Y., Lu, G. Y., and Cao, L. T. (2019). "Laboratory and numerical investigation of microwave heating properties of asphalt mixture." *Materials*, 12(1).
Wang, L. T., Gu, X. Y., Liu, Z., Wu, W. X., and Wang, D. Y. (2022). "*Automatic detection of asphalt pavement thickness: A method combining GPR images and improved Canny algorithm.*" Measurement, 196.
Xu, Z., Zhou, Q., Chen, G., Wang, Y., Yang, X., and Liu, Z. "*An Effectiveness Study of SPC Control Charts on the Quality Control for Construction in Asphalt Pavement.*" Proc., IOP Conference Series: Earth and Environmental Science, IOP Publishing, 012115.
Yang, Z. M., Mao, C. H., and Yang, J. (2004). "Matching characteristic of electro-magnetic parameters of microwave absorbing materials for low frequency (3 GHz)." *Chinese Journal of Rare Metals*.

Advances in Civil Engineering and Environmental Engineering – bin Yahya & Bin Daud (Eds)
© 2023 The Author(s), ISBN: 978-1-032-39397-1

Application of collaborative design in the maintainable and renewable of prefabricated buildings

Shenqi Gan* & Hong Zhang
School of Architecture and Urban Planning, Anhui Jianzhu University, Hefei, China
Key Laboratory of Built Environment and Health, Anhui Jianzhu University, Hefei, China
School of Architecture, Southeast University, Nanjing, China

ABSTRACT: With the development of information technology, BIM is being used more and more in the whole life cycle of prefabricated buildings. According to the construction characteristics of the prefabricated building, for the maintenance and renovation of the fabricated house, BIM technology can be used in three aspects: first, cooperative design pattern; second, computer coding technology; third, component information tracking and feedback technology. Among them, the cooperative design is not only of great significance to the construction and industrialization of the prefabricated residence in our country but also plays an important role in its maintenance and renewal.

1 INTRODUCTION

Collaborative design is to achieve a design goal, by two or more design subjects, through a certain mechanism of information exchange and mutual collaboration, different design tasks are used to accomplish this design goal (Zhang 2017). How to further apply the method of cooperative design in the development of prefabricated housing is a topic that we need to study seriously at present. The specific concept of the cooperative design of prefabricated housing is: the cooperative design is guided by the system cooperative theory and supported by the information technology, and the cooperative design theory and method are applied to the prefabricated housing design mode, and based on this study, how to cooperate and cooperate among different disciplines, different specialties and different team members in prefabricated housing design, how to optimize the process of prefabricated housing design, dispelling the prefabricated housing design conflict, then enhances the prefabricated housing design efficiency. Therefore, collaborative design is a comprehensive and integrated design methodology.

Cooperative design is the basis for the maintenance and renewal of prefabricated housing products (Gan 2020). As explained above, the development of maintenance and renewal of prefabricated housing will eventually form the scale of prefabricated operation and become an important part of the complete industrial chain of prefabricated housing construction. This should be the development direction of the maintenance and renewal of prefabricated housing. The maintenance and renovation of prefabricated housing are different from traditional housing, which involves various departments in the housing industry chain, only when all units, departments, and enterprises, such as design, manufacture, assembly, maintenance, and renewal, join forces and cooperate, can the prefabricated operation of maintenance and renewal be realized, the industrial functions of the industrial chain of Chinese prefabricated housing construction can be brought into full play. In addition, the cooperation model is also an important guarantee of computer coding, only through the cooperation of various departments, and ensures the real-time sharing of information, can complete the complex and cumbersome computer coding (Wang 2016). Therefore, the cooperative design pattern is an indispensable means for the maintenance and renewal of prefabricated housing.

*Corresponding Author: Gansq@ahjzu.edu.cn

DOI 10.1201/9781003349563-20

2 THE CHARACTERISTICS OF COLLABORATIVE DESIGN

2.1 *Characteristics of self-organizing systems*

As a self-organizing system, the collaborative design of prefabricated housing is more like a biological organism, which constantly knows itself and its surroundings, and can adapt to the changes in its surroundings to improve its performance (Li 2019). Self-organization doesn't mean "Out of control," it means collaborative, evolutionary design. Self-organizing systems contain multiple forms and levels of control and feedback, and transmit synergies and checks and balances to all affected project participants.

2.2 *Characteristics of comprehensiveness*

The integrated design approach is adopted in the collaborative design of prefabricated housing (Qin 2014). Collaborative design means integrated design, including the design of fabricated housing systems, subsystems, components, and components, all of which achieve the highest overall performance in a collaborative manner. Integrated design means that from the beginning of the design process to the final completion, all parts of the assembly of residential components keep as close contact as possible with the production enterprise, and the whole process is always accompanied by the production process from the parts of the components' feedback. The integrated design represents a collaborative approach in which problems of structure, installation, construction, and environmental performance are not left to others for final resolution, but are considered from the outset, from this, the progress of the project can be arranged in an orderly manner.

2.3 *Characteristics of informatization*

Collaborative design of prefabricated housing should be based on information, not on form. It is not concerned with the spatial division and modelling design of prefabricated housing, but with how to operate to achieve this state. The collaborative design of prefabricated housing is not a style, but an information-centered design method.

3 APPLICATION OF COLLABORATIVE DESIGN IN PREFABRICATED HOUSING

Cooperative design plays an important role in the maintenance and renewal of prefabricated housing, which is mainly reflected in three aspects: solving the collision problem, improving the quality of maintenance, and cooperative work in the operation and maintenance phase (Miao 2020). The following is a detailed description of these three aspects.

3.1 *Solve the collision problem*

The characteristics of the development of the construction industry are the individuation of the building appearance, the diversification of the building function, and the complexity of the building scale. In view of the fact that all kinds of structural components occupy a certain space in the building, under the premise of guaranteeing the use function and aesthetics of the building, supporting facilities and pipeline systems can only be "crowded" in the corridor ceiling or equipment room. Including water supply, drainage, fire water, sewage, rainwater, and other kinds of water supply and drainage pipes, as well as heating, air conditioning, ventilation, and other heating pipes are crowded together, inevitable "fight" situation, cross-collision problem is also common, to the construction of a great inconvenience, but also not conducive to future maintenance and updating. The computer is used to detect the collision of professional drawings. Although the accuracy of the design drawings is improved to a certain extent, it is still a manual operation based on the two-dimensional drawings, it has not got rid of the dependence on the designer's experience, and the conflicts in the 3D are still difficult to be detected and resolved, and the design drawings may still have a lot of conflicts such as "mistakes, omissions, bumps, flaws".

Although collision problems are inevitable, they can be detected ahead of time through the BIM collaborative design platform. As the most important part of the implementation of collaborative design, collision detection not only solves the problems of hard and soft collisions but also makes the design result more perfect.

Collision detection is the premise of maintenance and update work in the future, the existence of collision problems will bring great trouble to maintenance and update work, such as not solving such problems well, eventually, and it may even become impossible to maintain updates. Therefore, it is of great significance not only to improve the design efficiency but also to improve the quality of the building to solve the collision problems such as fault, leakage, collision, and shortage in a timely and predictable manner, it is also an important guarantee for the maintenance and renewal of prefabricated housing.

3.2 *Improvement of quality maintenance level*

Quality maintenance is the concrete embodiment of the maintenance and renewal of prefabricated housing in the production and construction stages (Liu 2016). It can effectively ensure the quality of the factory production of housing parts and improve the construction accuracy of housing, and the realization of a three-package maintenance service for prefabricated housing products will greatly facilitate the work in the later stage of operation and maintenance. Quality maintenance is an important link in the maintenance and renewal of prefabricated housing, and collaborative design is an important means to improve the quality maintenance level.

According to the research on the prefabricated housing system, the prefabricated housing system can be divided into four stages: the first stage of industrialization, the second stage of industrialization, the third stage of industrialization, and the field installation stage. Among them, the first-class factory is the standard parts production stage, its production goal is mainly to reduce the tolerance, through the production process progress and the organization management level promotion, in the end, the error in the manufacturing process of the component is reduced from 60 cm to 10 cm, and the second stage is to install the component and assemble the component into a whole, for example, the components are assembled into a relatively integral plate-shaped wall; the third stage of industrialization is the rapid assembly of components of prefabricated residential components, and the components are assembled into prefabricated residential components in the factory workshop, to facilitate the overall lifting from the workshop to the site; on-site installation phase is the prefabricated residential parts in the project site for the overall lifting.

The factory level of prefabricated housing products means the continuous improvement of the quality system of prefabricated housing products. The first level of industrialization is only the improvement of construction accuracy, while the third level of industrialization raises the prefabricated housing to the product level, prefabricated housing can be easily broken down into different components. The finished product of the prefabricated part means that the prefabricated residence can be like a common commodity, and can realize three guarantees, which symbolizes the substantial improvement of the quality maintenance level of the prefabricated residence.

To sum up, when the above systems and levels are complex enough, it can reflect the necessity of cooperative design of prefabricated housing, and the improvement of quality maintenance level is one of the goals of cooperative design of prefabricated housing, it can also be implemented at three levels:

(1) Establishing the quality guarantee system of the components of prefabricated housing

The quality assurance system of the components of prefabricated housing needs to be realized by digital tools. Digitalization is the conversion of a large amount of complex information into computer-processed binary digits. Digital tools are computers and other devices that store a large amount of related information and data. Based on this digital tool, the prefabrication of prefabricated housing components can be completed automatically by computer control, the construction error can be reduced, and the production and manufacturing precision of prefabricated housing components can be improved, to improve the quality of prefabricated housing. In addition, digital tools can transfer data information from individual components to components during the assembly stage of prefabricated housing components, which is helpful to improve the manufacturing precision of prefabricated housing components.

(2) Improvement of construction precision of prefabricated residence

Through the application of BIM tools, conflict detection and resolution can be carried out in the design and construction stages of prefabricated housing cooperative design, therefore, it can reduce the collision and conflict among the different specialties of design and the designers and constructors, and improve the construction precision of prefabricated residence greatly (Liu 2016). At the same time, by using BIM to discuss the construction plan and construction procedure among different kinds of work, the problems and hidden troubles in construction can be found and solved in advance, it is helpful to reduce the misunderstanding and disputes in the process of prefabricated residence construction, and it can also improve the construction precision of prefabricated residence.

(3) Establishing the quality tracing system for the whole life cycle of prefabricated residence

The improvement of construction accuracy is the first level of the improvement of the quality of prefabricated housing, and the establishment of a quality guarantee system for the components of prefabricated housing is the second level of the improvement of the quality of prefabricated housing, the cooperative design of prefabricated housing aims to improve the construction quality of the whole life cycle of prefabricated housing, the establishment of the quality traceability system of the whole life cycle of prefabricated housing should be regarded as the third level and ultimate goal of the quality maintenance and promotion of prefabricated housing (Wu 2017). On the basis of the realization of the first two levels, the data and information accumulated in the design and construction phase of cooperative design of prefabricated residences should be effectively transferred to the operation and maintenance phase of prefabricated residences through the application of digital tools, for example, by attaching RFID tags to the parts of prefabricated housing, we can identify the parts with quality problems in time and accurately and conduct quality tracing, so as to truly realize the transformation of prefabricated housing into a product, can also achieve the prefabricated residential products three-package maintenance services, improve the quality and efficiency of future maintenance and renovation work.

4 APPLICATION METHOD OF COOPERATIVE DESIGN FOR PREFABRICATED RESIDENCE IN THE MAINTENANCE STAGE

The operation and maintenance phase is always in the long-term use after the residential building is completed. This stage of the maintenance and renewal work cycle is long and difficult, there is certain human and material resource consumption, and it is the main stage of maintenance and renewal of prefabricated housing. Collaborative design is also important at this stage of the process (Gan 2019).

The completion of the collaborative construction and delivery of the project does not mean the end of the collaborative design of prefabricated housing. In the project delivery phase, a complete project completion BIM model must be submitted, linked to the property management plan, so that the operation and maintenance phase can achieve a lot of collaborative work, then greatly improve the operation and maintenance efficiency of prefabricated housing.

4.1 *The realization method of cooperative design of prefabricated housing*

The so-called collaboration cannot be separated from the cooperation of the participants, after the establishment of a consistent design goal, the participants in the BIM environment should follow the BIM design process to complete the tasks of the various stages (Figure 1).

(1) Preparation Stage: after the project is approved, it is necessary to establish an information-sharing platform, professional designers should strictly follow the BIM design process design. As

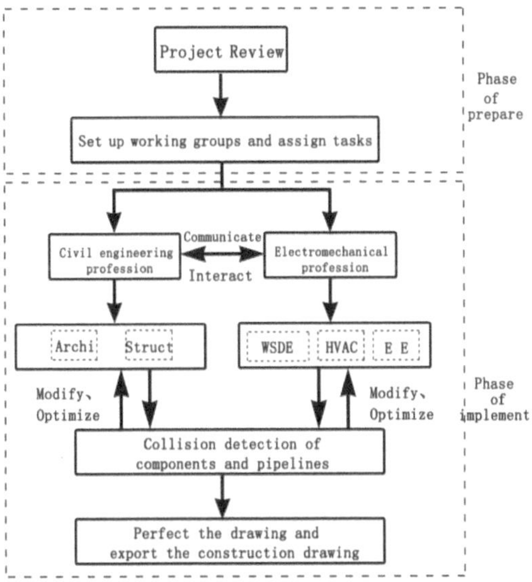

Figure 1. Basic flow of collaborative design.

the basis of collaborative design, the main work is to determine the quality delivery standard of the model, the accuracy of the model, modeling tools, coordinates, sample selection, etc., and according to the professional requirements to divide the work set to achieve design sharing and synchronization. As shown in Figure 2, it is the "Dream Housing" as an example of the BIM center document schematic diagram.

(2) Implementation phase: the implementation phase of collaborative design includes two parts: initial coordination and later coordination. Initial coordination means that the structural designer and electromechanical designer design according to the professional requirements on the basis of the building model established in the early planning and design stage. The main work of post-coordination is to deal with and optimize the collision problems of different majors.

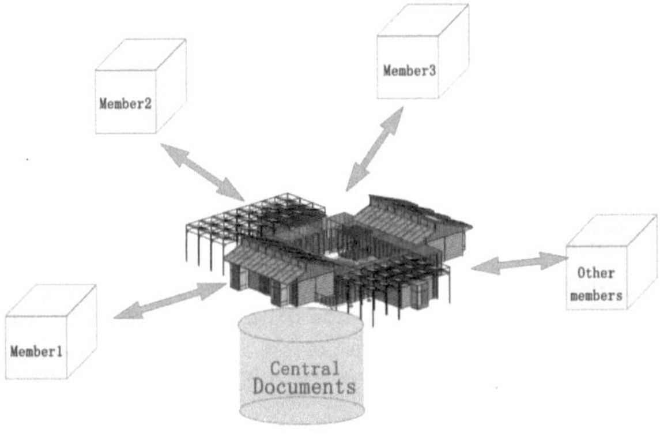

Figure 2. BIM centre document sketch map taking "Dream Housing" as an example.

4.2 *The application method of cooperative design of prefabricated residence*

4.2.1 *Space, customer information and energy management based on BIM*

The establishment of the BIM-FM system can effectively integrate the basic information of pre-fabricated housing and property equipment management after delivery (Wang 2015). FM Facility

Management, Equipment Management. The broad definition of FM also covers technical services such as PM (Property Management, Property Management) and AM (Asset Management, Asset Management). The BIM-FM system has the advantages of BIM technology in terms of spatial positioning of components and related information collection, as well as effective management of spatial information, customer information, and energy consumption:

1) In the aspect of space management, it can help property management to realize visual management and improve management efficiency;

2) At the level of customer information management, can achieve the effective integration of all user information;

3) At the energy management level, instruments in all areas can be connected to the BIM-FM system, and energy consumption data can be transmitted to the BIM-FM system in real time.

4.2.2 *Maintenance management based on BIM*

Based on the technology of BIM and RFID, it is possible to locate and maintain all the damaged areas of prefabricated housing accurately and quickly, mainly through the following steps:

1) First of all, the BIM information management platform quickly locates mechanical equipment and components when they are not working properly or are damaged;

2) Secondly, the system can help the maintenance manager to locate the position indicated by the RFID tag, and prompt the shortest and shortcut route to and from the maintenance area;

3) Thirdly, by scanning damaged equipment and components on a mobile device, you can quickly get detailed information about the damage, as well as drawing and manufacturer information from the BIM information management platform via remote technology, and the maintenance manual and other related maintenance information;

4) Finally, by reading the manufacturer's information in the RFID tag, we can quickly check whether there are replaceable components and equipment, if it cannot be repaired, we can contact the manufacturer according to the information in the tag, realization of repair and renewal of components of prefabricated housing.

It can be seen that the realization of the maintainable renewal of prefabricated housing products will involve the whole construction process of design, production, construction assembly, and future operation and maintenance, and collaborative design based on BIM technology plays a crucial role in this process. Therefore, collaborative design is the basis and important means of maintenance and renewal of prefabricated housing.

5 CONCLUSIONS

The cooperative design of prefabricated residences realizes the application transformation to cooperative construction through the cooperation of the designer and the builder, the coordination of virtual construction and actual construction, the coordination of the production and construction of components, and the coordination of different kinds of work in the construction link.

In introducing the fundamental role of cooperative design in the maintenance and renewal of prefabricated housing products, it is pointed out that, the prefabricated operation of the maintenance and renewal of prefabricated housing can be realized only when all participating units and related enterprises in the industrial chain, such as design, manufacture, assembly, maintenance, and renewal, are united and cooperate, which is of great significance to the formation and perfection of China's prefabricated housing industry chain. This paper studies the important role of cooperative design in the maintenance and renewal of prefabricated residences and considers that the cooperative design pattern is an important means of maintenance and renewal of prefabricated residences.

ACKNOWLEDGEMENTS

It is a project supported by The 2021 provincial natural science research emphases of universities in Anhui province "Research and application of prefabricated building maintenance update information technology based on BIM technology" (KJ2021A0613); The 2019 talents Introduction (doctor) Scientific Research Starting fund Project from Anhui Jianzhu University "Research on technical methods and application of industrial residential building renovation" (2019QDZ50). The corresponding author is Shenqi GAN.

REFERENCES

Gan Shenqi, Zhang Hong, Yang Yizhen. Research on the development of Chinese contemporary building maintenance, renovation, and renewal [J]. *Architecture and Culture*, 2019(08):180–183.

Gan Shenqi, Zhang Hong. Research on maintainability and renewal of industrial residential buildings [J]. *Housing Science and Technology*, 2020(09): 23–26+48.

Li Zhongfu, Jin Yuge, Li Long. Research on the selection of supporting structure and construction method of Si residence [J]. *Construction Economy*, 2019(06) : 94–99.

Liu Dongwei. SI Housing and Housing Construction Model, System, Technology, Illustration [M]. *Beijing: China Architecture and Building Press*, 2016.

Liu Dongwei. SI Housing and Housing Construction Model, Theory, Method, Case [M]. *Beijing: China Architecture and Building Press*, 2016.

Miao Q. A preliminary study on the design and type characteristics of KSI residential building [J]. *Acta Architecture Sinica*, 2020(05): 28–31.

Qin Shan, Wu Zhichao, Yu Lei. Research on the Development of KEP to KSI Interior Component System in Japan [J]. *Journal of Architecture*, 2014 (07): 17–23.

Wang Haining, Zhang Hong, Yin Jiang. Research and problem analysis on the application of building information model [J]. *Construction Science and Technology*, 2015(23): 47–49.

Wang Xiaomeng, Ma Tao. SI Residence Design – Building Centennial residence [M]. *Beijing: China Architecture and Building Press*, 2016.

Wu Xiuluan, Cheng Shuai. Study on basic theory and structure composition of CSI residential building [J]. *Architectural Design Management*, 2017(11): 88–90.

Zhang Hong, Cong Meng, Zhang Ruizhe. Design, development, improvement and application of a pre-assembled housing system-architectural product model and new architectural construction [J]. *New Architecture*, 2017 (02) : 19–23.

Advances in Civil Engineering and Environmental Engineering – bin Yahya & Bin Daud (Eds)
© 2023 The Author(s), ISBN: 978-1-032-39397-1

Research on influence and control of shield tunnel underpass construction on pile foundation settlement of neighboring buildings

Baofu Duan & Zhaowen Yu
Shandong University of Science and Technology, Shandong Provincial Civil Engineering Disaster Prevention and Mitigation Key Laboratory, Shandong Qingdao, China
Hubei Key Laboratory of Blasting Engineering of Jianghan University, Hubei Wuhan, China

Chunwu Zhang
China Railway Wuhan Survey and Design Institute Co, Ltd, Hubei Wuhan, China

Zongjun Sun*
Qingdao Ruihan Technology Group Co., Ltd., Shandong Qingdao, China

ABSTRACT: Passing the shield tunnel underground will cause settlement and deformation of the building, and in severe cases, it will cause damage to the pile foundation and even disasters. In order to study the settlement law of the pile foundation of adjacent buildings when the shield tunnel is under construction, taking the construction site of Shuxiang Primary School under the shield tunnel of Shenzhen Rail Transit Line 6 as an example, a three-dimensional rock-soil mass-tunnel-pile foundation was established in combination with the site. Through Numerical calculation model, the simulation results show that: When the stratum is not reinforced, the maximum settlement value of the pile foundation is 46.06mm, which exceeds the control standard after the stratum is grouted with the sleeve valve pipe and the rotary jet pile; when the stratum is reinforced, the maximum settlement value of the pile foundation is 11.07mm., which is about 75% less than the unreinforced case, meeting the building pile foundation control standard; the maximum settlement value of the pile foundation obtained by on-site monitoring is 10.93mm, which is similar to the simulation result. Numerical simulation and on-site construction results indicate that the above reinforcement method can effectively control the impact of shield tunnel and reduce the impact of shield tunnel on pile foundation settlement. The settlement values of the ground and pile foundation obtained by numerical simulation are similar to the settlement values obtained by actual monitoring, and the change trend is consistent, which can provide an effective soil reinforcement scheme for shield construction. Pile foundation settlement control has good reference value and guiding significance.

1 INTRODUCTION

It is well known that during tunnel construction, shield excavation will cause subsidence of the stratum, and at the same time, the pile foundation of buildings in the stratum will also be deformed and damaged. Therefore, it is of great significance to study the deformation law of the pile foundation and its reinforcement measures when the shield tunnel penetrates down.

At present, many scholars have done a lot of work on the research of shield tunnel passing through buildings. Morton et al.(Morton & King 1979) found that shield tunnel construction has a very large impact on the pile foundation of adjacent buildings through the shield tunnel simulation test, and the impact is greatest in soft soil; ONG et al. (2008) conducted a centrifuge test indoors

*Corresponding Author: 3519808967@qq.com

 DOI 10.1201/9781003349563-21

and concluded that the Change law of pile foundation of adjacent buildings during shield tunnel construction in clay; Chen et al. (Chen et al. 1999; Du et al. 2009) simulated the interaction between pile foundation and soil by superimposing the displacement of soil during shield tunnel on the pile foundation through numerical simulation, analyzing the internal force and displacement of the pile foundation during the construction of the shield tunnel; Zhao Min and Xu Zengwei (Zhao & Xu 2016) obtained the impact of the shield operation on the surface through the numerical simulation results of the Tianjin subway shield analyzed by ANSYS, and proposed a series of control measures, to enhance the safety of construction. Xiaotao Wu (Wu & Liu 2019) and others used finite element software to simulate the construction process of the lap joint, combined it with the field measured data, and finally concluded that reasonable reinforcement measures could be taken before the construction of the lap joint to effectively control the land subsidence. The above results discuss the influence of shield tunnel excavation on adjacent pile foundations, but there are few researches on the change law of pile foundation settlement during the construction of shield tunnels under buildings in complex strata and the verification based on on-site testing data. Based on this deficiency, this paper will analyze the pile foundation settlement caused by the double-track tunnel shield passing through the building at a short distance, and simulate and predict the settlement value of the pile foundation under the soil reinforcement and unreinforcement schemes around the pile foundation. The test data are compared with the pile foundation settlement obtained from the simulation to verify the effectiveness of the pile foundation reinforcement measures.

2 PROJECT OVERVIEW

2.1 *Project overview*

The second phase of Shenzhen Metro Line 6 has a total length of 11.77km. Thereinto, the most complicated construction is the section of Shuxiang Primary School, which has a total of 61 rings divided into left and right lines. There are two types of underground supports in Shuxiang Primary School: caps and prestressed pipe piles. The thickness of the bearing platform is 1.2m, and it is divided into four sizes. The top of the pile is 1.6m from the ground and the material of the pile is PHC-400AB-95, and the strength of the concrete is C80. The pile is 20~28 meters long, and the top is connected with strongly weathered granite. There are 5 pile foundations that may affect the cutter head, and the longest intrusion can reach 0.59m.

The soil layer structure around Shuxiang Primary School consists of fill layer, clay layer, strong, full and moderately weathered granite layers. Artificial filling and residual soil will affect the construction. The groundwater is dominated by stagnant water, phreatic water and fissure water in the upper layer. The geological profile is shown in Figure 1.

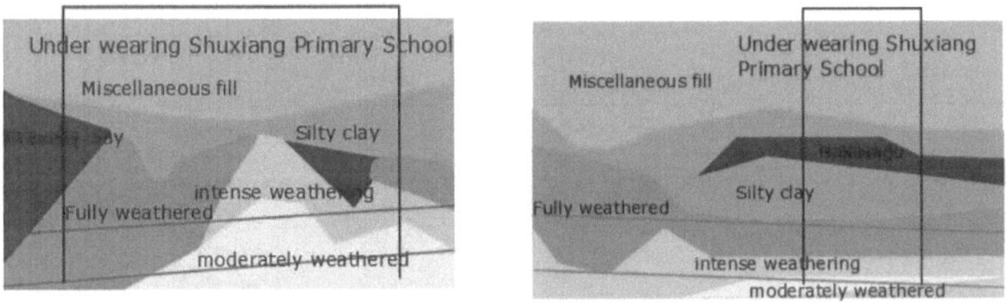

(a) Geological longitudinal section of the right line

(b) Geological longitudinal section of the left line

Figure 1. Longitudinal section of the detailed geological survey of Shuxiang Primary School.

3 COMPUTATION AND ANALYSIS

3.1 *Numerical computation model*

Based on MIDAS GTS NX, a finite element model is established for simulation. In order to better study the influence of shield tunnel penetration on the pile foundation, the model range can be 3~5 times of the hole, and twice the length of the pile in the vertical direction, that is, the calculation model is taken as 90m in the longitudinal direction of the tunnel (y direction), 70m in the transverse direction (x direction) of the tunnel, and 70m in the vertical direction (z direction). The tunnel is located 29m above the ground. The top of the cap is located on the ground surface with a height of 1.6m. The finite element model has a total of 111,225 elements and 52,089 nodes.

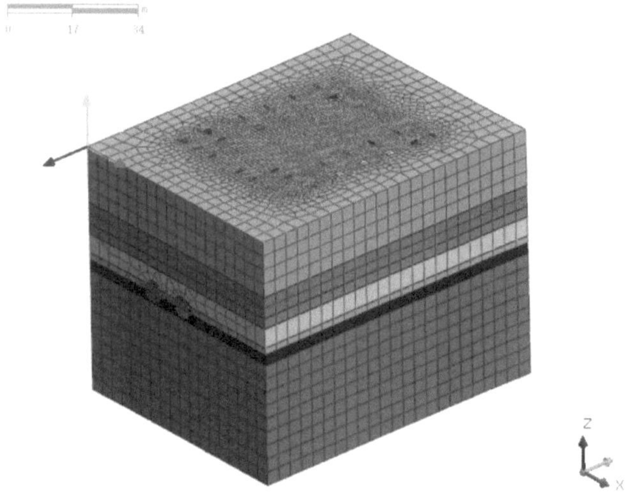

Figure 2. Meshing of finite element model.

3.2 *Calculation parameter selection*

Before the construction, the underground rock formation of Shuxiang Primary School is inspected in detail. The model calculation adopts the real formation parameters of the project. The formation parameters are shown in Table 1. The segment is equivalent to a homogeneous cylinder, and its stiffness reduction coefficient is 0.6~0.8 The mechanical parameters of the material are shown in Table 2.

Table 1. Physical and mechanical parameters of rock and soil layers.

Serial number	Formation name	Formation thickness m	Elastic Modulus E (*MPa*)	Test weight (KN/m^3)	Poisson's ratio μ	Cohesion (*KPa*)	Internal friction ϕ (°)
1	Miscellaneous fill	16	4.2	18.1	0.35	12	8.5
2	Silty clay	6	15.1	18.8	0.31	22	17.3
3	Zhongsha	3	30.5	20	0.3	1	30
4	Fully weathered granite	5.5	60	19	0.26	35	30
5	Strongly weathered granite	2	150	20	0.25	40	32.6
6	Medium weathered granite	3	2450	24	0.22	1000	40
7	Micro-weathered granite	34.5	32200	26.5	0.18	1500	46

Table 2. Mechanical parameters of structural materials.

Serial number	Material name	Elastic Modulus E (*MPa*)	Poisson's ratio μ	Test weight γ (KN/m^3)
1	Grouting reinforcement area	200	0.25	25
2	Grouting area	2000	0.23	20
3	Shield shell	210000	0.2	78.5
4	Segment	27600	0.2	26
5	C50 deck	34500	0.2	25
6	C80 pile	38000	0.2	25

When the relative displacement of the stratum changes, the relative position between the stratum and the pile foundation will also change. The pile interface in the finite element software MIDAS GTS is used to simulate the contact surface between the pile foundation and the soil. The parameters of the pile interface are obtained through calculation, and the shear force value, shear stiffness modulus and normal stiffness modulus of the pile interface are obtained by confirming with the field survey data.

3.3 *Analysis of calculation results*

In order to more advantageously analyze the influence of the shield tunnel passing through the pile foundation of the building, according to the construction of the shield, the stratum is cut into two monitoring sections S1 and S2, which are located at the 18th ring and the 35th ring respectively. The S1 section is located on the right side of the model and is 25.5m away from the model boundary, while the S2 section is located in the middle of the model and is 51m away from the model boundary. Ten groups of monitoring points are arranged on the monitoring section totally, and one group is arranged every 5m. According to the position distribution of the pile foundation, the pile foundation is divided into six groups, from bottom to top, they are group A, group B, group C, group D, group E and group F respectively. Pile groups A, E, and F are not penetrated by tunnels, while pile groups B, C, and D are penetrated by tunnels. The schematic diagram of monitoring section grouping is shown in Figure 3.

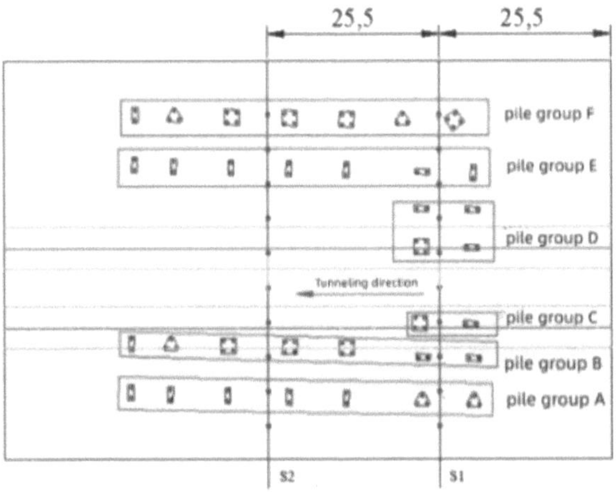

Figure 3. Schematic diagram of monitoring section grouping.

Figure 4 shows the settlement curve of the pile foundation of the building after the shield tunnel is constructed. It can be seen from the figure that the vertical displacement of pile group C is the largest one, because pile group C is located in the center of the tunnel underpass, and the maximum displacement is 46.06mm. The displacement of pile group D is 42.13 mm slightly smaller than that of pile group C. The displacement of pile group B is 27.94 mm, which is significantly lower than that of pile groups C and D. Since the pile groups A and E are far from the tunnel penetration position, the displacement is small. The pile group F is hardly affected by the shield construction because it is too far away from the tunnel.

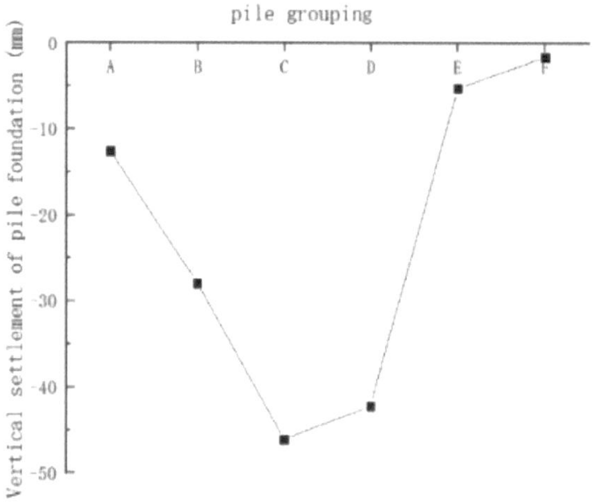

Figure 4. Settlement diagram of the pile foundation of the building after the completion of the shield.

4 REINFORCEMENT MEASURES OF PILE FOUNDATION SOIL UNDER SHIELD TUNNEL

Since the subsidence of the building, soil surface and pile foundation passed by the shield is higher than the maximum allowable subsidence value, it is necessary to reinforce the building soil of the project. The building soil passing through is reinforced by means of sleeve valve pipe and rotary jet pile grouting reinforcement.

According to the simulation data of the model when it is not reinforced, the whole model is divided, and a reasonable reinforcement scheme is designed. The model reinforcement scheme is shown in Figure 5.

4.1 *Soil reinforcement scheme*

The foundation near the tunnel is strengthened by grouting technology, which improves the strength and hardness of the soil near the tunnel, and reduces the impact on the surrounding soil structure, thereby reducing the impact of shield construction on the foundation and piles of adjacent buildings. The grouting reinforcement of the stratum with the sleeve valve pipe and the rotary jet pile improves the strength of the soil, and has the advantages of short construction period, good reinforcement effect, fast construction, and low price.

Sleeve valve pipe grouting is a method to put the pipe into the soil through pre-drilled small holes, and use the air pressure to evenly disperse the mud into the soil stratum to increase the strength of the cement body. This method has less pressure change and less interference factors, and is also conducive to strengthening the soil. The high-pressure rotary spray pile method uses a high-pressure rotating nozzle to spray the cement slurry into the soil layer and mix with the soil, thereby producing a continuous lapped concrete reinforcement.

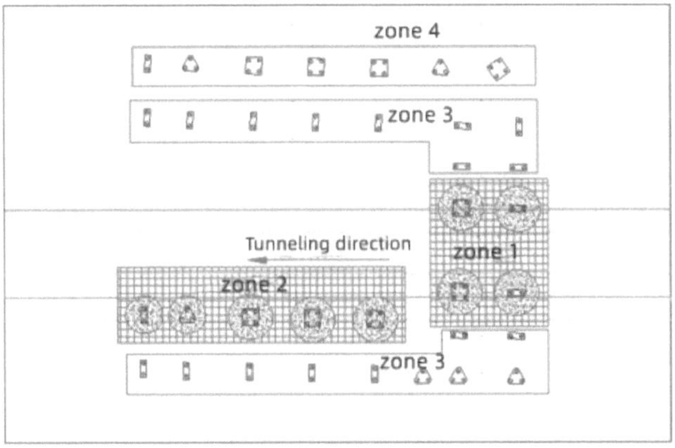

Figure 5. Schematic diagram of the underground reinforcement zone of Shuxiang Primary School.

Different reinforcement methods are adopted according to the different conditions of the reinforcement area. Because the grouting pressure of the sleeve valve pipe is large, the ground will bulge when the soil layer is shallow, so the reinforcement zone 1 and reinforcement zone 2 are only reinforced underground 3m. For soils in the middle to moderately weathered rock formations, the reinforcement range is the soil within 6m in diameter around the cap. Other parts are reinforced with double-pipe rotary spray piles. Reinforcement zone 1 ensures the safety of shield opening and cutting piles after reinforcement. The friction force between the pipe piles and the bearing capacity of the pile foundation are improved after reinforcement in the reinforcement zone 2. In reinforcement area 3, the sleeve valve pipe is used for tracking grouting to strengthen the soil strength and reduce settlement. The reinforcement area 4 is not reinforced. The mesh division after reinforcement is shown in Figure 6.

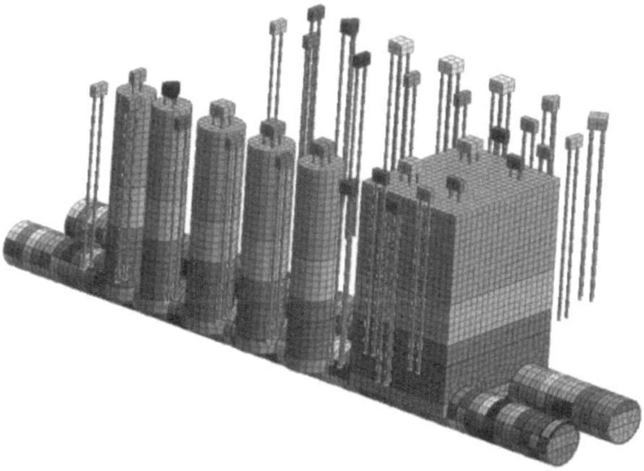

Figure 6. Schematic diagram of mesh division of pile foundation model after reinforcement.

4.2 *Numerical analysis of settlement of pile foundation after reinforcement*

Figure 7 is a cloud diagram of the vertical settlement and displacement of the pile foundation of the building after the soil around the pile is reinforced. From the cloud map, it can be found that the maximum settlement occurs at pile group B. The settlement of pile group B is 11.07mm, which is

34.99mm less than that without reinforcement, and the reduction range is 75%; and the maximum settlement is not high: less than 20mm, which meets the maximum settlement standard.

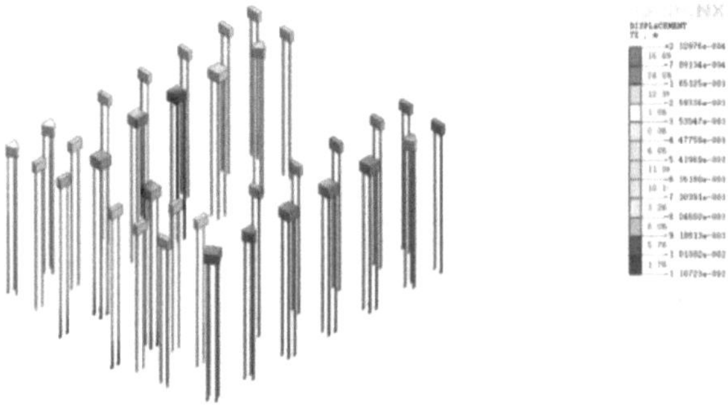

Figure 7. Cloud map of vertical displacement of pile foundation after reinforcement.

After grouting reinforcement with sleeve valve pipe and rotary jet pile, the maximum subsidence value of the pile foundation is shown in Table 3. After reinforcement, the settlement amplitude of each pile group decreases, and the change amplitude of pile group C and pile group D is largest with a reduction of about 78%. Pile group B and pile group A also have different degrees of reduction. It can be seen that the method of grouting and strengthening the ground with the sleeve valve pipe and the rotary jet pile to improve the strength of the soil body can effectively reduce the settlement of the pile foundation of the building.

Table 3. Maximum settlement of pile group after grouting reinforcement.

Pile group name	Pile group A	Pile group B	Pile group C	Pile group D	Pile group E	Pile group F
Unreinforcement	12.57	27.94	46.06	42.13	5.23	1.57
Reinforcement	7.25	11.07	8.93	8.46	4.71	1.55

4.3 *Numerical comparison of pile foundation settlement before and after reinforcement*

The comparison of the maximum settlement of each pile foundation before and after grouting reinforcement is shown in Figure 8. It can be seen from the figure that the influence of the pile foundation extending from the construction center of the shield tunnel to both sides is getting smaller and smaller. Therefore, the settlement change of the pile foundation after reinforcement is not obvious. Compared with the situation before reinforcement, the vertical subsidence of pile groups B, C, and D is reduced by about 75%. Strengthening the soil can improve the bearing capacity of the pile foundation and reasonably strengthen the control of the pile foundation subsidence. When pile group A is not reinforced, the vertical displacement of pile body is large; while after the soil around pile group A is reinforced, the settlement of pile group A is significantly reduced. Pile group C and pile group D also decrease in different extents after reinforcement, which shows that the control effect of pile foundation settlement is remarkable after reinforcement of soil below the neutral point.

5 ON-SITE CONSTRUCTION MONITORING AND ANALYSIS

5.1 *Monitoring and analysis of pile foundation settlement*

In the construction of Shenzhen rail transit line 6 under Shuxiang Primary School, according to the relationship between piles and tunnels, a number of monitoring points will be selected along

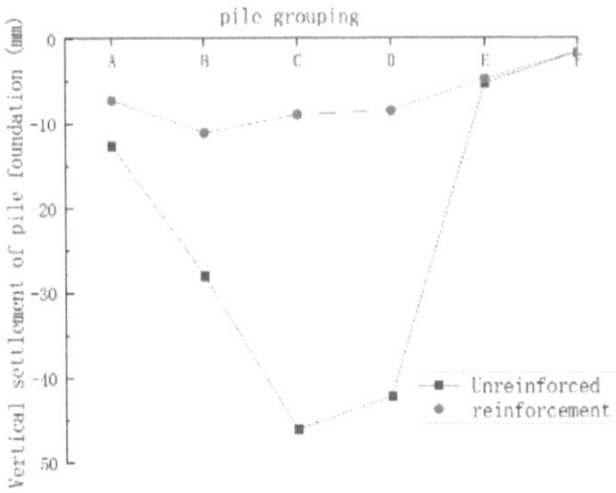

Figure 8. Comparison of settlement of pile foundation under reinforced and unreinforced conditions.

the central axis of the tunnel along the left and right lines, and a number of monitoring points will be selected around the pile foundation of the building along the tunnel transverse, and finally the measured data will be analyzed. The layout of each measuring point is shown in Figure 7. The maximum subsidence value of pile foundation monitored on site is shown in Table 4, and the foundation monitored on site is shown in Table 4.

Table 4. Maximum settlement monitoring table of pile group.

Pile group name	Pile group A	Pile group B	Pile group C	Pile group D	Pile group E	Pile group F
Pile foundation settlement (mm)	6.27	10.9	8.38	8.84	4.20	2.35

The relationship between the observed value of the maximum settlement of the pile foundation in each group and the simulated value is shown in Figure 9. The figure shows that the field detection values and numerical simulation results of all pile foundations are similar, and the change trends are consistent, indicating that the numerical simulation has good reference value and guiding significance for the construction scheme design and pile foundation settlement control of similar projects. The on-site detection value of some pile groups penetrated directly under the shield is smaller than the numerical simulation result, and the on-site monitoring value in the area three times the diameter of the hole is greater than the numerical simulation result.

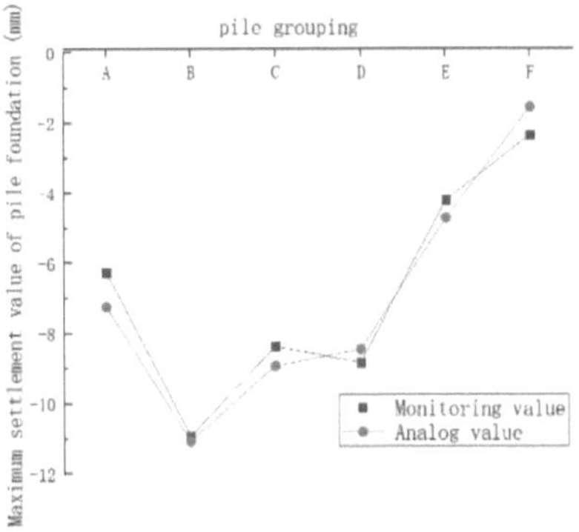

Figure 9. Comparison between the maximum settlement monitoring value and the simulated value of the pile group.

6 CONCLUSION

(1) Using MIDAS-GTS NX to establish a three-dimensional numerical model for analysis, it is concluded that the maximum settlement of the pile foundation during shield construction occurs at the pile group C, and the maximum settlement is 46.06mm. With the pile group C as the center, the settlement of each pile group extending to both sides becomes smaller and smaller. The pile group F is too far away from the construction and can be ignored.
(2) Combined with the field measured data and numerical simulation results, it is concluded that the maximum settlement value of the pile foundation is 11.07mm after grouting reinforcement of the stratum with sleeve valve pipe and rotary jetting pile, which is 75% lower than that without reinforcement, and the maximum settlement value is less than 20mm, which meets the building pile foundation control standard.
(3) The settlement value of ground and pile foundation obtained by numerical simulation is similar to the settlement value obtained by actual monitoring, and the change trend is consistent, indicating that the results obtained by numerical simulation have a good reference value for such the same construction scheme and guiding significance.

This paper only studies the effect of shield tunnel underpass construction on the settlement of ground surface and building pile foundation, and does not discuss the change of stress before and after pile foundation reinforcement, which can be further discussed in the next step.

ACKNOWLEDGMENTS

Funding: This research was funded by the Shandong Natural Science Foundation of China with grant number ZR2020ME096, and the work is supported by the Foundation of Hubei Key Laboratory of Blasting Engineering with grant number BL2021-24 and Qingdao West Coast New Area High-level Talent Team Project with grant number RCTD-JC-2019-06.

REFERENCES

Chen L T, Poulos H G, Loganathan N. Pile responses caused by tunneling [J]. *Journal of Geotechnical & Geoenvironmental Engineering*, 1999, 125(3):207–215.
Du Z L, Huang M S, Li Zao. DCM method of tunnel excavation effect on adjacent pile groups based on formation loss ratio [J]. *Rock and Soil Mechanics*, 2009, 30(10):3043–3047.
Morton J D, King K H. Effects of tunnelling on the bearing capacity and settlement of piled foundations. 1979.
Wu Xiaotao, Liu Zengwei. Numerical simulation of ground settlement caused by over-lapping tunnel shield construction and measures of stratum reinforcement [J]. *Journal of Physics: Conference Series*, 2019, 1176(5).
Zhao M, Xu Z W. Control of surface subsidence in construction of tianjin metro shield tunnel passing through helipad [J]. *Modern Tunnel Technology*, 2016, 53(01):180–186.
Zhu F B, Yang P, Ong C W. Numerical analysis of the influence of shield tunnel excavation on adjacent pile foundations [J]. *Chinese Journal of Geotechnical Engineering*, 2008(02):298–302.

Advances in Civil Engineering and Environmental Engineering – bin Yahya & Bin Daud (Eds)
© 2023 The Author(s), ISBN: 978-1-032-39397-1

Analysis on basic mechanical properties of manufactured sand concrete modified by super absorbent polymer

Zhaowei Cui*, Hao Chen & Ronggui Liu
Faculty of Civil Engineering and Mechanics, Jiangsu University, Zhenjiang, China

ABSTRACT: In order to study the influence of super absorbent polymer (SAP) on the basic mechanical properties of manufactured sand concrete, the experimental studies of SAP content (0%, 0.08%, 0.16%, 0.24%) on the cubic compressive strength, splitting tensile strength, flexural strength and axial compressive strength of manufactured sand concrete were carried out. The results reveal that the cube compressive strength and splitting tensile strength of manufactured sand concrete increased firstly, and then decreased with the increase of SAP content. The flexural strength and axial compressive strength of manufactured sand concrete decreased first, then increased and then decreased as SAP content increased. It could be concluded from the above experiments that SAP content of 0.08% played a positive role in the improvement of mechanical properties of manufactured sand concrete. The above research could provide support for the internal curing technology of manufactured sand concrete.

1 INTRODUCTION

With the comprehensive advancement of urbanization process in China and the overall improvement of infrastructure construction level, the amount of concrete is increasing year by year, which makes the demand for sand resources keep rising (Liu 2021; Wang 2022). According to incomplete statistics, the demand for sand resources in China has doubled compared with the previous decade (Liu 2021). As a result, the long-term exploitation has led to the depletion of river sand resources (Bendixen 2019; Li 2016). In order to ensure the sustainable development and utilization of river sand resources, local governments have issued a series of policies to limit the exploitation of river sand, resulting in the skyrocketing price of river sand (Bendixen 2019; Liu 2021; Li 2016; Wang 2022). Due to the prominent demand of sand used in infrastructure construction activities, the emergence of manufactured sand has been promoted, which has the advantages of controllable quality, high strength, local materials, low production cost, and lower environmental pollution compared with river sand. However, there are also some shortcomings, such as uneven gradation, high stone powder content, energy consumption and environmental pollution in the production process of manufactured sand (Li 2016; Liu 2021; Wang 2022). Therefore, in order to promote the market of manufactured sand, it is necessary to further study how to improve the characteristics of manufactured sand concrete.

In recent years, scholars at home and abroad have carried out a lot of research work on manufactured sand concrete, such as the influence of stone powder on mechanical properties of manufactured sand concrete (Abdulwahid 2020; Liu 2021; Yang 2018; Zhang 2017). Zhang et al. (2017) and Yang et al. (2018) studied the influence of stone powder on cement hydration process and hydration products in concrete by using different test methods, and found that stone powder could significantly promote cement hydration, and thus improve the strength of concrete. Abdulwahid et al. (2020) found that the addition of stone powder to concrete could significantly reduce the porosity of concrete, thus improving the strength of concrete. However, too much stone powder in manufactured sand will destroy the accumulation structure of concrete aggregate, increase water

*Corresponding Author: cuizw0961@163.com

DOI 10.1201/9781003349563-22

consumption and the probability of shrinkage cracking of concrete, and affect the mechanical properties and durability performance of concrete (Liu 2021). In addition, using reasonable curing method can also alleviate the mechanical properties and durability degradation caused by dry shrinkage cracking of concrete (Liu 2021; Wang 2022).

With the development of high performance curing technology in the 21st century, the appearance of internal curing technology using super absorbent polymer (SAP) provides a new idea to solve the hot issue of concrete cracking (Mignor 2016). SAP is a kind of multi-functional polymer material with unique characteristics of water absorption, storage and release, which can absorb and store water hundreds or even thousands of times its own weight in advance, and can timely release the pre-absorbed water for cement hydration when water is insufficient in the hydration reaction process (Mechtcherine 2012). SAP particle size, mixing amount, additional water diversion, mixing method, curing conditions and other factors will affect the mechanical properties of concrete, so there are great differences in the research results of domestic and foreign scholars. Han et al. (2014) made a comparative analysis of hydration degree between ordinary concrete and SAP internal curing concrete, and found that hydration degree and gel space of SAP internal curing concrete were significantly higher than ordinary concrete, so SAP internal curing concrete had higher strength. Lao et al. (2021) and Yu et al. (2021) analyzed the microstructure of SAP concrete, and found that SAP promoted the generation of ettringite in hydration products of concrete, and its internal structure was denser than ordinary concrete, which was the main reason for the improvement of concrete strength. Pourjavadi (2012) and Soliman (2013) found that the cavitation left inside concrete after SAP water release would reduce the compressive strength and flexural strength of concrete. Hu et al. (2021) found that the holes left inside concrete after SAP water release were the main reason for the decline of concrete strength. At present, the domestic and foreign researches mainly focus on the internal curing technology of high performance concrete to improve the performance of concrete, and there are few researches on manufactured sand concrete. Therefore, this study aims to improve mechanical properties of manufactured sand concrete by SAP internal curing technology, which can provide guidance for the practical application of manufactured sand concrete.

2 EXPERIMENT DETAILS

2.1 *Selection of test raw materials*

P·O42.5 ordinary Portland cement of Helin brand of Zhenjiang city was used in this work. Fine aggregate was divided into river sand and manufactured sand. River sand grand was medium, using the manufactured sand of tunnel excavation slag production of Sanming section of Puyan High-speed engineering site, the gradation curve was shown in the Figure 1. Continuous graded

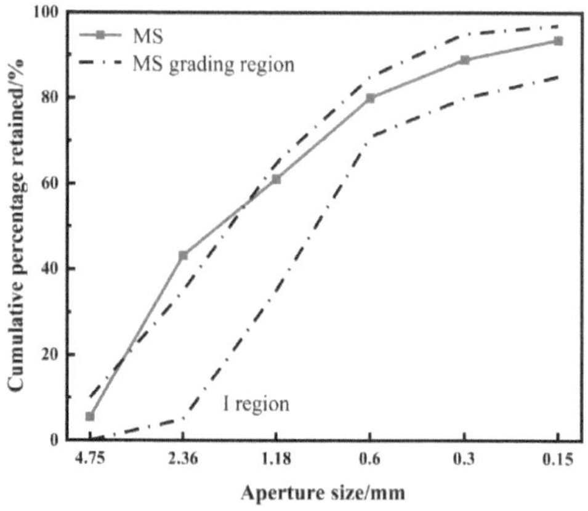

Figure 1. Grading curve of manufactured sand.

gravel with particle size of 5–20mm was used for coarse aggregate. Stone powder was the particle produced in the process of manufactured sand production, whose size less than 0.075mm. In this paper, 6% of the total weight of the cementified material was taken as the control group. Tap water was used in the experiment, water reducing agent was polycarboxylic acid concrete high efficiency water reducing agent, whose dosage was 0.2% of the total weight of the cementing material. The characteristics of SAP selected in this experiment are shown in Table 1.

Table 1. Characteristics of SAP.

Parameter	Water absorption ratio (g/g)	Salt water absorption ratio (g/g)	Pressurized water absorption ratio (g/g)	Bulk density (g/mL)	pH
Values	400–500	60	26	0.65–0.85	6.2

2.2 *Mix proportion design of concrete*

The proportion design of manufactured sand concrete was carried out in accordance with the Proportion Design Specification of Ordinary Concrete (MINISTRY OF HOUSING AND URBAN-RURAL DEVELOPMENT 2011), and a total of 8 working conditions were designed. The replacement rate of manufactured sand was 80%, which was the same as that of Sanming Section of Puyan Expressway, and the amount of cementing material replaced by stone powder was 6%. In this way, the influence of different SAP content (0%, 0.08%, 0.16%, 0.24%) on the basic mechanical properties of manufactured sand concrete were studied, and the detailed mix ratio was shown in Table 2.

Table 2. Mix ratio of manufactured sand concrete.

No.	SAP (%)	Stone powder (%)	Cement (kg/m^3)	River sand (kg/m^3)	Manufactured sand (kg/m^3)	Coarse aggregate (kg/m^3)	Water (kg/m^3)	Water reducing agent (kg/m^3)
SOFO	0.00	0	455.0	141	564.2	1057.8	182.00	0.91
S8F0	0.08	0	454.7	141	564.2	1057.8	191.09	0.91
S16F0	0.16	0	454.3	141	564.2	1057.8	191.09	0.91
S24F0	0.24	0	453.9	141	564.2	1057.8	191.08	0.91
SOF6	0.00	6	427.7	141	564.2	1057.8	182.00	0.91
S8F6	0.08	6	427.3	141	564.2	1057.8	190.55	0.91
S16F6	0.16	6	427.0	141	564.2	1057.8	190.54	0.91
S24F6	0.24	6	426.6	141	564.2	1057.8	190.53	0.91

Note: (1) SAP and stone powder were calculated by mass percentage of substituted cementitious materials. (2) SOFO means that SAP content was 0% and stone powder content was 0%. (3) The additional water diversion was calculated according to the Power model (SUN 2019).

2.3 *Preparation and test method of specimen*

The specifications of specimens required for each test were shown in Table 3. When preparing the specimen, a forced single horizontal shaft concrete mixer was used for mixing. SAP, stone powder, cement, river sand, manufactured sand and coarse aggregate were put into the concrete mixer for 30s in dry mixing. After SAP and stone powder were fully mixed with the cementing material, water and water reducing agent were added to continue mixing for 60s, and then the material was discharged and formed. The casting-shaped specimens were placed in the standard curing box (temperature 20±2°C, relative humidity 95%) for curing. After 24h, the molds were removed and numbered. Then the specimens were continued to be placed in the standard curing box for curing. All tests were carried out according to Standard for Test Methods of Physical and mechanical Properties of Concrete (GB/T 50081-2019).

Table 3. Specification of mechanical properties test specimen.

No.	Experiment content	Specimen size /mm	Standard curing age /d
1	Cube compressive strength test	$100\times100\times100$	3,7,28
2	Split tensile strength test	$100\times100\times100$	3,7,28
3	Axial compressive strength test	$100\times100\times300$	28
4	Flexural strength test	$100\times100\times400$	28

3 TEST RESULTS AND ANALYSIS

3.1 *Analysis of cube compressive strength test results*

The cube compressive strength test results of manufactured sand concrete with SAP internal curing are shown in Figure 2. The compressive strength of concrete at the period of 3d, 7d and 28d increased first and then decreased with the increase of SAP content. When SAP content was 0.08%, the compressive strength reached the maximum value, which was increased by 14.32%, 4.71% and 8.48% respectively compared with S0F0 group. It showed that a small amount of SAP

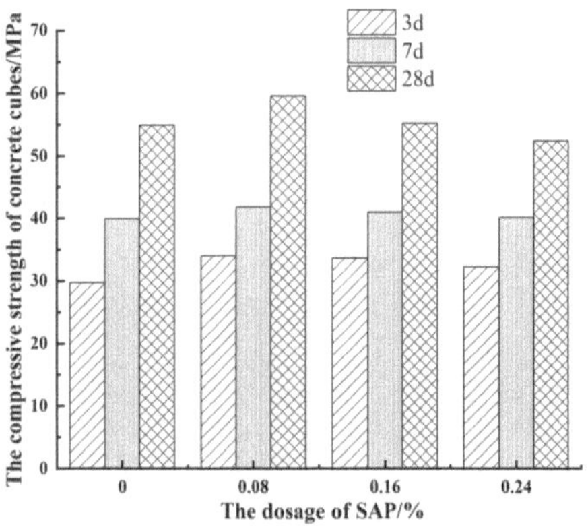

(a) The content of stone powder is 0%

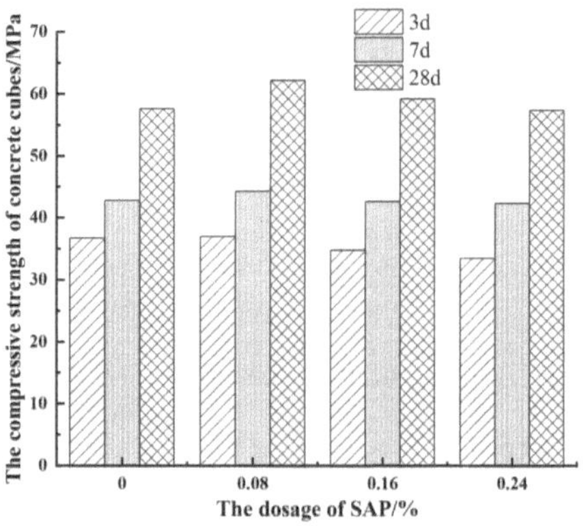

(b) The content of stone powder is 6%

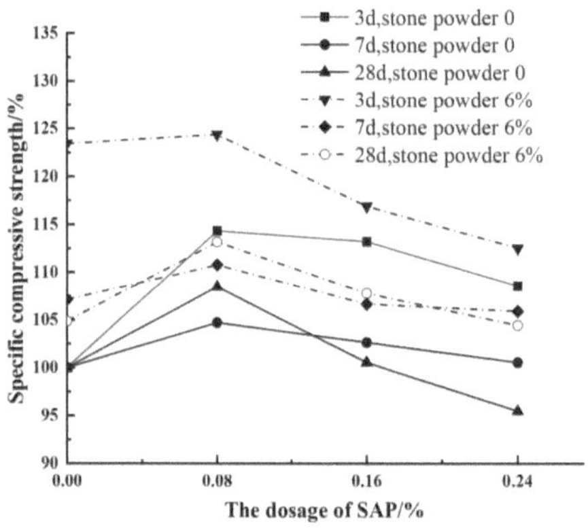

(c) Specific compressive strength of concrete

Figure 2. Influence of SAP on compressive strength of manufactured sand concrete.

could improve the compressive strength of manufactured sand concrete at various periods, mainly because SAP mixed into concrete absorbed water in slurry and reduced free water content in concrete, resulting in a smaller water-binder ratio, thus improving the strength and compactness of manufactured sand concrete. When SAP content exceeded 0.08%, the compressive strength of concrete began to decrease. In particular, the strength of 0.24% SAP specimen at 28d period was lower than that of S0F0 specimen, indicating that excessive SAP content had adverse effects on compressive strength. This was because the excessive SAP content leaded to the increase of the holes left in the concrete after SAP water release, causing the reduction of the compressive strength.

The addition of 6% stone powder could improve the compressive strength of concrete cubes at different periods. Stone powder had the greatest effect on the compressive strength at 3d age, followed by 7d and 28d. This was mainly due to the fact that the main component of stone powder was $CaCO_3$, which could promote the hydration of C_3S to generate C-S-H and $Ca(OH)_2$ in the very early stage of concrete, so the improvement effect of early compressive strength of concrete was obvious.

3.2 *Analysis of splitting tensile strength test results*

The splitting tensile strength test results of manufactured sand concrete with SAP internal curing are shown in Figure 3. It could be seen from the figure that the splitting tensile strength at 3d, 7d

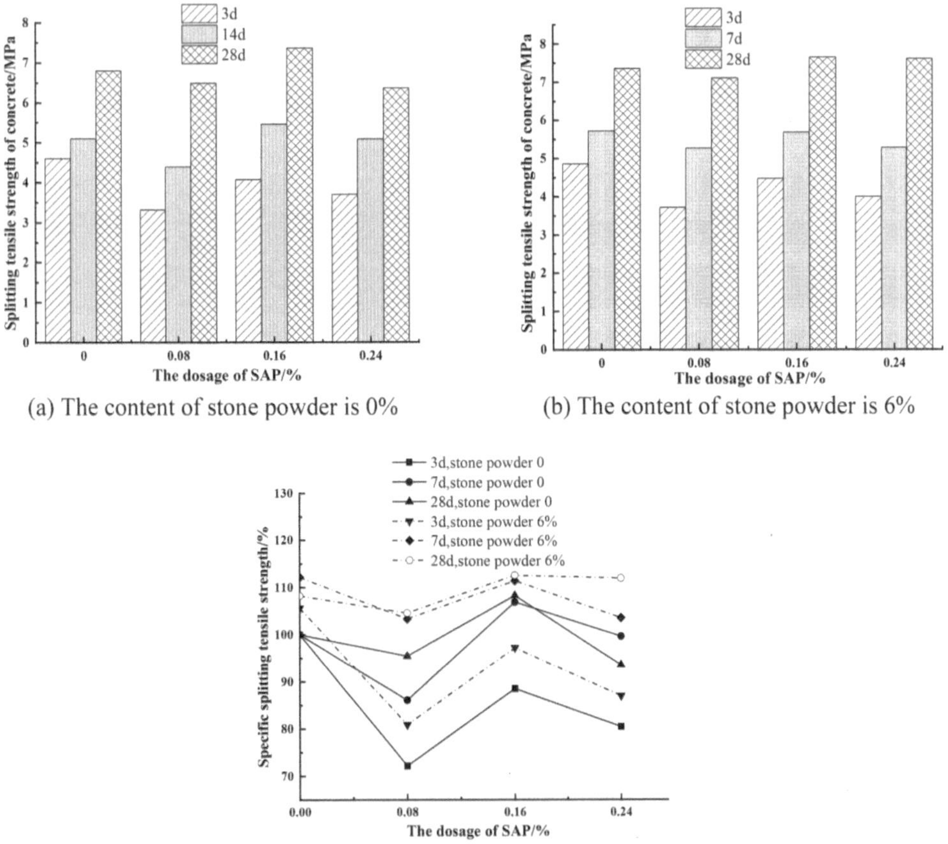

(a) The content of stone powder is 0%

(b) The content of stone powder is 6%

(c) Specific splitting tensile strength of concrete

Figure 3. Influence of SAP on splitting tensile strength of manufactured sand concrete.

and 28d ages decreased first, then increased and then decreased with the increase of SAP dosage. When SAP content was in the range of 0%-0.08%, the splitting tensile strength of the three ages showed a linear decreasing trend, which decreased by 27.83%, 13.92% and 4.56% respectively compared with S0F0 group. When SAP content was in the range of 0.08%-0.16%, the compressive strength of the three ages began to increase linearly again. At this time, the splitting tensile strength at 3d age was still 15.96% lower than that of S0F0 group, but the splitting tensile strength at 7d and 28d age was 6.86% and 8.23% higher than that of S0F0 group respectively. When the SAP content was in the range of 0.16%-0.24%, the splitting tensile strength at 3d, 7d and 28d ages decreased by 22.11%, 0.39% and 6.47% respectively, compared with the corresponding age S0F0 group. It could be seen that 0.16% SAP could improve the splitting tensile strength of manufactured sand concrete in the middle and later stage of hydration reaction.

When the stone powder content was 6%, the splitting tensile strength reached the maximum, which was increased by 5.65%, 12.16% and 8.24% respectively compared with S0F0.Too much or too little SAP content was detrimental to the splitting tensile strength, especially in the early stage of hydration reaction, the reduction of the splitting tensile strength was most obvious.

3.3 *Analysis of flexural strength test results*

As can be seen from the Figure 4, when the content of stone powder was constant, the flexural strength of manufactured sand concrete decreased first, then increased and then decreased with the increase of SAP content. When the stone powder content was 6%, the flexural strength of all SAP content groups reduced. Especially in the 0.08% SAP content group, its flexural strength was lower than that without stone powder, indicating that too much stone powder was unfavorable to the flexural strength and excessive or insufficient SAP would lead to the reduction of the flexural strength of mechanized sand concrete. The reasons might come from two aspects. On the one hand, because the SAP released water, making microporous hydration reaction fully around the hole wall and the formation of a large number of acicular AFt, which filled the micro cracks in concrete and micro pore and improved the strength of the concrete On the other hand, due to concrete matrix had been initially formed stable structure when the SAP pre-absorbed water was released, continuous ball-shaped holes were formed inside the concrete, causing the concrete to lose strength.

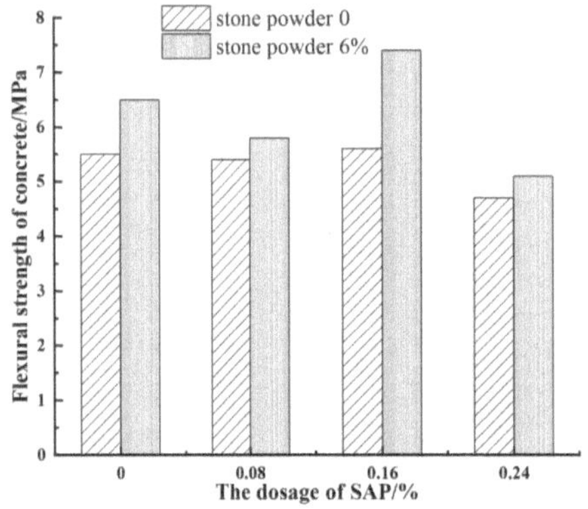

(a) The relationship between flexural strength and the dosage of SAP

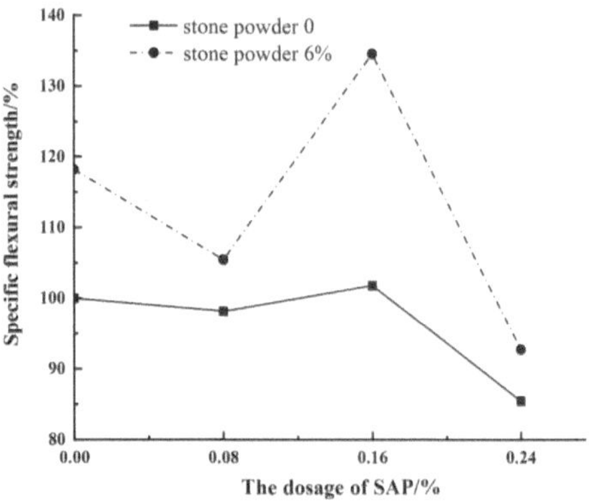

(b) The relationship between specific flexural strength and the dosage of SAP

Figure 4. Influence of SAP on the flexural strength of manufactured sand concrete.

3.4 *Analysis of axial compressive strength test results*

The relationship between axial compressive strength and the dosage of SAP is shown by the Figure 5. The axial compressive strength of manufactured sand concrete decreases first, then increased and

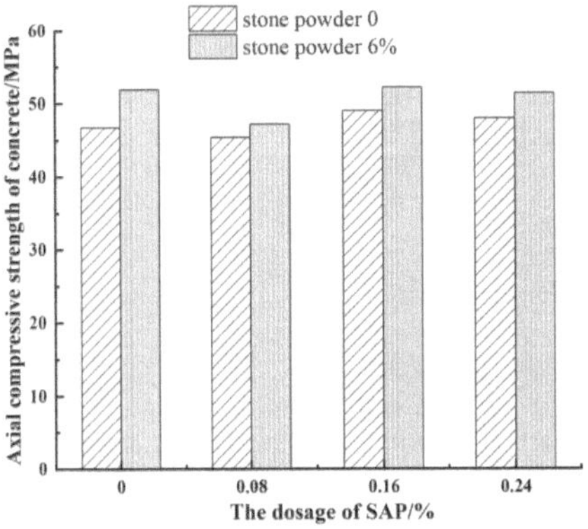

(a) The relationship between axial compressive strength and the dosage of SAP

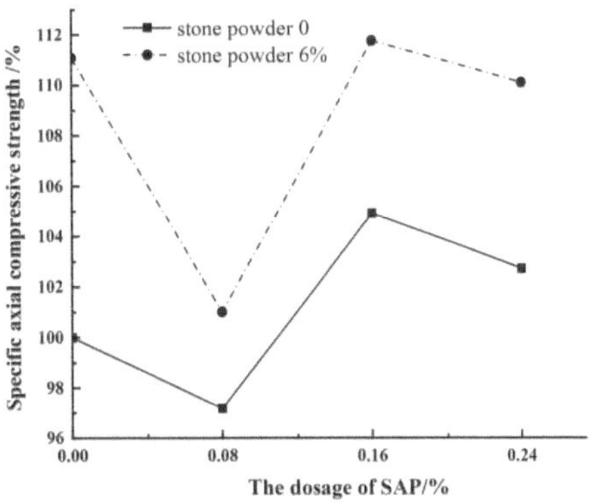

(b) The relationship between specific axial compressive strength and the dosage of SAP

Figure 5. Influence of SAP on the axial compressive strength of manufactured sand concrete.

then decreased with the increase of SAP content. The strength of S8F0, S16F0 and S24F0 specimens increased by –2.82%, 4.92% and 2.72% respectively, compared with S0F0 specimens. When the stone powder content was 6%, the axial compressive strength of different SAP content groups was larger than that of the group without stone powder. The main reason was the same as the flexural strength.

4 CONCLUSION

By studying the influence of SAP on the basic mechanical properties of manufactured sand concrete such as cube compressive strength, splitting tensile strength, flexural strength and axial compressive strength, the following conclusions are drawn:

(1) With the increase of SAP content, the cubic compressive strength of manufactured sand concrete increased first and then decreased, and the strength was the highest when SAP content was 0.08%.
(2) The splitting tensile strength of manufactured sand concrete at each age decreased first and then increased, then decreased with the increase of SAP content, and the strength was the highest when SAP content was 0.16%.
(3) The effect laws of flexural strength and axial compressive strength of manufactured sand concrete were basically the same. With the increase of SAP content, the strength decreased first, then increased and then decreased. When SAP content was 0.08%, the strength was the highest.

CREDIT AUTHORSHIP CONTRIBUTION STATEMENT

Zhaowei CUI: Investigation, Data curation, Writing-original draft. Hao CHEN: Experimentalize, Writing-review. Ronggui LIU: Project administration, Supervision.

DECLARATION OF COMPETING INTEREST

The authors declare that they have no known competing financial interests or personal relationships that could have appeared to influence the work reported in this paper.

ACKNOWLEDGEMENT

This study was supported by the National Natural Science Foundation of China (No. 51778272).

REFERENCES

Abdulwahid M. Influences of different stone powders on pervious concrete strength [J]. *Structural Concrete*, 2020, 22: 528–534.

Bendixen M, Best J, Hackney C, et al. Time is running out for sand [J]. *Nature*, 2019, 571(7763): 29–31.

Han Jianguo. Design and control shrinkage behavior of high-strength self-consolidating concrete using shrinkage-reducing admixture and super-absorbent polymer [J]. *Journal of Sustainable Cement-Based Materials*, 2014, 3(3–4):182–190.

Hu Shiyong, Zhang Gaozhan, Yang Jun, et al. Effect of super absorbent polymer (SAP) on the performance of low-shrinkage ult rahigh performance concrete (UHPC) [J]. *Concrete*, 2021(03):54–58.

Lao Jiarong, Huang Zhongcai, Guo Yinchuan, et al. Shrinkage and mechanical properties of internal curing pavement concrete with SAP [J]. *Bulletin of the Chinese Ceramic Society*, 2021, 40(02): 676–682.

Li H J, Huang F L, Cheng G Z, et al. Effect of granite dust on mechanical and some durability properties of manufactured sand concrete [J]. *Construction and Building Materials*, 2016, 109(4): 41–46.

Liu Minghui, Wang Xueyan, Liu Xuan, et al. Study on modification effect of calcium sulfate whiskers on stone powder content of manufactured sand concrete[J]. *China Civil Engineering Journal*, 2021, 54(05): 56–64.

Mechtcherine V, Reinhardt H W. *Application of super absorbent polymers (SAP) in concrete construction* [M]. Netherlands: Springer, 2012.

Mignon A, Snoeck D, Halluin K, et al. Alginate biopolymers: Counteracting the impact of super absorbent polymers on mortar strength [J]. *Construction and Building Materials*, 2016, 110:169–174.

Ministry of Housing and Urban-Rural Development, PRC. The design regulations of mix ratio of ordinary concrete (JGJ55-2011) [S]. *Beijing: China Architecture and Architecture Press*, 2011.

Pourjavadi A, Fakoorpoor S, Khaloo A , et al. Improving the performance of cement-based composites containing superabsorbent polymers by utilization of nano-SiO_2 particles[J]. *Materials and Design*, 2012, 42:94–101.

Soliman A, Nehdi M. Effect of partially hydrated cementitious materials and superabsorbent polymer on early-age shrinkage of UHPC [J]. *Construction and Building Materials*, 2013, 41:270–275.

Sun B, Wu H, Song W, et al. Design methodology and mechanical properties of Superabsorbent Polymer (SAP) cement-based materials [J]. *Construction and Building Materials*, 2019, 204(APR.20):440–449.

Wang Xueyan, Liu Minghui, Liu Xuan, et al. Study on mechanical properties and carbon emissions of desert sand and manufactured sand concrete [J]. *China Civil Engineering Journal*, 2022, 55(02): 23–30.

Yang Haifeng, Jiang Jiasheng, Li Dekun, et al. Analysis on basic mechanical properties and microstructure of concrete made with manufactured sand and recycled coarse aggregate [J]. *Bulletin of the Chinese Ceramic Society*, 2018, 37(12): 3946–3950.

Yu Bentian, Chen Yanfei, Wang Huan, et al. Experiment on control measures of shrinkage and cracking of high strength manufactured sand concrete containing a large amount of high absorbency stone powder [J]. *Acta Materiae Compositae Sinica*, 2021, 38(08): 2737–2746.

Zhang Zhihao, Shui Zhonghe, Yu Rui, et al. Effect of cement substitution by coral fillers on the workability and mechanical properties of UHSC[J]. *Bulletin of the Chinese Ceramic Society*, 2017, 36(12): 3993–3998.

Advances in Civil Engineering and Environmental Engineering – bin Yahya & Bin Daud (Eds)
© 2023 The Author(s), ISBN: 978-1-032-39397-1

Study on the proportion of soil similar materials in a model test of a deep foundation pit adjacent river

Zhenghong Su* & Boyang Huang*
College of Civil Engineering, Fuzhou University, Fuzhou, China

ABSTRACT: Based on the orthogonal experimental design method, using barite powder, quartz sand, gypsum, laundry detergent, bentonite, and water as raw materials, the soil similar materials of the foundation pit in the model test were configured, and the physical and mechanical indexes such as density, cohesion, and internal friction angle of similar materials with different proportions were obtained. The results show that (1) the physical and mechanical parameters of similar materials prepared according to the mixing plan have a wide distribution range, and the physical and mechanical parameters of similar materials prepared according to the mixing plan have a high degree of coincidence with the original rock; (2) the similar materials can lay a foundation for subsequent friction model tests.

1 INTRODUCTION

As more and more subway stations are built adjacent to rivers, more and more research is made on the mechanical properties of deep foundation pit construction. For this kind of problem, a similarity model test is often used to study to supplement the numerical simulation and theoretical derivation.

The selection and proportion of similar materials are the keys to the success of model tests. More and more scholars have carried out research on it through laboratory tests and theoretical analysis. Because of the simplicity and economy of the orthogonal test, a laboratory test is often carried out with this method. Wu (2006) and Dou (2006) used barite powder, gypsum, fine sand, laundry detergent, and water as raw materials to determine the mixture ratio of similar materials in the soft soil in subsequent model tests through direct shear tests, and conducted a sensitivity analysis on the physical and mechanical parameters. Miao (2016) determined the mixing ratio of similar soil materials using barite powder, clay, fine sand, and bentonite as raw materials. Guan (2016) determined the mix ratio of similar rock materials by using barite powder, iron ore powder, fine sand as aggregate, rosin and alcohol as a binder, and gypsum powder as a regulator. Jia (2022) introduced elastic energy index, urge index, and residual elastic energy index comparatively analyzed the influence of various influencing factors on these indexes, and formulated white sandstone materials that met the test requirements. Gao (2021) introduced a new grouting material AG-C solution to carry out orthogonal tests, conducted single-factor tests based on the test results, and obtained similar soft rock materials suitable for underwater shaking table tests.

Chen and Li (2018) conducted a sensitivity analysis on similar materials in the deeply buried rock mass and large-scale physical model test respectively to determine the significance of each influencing factor and established a fitting formula between each factor and physical quantity based on multiple linear regression analysis. Chai (2019) took slope in loess region as the research object and proposed the proportion method of shaking table test for large soil slope under six similar ratio conditions, and then obtained the material similarity evaluation system.

Based on previous studies, this paper designs the soil layer under the condition of similarity ratio through an orthogonal experiment. The test results can provide a basis for the follow-up

*Corresponding Authors: 1016375013@qq.com and boyoungh@163.com

model excavation test of the deep foundation pit along the river, and provide a reference for the preparation of similar soil materials in other similar model tests.

2 MODEL TEST SIMILARITY THEORY AND SIMILARITY RELATION

2.1 *Project background*

Based on the foundation pit engineering of A&F University station of Fuzhou Metro Line 5, the normal gravity model test was carried out based on the similarity theory. The safety level and complexity level of the supporting project are level 1, and the site is in an unfavorable area. The distribution and physical indexes of the soil layer refer to the representative section of the standard section of the foundation pit. The parameters of each soil layer are shown in Table 1.

Table 1. Main physical parameters of similar materials in the soil layer.

Soils	Thickness (m)	Density ($G \cdot cm^{-3}$)	Cohesive force (kPa)	The angle of internal friction	The compression modulus MPa
Silty clay	12	1.86	19.8	13.9	4.85
Silt soil	5.5	1.70	9.0	5.7	3.15
Residual sandy clay	11	1.79	19	16	4.96
Fully weathered granite	10	1.90	20	22	15
Strongly weathered syenite porphyry	20.5	2	23	25	40

2.2 *Similarity theory and similarity relation*

Based on the similarity theory, a reasonable similarity relation was set up to predict the engineering prototype by model test results. In the model test of foundation pit excavation, the material characteristics, geometric characteristics, and loads of the model and prototype are mainly considered. For the normal gravity model test adopted in the subject, the acceleration similarity ratio S_a is set to 1; therefore, when determining the similarity relationship between prototype and model, the similarity ratio conversion formula: S should be used; $S_E = S_\rho S_a S_L$. First determine the structural geometric similarity ratio S_L, the similarity ratio of elastic mode S_E, density similarity ratio S_ρ, and then deduce the similarity ratio of other physical quantities. To sum up, the similarity of the proposed physical quantities is shown in Table 2.

Table 2. Relationship of similar physical quantities of the model.

	Physical quantities	Relation between	Similar than	Note
Material properties	Strain epsilon	$S_\varepsilon = 1.0$	1	
	Stress sigma	$S_\sigma = S_E$	11/400	
	Elastic modulus E	S_E	11/400	Combined material control
	Poisson's ratio of mu	$S_\mu = 1.0$	1	
	The density of rho	$S_\rho = S_E/(S_a S_l)$	1.1	Combined material control
Geometric features	The length L	S_L	1/40	Model design control
	In S	$S_S = S_l^2$	1/1600	
	Linear displacement X	$S_X = S_l$	1/40	
	Angular displacement beta	$S_\beta = 1.0$	1	
Load	Concentrated force p	$S_p = S_E S_l^2$	11/640000	

Note: S is the similarity ratio of all physical quantities; P is the prototype structure; M is the model structure.

2.3 *Index of similar materials*

The model soil layer is divided into two layers: the upper layer is soft soil composed of silty clay and silty soil, and the lower layer is bedrock composed of residual sandy clay, fully weathered granite, and strongly weathered syenite porphyry. The weighted average of the overlying soft soil (the depth of each soil layer multiplied by the physical and mechanical parameters of the corresponding soil layer and divided by the total depth) was obtained, and the average value of the physical and mechanical parameters was converted by the similarity ratio to obtain the target value of the model soil, as well as the underlying bedrock. This target is used to find suitable similar materials, as shown in Table 3.

Table 3. Main physical parameters of similar materials.

Physical quantities	Depth (m)	Density ($g{\cdot}cm^{-3}$)	Cohesive force (kPa)	Angles of internal friction (°)	The compression modulus (MPa)
Similar than	$S_L = 1/40$	$S_\rho = 1.1$	$S_c = 11/400$	$S_\phi = 1.0$	$S_E = 11/400$
Upper soft soil (mean)	17.5	1.81	16.4	12.3	4.32
Similar materials (target value)	0.44	1.99	0.45	12.3	0.12
Lower hard soil (mean)	30.5	1.92	2.12	2.19	24.69
Similar materials (target value)	0.76	2.11	0.58	21.9	0.68

3 ORTHOGONAL TEST SCHEME OF SOIL SIMILAR MATERIALS

3.1 *Selection of similar materials*

According to the overall scheme of model test and parameter requirements of similar materials, this paper refers to Dou Yuanming et al. In the similar material scheme of soft soil, barite powder, quartz sand, gypsum, bentonite, laundry detergent, and water were selected as raw materials to configure the model soil. Barite powder and quartz sand are used as aggregate to adjust the density of the material. Gypsum and laundry detergent are used as cementing agents to adjust the compression modulus of materials to meet the needs of simulating soft soil. Bentonite is used as an additive to regulate the cohesion of materials. See Table 4 for raw material parameters.

Table 4. Raw material parameters of similar materials.

Material	Character	Main ingredients	Mesh	Density
Quartz sand	White fine-grained	SiO_2 99.5 to 99.9% or higher	40–70	1.6
Barite powder	White powder	$BaSO_4$> 98%	400	4.2
Bentonite	Yellow powder	Sodium silicate aluminate	325	1.5
Gypsum	White powder	$CaSO_4$> 90%	–	1.5
Laundry detergent	Blue-green viscous liquid	Surfactants and water softeners	–	1.1

3.2 *Orthogonal experimental design*

To select the appropriate mix ratio with the least number of tests, L25(5) is designed according to the requirements of the orthogonal test (See Table 5 for the design arrangement of the orthogonal test). The following five factors were mainly considered in the test (5 levels were set for each factor): bone glue ratio (aggregate/binder), washing paste ratio (laundry detergent/gypsum powder), barite powder content (barite powder/aggregate), bentonite dosage (bentonite/total mass), water content (water/total mass). The above factors are all mass ratios, and the matching scheme and results are shown in Table 6.

Table 5. Arrangement of orthogonal test design.

Level	Bone glue	Wash cream	Barite content (%)	Bentonite content (%)	Water content (%)
1	5-1	3:7	0	5	7
2	5-1	4:6	10	10	9
3	5-1	5:5	20	15	11
4	5-1	6:4	30	20	13
5	5-1	Mixed	40	25	15

Table 6. Orthogonal experimental results.

	Factors affecting the					Basic physical parameters		
Serial number	Bone glue	Wash cream	Barite content (%)	Bentonite content (%)	Water content (%)	Density (right to m^{-3})	Cohesion (kPa)	Internal friction Angle (°)
1	5:1	3:7	0	5	7	1.82	1.29	32.54
2	5:1	4:6	10	10	9	1.99	4.77	28.77
3	5:1	5:5	20	15	11	1.95	0.27	10.59
4	5:1	6:4	30	20	13	1.90	1.33	3.03
5	5:1	7:3	40	25	15	1.85	5.16	0.34
6	**7:1**	**3:7**	**10**	**15**	**13**	**2.01**	**0.61**	**12.49**
7	7:1	4:6	20	20	15	1.90	0.10	5.14
8	7:1	5:5	30	25	7	1.57	19.99	25.36
9	7:1	6:4	40	5	9	1.97	0	27.59
10	7:1	7:3	0	10	11	1.75	0	31.67
11	9:1	3:7	20	25	9	1.61	6.78	26.89
12	9:1	4:6	30	5	11	2.01	0.33	36
13	9:1	5:5	40	10	13	2.01	0	27.52
14	9:1	6:4	0	15	15	1.79	0.18	30.28
15	9:1	7:3	10	20	7	1.93	15.48	24.13
16	11:1	3:7	30	10	15	2.01	1.61	32.17
17	11:1	4:6	40	15	7	1.76	27.09	29.94
18	11:1	5:5	0	20	9	1.93	1.93	30.46
19	11:1	6:4	10	25	11	1.98	28.05	7.97
20	11:1	7:3	20	5	13	1.87	0	33.86
21	13:1	3:7	40	20	11	1.77	8.06	23.65
22	13:1	4:6	0	25	13	2.03	10.87	12.73
23	13:1	5:5	10	5	15	1.88	0.53	33.58
24	13:1	6:4	20	10	7	1.97	0.86	30.90
25	**13:1**	**7:3**	**30**	**15**	**9**	**2.12**	**1.05**	**22.75**

3.3 *Orthogonal test process*

While preparing the sample, join in turn define the quality of quartz sand, barite powder, gypsum powder, and bentonite powder dry mixing evenly, then stir in the secondary regulation quality laundry detergent and water, stir well until the material group that has no obvious particles after the material into the cutting ring, weighing of each specimen, take the average to obtain the group sample density. Five samples were prepared for each ratio, four of which were subjected to a direct shear test to measure the cohesion and internal friction Angle of soil samples, and the remaining one was reserved, as shown in Figure 1.

Figure 1. Sample of direct shear test.

Figure 2. Direct shear instrument.

Use the direct shear instrument as shown in Figure 2. Four samples were taken from each group at 0.8 mm·min at vertical pressures of 100, 200, 300, and 400 kPa, respectively. The shear displacement L and shear strength τ relation curves of A6 and A25 samples under various vertical pressure σ are shown in Figure 3. The maximum shear stress at each vertical pressure was further selected and the relationship between vertical pressure σ and shear strength τ was obtained by linear fitting, as shown in Figure 4. According to the curves, the cohesion of samples A6 and A25 were 0.61 kPa and 1.05 kPa, respectively, and the internal friction angles were 11.49° and 22.75°, respectively.

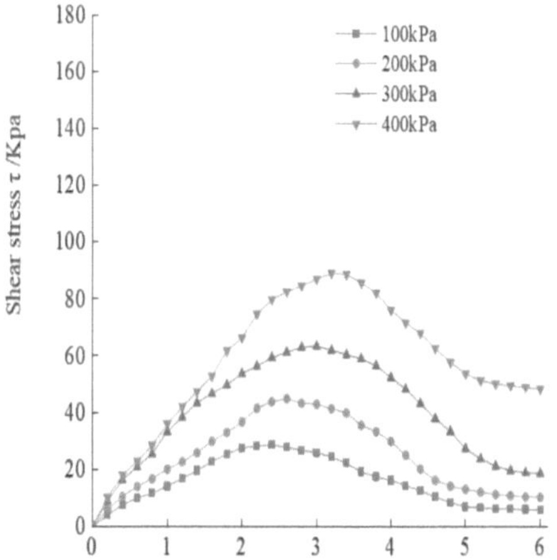

A) Sample A6

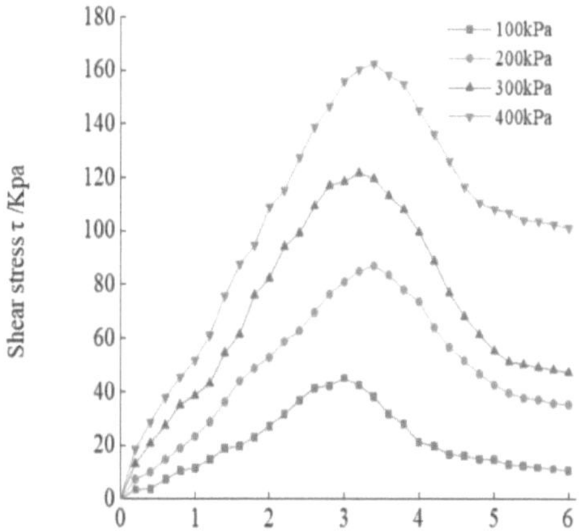

B) Sample A25

Figure 3. L-τ relation curve.

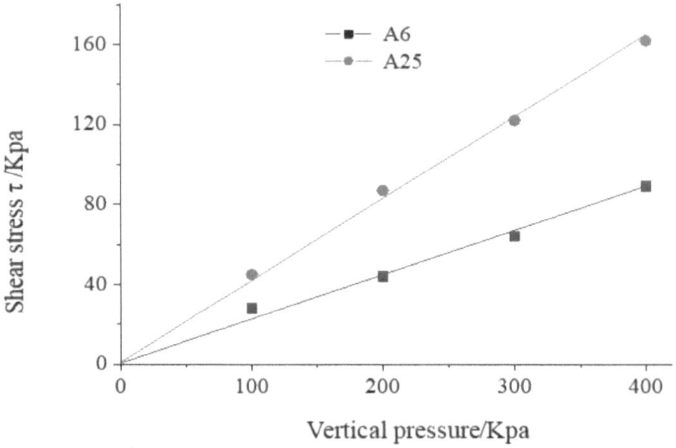

Figure 4. σ − τ relation curve.

4 CONCLUSIONS

Direct shear tests were carried out for 25 groups of similar soil materials with different ratios (the average of each group sample was taken from three tests), and the cohesion, internal friction Angle, and density of each ratio were finally measured, as shown in Table 6. The physical parameters of similar materials with different proportions are distributed in a wide range, with the density ranging from 1.57 g/cm to 2.12 g/cm^3, the cohesion is distributed between 0 and 28.05 kPa, and the internal friction. Angle is distributed between 0° and 33.86°, which can meet the requirements of most model tests for similar materials in soft soil.

According to the orthogonal experiment results, the A6 soil sample meets the requirements of soft soil in the upper layer, and its ratio is quartz sand: barite powder: bentonite: gypsum: laundry detergent: water = 567:63:153:63:27:130. In addition, the A25 soil sample is close to the physical property parameters required by the underlying bedrock, and its ratio is quartz sand: barite powder: bentonite: gypsum: laundry detergent: water = 494:212:150:16:38:90.

REFERENCES

Chai Shaofeng, Wang Ping, Guo Haitao, et al. (2019). *Journal of Earthquake Engineering*, 41(05): 1308–1315.

Chen Zhuoli, Zhu Xunguo, Zhao Deshen. (2018). Multiple regression analysis of similar material ratio in deep tunnel model test. *Journal of guangxi university* (natural science), 43(1): 379–389.

Dou Yuanming, Wang Jianning, Tian Guizhou, et al. (2018). *Journal of Chongqing Jiaotong University* (Natural Science), 37(4): 65–71.

Gao Huihui, Pei Xiangjun, Ma Zhigang, et al. (2021). Experimental study on similar Material Ratio of soft saturated interlayer. *Water Resources and Hydropower Technology*, 52(09): 156–164.

Guan Zhenchang, Gong Zhenfeng, Chen Renchun, et al. (2016) *Journal of Highway and Transportation Technology*, 33(9): 92–98, 111.

Jia Baoxin,Yuan Wenya, Ma Chaoyi, et al. (2022). *Chinese Journal of Underground Space and Engineering*, 18(02): 503–512.

Li Guang, Ma Fengshan, Guo Jie, et al. (2020). Experimental study on Similar Material Ratio in Large-scale Engineering Model Test. *Journal of Northeastern University* (Natural Science), 41(11): 1653–1660.

Miao Yuanbing, Wei Wen, Chen Renchun, et al. (2016). *Journal of Fuzhou University* (Natural Science), 44(4): 132–138.

Wang Hanpeng, Li Shucai, Zhang Qiangyong, et al. (2006). *Chinese Journal of Rock Mechanics and Engineering*, 25(9): 1842–1847.

Wu Botao, Zhu Hehua, Xu Weiwei, et al. (2013). Experimental study on similar materials of grade iv soft rock. *Rock and Soil Mechanics*, 34(SI): 109–116.

Advances in Civil Engineering and Environmental Engineering – bin Yahya & Bin Daud (Eds)
 ISBN: 978-1-032-39397-1

Discussion on calculation method of bearing capacity of super-soft soil foundation reinforced by bamboo grid and sand cushion

DongKui Zhao, Shaolin Yue & Jihui Ding*
CCCC Road & Bridge Special Engineering Co., Ltd. Wuhan, Hubei, China

Zenghui Yu & Haochen Wang
China Construction Second Bureau Eighth Engineering Co., Ltd, Jinan, China

ABSTRACT: When the hydraulic fill soft soil foundation is a deep super soft soil foundation and the surface layer is super soft soil, the bearing capacity of the surface layer is very small or even zero, and it is difficult for manual and construction equipment to work on it. The existing calculation theory of the bearing capacity of super soft soil foundations reinforced by a bamboo grid and sand cushion is not perfect. (1) Taking bamboo net and super soft soil foundation, the ultimate bearing capacity of bamboo grid and soft soil foundation is this paper determined, and sand cushion, bamboo grid and soft soil layer are taken as a double-layer foundation and deduced the formula of ultimate bearing capacity of bamboo grid and sand cushion reinforced super soft soil foundation. (2) For the field load test results of Yang and Cho, through the bearing capacity calculation method based on the diffusion of the bamboo grid and sand cushion, the fitting relationship between the diffusion angle and the thickness width ratio and bearing ratio is given. The average relative error of the bearing capacity of a composite foundation is 3.84%. (3) According to the double-layer theory, the fitting relationship between the impact coefficient and the thickness width ratio and bearing ratio is given. The relative error of the bearing capacity of the composite foundation is 6.94%. (4) The relative error of the calculation results based on the surface bearing capacity of the bamboo grid is the largest. When the angle between the bamboo net tension and the horizontal is taken as 10°, the calculation results are generally small, with the maximum size of 34.42%; and that is taken as 20°, the calculation results are generally large, with the maximum size of 49.95%.

1 INTRODUCTION

The site with thick silt or soft soil layer has high moisture content and low bearing capacity, and the construction machinery cannot operate. The use of reinforced cushion can improve the bearing capacity of silt and soft soil foundation. Bamboo has the advantages of a short growth period, high tensile strength and bending stiffness. Strengthening silt or soft soil with the bamboo grid as reinforcement has the advantages of stable structure, low price, convenient construction, energy conservation, and environmental protection. The bearing capacity of soft soil foundations reinforced with bamboo grids after surface treatment has been significantly improved (Chen 2007; Dang 2015; Yuan 2014). However, there are few research results on how to determine the deformation of the bamboo net and the calculation of the bearing capacity of bamboo grid reinforcement in practical projects. The existing calculation methods of bamboo grid reinforced soft soil bearing capacity are all based on the calculation methods of geotextile-reinforced soft soil bearing capacity. The calculation methods of geotextilereinforced soft soil bearing capacity are summarized, including empirical methods, calculation methods considering the diffusion of sand cushion and calculation methods based on the surface bearing capacity of the bamboo grid (Lin 1992). The calculation results of various calculation formulas vary greatly. One of the main reasons is that the

*Corresponding Author: dingjihui@126.com

DOI 10.1201/9781003349563-24

calculation formula itself is not perfect, such as the assumption of calculation mode, the factors affecting bearing capacity and the selection of calculation indicators. The second is that various formulas have specific applicable conditions, such as the layout type, laying range and loading characteristics of bamboo grids. Based on the calculation of the bearing capacity acting on the surface of the bamboo grid, a bearing capacity method is proposed for reinforcing soft soil foundations with bamboo grids, and the calculation result is too large (Yamanouchi 1985). A calculation method of bearing capacity considering the diffusion of the sand cushion is proposed that of bearing capacity of reinforced soft soil foundation under three-dimensional stress state of geotextile (Bai 1994). Based on the field load test, the expression of the relationship between the bearing ratio of bamboo grid reinforced soft soil foundation and the thickness of the sand cushion and the space ratio of the bamboo grid is given, but it does not apply to the case that the bearing capacity is less than 12.0 kPa or even equal to 0.0 (Yang 2011). Therefore, this paper improves the existing calculation method of bearing capacity of super soft soil composite foundation with bamboo grid sand cushion based on analysis and provides a basis for the treatment design of similar soft soil foundation.

2 CALCULATION OF BEARING CAPACITY OF BAMBOO GRID REINFORCED SUPER SOFT SOIL FOUNDATION

2.1 *Empirical method*

Yang (2011) gave the expression of the relationship between the bearing ratio of bamboo grid reinforced soft soil foundation and the thickness of the sand cushion and the space ratio of the bamboo grid through field load test:

$$BCR = f_u/f_{u0} = 4.56\,(H/l)^{0.644} \tag{1}$$

Where CBR is the bearing ratio; f_u is the ultimate bearing capacity of the foundation after reinforcement (kPa); f_{u0} is the ultimate bearing capacity of the foundation before reinforcement (kPa), in which $f_{u0} = C_u N_c$, C_u is the undrained strength of saturated soft clay (kPa); N_c is the bearing capacity coefficient, that is 5.3; H is the thickness of sand cushion (m); l is the space of bamboo grid (m).

2.2 *Calculation method of bearing capacity based on sand cushion diffusion*

When a certain thickness of the sand cushion is laid on the bamboo grid, the sand cushion and the bamboo grid have the effect of stress diffusion. When the bamboo grid, sand cushion and soft soil composite foundation reach the limit state, currently, the bamboo grid tension T_u is the tensile strength of the bamboo grid, and the vertical load is the ultimate bearing capacity f_u (Bai 1994):

$$f_u = [C_u N_c + K T_u \sin\theta_{hu}/a](B + 2H\tan\alpha)(L + 2H\tan\alpha)/(BL) \tag{2}$$

Where B is the width of the load action surface (m); L is the length of the load action surface (m); θ_{hu} is the angle between T_u and the horizontal plane; α is the pressure diffusion angle; K is the coefficient related to the foundation shape, and K is generally 1.0 to 2.0; if $L/B \geq 10$, then $K = 1$; if $L/B = 1$, then $K = 1$.

2.3 *Calculation method based on bamboo grid surface bearing capacity*

Yamanouchi (1985) regards the bamboo grid reinforced soft soil foundation as that the upper layer is covered soil and the lower layer is saturated soft soil foundation. When the sand cushion is damaged, the bearing capacity of the foundation is:

$$f_u = 0.5\gamma_1 B N_{\gamma 1} + \gamma_1 D_f N_{q1} \tag{3}$$

When the lower saturated soft soil layer is damaged, the bearing capacity of the foundation is:

$$f_u = (1 + H/B)\,(C_u N_c + 2T_u \sin\theta_{hu}/B) \tag{4}$$

In Formulas (4) and (5), γ_1 and ϕ_1 are the gravity and internal friction angle of sand cushion respectively; D_f is the load plate settlement; $N_{\gamma 1}, N_{q1}$ is the bearing capacity coefficient related to the internal friction angle ϕ_1 of sand cushion; T_u is the ultimate tension of bamboo grid and is mainly composed of the bamboo net and geotextile (geotextile has low strength, so it is not considered); θ_{hu} is the angle between T_u and the horizontal plane.

In Formula (4), the effects of S_a (friction at the bamboo mesh interface), $1/r$ (curvature of the imaginary circle), and D_f (settlement of the loaded plate) are ignored. S_a, $1/r$ and D_f are the coefficients related to the deformation shape of the load pattern. So far, the values of S_a, $1/r$ and D_f have no definite conclusion in theory, and there are few quantitative test results. At present, most of them use speculative values.

2.4 *Calculation method of double layered foundation based on upper hard and lower soft*

The thickness h of the overlying sandy soil layer on the soft soil layer is smaller than the width of the load action surface. The research results of Meyerhof and Hanna show that the upper soil layer will have punching shear failure, and the lower soil layer will have overall shear failure. The foundation bearing capacity of the upper hand and lower soft soil layers consists of the upper and lower layers, namely (Qian 2007):

$$f_u = f_{u0} + \gamma_1 H^2 k_s \tan\varphi_1/B - \gamma_1 H \tag{5}$$

Considering the influence of the bamboo grid, the bearing capacity of the lower soft soil is:

$$f_{u0} = C_u N_c + \gamma_1 H + 2T_u \sin\theta_{hu}/B \tag{6}$$

Substituting Formula (6) into Formula (5), we get:

$$f_u = C_u N_c + 2T_u \sin\theta_{hu}/B + \gamma_1 H^2 k_s \tan\varphi_1/B \tag{7}$$

Considering the influence of foundation shape, Formula (8) is modified, including:

$$f_u = (1 + B/L)\,\gamma_1 H^2 k_s \tan\varphi_1/B + (1 + 0.2B/L)\,(C_u N_c + 2T_u \sin\theta_{hu}/B) \tag{8}$$

In Formulas (5) to (8), k_s is the impact coefficient, and is related to the bearing ratio of sand cushion and soft soil layer and the thickness width ratio of sand cushion and foundation width; Other variables are the same as before. Formula (9) for the ultimate bearing capacity of soft soil composite foundation with bamboo grid and the sand cushion is composed of three parts: sand cushion, bamboo grid, and soft soil foundation. The bearing capacity of a sand cushion considers the influence of the internal friction angle of sand and the thickness of the sand cushion.

3 COMPARISON OF BEARING CAPACITY RESULTS

3.1 *Field test results*

Yang (2011) conducted the bamboo grid reinforced super soft soil test on site, and the size of the load plate is 1.0 m ×1.0 m, and the tensile strength of the bamboo grid is 22.1 kN/m. The unit weight of the upper sand cushion is 20 kN/m^3 and has a friction angle of 30°. Table 1 shows the results of the composite bearing capacity of soft soil with bamboo grid and sand cushion in the field experiment using the bamboo grid laying scheme [3]. According to Table 1, Yang and Cho gave the empirical formula (1).

Table 1. Comparison of different bamboo net spacing and sand cushion thickness.

Size /m	H /m	C_u t/m^2	Bearing capacity of the test (t/m^2)	Bearing capacity of soft soil (t/m^2)	BCR
0.5×0.5	1.00	0.23	3.61	0.61	5.92
0.7×0.7	0.40	0.27	2.09	0.87	4.15
1.0×1.0	0.60	0.46	3.97	0.87	5.72
1.0×1.0	0.60	0.31	3.89	0.98	4.00
1.0×1.0	0.95	0.33	3.63	0.98	5.10
1.0×1.0	1.45	0.33	5.35	1.22	3.25
1.0×1.0	1.95	0.33	6.75	0.82	4.74
1.25×1.25	0.95	0.37	3.92	0.72	2.92
1.25×1.25	1.25	0.37	5.00	0.72	4.81

3.2 *Calculation results of bearing capacity based on sand cushion diffusion*

According to the measured results in Table 1, the diffusion angle of the sand cushion bamboo grid is calculated according to Formula (2), as shown in Table 2. The diffusion angle of the bamboo grid and sand cushion is 22° to 40°, with an average of 27.44°. The diffusion angle of the bamboo grid and sand cushion is fitted by a genetic algorithm, and the fitting expression is:

$$\alpha = p_1 + p_2 y + p_3 \exp\left(-0.5((x - p_4)/p_5)^2\right) \tag{9}$$

Where $x = H/B, y = q_2/q_1$; q_1, q_2 are the bearing capacity of sand cushion and soft soil layer, respectively. See Table 2 for fitting parameters. The correlation coefficient, R-square, and determination coefficient both are greater than 0.95.

Table 2. Fitting parameters of expression α.

p1	p2	p3	p4	p5
42.23774399	−191.295941	13.64138041	0.633569144	0.139173962

The bearing capacity is calculated by using the fitted diffusion angle, and the calculation results are shown in Table 3. Compared with the measured results, the relative error is 0.01% ~10.04%, and the average relative error is 3.84%. The large error is mainly due to the large thickness of the sand cushion. For example, when H/B = 1.95, the error is 9.99%. At this time, H/B = 1.95, the bearing capacity of soft soil composite foundation with bamboo grid sand cushion should be determined by the sand cushion.

Table 3. Calculation results of bearing capacity based on sand cushion diffusion.

l/B	H/B	q_2/q_1	α(°)	Fitting α(°)	Bearing capacity (kPa)	relative error (%)
0.5	1	0.066	30.13	30.04	35.37	0.40
0.7	0.40	0.078	30.65	30.65	20.48	0.01
1.0	0.60	0.133	29.33	30.05	45.84	2.34
1.0	0.60	0.089	39.18	38.46	32.18	2.56
1.0	0.95	0.095	24.11	25.09	35.73	4.11
1.0	1.45	0.095	23.86	24.06	51.38	1.04
1.0	1.95	0.095	22.32	24.06	67.93	9.99
1.25	0.95	0.107	23.73	22.80	38.53	4.11
1.25	1.25	0.107	23.68	21.77	48.44	10.04

3.3 *Calculation results based on the surface bearing capacity of bamboo grids*

Table 4 shows the calculation of the surface bearing capacity of bamboo grids. The calculation results show that the included angle between bamboo grid tension T and horizontal plane $\theta = 10°$, the calculation results of bamboo grid sand cushion and soft soil composite foundation are relatively small, and the relative error is 0.30%~ 34.42%; when $\theta = 20°$, the calculation result is larger, and the relative error is 4.78% ~ 49.50%.

Table 4. Calculation results based on the surface bearing capacity of bamboo grids.

l/B	H/B	Bearing capacity (kPa) $\theta = 10°$	Relative error (%) $\theta = 10°$	Bearing capacity (kPa) $\theta = 20°$	Relative error (%) $\theta = 20°$
0.5	1	27.10	−23.41	41.80	18.14
0.7	0.40	20.42	−0.30	30.71	49.95
1.0	0.60	31.23	−19.72	42.99	10.51
1.0	0.6	25.00	−34.42	36.76	3.57
1.0	0.95	31.48	−11.50	45.82	28.79
1.0	1.45	39.55	−24.56	57.56	9.79
1.0	1.95	47.63	−28.00	69.31	4.78
1.25	0.95	33.51	−12.78	47.84	24.54
1.25	1.25	38.66	−21.10	55.20	12.66

3.4 *Calculation results based on upper hard and lower soft double-layer Foundation*

Table 5 shows the impact coefficient ks derived from the bearing capacity calculation Formula (8) given by the bearing capacity of the upper and lower soft double-layer foundation. It can be seen from Table 5 that the impact coefficient is 0.57 ~ 2.73. The impact coefficient is fitted by a genetic algorithm, and the fitting function is:

$$k_s = p_1 + p_2 y + p_3/(1 + ((x - p_4)/p_5)^2) \tag{10}$$

Where $x = H/B, y = q_2/q_1$; q_1, q_2 are the bearing capacity of sand cushion and soft soil layer, respectively. See Table 5 for fitting parameters. The correlation coefficient, R-square, and determination coefficient both are greater than 0.96.

Table 5. Fitting parameters of expression k_s.

p1	p2	p3	p4	p5
0.507135618	1.679542924	32.779410383	0.69762898	−0.023863164

According to the above fitting coefficient ks. The calculated bearing capacity of the foundation is shown in Table 6. The relative error between the calculated bearing capacity and the measured bearing capacity is 0.84% ~ 14.12%, and the average relative error is 6.94%.

Table 6. Calculation results of upper hard and lower soft double-layer Foundation.

l/B	H/B	q_2/q_1	ks	Fitting ks	Bearing capacity (kPa)	Relative error (%)
0.5	1	0.066	0.64	0.82	39.65	12.07
0.7	0.40	0.078	0.80	0.85	20.66	0.88
1.0	0.60	0.133	2.36	2.58	40.73	4.70

(continued)

Table 6. Continued.

l/B	H/B	q_2/q_1	ks	Fitting ks	Bearing capacity (kPa)	Relative error (%)
1.0	0.60	0.089	2.73	2.50	36.23	4.98
1.0	0.95	0.095	0.94	0.96	35.87	0.84
1.0	1.45	0.095	0.75	0.70	49.88	4.85
1.0	1.95	0.095	0.57	0.68	75.49	14.12
1.25	0.95	0.107	1.10	0.98	35.86	6.66
1.25	1.25	0.107	0.93	0.75	42.47	13.32

4 CONCLUSIONS

(1) Taking bamboo grid and soft soil foundation, the ultimate bearing capacity of bamboo grid and soft soil foundation is determined. Taking sand cushion and bamboo grid soft soil as a double-layer foundation, the formula of ultimate bearing capacity of bamboo grid reinforced soft soil foundation with sand cushion is derived. The bearing capacity formula is composed of three parts: sand cushion, bamboo grid, and soft soil foundation bearing capacity. The bearing capacity of a sand cushion considers the internal friction angle of sand and the thickness of the sand soil layer.

(2) Different calculation methods of bearing capacity of the bamboo grid, sand cushion, and soft soil composite foundation are adopted to calculate and compare the field load test results given by Yang and Cho. The results show that the maximum relative error of the bearing capacity calculation method based on sand cushion diffusion is 10.04%, and the average relative error is 3.84%. The relative error of the calculation result based on the surface bearing capacity of the bamboo grid is the largest. When the angle between the bamboo grid tension and the horizontal plane is 10°, the calculation result is generally small, with a maximum size of 34.42%. When the angle is 20°, the calculation result is generally large, with a maximum relative error of 49.95%. Based on the calculation method of upper hard and lower soft bearing capacity, the maximum relative error is 14.12%, and the average relative error is 6.94%.

REFERENCES

Bai Bing (1994). Calculation of bearing capacity of reinforced soft foundation under three-dimensional stress state of Geotextile[J]. *E ast China Highway*. 91(6):54–58.

Chen Chunan, XU Jinming (2007). Application of bamboo grid for silt surface treatment of ultra- soft ground [J]. *Port & Waterway Engineering*. (10): 75–77.

Dang Fa-ning, LIU Hai-wei, Wang Xue-wu. Application of bamboo as tensile reinforcement to strengthening of embankment of soft soils [J]. *Chinese Journal of Geotechnical Engineering*. 2013, 35(S2):44–48.

Lin Kaiqiu, Bao Chenggang (1992). A new practical calculation method and test verifications of geotextile reinforced soft foundation [J]. *Journal of Yangtze River Scientific Research Institute*. (01):47–56.

Qian Jiahuan, Yin Zongze (2007). Geotechnical principles and calculations (Second Edition) [M] *Beijing: Water Resources and Electric Power Press*. 2007, 374–378.

Yamanouchi T(1985). Recent development in the used synthetic geofabrics and geogrids [C]. *Symposium on Recent Developments in Ground improvement techniques*, Bangkok. 205–224.

Yang K S, Cho S D, Ham T G, et al. (2011). A bearing capacity characteristic of very soft ground reinforced by bamboo net [J]. *International Symposium of Korean Geosynthetics Society Seoul*, South Korea , 185–190.

Yuan Man, Ding Jihui, Cao Yanliang (2014). The application of bamboo network reinforcement technology on hydraulic fill soft-soil foundation treatment [J]. *World Journal of Engineering & Technology*, 02(2): 68–72.

Advances in Civil Engineering and Environmental Engineering – bin Yahya & Bin Daud (Eds)

Study on displacement warning standard value of large section highway tunnel extension construction

Jiangling He*
Guangdong Province Freeway Co. Ltd, Tianhe, Guangdong, China

Yao Shi*
Chongqing Expressway Group Co. Ltd, Chongqing, China

Lianchao Ye*
School of Civil Engineering, Chongqing Jiaotong University, Chongqing, China

Xinghong Jiang*
China Merchants Chongqing Communications Technology Research and Design Institute Co., Ltd Chongqing, China

ABSTRACT: In view of the construction of large section highway tunnel reconstruction and expansion, the article takes the Shenzhen-Shantou west reconstruction and expansion project as the research object, and systematically expounds on the construction safety and early warning standard of large section highway tunnel reconstruction and expansion. In this paper, the mechanical parameters of grade IV and V surrounding rock of Shen-Shan-West tunnel are determined by a uniaxial compression test. The ultimate shear strain and ultimate principal tensile strain of grade IV and V surrounding rock are determined by the numerical analysis method. The finite element limit analysis method is adopted to determine the construction early-warning index of the reconstruction and expansion of the Shenzhen-Shantou west highway tunnel under the influence of the existing tunnel under the condition of different net distances by taking the plastic strain transfixion in the rock and soil as the criterion, which guides the subsequent construction and technical reference for similar projects.

1 INSTRUCTIONS

In some economically developed areas and super-large urban areas, two-way four-lane or even two-way six-lane road tunnels can no longer adapt to the increasing demand of traffic volume (Wang 2001), and there are increased projects to reconstruct two-way eight-lane road tunnels based on the original highway or road standard. The main types of reconstruction and extension of large-section highway tunnels include in-situ extension, parallel extension, and combined extension (Hu 2010). In-situ expansion from the small section tunnel formed by the original tunnel excavation to the large section tunnel excavated by the demolition of the original structure (Li 1996), the stress state of tunnel surrounding rock is complicated, and the stability of surrounding rock is greatly affected by the excavation method (Yang 2010). In addition, the surrounding rock pressure of the newly expanded tunnel with a large section is essentially different from that of the tunnel with a large section directly excavated under the original topography (Liu 2010), so the design and calculation of the support structure are different from that of the ordinary tunnel with large section. However, at the present stage, the development of tunnel reconstruction and expansion in China is not particularly comprehensive. There is a lack of guidelines or specifications for the risk research

*Corresponding Authors: hejl0620@126.com; 357050382@qq.com; 1036731016@qq.com and jiangxinghong@cmhk.com

DOI 10.1201/9781003349563-25

of existing road tunnel reconstruction and expansion, and academic research on construction early warning of large section road tunnel reconstruction and expansion under the influence of existing tunnels is also lacking. Based on the existing tunnel reconstruction and expansion project, Li Jian et al. (Li 2013) studied the construction structure safety of tunnel reconstruction and expansion projects using numerical simulation and theoretical analysis. Chen Xiao (Chen 2012) Yang analyzed the reconstruction and expansion construction method, mechanical properties of surrounding rock, and supporting structure of urban tunnel with large span and small clean distance. Jin Bo et al. (Jin 2008) qualitatively summarized the solutions to the problems involved in tunnel reconstruction, such as renovation schemes, construction safety and operational safety. The above scholars have analyzed the construction scheme and mechanical properties of highway tunnel reconstruction and expansion, but lack research on safety warnings in the construction process of large section highway tunnel expansion. Although the above research has a deep impact on the reconstruction and expansion of China's highway tunnel, it also has limitations.

At present, under the influence of existing tunnels in China, there are few engineering practices and a relative lack of experience in newly built highway tunnels with large sections. In addition, large section tunnels have a high flattening rate and more complex stress conditions than two-lane small section tunnels. Therefore, there is an urgent need for early warning indexes to guide the construction of large-section tunnel expansion to ensure the safety of construction and avoid huge losses to society. In this paper, the displacement values of each key point during the instability failure of the tunnel with a large section under the condition of the soft surrounding rock are studied by using the method of combining experiment and numerical simulation.

2 PROJECT PROFILE

Shenzhou-Shantou Tunnel is a double-hole eight-lane expressway tunnel with a design speed of 120 km/h. The single hole excavation width of the tunnel is about 22.6 m, and the excavation height is about 14.5 m, which belongs to the super-large span tunnel and the tunnel with a large section and small clear distance. The overall form of the tunnel is the hidden ridge tunnel + the open-cut sinking tunnel, among which YK56+334~YK56+950 are the hidden ridge tunnel section, with a maximum buried depth of about 67 m. According to the drilling results, the excavation section is mainly composed of fully to strongly weathered sandstone, with well-developed joints and fractures, extremely broken rock mass, low strength, easy to soften in water, and low-grade surrounding rock. The tunnel body lining adopts compound lining according to the new Austrian method principle. The secondary lining is molded concrete lining, the lining adopts curved wall lining, the concrete strength grade is C30, and the invert concrete grade is C20. The initial support adopts bolt-shotcrete support because the steel arch and shotcrete are integrated into a whole and bear the pressure of surrounding rock, Table 1 shows the physical and mechanical parameters of concrete materials.

Table 1. Mechanical parameter of materials.

Material parameter	Elasticity modulus /GPa	Elasticity modulus	Density kg/m^3	Cohesion /MPa	Frictional angle /°
C25	26.94	0.2	2250	\	\
C30	30	0.2	2500	\	\
C20	25.5	0.2	2500	\	\

3 TEST OF MECHANICAL PROPERTIES OF SURROUNDING ROCK

3.1 *Testing program*

Through uniaxial compression deformation test can measure IV, V surrounding rock elastic modulus E and Poisson's ratio of μ, IV, V surrounding rock under uniaxial compression deformation test specimen size Φ = 50 mm diameter, high H = 100 mm standard cylinder, IV, V surrounding rock mineral particles less than 2 mm in diameter, and largest specimen diameter greater than 10 times the diameter of the largest mineral grains. The accuracy of specimen preparation, in the whole height of the specimen, the diameter error shall not exceed 0.3 mm. The maximum non-parallelism of the two ends shall not exceed 0.05 mm. The end face shall be perpendicular to the axis of the specimen with a maximum deviation not exceeding 0.25 degrees.

To measure the elastic modulus E and Poisson's ratio μ of grade IV and V surrounding rock, combined with the existing test conditions, the displacement control loading method was adopted, the loading speed was set as 0.01mm/min, and the confining pressure was set as 0 MPa. To improve the test accuracy, a total of 2×3 specimens were prepared.

3.2 *Test instrument*

Taw-200 multifunctional rock triaxial test system (Figure 1), the main parameters of the instrument are as follows: specimen size: Φ50 × 100 mm; maximum axial force: 1,000 kN; Maximum ambient pressure: 60 MPa; maximum pore pressure: 60 MPa; heating temperature: –30 to 80°C; axial deformation: 0 to 20 mm; radial deformation: 0 to 5 mm; axial loading speed: 0.01 to 10 kN/s. Marker, screwdriver, wrench, rubber hammer rag, chipless paper, etc.

Figure 1. Taw-200 multifunctional triaxial rock testing system.

3.3 *Test procedure and result*

At the construction site, rock samples of all levels of surrounding rock were obtained. Through the standard production process, the test specimen was made into a standard cylinder with a diameter of Φ = 50mm and a height of H = 100mm. After the specimen reaches the test standard, a uniaxial compression test is carried out on the experimental specimen through the Taw-200 multifunctional triaxial rock testing system.

Table 2. Summary of test results.

Class IV surrounding rock specimen	Deformation parameter	Elasticity modulus E/MPa	3470
		Poisson ratio u	0.33
	Intensity parameter	Cohesion c/MPa	0.51
		Internal friction angle φ/°	30.0°
Class IV surrounding rock specimen	Deformation parameter	Elasticity modulus E/MPa	1150
		Poisson ratio u	0.4
	Intensity parameter	Cohesion c/MPa	0.13
		Internal friction angle φ/°	22.50°

According to the data recorded by the Taw-200 multifunctional triaxial rock testing system, the test results of each specimen were checked, the suspicious data were discarded, and the relationship curves of axial stress, axial strain, and radial strain were drawn. The peak value on the relation curve between axial stress and axial strain is taken as the uniaxial compressive strength. The uniaxial compression test results of grade IV and V surrounding rock remolded soil are shown in Table 2.

4 STUDY ON EARLY WARNING OF RECONSTRUCTION AND EXTENSION OF LARGE SECTION TUNNEL

4.1 *Finite element limit analysis*

Numerical limit analysis method solving process is different from the traditional method, calculated with strict ideal elastic-plastic numerical solution method, the process of numerical calculation, through continuously reduce the strength of materials (in the same proportion reduce soil cohesion and internal friction factor c tan phi) or increasing load (Zheng 2004; Zhou 2010; Zhao 2002), make it eventually reached damage state in the numerical calculation, At this point, the failure surface is automatically generated and the failure information is sent. Therefore, it does not need to assume the failure surface in advance, the strength reduction factor at the time of failure is the stable safety factor, and the load at the time of failure is the ultimate load. In this method, when the calculation reaches the failure state, the software will send out the failure information and automatically generate the failure surface, without knowing the failure surface in advance. The traditional limit analysis method only requires the failure surface to be known in advance but does not require the calculation to reach the failure state. This is the difference between the two limit analysis methods, but their principles are the same.

The numerical limit analysis does not need to know the sliding surface in advance, nor need to require the sliding force and anti-sliding force on the sliding surface, and can directly obtain the ultimate load and stability safety factor. However, the numerical limit analysis method needs to find the effective criterion for rock mass failure in the calculation process. If such a criterion cannot be found, even if rock mass failure has occurred, the solver does not know it. In this paper, the plastic strain is used as the criterion of failure, that is, the plastic zone is used as the criterion of failure. However, transfixion of the plastic zone only means the yield state, not necessarily the whole failure state of the soil.

4.2 *Safety limit strain control standard*

According to the mechanical parameters of grade IV and V surrounding rock, the limit strain value of equivalent rock block is numerically analyzed, and the value of equivalent rock block is obtained by numerical analysis. FLAC3D was used to simulate the equivalent rock block uniaxial compression test of grade iii surrounding rock. The calculation model was a cubic specimen with a size of 150 mm × 150 mm × 150 mm. Constraints were applied on the bottom surface, and a uniform vertical downward load was applied on the top surface, but friction on the top surface was not considered. In the equivalent block ultimate strain analysis, the central point displacement is monitored by increasing the load step by step. When the central point displacement of the block does not converge under the action of a certain level of load, the block will be destroyed, which is the equivalent ultimate load of the block. The strain value corresponding to the upper load is the ultimate strain. The vertical displacement of the center point of the equivalent block under different loads is shown in Figure 2.

The results show that the amplitude of vibration velocity change after blasting in the Y and Z directions is larger than that in the X direction, and the vibration velocity of the initial blasting decreases gradually with the increase of the distance from the face of the hand, and the influence of blasting on the initial support also decreases gradually. Primary support the vibration velocity of the closer distance constraints, the greater the vibration velocity values, the change of amplitude

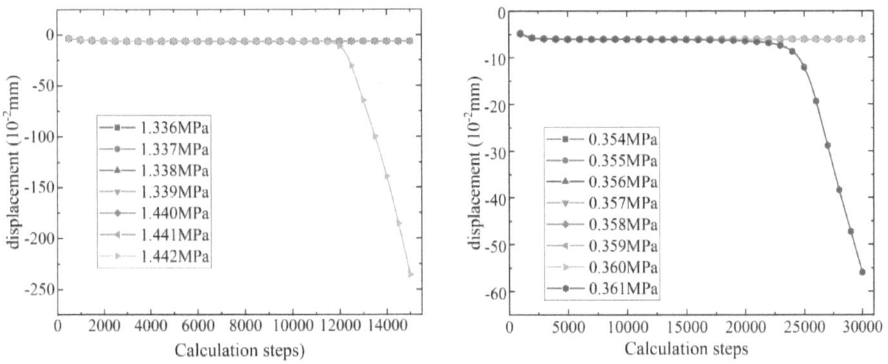

(a) Four surrounding rock (b) Five surrounding rock

Figure 2. The vertical displacement variation rule of equivalent block center point.

is, the greater the through the study on the mechanical behavior of primary support above, we know that at the beginning of the physical and mechanical properties and its curing ages have great correlation, where the nearer the distance constraints, the early support of curing age is shorter, not fully meet the design strength, This also makes the blasting vibration velocity relatively larger in the place closer to the palm surface.

Figure 2 shows that when the vertical load is 1.442 MPa and 0.361 MPa, the vertical displacement of the center point of the equivalent rock block does not converge and the rock block is damaged. The strain value of the upper load is extracted as the ultimate strain value, namely 1.441 MPa and 0.361 MPa. Finally, the ultimate shear strains of equivalent rock blocks of grades IV and V are 1.08‰ and 0.96‰, respectively, and the ultimate principal tensile strains are 0.28‰ and 0.27‰, respectively.

4.3 *Numerical analysis*

This paper aims to study the construction early-warning standard of a new four-lane tunnel with the large sections under different clear distances under the influence of existing tunnels and provide guidance for the follow-up construction of extended tunnels. Under the conditions of grade IV and V surrounding rock, buried depth of 60 m, and construction method of double side wall diversion pit, 5 working conditions under different net distances are proposed respectively, as shown in Table 3.

The parameters of surrounding rock in the numerical simulation are selected in Table 1 and Table 2, to more truly reflect the deformation of surrounding rock in actual construction, improve the authenticity of data, and provide a reference for subsequent construction.

Table 3. Numerical simulation condition.

Surrounding rock classification	Burial depth (m)	Construction method	Distance
IV	60	Double side drift method	0.5D
			0.75D
			1D
			1.25D
			1.5D
V			0.5D
			0.75D
			1D
			1.25D
			1.5D

4.4 *Interpretation of result*

In this paper, finite element limit analysis method, which transfixion of plastic strain in the rock and soil body is as a criterion, namely, which transfixion of plastic zone from the inside to the ground or airport surface as a failure criterion, get all the plastic zone in the vault under the clear distance tunnel through strength reduction factor, now think instability and failure of the tunnel and argues that the critical safety coefficient of the tunnel at the next higher level for the reduction factor. The displacement curves at key positions of the tunnel as a function of reduction coefficients were extracted and drawn, as shown in Figure 3. Conclusions can be drawn from the figure: (1) as the reduction coefficient increases, the displacement values of each key position of the tunnel also increase; (2) when the safety factor varies between 1.0 and 1.5, the displacement of each key position increases slowly; when the safety factor varies between 1.5 and 2.0, the displacement of each key position increases sharply. (3) The reduction coefficient is not much when the tunnel is connected at different net distances.

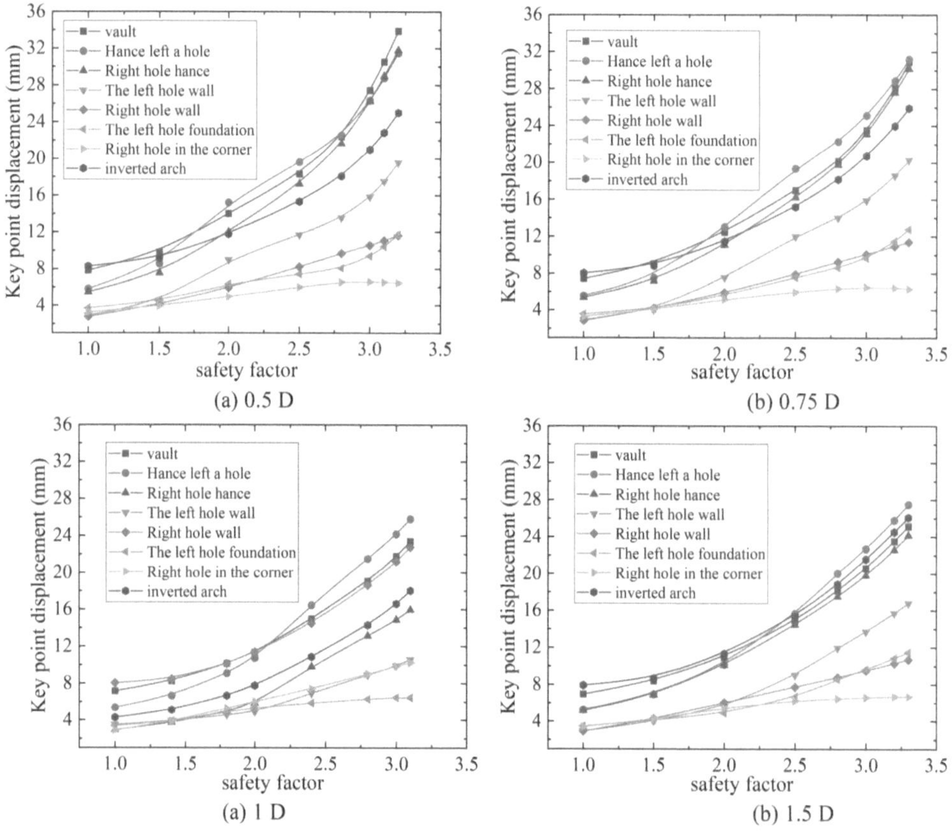

Figure 3. Curve of displacement value of key parts of tunnel changing with the safety factor.

The above chart only lists the variation rule of displacement of each key position with safety factor under the condition of grade IV surrounding rock, and the rule of grade V surrounding rock is familiar with it, but the value is different. Through the analysis of displacement law and size of each key point under a different surrounding rock and different net distance. The construction early-warning standards of large section four-lane highway tunnels under different net distances under the influence of existing tunnels are obtained, as shown in Table 4.

Table 4. Safety warning standard for tunnel construction under 60m buried depth.

The surrounding rock grade	Channel spacing	Danger classes	The vault of the tunnel settled /mm
IV	0.5D	Instability	U>30.53
		Dangerous	20<U≤31
		Abnormal	10<U≤20
		Normal	U≤10.18
	0.75D	Instability	U>28.13
		Dangerous	19<U≤28
		Abnormal	9<U≤19
		Normal	U≤9.38
	1D	Instability	U>21.77
		Dangerous	15<U≤22
		Abnormal	7<U≤15
		Normal	U≤7.26
	1.25D	Instability	U>23.55
		Dangerous	16<U≤24
		Abnormal	8<U≤16
		Normal	U≤7.85
	1.5D	Instability	U>22.51
		Dangerous	15<U≤23
		Abnormal	8<U≤15
		Normal	U≤7.5
V	0.5D	Instability	U>45.79
		Dangerous	31<U≤46
		Abnormal	15<U≤31
		Normal	U≤15.26
	0.75D	Instability	U>42.19
		Dangerous	28<U≤42
		Abnormal	14<U≤28
		Normal	U≤14.06
	1D	Instability	U>32.65
		Dangerous	22<U≤33
		Abnormal	11<U≤22
		Normal	U≤10.88
	1.25D	Instability	U>35.32
		Dangerous	24<U≤35
		Abnormal	12<U≤24
		Normal	U≤11.77
	1.5D	Instability	U>33.76
		Dangerous	23<U≤34
		Abnormal	11<U≤23
		Normal	U≤11.25

5 CONCLUSIONS

(1) By analyzing the failure mechanism of geotechnical materials, the finite element limit analysis method is determined to carry out numerical simulation; The mechanical properties of rock blocks in the Shenzhen-Shantou-West tunnel are analyzed experimentally, and the deformation parameters and strength parameters of surrounding rock IV and V are determined. Numerical simulation is carried out on selected parameters, and the surrounding rock deformation of the Shenzhen-Shantou West tunnel is simulated and analyzed. Finally, through data analysis and

comparison, it is concluded that the ultimate shear strain of equivalent rock block of grade IV surrounding rock is 1.08‰, and the ultimate principal tensile strain is 0.28‰. The ultimate shear strain and ultimate principal tensile strain of the equivalent rock block in class V are 0.96‰ and 0.27‰ respectively.

(2) the finite element limit analysis method, which transfixion of plastic strain in the rock and soil body is as a criterion, get the clear distance tunnel under the plastic zone in the vaults of the strength reduction factor, the maximum safe coefficient corresponding to the displacement of each point is the instability of surrounding rock displacement, through to the displacement of each point fitting. In this paper, the safety early-warning indexes of different spacing under the conditions of 60m depth and IV and V surrounding rock of the Shenzhen-Shanxi tunnel are obtained.

ACKNOWLEDGMENTS

This paper is supported by the Science and Technology Project of the Department of Transportation of Guizhou Province (Grant No. 2020-122-031).

REFERENCES

Chen Xiaoyang. Construction Mechanics Analysis of urban large-span small-distance tunnel in-situ Expansion Project [D]. *Chongqing: Chongqing University*, 2012.

Hu Juyi, Chen Libiao, Huang Lunhai. Study on the extension form of 2-lane tunnel to 4-lane tunnel J1, *Highway Traffic Technology*. 2010, (5): 93–97.

Jin Bo, Han Changling, Wang Wanwan, et al. Safety risks and countermeasures of existing tunnel reconstruction construction []. *Highway*, 2008, (7): 269–27L.

Li Jian, Li Xin, Li Hua. Research on structural safety of existing tunnel reconstruction and extension engineering [J]. *Highway Tunnel*, 2013, (1): 7–11.

Li Yuanfu. Reconstruction and extension construction technology of guancun ba tunnel [J]. *Railway Construction Technology*, 1996, (4): 22–26.

Liu Zhengxiong, LUO Chongjing, Zhou Yukui. Control blasting technology of guancun ba tunnel extension bell mouth engineering [J], *Blasting*, 1999, 16(3): 40–44.

Wang meng-shu. Suggestions on China's tunnel engineering construction in the 21st century [J]. *Modern Tunnel Technology*, 2001, 38(1): 2–4.

Yang Xiaoli, Wang Zuowei, Chen Jie. Study on stability of existing tunnel and lining [J]. *Transportation Science and Engineering*, 2010, 26(1): 49–52.

Zhao Shangjiao, Zheng Yingren, Shi Weimin, et al. Calculation of slope stability safety factor by finite element strength reduction method []. *Chinese Journal of Geotechnical Engineering*, 2002, 24(3): 343–346.

Zheng Yingren, Zhao Shangyi. *Chinese Journal of Rock Mechanics and Engineering*, 2004, 23(19): 3381–3388.

Zhou Guiyun. Finite element iterative method for slope stability safety factor based on strength reduction [J]. *Progress in Science and Technology of Water Resources and Hydropower*, 2010, 30(3): 5861.

Advances in Civil Engineering and Environmental Engineering – bin Yahya & Bin Daud (Eds)
© 2023 The Author(s), ISBN: 978-1-032-39397-1

Study on deformation characteristics of dam rockfill materials in a hydropower station

Xi Lu, Wei Wang* & Jiayuan Wang
Power China Northwest Engineering Corporation Limited, Xi'an, Shaanxi, China

ABSTRACT: The analysis of deformation characteristics of dam materials is an unavoidable technical difficulty in the engineering design of CFRDs (concrete face rockfill dams). Based on the compression test results of five kinds of rockfill materials with different lithologies in a large-scale hydropower station, this paper analyzes the porosity, breakage, immersion deformation and compression modulus of rockfill materials, the deformation characteristics and rules of different dam materials are summarized, which can be used for reference in the design of dam materials for similar CFRDs.

1 GENERAL INSTRUCTIONS

The main dam types of water retaining structures in hydropower projects include concrete dams and earth-rock fill dams, and the CFRD is one of the earth-rock fill dam types, it is widely used in dam engineering because of its good adaptability to topographic and geological conditions. Its characteristic is that the deformation problem of dam rockfill granular material is prominent (Gao 2002; Li 2007), which directly determines the feasibility and economy of a CFRD project. There is a lot of literature on the study of dam materials. For example, Lan Daoyin et al. (Lan 2008) have systematically studied different dam materials based on Shuibuya Dam engineering, the contents of the test include particle size test, compaction test, compression test, triaxial shear test, permeability test. Tang Hongjie et al. (Tang 2019) pointed out that there is a certain difference in deformation law between gravel and blasting material, for the mixed dam, it is important to pay attention to the deformation coordination. Li Weishu and Wei Wei et al. (Li 2021; Wei 2012) have carried out an experimental study on soft rock such as sandy mudstone and sandy slate, which provides valuable information for designing and constructing a dam in soft rock. Wei Ran et al. (Wei 2020) studied the scale effect of deformation characteristics of dam materials, and revealed the influence rule and internal mechanism of 5 external factors, such as lateral limit condition, dry density, edge breakage and load grade, on the scale effect. Han Huaqiang and Zhu Sheng et al. (Gao 2019; Han 2015; Zhu 2010) have done a lot of research on the dynamic characteristics of dam materials, which provides a reliable theoretical basis for the further study of super-high earth-rock dams in meizoseismal areas.

To sum up, the analysis of dam material characteristics based on the experimental study is an important research content in the design of CFRDs. In this paper, combined with the compression tests of rock-fill and sand-gravel materials of a CFRD of a hydropower station, proceeding from a unique angle, the compression deformation characteristics of different dam materials (single lithology, mixed lithology, and sand-gravel) are analyzed and studied.

*Corresponding Author: wwtiger@126.com

DOI 10.1201/9781003349563-26

2 COMPRESSION TEST OF DAM MATERIALS

A hydropower station is a large-scale hydro-junction project which mainly produces electricity. The water-retaining structure is a CFRD, the maximum dam height is 211m, the reservoir capacity is 1.26 billion m^3, and the installed capacity of the power station is 1500MW. The main dam rockfill materials of the face-slab dam are sand stone, monzonite, and natural sand-gravel. The mechanical properties of monzonite and sand stone are basically the same, compact, and hard, uniaxial saturation compressive strength 60~177MPa, softening coefficient 0.8. The parent rock of natural sand-gravel material mainly consists of medium-fine sand stone, slate and a small amount of granite. The maximum grain size is 150mm and the needle-like content is high, up to 30%.

Compression tests were carried out on five main rockfill materials to be used, namely, weakly weathered sand stone, weakly weathered monzonite, weakly weathered sand stone and monzonite (1:1 mixed ratio), weakly weathered sand stone and monzonite (3:7 mixed ratio), and natural sand-gravel, the study on compression deformation characteristics of dam materials is carried out. Before the compression test, the maximum dry density of the five kinds of dam materials was carried out to determine the sample-making density. The results of the maximum dry density that can be achieved in the laboratory are shown in Table 1.

Table 1. Maximum dry density test results of various dam materials.

No.	Main rockfill materials	Specific gravity	Density (g/cm^3)
1	Weakly weathered sand stone	2.78	2.25
2	Weakly weathered monzonite	2.70	2.19
3	Sand stone and monzonite 1:1 mixed	2.74	2.22
4	Sand stone and monzonite 3:7 mixed	2.72	2.21
5	Natural sand-gravel	2.71	2.34

The size of the compression test sample is Φ450 × 300 mm. Except for the sand-gravel material, which is made according to the relative density of 0.98, the rest of the rockfill material is made according to the maximum dry density in Table 1. The samples are in two conditions of dry and saturation, the vertical load was applied in seven grades: 100, 200, 400, 800, 1600, 3200 and 6000 kPa. The test results of various materials are shown in Tables 2–6.

Table 2. Compression test parameters of weakly weathered sand stone.

Vertical load	Void ratio e		Compression modulus E_s (MPa×10^2)		Compressibility coefficient α_v ($MPa^{-1} \times 10^{-2}$)		Volume compression coefficient m_v ($MPa^{-1} \times 10^{-2}$)	
P (kPa)	Dry	Saturation	Dry	Saturation	Dry	Saturation	Dry	Saturation
0.0	0.235556	0.235556						
105.5	0.234444	0.234136	1.037	0.866	1.191	1.427	0.964	1.155
203.0	0.233496	0.232837	1.350	0.965	0.915	1.280	0.741	1.036
401.9	0.232041	0.230784	1.706	1.209	0.724	1.022	0.586	0.827
799.7	0.229899	0.228005	2.283	1.769	0.541	0.698	0.438	0.565
1599.4	0.226522	0.223656	2.926	2.272	0.422	0.544	0.342	0.440
3202.8	0.216638	0.206471	2.004	1.151	0.616	1.073	0.499	0.869
5999.6	0.192393	0.171677	1.425	0.994	0.867	1.243	0.702	1.006

Table 3. Compression test parameters of weakly weathered monzonite.

Vertical load P (kPa)	Void ratio e		Compression modulus E_s (MPa×10^2)		Compressibility coefficient α_v (MPa^{-1}×10^{-2})		Volume compression coefficient m_v (MPa^{-1}×10^{-2})	
	Dry	Saturation	Dry	Saturation	Dry	Saturation	Dry	Saturation
0.0	0.232877	0.232877						
105.5	0.231726	0.231452	1.131	0.878	1.090	1.404	0.884	0.884
203.0	0.230904	0.230301	1.461	1.087	0.844	1.134	0.684	0.684
401.9	0.229616	0.228685	1.904	1.517	0.647	0.813	0.525	0.525
799.7	0.227726	0.226384	2.594	2.131	0.475	0.579	0.385	0.385
1599.4	0.224658	0.222521	3.213	2.552	0.384	0.483	0.311	0.311
3202.8	0.215534	0.207397	2.167	1.305	0.569	0.944	0.462	0.462
5999.6	0.192712	0.177288	1.511	1.146	0.816	1.076	0.662	0.662

Table 4. Compression test parameters of sand stone and monzonite (1:1 mixed ratio).

Vertical load P (kPa)	Void ratio e		Compression modulus E_s (MPa×10^2)		Compressibility coefficient α_v (MPa^{-1}×10^{-2})		Volume compression coefficient m_v (MPa^{-1}×10^{-2})	
	Dry	Saturation	Dry	Saturation	Dry	Saturation	Dry	Saturation
0.0	0.234234	0.234234						
105.5	0.232973	0.232685	0.973	0.792	1.269	1.558	1.028	1.262
203.0	0.231999	0.231409	1.286	0.982	0.959	1.257	0.777	1.018
401.9	0.230641	0.229626	1.827	1.391	0.676	0.887	0.547	0.719
799.7	0.228721	0.227254	2.570	2.070	0.480	0.596	0.389	0.483
1599.4	0.225595	0.223277	3.149	2.482	0.392	0.497	0.318	0.403
3202.8	0.215830	0.207177	2.024	1.228	0.610	1.005	0.494	0.815
5999.6	0.193038	0.171055	1.516	0.956	0.814	1.291	0.660	1.046

Table 5. Compression test parameters of sand stone and monzonite (3:7 mixed ratio).

Vertical load P (kPa)	Void ratio e		Compression modulus E_s (MPa×10^2)		Compressibility coefficient α_v (MPa^{-1}×10^{-2})		Volume compression coefficient m_v (MPa^{-1}×10^{-2})	
	Dry	Saturation	Dry	Saturation	Dry	Saturation	Dry	Saturation
0.0	0.230769	0.230769						
105.5	0.229634	0.229306	1.078	0.837	1.141	1.471	0.927	1.195
203.0	0.228759	0.228034	1.427	0.982	0.862	1.253	0.701	1.018
401.9	0.227405	0.226297	1.845	1.424	0.667	0.864	0.542	0.702
799.7	0.225491	0.223863	2.557	2.011	0.481	0.612	0.391	0.497
1599.4	0.222386	0.219966	3.171	2.525	0.388	0.487	0.315	0.396
3202.8	0.213374	0.203747	2.184	1.215	0.563	1.013	0.458	0.823
5999.6	0.191221	0.169573	1.555	1.008	0.792	1.221	0.643	0.992

Table 6. Compression test parameters of natural sand-gravel.

Vertical load P (kPa)	Void ratio e		Compression modulus E_s (MPa$\times 10^2$)		Compressibility coefficient α_v (MPa$^{-1}\times 10^{-2}$)		Volume compression coefficient m_v (MPa$^{-1}\times 10^{-2}$)	
	Dry	Saturation	Dry	Saturation	Dry	Saturation	Dry	Saturation
0.0	0.183406	0.183406						
105.5	0.182540	0.182184	1.359	0.963	0.871	1.229	0.736	1.038
203.0	0.181986	0.181389	2.125	1.481	0.557	0.799	0.471	0.675
401.9	0.181091	0.180261	2.658	2.150	0.445	0.550	0.376	0.465
799.7	0.179543	0.178415	3.057	2.537	0.387	0.466	0.327	0.394
1599.4	0.176519	0.174791	3.129	2.611	0.378	0.453	0.320	0.383
3202.8	0.170000	0.166671	2.911	2.337	0.407	0.506	0.344	0.428
5999.6	0.159052	0.152945	3.021	2.411	0.392	0.491	0.331	0.415

3 ANALYSIS OF COMPRESSIBILITY OF DAM MATERIALS

Referring to above mentioned Tables 2~6, the compression coefficients of the five kinds of dam dry materials are between 0.00378~0.01269 MPa^{-1}, and that of the saturated materials are between 0.00453~0.01558 MPa^{-1}. They are all less than 0.1 MPa^{-1} and belong to low compressibility materials. The compressibility is related to the lithologic strength of the material, the density of rockfill and the environmental conditions.

3.1 *Comparative analysis of porosity of dam materials*

Porosity is an index to reflect the compactness of dam filling. The porosity of a dam increases with increase of the compression deformation. According to the test results, the relation curves of pressure and porosity of various dam materials are shown in Figure 1.

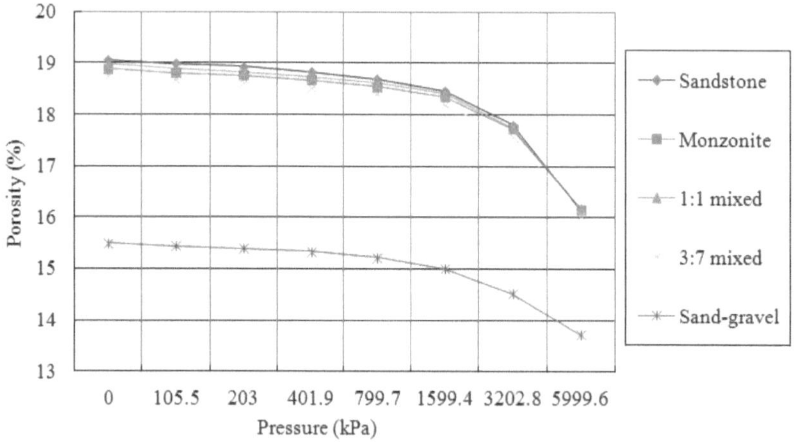

Figure 1. Relation curves of pressure and porosity of various dam materials.

Referring to Figure 1, the porosity of sand-gravel material is the smallest, ranging from 0.137 to 0.155, while the porosity of the other four dam materials has little difference, ranging from

0.161 to 0.191, but the void ratio between the other four kinds of dam materials and sand-gravel material is quite different, and the minimum difference value is more than 0.023. It shows that the compression deformation of sand-gravel material is smaller than that of other four kinds of dam materials, and the dam settlement of sand-gravel material is smaller than that of other four kinds of dam materials.

In addition, when the pressure is less than 1.6MPa, the slope of several curves is basically the same and gentler. When the pressure is more than 1.6MPa, the slope of several curves becomes steeper and the slope increases, but in contrast, the change range of sand-gravel is the smallest, about 0.01, the porosity of the other four dam materials decreased by 0.02. This phenomenon shows that, firstly, for extra-high dams above 200m, the sand-gravel material has the obvious advantage of less denaturation, and secondly, for high dams, with the rise of the fill height and the water level in front of the dam, the compression deformation of sand-gravel material is smaller than that of other four kinds of dam materials, and the late compressibility of sand-gravel material is smaller than that of rubble stone material. According to the test results of porosity, as the rock filling materials of High Dams, sand-gravel material is the most suitable, and there is no great difference between other dam materials, but there is a certain gap between sand-gravel material.

3.2 *Breakage analysis of various dam materials*

Under the action of self-weight or water load, rockfill will have a certain degree of breakage, rockfill breakage will cause compression deformation, the greater the amount of breakage, the greater the deformation. For this characteristic of rockfill, it can be reflected by the volume compression coefficient. The relation curves of pressure and volume compression coefficient of various dam materials are shown in Figure 2.

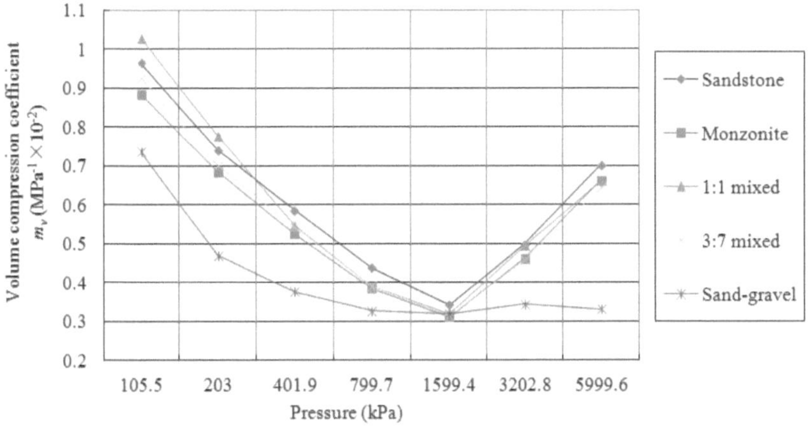

Figure 2. Relation curves of pressure and volume compression coefficient of various dam materials.

As can be seen from Figure 2. When the pressure is less than 1.6Mpa, the volume compression coefficient decreases gradually with the increase of the pressure, and there is no sudden change, indicating that the dam material particles have not changed greatly at this time the dam material has not broken. However, 1.6Mpa is a turning point. When the pressure is greater than 1.6Mpa, the curve begins to turn around and reverse, especially for the other four dam materials except sand-gravel, the curve is always upward and the slope is large, the results show that the particle breakage of the four dam materials is serious under high pressure, but the curve of sand-gravel material rises slightly between 1.6Mpa and 3.2Mpa, and then begins to decline, there is particle breakage, but because of the small warpage, the degree of breakage is less, after the pressure, the curve is flat, the particle breakage is not obvious. In theory, the breaking rate of sand-gravel material should be very low, but because of the high needle-like content of sand-gravel material in this project, the needle-like gravel is easy to break, which causes the particle breaking phenomenon of sand-gravel material under great pressure.

3.3 *Analysis of water-immersion deformation characteristics of various dam materials*

The experimental results show that the porosity of the five dam materials in the saturation condition under the same pressure is smaller than that in the dry condition, which indicates that the particle system of rockfill is more closely arranged and the rockfill body is more compact under the saturation condition, it also shows that water saturation is more favorable for dam filling. For accuracy explanation of the change of porosity after water immersion, the difference between dry and saturation porosity of five dam materials is analyzed, refer to Figure 3.

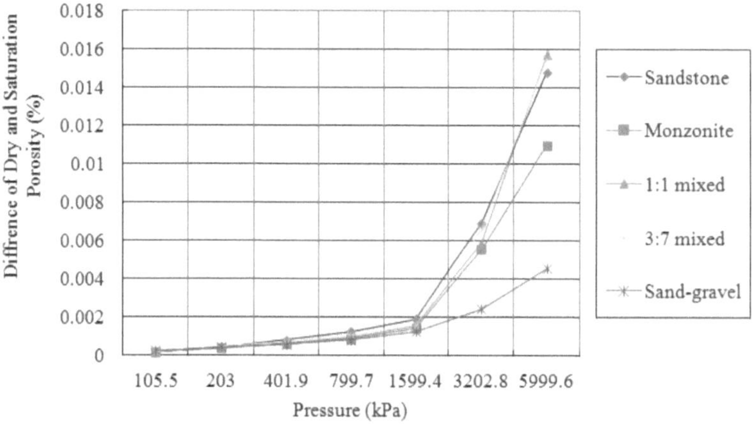

Figure 3. Relation curves of pressure and difference between dry and saturation porosity of various dam materials.

From Figure 3, it can be seen that when the pressure of these dam materials is less than 1.6MPa, the increment of compression deformation of saturation condition and dry condition is not much different, but when the pressure is more than 1.6MPa, the difference between sand-gravel material and other four materials opens up, the curve of sand-gravel material is slightly raised, while the curve of other four materials is steeper and the slope is larger, which shows that the increment of compression deformation of sand-gravel material is smaller after water-immersion, and that of other rubble stone material is larger after water-immersion.

In addition, the volume compression coefficient of the five dam materials under the saturation condition under the same pressure is greater than that under the dry condition, which indicates that the rockfill under the saturation condition has further compression settlement, the dam body has a process of aggravation of settlement, which also provides a basis for some CFRDs to make the bottom rock fill soaked in water in the early stage. The difference of volume compression coefficient between saturated and dry materials can show the settlement of dam materials under saturated and dry conditions, refer to Figure 4.

As can be seen from Figure 4, when the pressure is less than 1.6MPa, with the increase of the pressure, the difference curve slope of the volume compression coefficient between the saturated and the dry materials has little difference, the results show that the deformation increment of these materials is similar with the change of pressure, but the separation degree of the curve is larger, which shows that the deformation increment of different dam materials is different, and the increment of sand-gravel material is the smallest under the same pressure condition. When the pressure is greater than 1.6MPa, the curves of four kinds of rubble stone materials rise rapidly and the slope increases, which shows that the saturated material has a larger deformation increment than the dry material, and the moisture in the dam material replaces the good lubrication effect, it

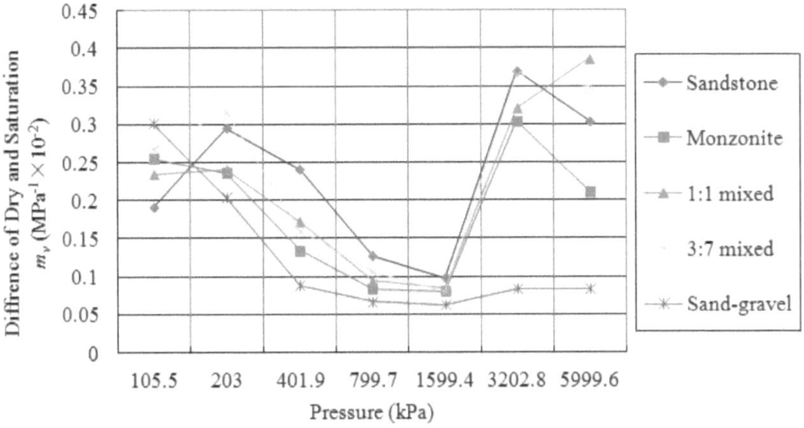

Figure 4. Relation curves of pressure and difference between dry and saturation volume compression coefficient of various dam materials.

also shows the advantage of adding water in the process of rockfill compacting, and the particle breakage of rockfill under high pressure, the degree of particle rearrangement and combination becomes larger, which aggravates the settlement deformation. The curve of sand-gravel is flat, which means that the deformation increment is small when filling water, and the particle breakage is better, so the degree of particle rearrangement is not very big, so the sedimentation increment is small.

3.4 *Analysis of compressive modulus of various dam materials*

In this paper, the compressive modulus of rockfill materials is analyzed from the relation between compressive modulus and compressive stress, environmental conditions and so on. Of course, the compression modulus is related to the dry density and the shape of river valley. The dry density of our dam materials is basically the same, so we do not study it from this aspect. The compressive modulus of rockfill body changes with the change of pressure value. The relationship curves of pressure and compressive modulus of various dam materials are shown in Figure 5.

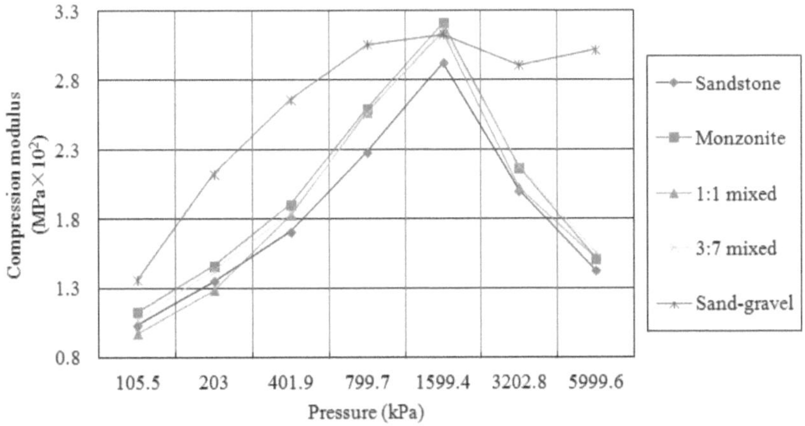

Figure 5. Relation curves of pressure and compression modulus of various dam materials.

It is obvious from Figure 5 that the peak values of compression modulus of the five kinds of dam materials are all at the pressure of 1.6MPa, and the peak values are not much different. When the pressure is less than 1.6Mpa, the compressive modulus of the five dam materials is basically the same as that of the other four dam materials, but the compressive modulus of sand-gravel is much higher than that of the other four dam materials, and when the pressure reaches 1.6Mpa, the compressive modulus of sand-gravel is much higher than that of the other four dam materials, the compressive modulus of the dam materials approaches the peak value. When the pressure is more than 1.6Mpa, the curve of sand-gravel deviates completely from that of the other four kinds of dam materials. The curve of sand-gravel material turns down slightly and rises gradually after 3.2Mpa, which indicates that the later settlement will be very small. But the other four kinds of dam materials turn down at a steep slope and extend all the time, and there is no reverse upward trend, which indicates that the deformation is larger in the later stage. The settlement of rockfill is very important in the later stage, which will have a far-reaching influence on the stress-strain of concrete face slab and the deformation of peripheral joints. From this point of view, we hope that the later deformation as small as possible.

Under the same other conditions, the compression modulus of saturated and dry rockfill is different. The compression modulus under saturation condition is lower than that under dry condition, which is also verified in the test results. Figure 6 shows the pressure and compression modulus difference curves of various rockfill materials.

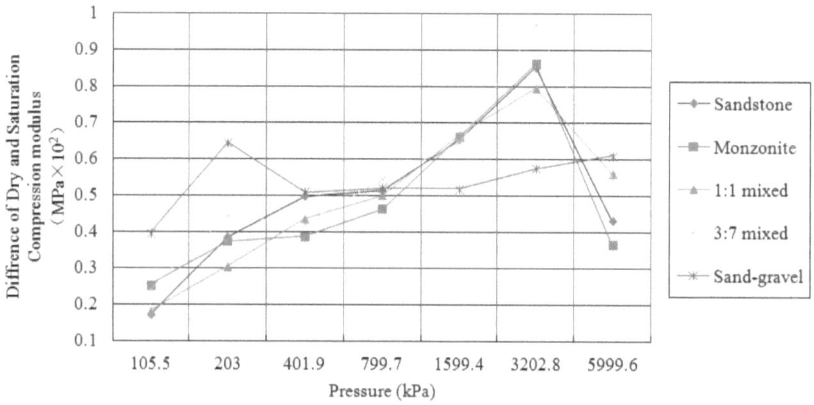

Figure 6. Relation curves of pressure and difference between dry and saturation compression modulus of various dam materials.

As can be seen from Figure 6, for sand-gravel, the difference of compression modulus changes obviously with the increase of pressure, but tends to smooth very quickly. When the pressure is greater than 0.4MPa, the difference of compression modulus changes very little, and the later change is more stable. This shows that the deformation of the dam material will be much smaller once it is saturated. And the pressure required to complete the saturation deformation is not large. For other dam materials, when the pressure is less than 3.2 MPa, the modulus difference shows an upward trend, and when the pressure is more than 3.2 MPa, it shows a downward trend. The curve changes sharply with the increase of pressure, and the range of change is also wide, especially in the later stage, the settlement of rockfill material has a greater change. It also provides some basis for us to soak the rockfill body in water before casting the concrete face slab. In this respect, the sand-gravel material still has a big advantage, but generally put the sand-gravel material in the dry area, that is from the seepage stability and drainage, but from the compression modulus aspect, its characteristics are better than rubble stone.

4 CONCLUSIONS

According to the dam material of a hydropower station in this paper, the following conclusions can be drawn from the analysis of the test results.

(1) No matter it is rubble stone material or sand-gravel material, they are all low compressibility material, and can be used as rockfill materials of dam body of CFRDs.
(2) The vertical load of 1.6MPa is the turning point of the engineering characteristics of various dam materials. When the vertical load of dam materials is greater than 1.6 MPa, the compression characteristics of dam materials will change. It is proved that the deformation characteristics of high dams are completely different from those of low dams.
(3) The particle of rubble stone is easy to break under high pressure, the deformation after water-immersion is large, namely the rheology is large, but the particle of sand-gravel is not easy to break, and the deformation after water-immersion is small, namely the rheology is small, therefore, the deformation characteristics of sand-gravel in the hydropower station are better than that of rubble stone.

REFERENCES

Gao Deqing (2019). Test and particle flow simulation of dynamic deformation characteristic of dam materials [J]. *China Earthquake Engineering Journal*, 41(06):1615–1622.

Gao Lianshi, Song Wenjing, Zhang Zongliang, et al. (2002). 3-D feedback analysis based on the observational deformation of concrete face rockfill dam in Tianshengqiao Project [J]. *Journal of Hydraulic Engineering*, (03): 26–31.

Han Huaqiang, Chen Shengshui, Fu Hua, et al. (2015). Effect of previous cyclic loading on deformation characteristics of dam materials [J]. *Chinese Journal of Geotechnical Engineering*, 2015, 37(02):263–268.

Lan Daoyin, Huang Bin, Li Hongli (2008). Statistical analysis of engineering characteristics of Shuibuya CFRD [J]. *Yangtze River*, (06):71–73.

Li Nenghui (2007). New concept of design for high concrete face rockfill dams [J]. *Chinese Journal of Geotechnical Engineering*, 29(8): 1143–1150.

Li Weishu, Zhang Feng, Zhi Keting (2021). Experimental study on the influence of soft rock on compressive modulus of filling materials of rockfill dam [J]. *Chinese Journal of Underground Space and Engineering*, 17(S2):703–709.

Tang Hongjie, Yang Zhengquan (2019). Understanding and thinking of material characteristics of high earth-rock dam [J]. *Water Resources Planning and Design*, (01):125–129.

Wei Wei, Shen Junhui, Miao Chao, et al. (2012). Feasibility analysis on weathered rock of a hydropower as rock-fill dam materials [J]. *Yellow River*, 34(07):127–129.

Wei Yan, Wu Shuaifeng, Wang Xiaogang, et al. (2020). Theoretical basis and application verification of scale effects of deformation characteristics of rockfill [J]. *Chinese Journal of Geotechnical Engineering*, 42(S1):161–166+213.

Zhu Cheng, Zhou Jianbo (2010). Deformation behaviour of coarse grained materials under cyclic loading [J]. *Rock and Soil Mechanics*, 31(05):1375–1380.

Advances in Civil Engineering and Environmental Engineering – bin Yahya & Bin Daud (Eds)
© 2023 The Author(s), ISBN: 978-1-032-39397-1

Study on the seismic performance of rusted stirrup concrete columns strengthened with externally wrapped shaped steel

Qiang Li* & Diwen Chang
Department of Architecture and Civil Engineering, Xi'an University of Science and Technology, Xi'an, China

ABSTRACT: To investigate the effect of strengthened externally wrapped shaped steel reinforcement on the seismic performance of rusted stirrup reinforced concrete columns, eight concrete columns were designed and fabricated, and accelerated rusting was used to obtain six rusted concrete columns. By controlling the stirrup spacing, corrosion rate, and material bonding method, the effect of strengthened externally wrapped shaped steel on the seismic performance improvement under different conditions was investigated. After a low reversed cyclic loading test, it was concluded that the different binders had almost no effect on the seismic performance, that the reinforcement effect was better for specimens with small stirrup spacing than for specimens with large stirrup spacing. The maximum ductility improvement of reinforcement with externally wrapped shaped steel was about 100% and that the energy consumption capacity was about 55% higher. The strengthened with externally wrapped shaped steel is more effective and can provide a reference role for improving the seismic performance of rusted reinforced concrete structures.

1 INTRODUCTION

The problem of durability of concrete structures has been the focus of research, as early as 1991, Professor Mehta pointed out that the causes of damage to concrete structures in increasing order of importance are: physical and chemical effects of aggressive environments, frost damage in cold regions, and reinforcement corrosion (P.K. Mehia 1991). Since then, scholars at home and abroad have been studying the durability of reinforced concrete structures, and the industry has agreed that the main cause of durability damage of reinforced concrete is caused by reinforcement corrosion (Dunker 1993; Wang 2000; Zhou 2019). The corrosion of reinforcing steel produces rust pits and products such that rust not only reduces its cross-section, but also its mechanical properties, which leads to rust swelling or cracking or even loosening of the concrete protection layer, which not only seriously affects the normal use of concrete structures but also greatly reduces the bearing capacity of concrete structures (G.C. Hoff 1999; Li 2015; Xin 2019). Many research results on concrete damage factors also show that the corrosion of reinforcement is the primary factor (Ting 1991; Zheng 2019). The seismic performance of the structure is reduced after rusting, and the study of the seismic performance of rusted reinforced concrete reinforced columns is one of the mainstream contents of seismic strengthening research in recent years. In this paper, low reversed cyclic loading tests were conducted on columns strengthened with externally wrapped shaped steel to study the effect of the strengthened with externally wrapped shaped steel method on the seismic performance.

2 TEST OVERVIEW

2.1 *Test specimen design and fabrication*

In this test, eight concrete specimens were designed and fabricated, considering two variables: stirrup spacing and maximum mass loss rate of rusted stirrup, of which six specimens were rusted

*Corresponding Author: liguanqiang2005@126.com

DOI 10.1201/9781003349563-27

concrete columns and two specimens were uncorroded comparison concrete columns. The concrete columns were reinforced by wrapping steel sections with a width of 50 mm and setting a batten plate at the root of the column and the foundation, with epoxy resin bonding and cementitious grout bonding. The concrete column size is shown in Figure 1, and the grouping table of each test column and reinforcement method is shown in Table 1.

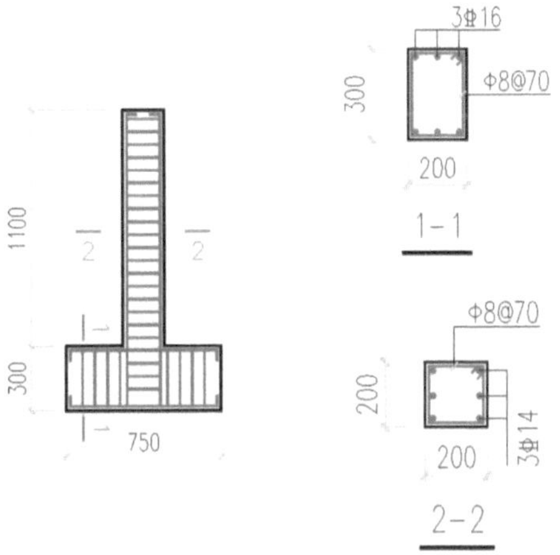

Figure 1. Specimens design (size: mm).

Table 1. Concrete specimen parameters table.

Specimen number	Stirrup spacing	Corrosion rate	Bonding method	Specimen number	Stirrup spacing	Corrosion rate	Bonding method
J-1	70	45%	Epoxy resin adhesive	J-5	90	45%	Epoxy resin adhesive
J-2	70	45%	Cement-based grout	J-6	90	45%	Cement-based grout
J-3	90	30%	Epoxy resin adhesive	R-1	70	0	
J-4	90	30%	Cement-based grout	R-2	90	0	

The test was conducted under natural conditions. To speed up the test process and improve the rusting effect when rusting by electricity, wet salt sand was mixed in place of salt solution to speed up the rusting of the reinforcing bars based on the close natural state.

First, four reinforcement cages with 70 mm stirrup spacing and five with 90 mm stirrup spacing were tied, and the insulating tape was pasted at the intersection of longitudinal bars and stirrup in the test concrete column to ensure that only the stirrup would be rusted. The strain gauges are wrapped tightly with epoxy resin glue after the paste is applied, and the concrete is ready to be molded, poured, pounded, and maintained.

2.2 *Accelerated corrosion test for stirrup reinforcement*

To meet the corrosion rate required for the experimental design, the current and duration of the test were determined according to Faraday's law calculation and references (Zeng 2009). 500 $\mu A/cm^2$ of current density was adopted for a single stirrup, and the specimens were placed in a 5% sodium chloride solution for corrosion after energization was completed, with each specimen connected in parallel, the positive pole connections to a DC power source and the negative pole connections a copper plate. After the corrosion of each specimen is completed, the specimens are dried naturally. Then the protective layer is chiseled and the stirrup strain gauges are pasted.

After the completion of the low reversed cyclic loading, the specimens are broken to determine the actual corrosion rate of each specimen. According to the method described in the specification (JGJ/T 101—2015), the actual corrosion rate was obtained, and the actual corrosion rate of stirrup reinforcement for each specimen is shown in Table 2.

Table 2. Corrosion rate of stirrups of each specimen.

Specimen number	Average weight loss rate/%	Maximum weight loss rate/%	Specimen number	Average weight loss rate/%	Maximum weight loss rate/%
J-1	29.01	50.10	J-5	29.71	44.51
J-2	30.49	49.34	J-6	22.65	39.42
J-3	18.93	26.72	R-1	0	0
J-4	16.27	25.17	R-2	0	0

2.3 *Low reversed cyclic loading test*

After the completion of the reinforcement of the externally wrapped shaped steel of the concrete column, the low reversed cyclic loading test was carried out, and the test equipment was as shown in the figure. Displacement-controlled loading was adopted First the axial force was loaded, keeping it constant after the axial force reached 200kN. Then the repeated load was applied, within 10 mm of displacement. 2 mm of each stage is reciprocated once. The displacement of 10 mm or more and 10 mm of each stage, with three cycles of each loading cycle, ended the test. The load was reduced to 70% of the peak load

Figure 2. The loading device.

3 TEST RESULTS AND ANALYSIS

3.1 *Specimen damage process and damage analysis*

As shown in Table 3, each reinforced specimen undergoes a period of plastic deformation after the yielding of the longitudinal reinforcement before the complete failure of the reinforcement material, showing bending damage characteristics, but eventually shear damage occurs due to the failure of the reinforcement material. The damage of the stirrup reinforcement of the specimen reinforced by the external wrapped steel is 1~2 fractures at the bottom, the concrete crushing damage in the core area and the shear strength decreases due to the external bulge or fracture of the angle steel, and finally, shear damage occurs. Due to the restraint of the concrete by the reinforcement material, the concrete protective layer was first crushed, and then the concrete crushed and squeezed the reinforcement material leading to its external drum, which eventually led to the fracture of the reinforcement material and the concrete falling off in large pieces. As the bonding material has failed in the damage stage, only relying on the outer cladding steel to provide restraint, the final damage form is consistent across different bonding methods.

Figure 3. Diagram of the test process of the externally wrapped shaped steel reinforcement specimen J-1.

Table 3. Failure modes of each specimen.

Specimen number	Failure mode	Damage of reinforcement materials	Stirrup failure	Failure of longitudinal reinforcement
J-1	Shear failure	The bottom anchor angle steel is broken on the east and west side	The 1st ~ 3rd stirrup at the bottom is fracture	Bending of 1 ~ 3 longitudinal bars at the bottom
J-2	Shear failure	Bottom anchor angle steel weld seam fracture	The 1st ~ 3rd stirrup at the bottom is fracture	Bending of 1 ~ 4 longitudinal bars at the bottom
J-3	Bending and shearing failure	The bottom anchor angle steel is broken on the east side	The 1st ~ 2nd stirrup at the bottom has bulged	Bending of 1 ~ 2 longitudinal bars at the bottom
J-4	Bending and shearing failure	The four corners of the bottom anchoring angle are broken, and the concrete connecting the specimen to the base is crushed	The first stirrup at the bottom is broken	Bending of 1 ~ 3 longitudinal bars at the bottom
J-5	Shear failure	Bottom anchor angle weld steel fracture at all four corners	The 1st ~ 2nd stirrup at the bottom is fracture	Bending of 1 ~ 4 longitudinal bars at the bottom
J-6	Shear failure	Bottom anchor angle weld steel fracture at all four corners	The 1st ~ 2nd stirrup at the bottom is fracture	Bending of 1 ~ 4 longitudinal bars at the bottom
R-1	Bending failure	——	Slight deformation	Bending of 1 ~ 3 longitudinal bars at the bottom
R-2	Bending failure	——	No fracture, no obvious deformation	stirrup at the bottom is yield

3.2 *Hysteresis curve*

The hysteresis curves can either be used to directly analyze the seismic performance of such specimens or to determine their restoring force models. The hysteresis curves of this test specimen are shown in Figure 4 after compiling the data collected in the test.

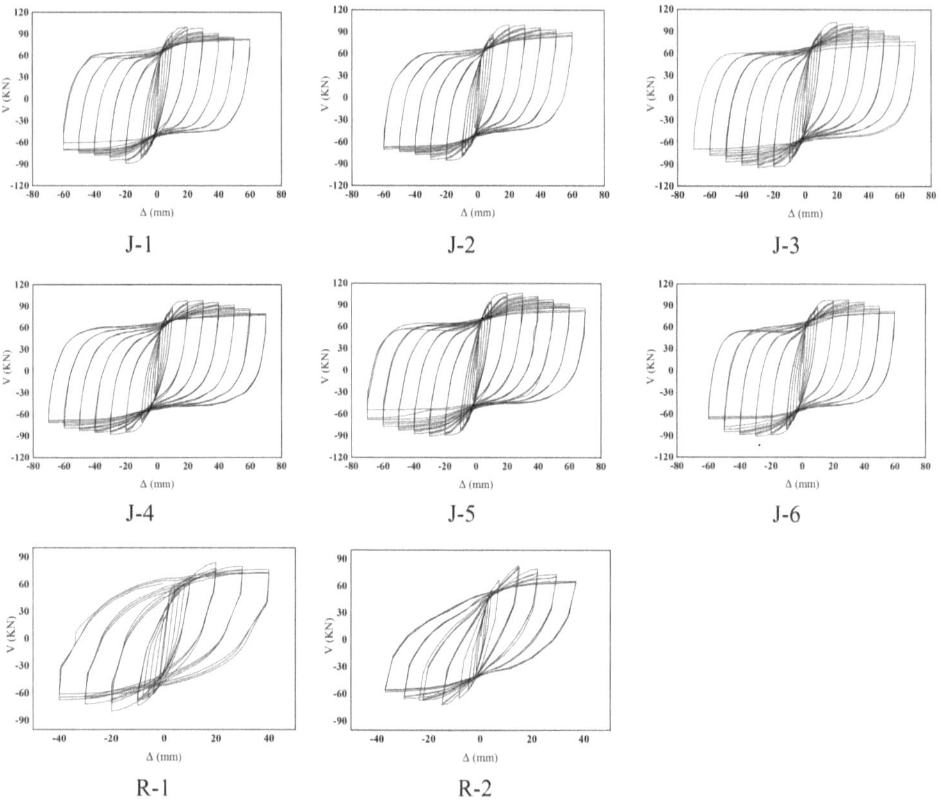

Figure 4. Hysteretic curve of each specimen.

The hysteresis rings of the reinforced specimens were full, and the hysteresis ring area was larger than that of the uncorroded specimens. The hysteresis loop area of the epoxy resin bonded reinforced specimens is larger than that of cementitious grout bonded specimens, and the greater the degree of corrosion, the greater the difference in the area. Reinforcement can effectively improve the ductility of rusted stirrup concrete columns. The hysteresis curves of moderately and severely corroded specimens were more pinched than those of uncorroded specimens, and the peak load under each displacement cycle of reinforced specimens was greater than that of uncorroded specimens.

3.3 *Load displacement skeleton curve*

The skeleton curve can reflect the seismic characteristics of the specimen, from which the direct influence of different parameters on the seismic performance of the specimen can be analyzed.

Reinforcement can restore the deformation characteristics of the specimen to be comparable with those of the intact specimen, and the ductility has been improved to a certain extent. The external clad steel reinforcement of moderately and heavily corroded specimens can significantly improve the deformation characteristics of corroded columns by about 29% as compared to intact specimens. The peak load of epoxy resin adhesive-bonded external clad steel reinforcement specimens is greater than that of cementitious grout bonding, and the smaller the degree of corrosion, the greater the peak load. The peak load of epoxy resin bonded specimens with moderate corrosion is about 8% higher than that of cementitious grout bonded specimens, and that of heavily corroded specimens is about 6% higher.

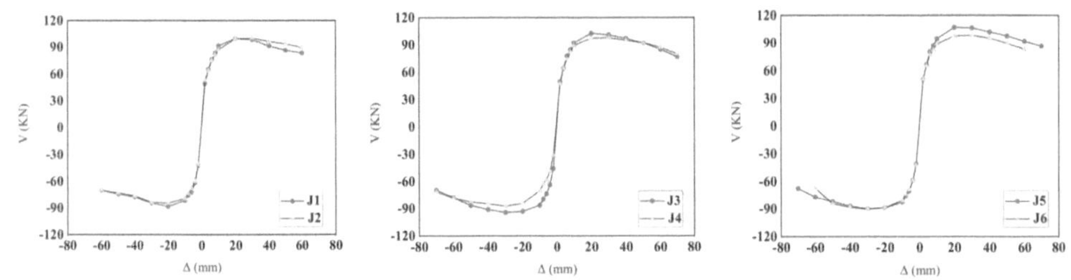

(a) Heavy corrosion, stirrup spacing 70 mm

(b)Moderate corrosion, stirrup spacing 70 mm

(c) Moderate corrosion, stirrup spacing 90 mm

Figure 5. Load displacement skeleton curve of each specimen.

3.4 *Ductility analysis*

The size of ductility directly reflects the form of damage to the member. This paper uses the ductility coefficient method to comprehensively analyze the impact and size of ductility, using the ratio of ultimate displacement to yield displacement to quantify ductility, and its calculation formula is as follows:

$$\mu = \frac{\Delta_u}{\Delta_y} \tag{1}$$

The relative deformation value is Δ / H , Where Δ is the lateral displacement of the column horizontal load loading point, H is the distance from the horizontal load loading point to the base on the bottom. The measured characteristic points of the specimens with different stirrup spacing are shown in Table 4.

Table 4. Specimen actual measurement characteristic point.

Specimen number	Ductility coefficient	Relative deformation	Specimen number	Ductility coefficient	Relative deformation
J1	12.92	1/21	J5	12.59	1/20
J2	16.66	1/19	J6	14.40	1/20
J3	14.92	1/18	R1	10.34	1/32
J4	13.59	1/17	R2	6.95	1/34

Analysis of the data is shown in Table 4. Reinforcement can significantly improve the ultimate deformation capacity and ductility of rusted reinforced concrete columns. The displacement ductility coefficients of this test specimen range from 6.95 to 16.66, all of which have good ductility, and the external clad steel reinforcement has a maximum improvement of about 100% for ductility. The stirrup corrosion rate is comparable, the smaller the stirrup spacing, the greater the ductility, and the greater the ultimate displacement angle. The stirrup spacing is the same, the bonding method is the same, and the greater the corrosion rate the smaller the ductility coefficient and the smaller the ultimate displacement angle. The specimens bonded with cement-based grout and the specimens bonded with epoxy resin adhesive both reach yield load at 4 mm displacement cycle and peak load at 20 mm displacement cycle and the epoxy resin adhesive bonded specimens with moderate corrosion are about 8% higher than the cement-based grout bonded peak load and the specimens with heavy corrosion are about 6% higher; at the ultimate load, the bonding materials between the reinforcement materials are disengaged from the work and the specimens are is restrained only by the reinforcing material itself.

3.5 *Strain analysis of reinforced materials*

It can be observed in the test that when the binder cracked, the reinforcement material failed early due to peel damage before the ultimate damage occurred. The measured strains of the reinforcing material are given in Table 5.

Table 5. Utilization of reinforcement materials.

Specimen number	1	2	3	4	5	6	Average value	Utilization rate
J1	6261	644	349	221	156	160	1299	55%
J2	6127	493	296	153	129	105	1217	55%
J3	8724	1262	369	290	242	164	1842	77%
J4	9998	795	388	378	256	116	1989	84%
J5	5518	878	438	336	332	272	1296	54%
J6	4873	2506	365	240	133	108	1371	58%

Note: 1-6 indicates the average value of six measurement points and two surface strains from bottom to top.

As shown in Table 5, the effective utilization rate of the outer clad steel is all above 50%. The strains from the root upward at each measurement point decreased significantly for the specimens reinforced with outsourced sections, exceeding the material limit strain value. This is mainly because as the test proceeds, the root concrete is crushed and extruded, resulting in extrusion of the root reinforcement material and damage such as tearing and fracture, so it is difficult to effectively provide tensile strength. The utilization rate of the two bonding methods for the specimens reinforced with external wrapped steel did not differ significantly, with the maximum difference being 7% higher for the cement-based grout bonded specimens with external wrapped steel for the moderately rusted specimens.

4 CONCLUSIONS

(1) Under the low circumferential reciprocal load, the reinforced specimen damage has a distinctly ductile nature, with shear damage dominating the damage pattern of the moderately and heavily corroded specimens. The damage was also more serious due to the larger corrosion rate of the original specimen.
(2) The external clad steel reinforcement can effectively improve the seismic resistance of the rusted column. According to the phenomenon of low reversed cyclic loading test and the hysteresis curve obtained from the test results, the specimens reinforced by clad steel have a good energy dissipation capacity with a maximum ductility increase of about 100% and an energy dissipation capacity increase of about 55%.
(3) By observing the low circumferential reciprocal load test phenomena and comparing the hysteresis curve, skeleton curve, ductility analysis, and material utilization rate, the moderately rusted specimens reinforced with epoxy resin adhesive bonded with clad steel are slightly better than cement-based grout bonding, and the epoxy resin adhesive bonded reinforcing material works better with the specimens, while for heavily rusted specimens The two bonding methods have almost no effect on the seismic performance.

ACKNOWLEDGEMENT

This project is financially supported by the Natural Science Basic Research Program of Shaanxi Province of China (2017JM5054).

REFERENCES

Dunker, Kenneth F, Rabbat, Basile. G. Why American bridges are crumbling [J]. *Scientific American*, March, 1993.

G.C. Hoff. Integrating durability into the design Process [J]. *International Congress on Creating with Concrete Dundee*, Scotland U. K.6-10, September 1999:101–114.

Lifumin, Dengtianci, Wangjianghao et.al Review of research on durability of prestressed concrete structures [J]. *Journal of Architecture and Civil Engineering*, 2015, 32(02):1–20.

P.K. Mehia. Durability of concrete-fifty years of Progress [A]. G.M. Idorn International Symposium [C], SP-126, *America Concrete Institute*, Detroit, 1991:1–3.

Shuenn-Chern Ting, Andrzej S. Nowak. Effect of reinforcing steel area loss on flexural behavior of reinforced concrete beams [J]. *ACI Structural Journal*, 1991, 88(3):309–314.

Specification for seismic test of buildings: JGJ/T 101-2015[S] Beijing: China Architecture & Building Press, 2015:6–9 (In Chinese)

Wangyuanli, Yaoyan. Research and engineering application of concrete durability in key projects [M]. *Beijing: China Architecture& Building Press*, 2000.

Xinjingzhou, Suxin, Zhoujianting. Review of resistance decay of corroded reinforced concrete structures (II): component level [J]. *Journal of Jiangsu University (Natural Science Edition)*, 2019, 40(6):719–726.

Zeng Yanhong, GU Xianglin, Zhang Weiping, et al. Accelerated corrosion technique for reinforcement steel bars in concrete [J]. *Structural Engineers*, 2009, 25, (01): 101–105.

Zhengyongbo, Baitinghui, Lixiaojun Influence of reinforcement corrosion on bending mechanical property of longitudinal joint in shanghai metro shield tunnel [J]. *Tunnel Construction*, 2019, 39(11):1823–1831.

Zhou jianting, Suxin, Xinjingzhou. Review of resistance decay of corroded reinforced concrete structures (I): material level [J] *Journal of Jiangsu University (Natural Science Edition)*, 2019, 40(4):458–464, 471.

Study on the mechanical characteristics of CRD reconstruction and expansion of oversized four-lane tunnel

Min Liu & Jiangling He
Guangdong Expressway Co., Ltd., Guangdong, China

Fang Wang*
School of Civil Engineering, Chongqing Jiaotong University, Chongqing, China

Xinghong Jiang
China Merchants Chongqing Traffic Research and Design Institute Co. LTD, Chongqing, China

Langping Xiong
School of Civil Engineering, Chongqing Jiaotong University, Chongqing, China

ABSTRACT: In view of the complicated process and low work efficiency of the traditional excavation method of the super large section four-lane tunnel, based on the project of the reconstruction and expansion of the large section four-lane deep Shanxi tunnel, the reconstruction and expansion construction method of the CRD of the super large section four-lane tunnel is studied by combining numerical simulation and theoretical analysis, and on this basis, the mechanical characteristics of the reconstruction and expansion construction of the CRD are further studied for the CRD construction method. The results show that the CRD method is applicable to the IV. and V. surrounding rock highway tunnels of the oversized four-lane tunnel; when the CRD method is used for the shallow buried surrounding rock of the V. class, the recommended upper step height is 3/5B tunnel height; before the demolition of the next door, the settlement of the left and right guide pit vaults in the tunnel excavation process is first increased, and then basically unchanged; when the next door is demolished, the sedimentation volume suddenly increases, and then the second lining is applied, and the settlement basically no longer changes. The research results can provide reference and reference for the reconstruction and expansion of the super-large cross-section tunnel.

1 GENERAL INSTRUCTIONS

At present, China's transportation demand is growing, road transportation capacity is becoming increasingly saturated, and the construction of super-large cross-section tunnels is imminent. Commonly used excavation methods are double-sided wall pit method, CRD method, CD method, etc., of which CRD method is also known as cross-in-the-middle next-door method, focusing on solving technical problems such as large span, ultra-shallow burial, CRD method is generally applied to three-lane and four-lane weak surrounding rock construction (Hou 2017; Shen 2012). Due to the difficulty of controlling the deformation of weak surrounding rock, the safety risk is high, and it is easy to cause safety accidents, so the study of the mechanical characteristics of CRD method construction is particularly important.

Some scholars at home and abroad have conducted certain research on the construction characteristics of the CRD method of large-section highway tunnels. Zhang Tiezhu (Zhang 2015) Based

*Corresponding Author: 3277909052@qq.com

 DOI 10.1201/9781003349563-28

on the on-site tunnel deformation and internal force monitoring data, the mechanical response of the CRD method of the single-span four-lane expressway tunnel in Dadongshan, Dalian was studied; Wu Mengjun et al. (Wu 2006) conducted similar model tests and numerical simulations on the four-lane extra-large section highway tunnel, and studied the dynamic construction mechanical behavior under different construction methods; Liu Cong et al. (Liu 2018) relied on the Beijing-Shanghai Expressway Jiganggou Tunnel. The mechanical evolution law of the construction process of the super large section tunnel crossing the fracture zone is studied by the model test method.

At present, there have been a lot of research on large-section highway tunnels, but the research direction mainly focuses on the optimization of excavation methods, safety control, etc. For the construction conditions of large cross-sectional four-lane tunnels, the research on the mechanical characteristics of CRD method construction lacks specificity and systematicness, which brings great challenges to the construction of similar projects. Based on the research results of predecessors, this paper relies on the reconstruction and expansion project of Shenzhen-Shanxi Tunnel to study the construction mechanical characteristics of CRD method of large-section highway tunnels, which provides a reference for subsequent engineering construction.

2 PROJECT OVERVIEW

Based on the double-hole eight-lane deep Shanxi Tunnel, the YK56+950~YK58+137 (ZK56+918~ZK58+105.28) section adopts the sunken tunnel scheme due to the ground planning and landscape requirements of the cooperation area, and the overall form of the tunnel is the underground excavation mountain tunnel + open-cut sunken tunnel, and the stratigraphic lithology of the tunnel site area is slope residual silty clay, alluvial silty clay, Jurassic sandstone and its weathering layer; the stratigraphic lithology of the dark excavation tunnel site is slope residual silty clay. Jurassic sandstone and its regolith. According to the regional geological data, mapping results and drilling results, combined with the work experience of similar projects nearby, the rock structure, rock strength and integrity of the tunnel are divided into V. grades.

3 MODEL BUILDING

In this paper, based on the finite difference software FLAC3D, the numerical model is established, and the planar strain numerical calculation method is used, as shown in Figure 1, according to the principle of Saint-Vinant, to reduce the influence of boundary conditions on the model, the distance between the left and right and lower boundaries of the model to the surrounding rock around the tunnel should be taken as 3 to 5 times the tunnel span. The constraints are: the top of the model is a free boundary without any constraints; the left and right borders are constrained by the horizontal displacement, and the bottom boundary of the model is constrained by the vertical direction displacement.

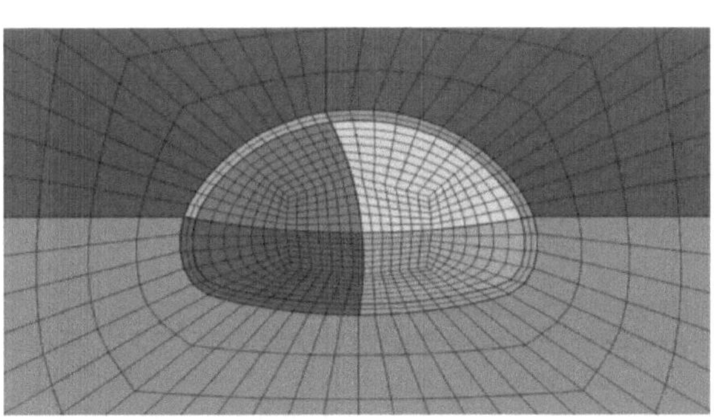

Figure 1. Model presentation.

Table 1. Variation curve of arch crown settlement with buried depth under different surrounding rocks by CRD method.

III level of surrounding rock

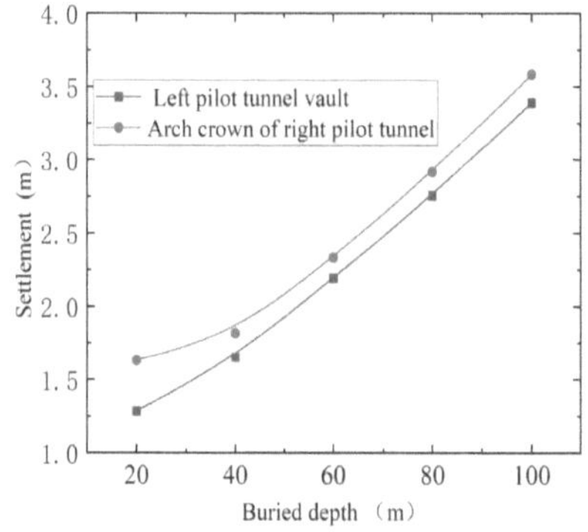

IV level of surrounding rock

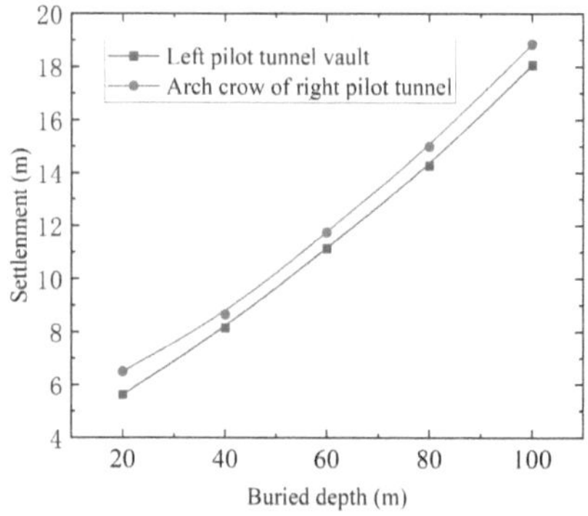

V level of surrounding rock

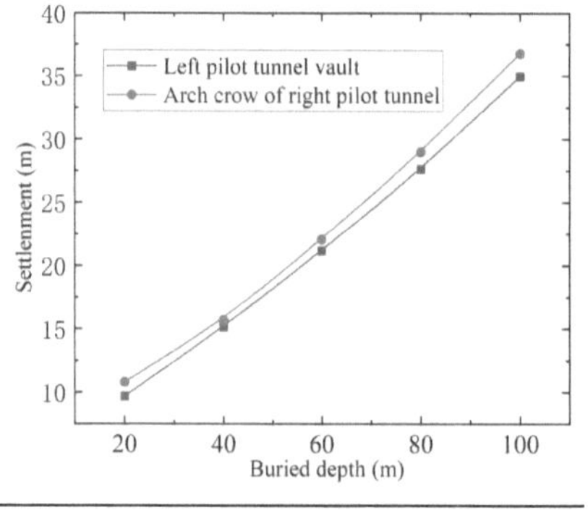

4 ANALYSIS OF RESULTS

4.1 *Stratigraphic suitability*

As shown in Table 1, under the condition of class III surrounding rock, the arch settlement increases with the increase of burial depth, and the maximum settlement value is about 3.8mm. Affected by the disturbance of the left guide hole excavated first, the settlement of the arch of the right guide hole excavated after the excavation is larger than that of the left guide hole, and the law is not related to the burial depth; it can be seen from Figure 2(b) and (c): the maximum vertical settlement after the tunnel excavation is located at the vault, and the arch settlement value increases with the increase of the burial depth, and the maximum settlement value under the surrounding rock conditions of class IV. And V. is about 19mm and 38mm, respectively, and the surrounding rock deformation of the settlement value at the arch is not large, which meets the engineering requirements.

4.2 *Step height analysis*

Relying on the hole shape and span of the Shenzhen-Shanxi Tunnel, the height of the pit steps on the three CRD methods was calculated, namely (1) h=1/3H (H is the tunnel excavation height), (2) h=1/2H, and (3) h=3/5H. Table and Table 3 were analyzed, and the step height was studied by comparing the two variables of surro2unding rock deformation and supporting stress.

When the height of the steps on the sidewall pit is reduced from 1/2 tunnel height to 1/3 tunnel height, the settlement of the left and right arches increases slightly, but the increase is not significant, and the horizontal displacement also increases or decreases, but the increase or decrease is small; when the height of the steps on the side wall guide pit increases from 1/2 tunnel height to 3/5 tunnel height, the settlement of the left and right arches is slightly reduced, the reduction is small, the horizontal displacement is increased, and the increase is not large; The height of the steps on the sidewall pit has a certain impact on the initial support spray concrete, and the specific law is that with the increase of the upper step height, the specific law is that with the increase of the upper step height, The initial supporting shotcrete stress of the ring increases, and the support stress of the middle wall also increases. In general, the increase in size does not affect the safety of support.

Table 2. Comparison of surrounding rock deformation under three working conditions.

	Crown settlement/mm		Horizontal displacement/mm				Plastic zone/m	
			Left pilot tunnel		Right pilot tunnel			
Working condition	Left vault	Right vault	Step up	Step down	Step up	Step down	Left wall	Right wall
1 h=1/3H	10.65	11.74	0.66	0.06	0.13	0.48	4	4
2 h=1/2H	10.40	11.71	0.32	0.70	0.46	0.42	5	5
3 h=3/5H	9.68	10.80	0.54	0.77	0.55	0.37	5~6	4

Note: The displacement values in the table are the total displacement values, including the displacements before they were unsupported.

Table 3. Comparison of support stresses for three operating conditions.

	Initial support stress/MPa								Middle partition/MPa	
	Left				Right					
Working condition	Cave top	Arch bottom	Middle of side wall	Wall bottom	tunnel vault	Arch bottom	Middle of side wall	Wall bottom	Up	Down
h=1/3H	1.71	2.72	3.75	3.71	1.07	0.97	1.72	1.81	0.92	2.79
h=1/2H	1.58	3.20	3.81	3.75	0.99	1.23	1.51	1.59	2.33	2.69
h=3/5H	2.06	3.43	3.50	3.50	1.02	1.38	1.37	1.46	2.83	2.11

4.3 *Construction parameterization*

For ease of comparative analysis, the horizontal displacement of the right monitoring point takes the absolute value, and the horizontal convergence indicates the relative horizontal displacement of the left and right monitoring points. As can be seen from Figure 2 and Figure 3, the horizontal displacement of the left and right monitoring points increases first and then decreases slowly before the next door is removed. The excavation of the left guide pit has a certain impact on the horizontal displacement of the right monitoring point; when the next door is demolished, the horizontal displacement of the left and right monitoring points is suddenly reduced, and after the second lining is applied, the horizontal convergence basically no longer changes. The maximum value of horizontal convergence (about 5.0 mm) occurs during tunnel excavation, rather than during the later application of the double lining.

Overall, before the demolition of the middle next door, during the tunnel excavation process, the settlement of the arch of the left and right guide pits first increased, and then basically unchanged. The vault settlement of the right guide pit is greater than the settlement of the arch of the left guide pit, because the excavation order is the left guide pit in front, the right guide pit in the back, the left guide pit in front of the right guide pit by 15m, and the settlement measurement point is 10m away from the edge of the model. Therefore, when the palm surface of the left guide pit reaches

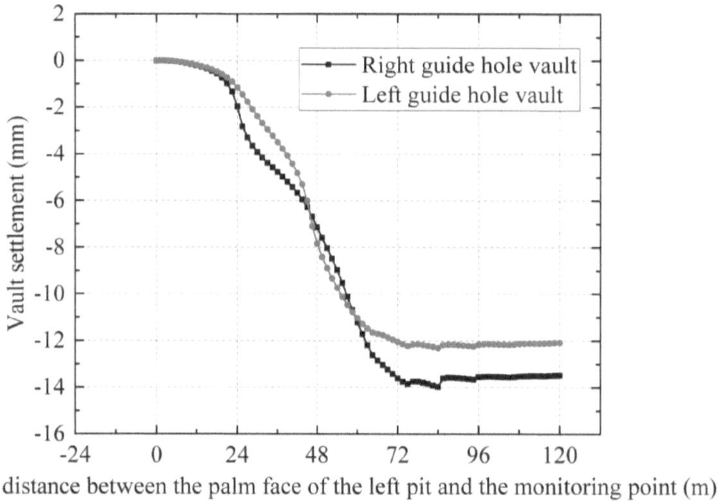

Figure 2. Settlement curve of arch crown of left and right heading of CRD construction method tunnel.

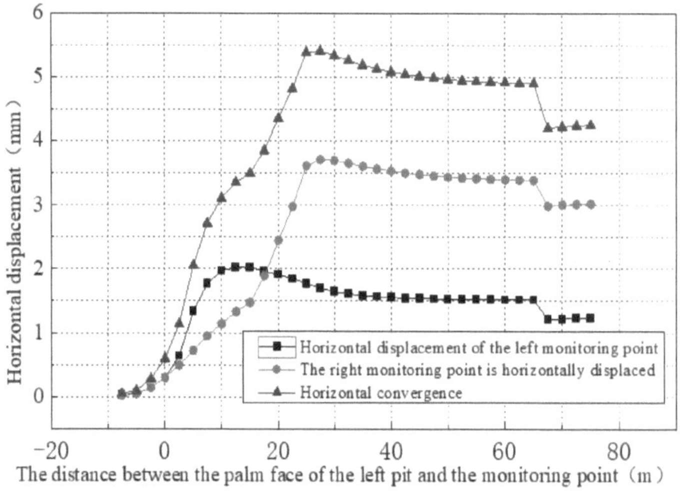

Figure 3. Horizontal displacement curve of CRD construction method tunnel.

the settlement measurement point, the right guide pit has not yet been excavated, so the initial settlement of the vault of the left guide pit is greater than the settlement of the vault of the right guide pit.

5 CONCLUSIONS

1) The results converge during the numerical calculation of the applicability of the formation, and the surrounding rock of the tunnel is in a stable state. According to empirical displacement criteria, the CRD method is applicable to oversized four-lane tunnels around rock road tunnels of Class IV and Class V.
2) Considering the mechanical effect, it should be recommended that the size of the sidewall guide pit be 1/3 span height or 3/5 span height, but from the construction space analysis, to facilitate the use of mechanized excavation and transportation tools, improve construction efficiency, cannot make the span of a single step too small. Theoretical analysis shows that the height of the three steps is not large, and the construction work space should be mainly considered, so when the CRD method is used for the V. shallow buried surrounding rock, the recommended upper step height is 3/5B tunnel height.
3) Before the demolition of the middle next door, during the tunnel excavation process, the settlement of the arch of the left and right guide pits is increased first, and then basically unchanged.
4) When the middle next door (abscissa coordinate 65.0m) is demolished, the sedimentation volume suddenly increases, and after the second lining is applied, the settlement basically no longer changes.

REFERENCES

Hou Fujin, Sun Keguo, Zhao Ran, et al. Applicability study on construction method of super-large section neighborhood tunnel [J]. *China Civil Engineering Journal*, 2017, 50(S1):111.

Liu Gong, Li Shucai, Zhou Zongqing, at al. Model test study on mechanical characteristic of surrounding rock during construction process of super-large section tunnel under complex strata [J]. *Rock and Soil Mechanics*, 2018, 39(9):3495.

Qu Haifeng. An overview on construction and research of super-large cross-section tunnels with low height-span ratios [J]. *Tunnel Construction*, 2009, 29(2):166.

Shen Lingjun. Construction scheme optimization and technology research of soft ground large section tunnel [D] *Changsha: Central South University*, 2012.

Wu Mengjun, Huang Lunhai. Research on dynamic construction mechanics of four-lane highway tunnel [J] *Chinese Journal of Rock Mechanic and Engineering*, 2006(S1):3057.

Zhang Junru, Wu Jie, Yan Congwen, et al. Construction technology of super-large section of highway tunnels with four or more lanes in China [J]. *China Journal of Highway and Transport*, 2020, 33:14.

Zhang Tiezhu. Analysis of mechanical response of four-lane small clear spacing highway tunnel with super-large cross-section [J]. *China Civil Engineering Journal*, 2015, 48(S1):302.

Advances in Civil Engineering and Environmental Engineering – bin Yahya & Bin Daud (Eds)
© 2023 The Author(s), ISBN: 978-1-032-39397-1

Design study on floor structure of large hole cantilever patio

Xinming Guo
CCCC Third Highway Engineering Bureau Co., Ltd., Binzhou, Shandong

Zhiyue You*
School of Civil Engineering, Hainan University, Haikou, Hainan

Weiwei Liao, Zhenyu Zhang & Quan Jin
CCCC Third Highway Engineering Bureau Co., Ltd., Binzhou, Shandong

ABSTRACT: The design and calculation of the suspended patio floor structure with a span of 25.2 m and an opening diameter of 20.8 m of Shandong Bohai advanced technology research institute are carried out. To study the stress law of the large opening structure of the floor slab, the finite element analysis of the mechanical indexes of the round hole diameter and the floor structure layer thickness is carried out by using ANSYS finite element software. The results show that the stress at the hole edge beam and support is larger when the opening is smaller, and the stress level at the hole edge beam and support decreases significantly with the increase of the opening diameter. Cross-sloped beams with large open edges were studied using steel or reinforced concrete beam-slab structures. The bearing capacity and deflection need to meet the design requirements.

1 INTRODUCTION

The Shandong Bohai Research Institute of Advanced Technology (Phase I) project is located in the Binzhou Economic and Technological Development Zone, which is a key project of Shandong Province. The project has 4 floors above ground and 1 floor underground. The construction area is about 80,000 square meters. The main height of the building is 25.1 m, and the whole building is in C shape. The project has a giant round hole of 20.8 m in diameter on the second floor of the core area of the building, the surrounding beams and plates are cantilevered inward, the vertical stressed components are steel and concrete composite columns, and the preliminary plan of the floor structure is steel structure or cast-in-place concrete structure. The structure of the steel structure floor scheme adopts the rigid connected H-type steel main and secondary beam system, and the steel plate material used for welding the H-type steel beam is Q355B, and the floor plate adopts the pressed steel plate-concrete combination floor plate. The concrete floor scheme is the main secondary beam floor structure, with concrete C30 and reinforcement HRB400. ANSYS is used to check the two schemes of the structure for practical design.

2 PLANE STRUCTURE LAYOUT AND CROSS-SECTION CHARACTERISTICS

Since the second floor of the building is a circular patio with a 20.8 m diameter, a single-story concrete steel tube column cantilever structure is adopted. Above the patio is a large-span giant steel structure truss (Figure 1), supported by 30 square concrete-filled steel tube columns (500 mm

*Corresponding Author: 1243262236@qq.com

 DOI 10.1201/9781003349563-29

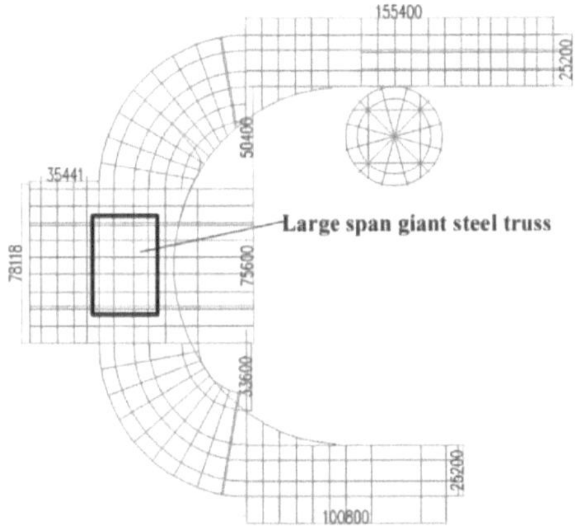

Figure 1. Plane position of large opening structure.

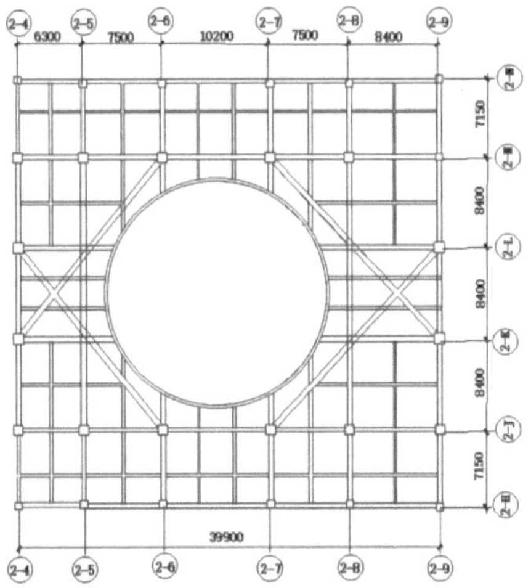

Figure 2. Plane layout of large opening structure.

× 500 mm × 30 mm). The giant round hole floor structure span 39.9 m × 25.2 m and the plan is shown in Figure 2.

Scheme 1: Floor main beam and secondary beam are made of H-shaped steel, overburden type steel plate concrete composite floor. The floor adopts noncomposite pressure steel plate concrete composite floor plate (1.2mm thick pressure steel plate only for formwork), the concrete uses C30, plate bottom with C10@185 reinforcement, plate top with C10@150 reinforcement, and plate thickness of 150mm. Scheme 2: Using reinforced concrete main and secondary beam structure, concrete-filled steel tube columns are equipped with a short leg and concrete beam longitudinal stressed steel welding, the floor is a cast-in-place reinforced concrete structure, concrete using C30, and the bottom and the top are equipped with C10@150 steel, with the thickness of 120mm. Section dimensions of the components are shown in Table 1.

Table 1. Part number and section size.

Plan 1		Plan 2	
Part number	Sectional dimension (mm)	Part number	Sectional dimension (mm)
Main beam (side)	H700×600×30×40	Main beam (side)	850×400
Main beam (middle)	H1000×600×30×40	Main beam (middle)	1500×500
Secondary beam	H600×600×36×40	Secondary beam	850×400
Secondary beam (II)	H500×500×26×30	Secondary beam (II)	800×300
Kong edge beam	H600×300×12×25	Kong edge beam	800×250

3 FINITE ELEMENT ANALYSIS OF STRESS PERFORMANCE OF LARGE OPEN FLOOR STRUCTURE

The floor structure of the building bears vertical loads, and its internal force is mainly characterized by bending. When the building function requires the floor plate to be opened, the structure of the hole edge needs to be strengthened for the smaller opening, and the larger hole needs to be conducted for structural design and verification. This is because the size of the floor opening greatly affects the force of the floor structure by the finite element method. At the same time, the steel structure system and concrete structure system are analyzed.

3.1 *Parameter analysis of open floor slab*

Plate opening stress analysis is a classic problem of elastic mechanics (Wang & Xiong & Huang 1982; Yang 1980), but the existing theoretical analysis mainly focuses on the infinite plate small opening stress under tensile stress concentration. ANSYS finite element is used to verify the classical solution of the relevant literature (Zhu & Tan 2004), proving that the finite element method analysis plate opening problem can meet the precision requirements. The open floor structure is a finite-size rectangular plate model. The thickness and aperture of the plate to examine the parameter change to the opening plate should be force and deformation, the ratio of the parameter hole diameter D to the rectangular plate side length B is from 0~0.8, and the scores of the plate thickness (h) and the rectangular board side length (B) are 1/20, 1/30, and 1/The SHELL181 cell in the ANSYS finite element program was used for modeling (Figure 3a), with a slab span of 10m. An elastic modulus for the concrete material of $E_c = 20000$ N/mm^2 and a Poisson ratio of 0.2 were used. For the parameter analysis, the unit load 1 was applied to the surface of the circular hole square plate and calculated by the 1/4 model based on the symmetry model. The hole edge stress, plate edge support stress, and deflection under different parameters are listed in Table 2. Figure 4 shows the edge stress of the upper- and lower-hole surfaces and the upper- and lower-hole surfaces at the support position at h/B=0.05. At h/B=0, the hole edge stress is the central stress of the plate. As can be seen from the curve in the figure, the value of the stress concentration of the edge of the plate exceeds the value of the plate center stress without the hole opening. Then with the increase of the opening diameter, the stress of the hole and the side of the support is gradually reduced. Figures 5 and 6 respectively show the relationship between unopen plate stress and D/B=0.2 open plate stress with plate thickness. The stress of the thickness of the plate decreases gradually with the ratio h/B. The changing pattern of hole edge deflection with D/B is roughly the same as the maximum stress of the plate.

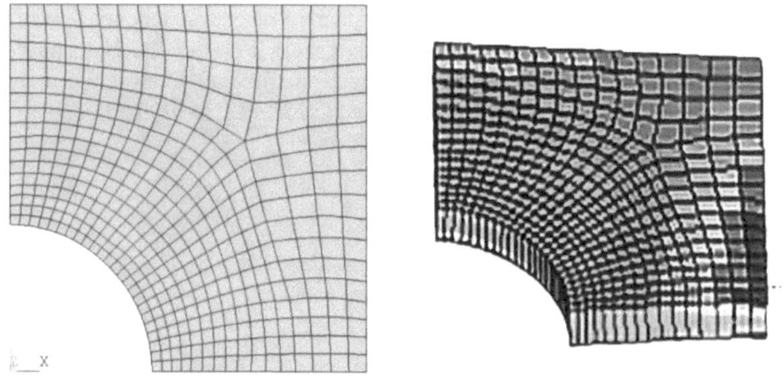

(a) (b)

Figure 3. Finite element model and stress distribution.

3.2 *Finite element analysis of large open floor structure of main and secondary beams*

According to the specific practice of the floor structure layer and the protection (fire prevention, corrosion prevention) layer, the constant load standard value is 6.495 KN/m^2 after pooling, and the variable load of the floor is 2.0 KN/m^2. Load sub-component factor (GB50009-2019) is 1.3 (constant load) and 1.5 (live load), respectively. General finite element software ANSYS18.0 is used to analyze the stress performance of steel structure and concrete structure, respectively, and to check the stress state and deflection deformation of the main components of the two schemes. The beam189 unit is used to simulate the H-shaped steel beam and the concrete section beam, whose dimensions are shown in Table 1. Because the design requires that the material cannot enter the

Table 2. Stress and deflection of perforated plate under different parameters.

D/B	h/B	Upper hole edge stress (N/mm^2)	Lower hole edge stress (N/mm^2)	Upper edge stress on the support seat (N/mm^2)	Lower edge stress of the support seat (N/mm^2)	Deflection (mm)
0	1/20	–51	51	112	–112	60.8
	1/30	–114.5	114.5	253.5	–253.5	201.2
	1/40	–203	203	450	–450	471.6
0.2	1/20	–69.4	69.4	103.6	–103.6	60.9
	1/30	–156.3	156.3	234.1	–234.1	201.7
	1/40	–277.4	277.4	415.5	–415.5	473.3
0.4	1/20	–28.9	28.9	88	–88	38.5
0.6	1/20	–12.5	12.5	53.8	–53.8	15
0.8	1/20	–6.6	6.6	23.8	–23.8	4.3

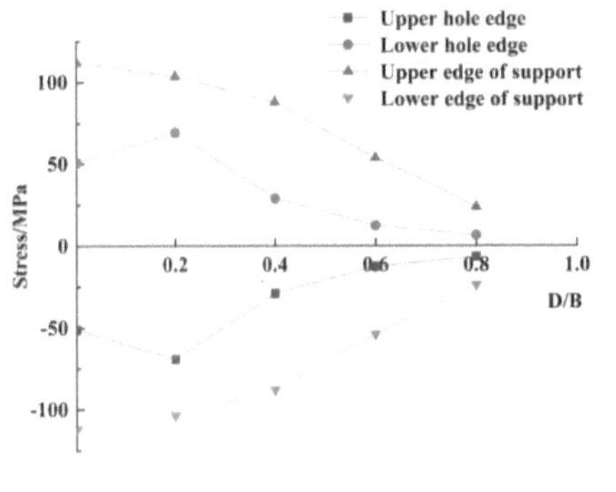

Figure 4. Stress of plate with holes when h/B=0.05.

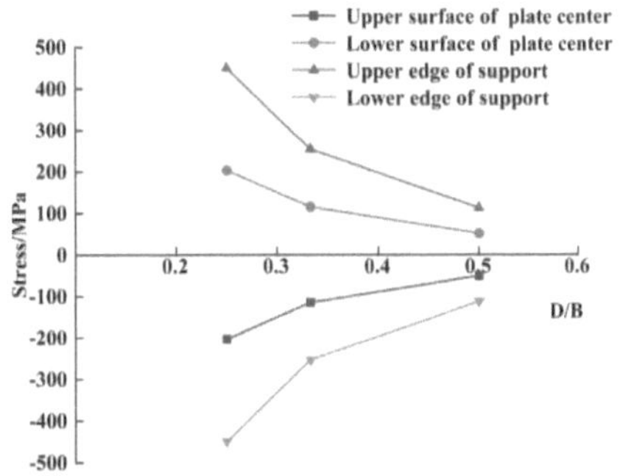

Figure 5. Stresses of unperforated plates.

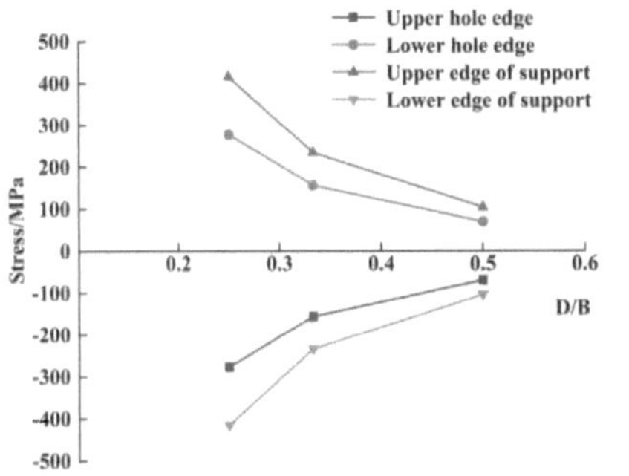

Figure 6. Stress of the plate with holes when D/B=0.2.

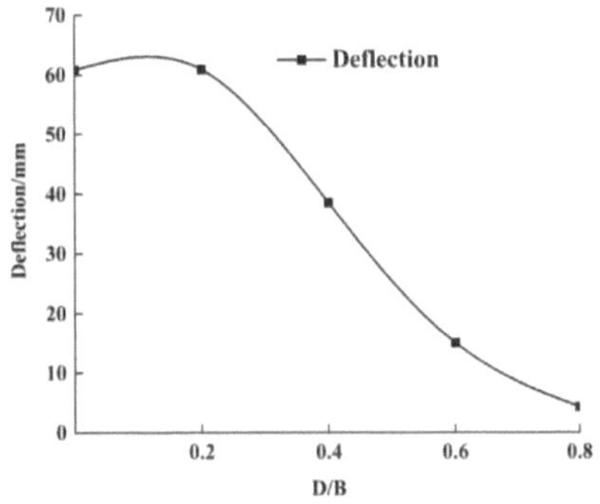

Figure 7. Deflection of plate when h/B=0.05.

yield state, the analysis uses the elastic analysis. The elastic modulus and Poisson ratio of steel are $E_s=2.06\times10^5$ N/mm^2, $u_s=0.3$; the elastic modulus and Poisson ratio of concrete are $E_c=2.0\times10^4$ N/mm^2, $u_c=0.2$. The contribution of the floor of the analysis to the bearing capacity of the beam is not considered, and the floor load and the node load of the main beam are applied to the calculation model. The nodes of the main and secondary beams are rigid connections, and vertical constraints are placed at the top of the column.

The analysis results show that the stress of the main and secondary beams of the steel structure is relatively small (Figure 8), and the deflection value is too large. The maximum radial stress of upper and lower edge stress at the support seat is 68.9 N/mm^2; 67.7 N/mm^2 of upper and lower edge stress of edge beam at support seat; 68.1 N/mm^2 of maximum stress of hole edge beam; 54.5 N/mm^2 of maximum stress of inner main beam span. The maximum vertical displacement occurs at the position of the large-open hole side beam 1, with a value of 35.1mm, and exceeds the deflection limit of the steel structure beam of (GB50017-2017) [L/500] =10600/500=21.2 mm. Scheme 2: the stress of the reinforced concrete primary and secondary beams is less than the design value of the compressive strength of C30 concrete (Figure 9). The maximum radial stress of the upper and lower edge stress is 15.1 N/mm^2; the stress of the hole edge column top beam is 17.5 N/mm^2; at the upper and lower edge beam is 13.4 N/mm^2; the stress of the maximum stress of the hole edge beam is 6.3 N/mm^2; and the maximum stress is 10.0 N/mm^2 in the inner girder span. Maximum vertical displacement is 15.2 mm (Figure 9 large open hole edge beam 1 position), which is less than the deflection limit of the concrete beam (GB 50010-2019): [L/400] = 10600/400=26.5 mm.

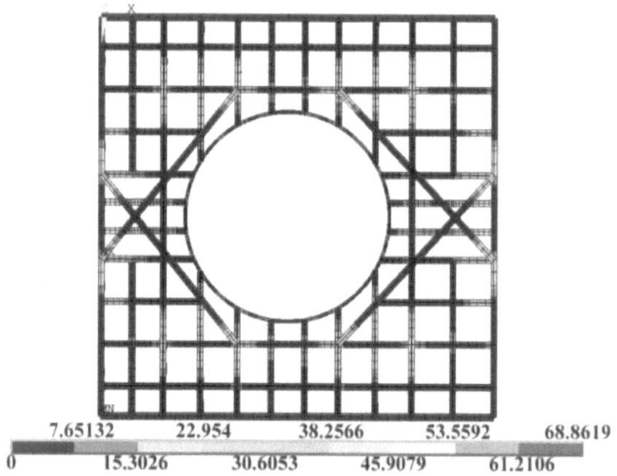

Figure 8. Steel structure floor beam stress.

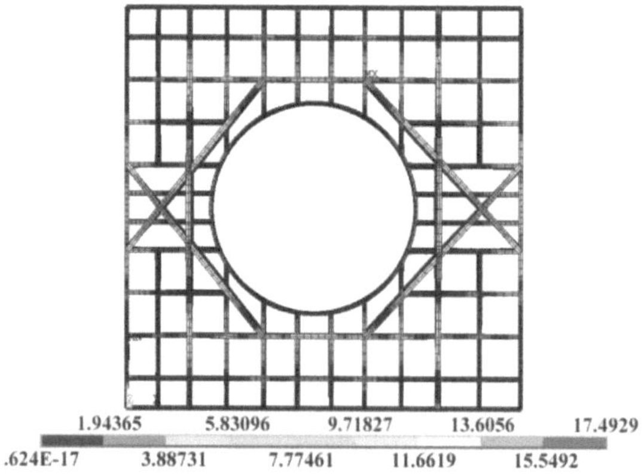

Figure 9. Concrete floor beam stress.

Under the action of the design load, the steel structure floor system is completely in the elastic state, and the upper and lower surface stress of each component is small when the deflection plays a control role. The maximum stress of some components of the concrete floor system exceeds 30% of the concrete strength, indicating that the stress area has entered a nonlinear state. Because the concrete has a certain compression area height, the actual surface stress of the actual component is less than the elastic analysis stress value. Concrete structure mostly adopts the elastic method in calculating the stiffness and deflection test, which is consistent with this analysis model. The finite element analysis process of the concrete structure system does not consider the favorable influence of the reinforcement on the strength increase and the deflection test calculation. The specific section design can be calculated according to the actual reinforcement of the beam end and the span control section. In actual engineering, the floor has a great contribution to the bearing capacity and deflection control of the floor structure system, and the beneficial contribution of the floor is ignored in the finite element analysis, so the comparative analysis of these two schemes is biased to safety. The results of the deflection test of the two systems show that using the concrete structure will make the floor system have greater stiffness and be more beneficial to the deformation control.

4 CONCLUSIONS

The large-opening cantilevered patio floor structure of Shandong Bohai Institute of Advanced Technology as a research object, the finite size open floor finite element parameter analysis, get hole size, plate thickness on the influence of structure component stress rule, and establish the foot steel structure and concrete structure two kinds of primary and secondary beam system model, the conclusion of this research is as follows:

The floor hole opening is small (D/B $\leq$0.2), due to the stress concentration, the hole edge stress is large (exceeds the unopen plate center stress), and the hole edge stress decreases rapidly with the increase of the aperture.

The stress of the support member of the large hole floor decreases with the increase of the aperture, and decreases with the increase of the floor structure, and its stress performance is gradually close to the overhanging structure.

When the large span and large hole floor structure system adopts a steel structure, the deflection plays a control role, and the steel strength has not been fully utilized.

Due to the good integrity of cast-in-place beam and slab, and its large stiffness, using large hole floor structure can give full play to the overall working performance of each component, its strength and deflection are easy to achieve the optimal design goal of carrying capacity limit state and normal use limit state.

REFERENCES

GB 50009-2019 (2019). Load code for the design of building structures. S. *Beijing: China Construction Industry Press*.

GB 50010-2019 (2019). Code for design of concrete structures. S. *Beijing: China Construction Industry Press*.

GB 50017-2017 (2017). Standard for design of steel structures. S. *Beijing: China Construction Industry Press*.

Wang, R. & Z. H. Xiong & W. B. Huang (1982). Fundamentals of plastic mechanics. M. *Science Press*, 343–352.

Yang, G. T. (1980). Elastoplastic mechanics. M. *People's Education Press*, 148–163.

Zhu, S. D. & Q. D. Tan (2004). Stress concentration analysis of thin plates with circular holes based on ANSYS platform. J. *Journal of Suzhou University* (Engineering Edition), 51–53.

Advances in Civil Engineering and Environmental Engineering – bin Yahya & Bin Daud (Eds)
© 2023 The Author(s), ISBN: 978-1-032-39397-1

Preliminary determination and inversion of rock mass parameters of underground cavern of Kaluma hydropower station

Hua Li*, An Liu* & Ting Shi*
PowerChina Huadong Engineering Corporation Limited, Huadong, China

ABSTRACT: The dam site of Kaluma Hydropower Station is located on the Nile River in northern Uganda. It is a diversion power station with a total installed capacity of 600MW. The surrounding rock deformation and block instability in the plant area of the hydropower station are the key factors affecting the stability of the underground cavern. Determining the mechanical parameters of rock mass is the primary task of the cavern stability evaluation and calculation. Based on the actual geological conditions revealed by the Kaluma Hydropower Station on-site, and on the basis of mastering the deformation and stress of the surrounding rock of the underground cavern group and the stress characteristics of the supporting structure, this paper systematically analyzes the on-site monitoring data, adopts the appropriate feedback analysis method to modify the three-dimensional numerical calculation model, simulates the complex geological conditions on site, and carries out the inverse analysis of the relevant parameters of rock mechanics and in-situ stress. The results show that the calculated response characteristics of surrounding rock excavation of underground powerhouse caverns are consistent with the displacement and stability characteristics of surrounding rock revealed by the monitoring data. The rock mass parameters obtained by inversion are also close to the empirical estimation results, indicating that the empirical prediction is consistent with the actual excavation exposure.

1 INTRODUCTION

The determination of mechanical parameters of rock mass is one of the keys to the calculation, but it is very difficult. In the early stage, it was popular to do some large-scale in-situ tests, but it is generally believed that the discreteness of large-scale test results is too large, and the cost and cycle constraints are large, so it is difficult to obtain good results (Jiang et al., 2011).

The failure mode of rock mass is completely different from that of rock block, which is mainly controlled by structural plane and shows nonlinearity, that is, it is affected by confining pressure. In general, the rock types and rock structures are ever-changing, so it is difficult to summarize a simple formula. It is generally believed that the main factors that need to be considered are the current RMR and Q system scores (Zhang et al., 2007).

The domestic engineering circles tend to adopt the empirical method, that is, look up the table from the classification results of the surrounding rock. This method is qualitative in general and combined with the correction and supplement of thousands of engineering cases at home and abroad (Zhu et al., 2013). Therefore, it is reliable and reflects strong practicality.

Based on the above empirical method of surrounding rock classification, some rock mechanics experts have been trying to find some quantitative methods. At present, the engineering community tends to derive empirical formulas from some rock mass classification systems (Deng et al., 2008) and the factors considered include rock block strength, rock mass integrity, structural plane

*Corresponding Authors: li_h5@hdec.com, liu_a2@hdec.com and shi_t@hdec.com

 DOI 10.1201/9781003349563-30

properties, groundwater, in-situ stress, etc. The Hoek-Brown failure criterion is an internationally recognized formula, and this method has been revised and improved.

$$\sigma_1' = \sigma_3' + \sigma_{ci}\left(m_b \frac{\sigma_3'}{\sigma_{ci}} + s\right)^a \tag{1}$$

In view of the above difficulties, some rock mechanics experts also put forward the idea of parameter inversion, that is, comparing the numerical calculation results with the on-site monitoring results to deduce the rock mechanics parameters. In general, this method is an important direction for future theoretical progress, but it also has its disadvantages. On the one hand, it tests the reliability of on-site monitoring results, on the other hand, it cannot be applied to the engineering design stage.

1.1 *Empirical method*

The surrounding rock in the underground powerhouse area is mainly slightly weathered granite gneiss, the rock mass is hard and relatively complete, the structural plane is slightly developed, most of the caverns are dry, and only local seepage and dripping water are available. Therefore, it is mainly class II surrounding rock, but the local tunnel section passing through the alteration zone is class III surrounding rock. According to the rock test results and engineering experience, the physical and mechanical parameters of surrounding rock in underground plant area are proposed in Table 1.

Table 1. Recommended parameters of rock mass.

	Uniaxial saturated compressive strength	Bulk density	Shear strength (rock mass)		Deformation modulus	Poisson's ratio
Rock mass classification	RC MPa	γ kN/m^3	C' MPa	f' °	Em GPa	μ /
II	60~90	26.5~28.0	1.5~1.7	1.1~1.2	13~15	0.17~0.2
II	40~60	25.5~26.5	1.0~1.1	0.9~1.0	6~8	0.20~0.23

1.2 *Hoek Brown failure criterion method*

Rock mass usually shows obvious anisotropy and is strongly affected by discontinuous structural planes. The strength of the rock can be sampled and tested in the laboratory for various strength and deformation, but the rock on site is affected by various weathering, groundwater, discontinuous structural plane and in-situ stress, and its contact surface may change from closure to clay filling, straightness to wavy fluctuation, thus showing obvious differences.

The strength of on-site rock mass is estimated from the on-site observation of rock block test, in-situ stress results, rock mass integrity, structural plane properties, and groundwater. In this regard, Hoek-Brown destruction criterion is widely recognized internationally. Hoek-Brown failure criterion is obtained by fitting the statistical results from many test data.

Relevant parameters are as follows:

$$\begin{cases} m_b = m_i \exp\left(\dfrac{GSI - 100}{28 - 14D}\right) \\ s = \exp\left(\dfrac{GSI - 100}{9 - 3D}\right) \\ a = \dfrac{1}{2} + \dfrac{1}{6}\left(e^{-GSI/15} - e^{-20/3}\right) \end{cases} \tag{2}$$

Among them, σ_{ci} is the uniaxial compressive strength of rock; D is between 0 and 1, which reflects the influence of external factors such as blasting and internal factors such as stress release on rock mass properties. The better the blasting quality is, the smaller the D value is; m_i is a mathematical parameter for compressive strength test fitting, which is mainly related to lithology; GSI is the strength index of rock mass, which is related to the development and properties of discontinuous structural planes. Since most engineers are more familiar with the Mohr-Coulomb model. A method is proposed to obtain the cohesion c and friction angle φ applied to the Mohr Coulomb model, as shown in Figure 1.

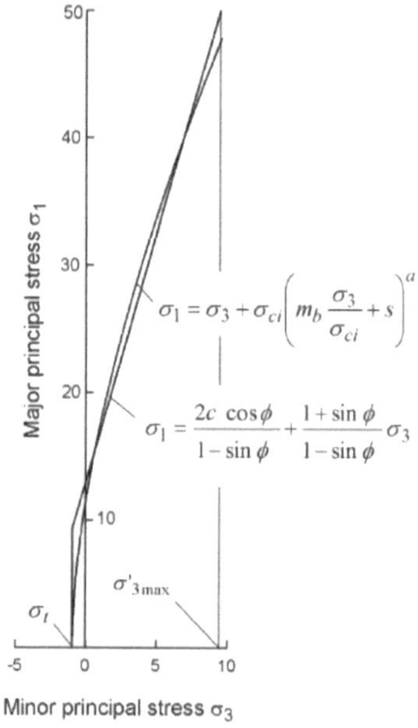

Figure 1. Hoek-Brown and Mohr-Coulomb criteria.

Marinos and Hoek (2000) put forward the empirical formula of rock mass elastic modulus as follows:

$$s\begin{cases} E_m = \left(1 - \frac{D}{2}\right)\sqrt{\frac{\sigma_{ci}}{100}} \cdot 10^{(GSI-10)/40} & \sigma_{ci} < 100MPa \\ E_m = \left(1 - \frac{D}{2}\right) \cdot 10^{(GSI-10)/40} & \sigma_{ci} > 100MPa \end{cases} \quad (3)$$

According to the results of the uniaxial compression test (with size correction), GSI system is used to evaluate the rock mass, Hoek-Brown empirical formula is used, and the rock mass mechanical parameters calculated by RocData software are shown in Table 2.

2 INVERSION OF ROCK MASS PARAMETERS

According to the rock mechanics test results and the on-site surrounding rock quality score, the Hoek-Brown experience system is used to estimate the mechanical parameters of class II rock mass. The inverse analysis takes the estimated rock mass parameters as the standard, and changes within a certain range. Through multiple trial calculations, the rock mass parameters that are most consistent with the monitoring data are obtained.

The range of deformation and strength parameters of II surrounding rock calculated by inversion trial is as follows: deformation modulus: 14~16 GPa; Cohesion: 0.9~1.1MPa; Friction angle:

Table 2. Mechanical parameters of rock mass in underground plant area.

Rock mass classification	Compressive strength of intact rock block/σ_{ci}	GSI	m_i	Mohr-Coulomb strength criterion (fitting): Cohesion/c	Mohr-Coulomb strength criterion (fitting): Internal friction angle/ϕ	Deformation modulus/ E_m	Compressive strength of rock mass σ_{cm}
	MPa			MPa	°	GPa	MPa
II	71	70	28	1.54	63.4	26.6	30.7
III	47	50	28	0.69	57.3	6.9	13.5

45° ~60°. After many tests, it is found that the adopted surrounding rock parameters are: Em= 15 GPa, Poisson's ratio 0.25, cohesion c=1.0, friction angle 48°, and the numerical analysis results are in good agreement with the field measured results.

The numerical simulation results show that the surrounding rock displacement occurs in all buried parts of the crown after the excavation of the third floor of the powerhouse. Since the monitoring data reflects the surrounding rock deformation increment of each part of the plant roof during the excavation of layer II and layer III, the relative displacement of layer II and layer III after excavation is selected for comparative analysis in the numerical simulation. The comparative results are as follows:

(1) In the monitoring data of the A section (PH 0+115), the displacement of the crown is not obvious, and the actual monitoring displacement of the crown is slightly larger than the numerical simulation results; There is asymmetric relative displacement on the upstream and downstream arch sides of the numerical simulation, and the displacement on the upstream side is generally larger, with a depth of 1.5 m reaching 2.88 mm. However, the displacement on both sides of the actual monitoring results is relatively symmetrical and the degree of displacement is less than the numerical simulation results, as shown in Table 3.

Table 3. Comparison between monitoring results and numerical calculation results of Section A (PH 0+115).

Depth of measuring point	Upstream side spandrel: Actual value	Upstream side spandrel: Calculated value	Crown: Actual value	Crown: Calculated value	Downstream side spandrel: Actual value	Downstream side spandrel: Calculated value
1.5m	0.49	2.88	1.21		0.51	0.87
4.5m	0.46	2.43	1.00	0.09	0.22	0.41
9.5m	0.59	1.92	−0.13	0.16	0.24	0.07
24.5m	0.40	1.07	0.39	0.17	0.05	0.46

(2) In the monitoring data of Section B (PH0+46), similar to Section A, the displacement of the crown is not obvious, with a rebound of 0.04 mm at the depth of 1.5 m, while the actual monitoring displacement of the crown at the depth of 1.5 m reaches 1.56 mm. The numerical simulation results show that there is asymmetric relative displacement on the upstream and downstream sides, and the upstream side is generally larger, with a maximum depth of 1.5 m up to 2.89 mm. The actual monitoring results show that the displacement of the surrounding rock on the upstream side is slightly larger than that on the downstream side (Table 4).

Table 4. Comparison between monitoring results and numerical calculation results of Section B (PH0+46).

Depth of measuring point	Upstream side spandrel		Crown		Downstream side spandrel	
	Actual value	Calculated value	Actual value	Calculated value	Actual value	Calculated value
1.5m	1.56	2.89	1.56	−0.04	1.04	1.03
4.5m	0.69	2.41	0.69	0.04	0.19	0.53
9.5m	0.46	1.99	0.46	0.12	−0.68	0.2
24.5m	0.03	1.11	0.03	0.15	0.29	0.47

(3) In the monitoring data of Section C (PH0+0), the numerical simulation results are still similar to that of Section A. There is a certain degree of rebound at the crown, with a rebound of 0.37 mm at the depth of 1.5 m, a displacement of surrounding rock of 2.48 mm at the depth of 1.5 m at the upstream side, and a displacement of surrounding rock of 1.83 mm at the depth of 1.5 m at the downstream side. Because the actual monitoring instrument is buried late and the monitoring data is too small, it is not suitable for the inversion analysis of surrounding rock parameters, as shown in Table 5.

Table 5. Comparison between monitoring results and numerical calculation results of Section C (PH0+0).

Depth of measuring point	Upstream side spandrel		Crown		Downstream side spandrel	
	Actual value	Calculated value	Actual value	Calculated value	Actual value	Calculated value
1.5m	0.05	2.48	0.65	−0.37	0.17	1.83
4.5m	−0.11	1.96	−0.04	−0.4	−0.23	1.83
9.5m	−0.22	1.58	−0.83	−0.31	−0.16	1.36
24.5m	−0.14	0.9	0.30	−0.21	−0.44	0.93

Mcf-(x)-4 and Mcf-(x)-5 multipoint displacement meters at the rock beam mainly reflect the deformation characteristics of surrounding rock during the excavation of layer III of the powerhouse. Due to the lag of embedding, the measured value of monitoring data is small: the deformation of surrounding rock at the depth of 1.5 m is basically after the excavation of layer III of the powerhouse. As of March 14, 2016, the deformation of surrounding rock at the depth of 1.5 m at the rock beam is 0.32~0.76 mm, reflecting the deformation increment of surrounding rock at this part caused by the excavation of layer III of the powerhouse. Because the monitoring deformation is small, it is even difficult to judge whether the dominant factor of this deformation is the influence of blasting relaxation or stress adjustment. Therefore, the monitoring data at the rock beam at the current stage has a relatively small reference value for inversion, as shown in Table 6.

3 CONCLUSION

Because the surrounding rock condition of the underground powerhouse of Kaluma Hydropower Station is good, and the rock mass category is mainly class II, the deformation of the surrounding rock monitored by the multipoint displacement meter after excavation is generally very small, generally no more than 2 mm. Because the deformation is too small, it objectively brings some difficulties to the inversion of rock mass parameters.

Table 6. Comparison between monitoring results and numerical calculation results of Mcf-(x)-4 and Mcf-(x)-5.

			Deformation increment/mm			
Monitoring position			1.5 m	4.5 m	9.5 m	20.5 m
A (PH0+115)	Upstream side rock beam	Actual value	0.46	0.24	0.13	−0.01
		Calculated value	2.79	2.66	2.28	1.53
	Downstream side rock beam	Actual value	0.76	0.13	0.14	−0.08
		Calculated value	2.06	1.82	1.36	0.43
B (PH0+46)	Upstream side rock beam	Actual value	0.32	0.34	0.17	−0.20
		Calculated value	2.87	2.69	2.33	1.53
	Downstream side rock beam	Actual value	0.76	0.13	0.14	−0.08
		Calculated value	2.07	1.78	1.38	0.49
C (PH0+0)	Upstream side rock beam	Actual value	0.17	−0.23	−0.16	−0.44
		Calculated value	2.62	2.5	2.11	1.4
	Downstream side rock beam	Actual value	–	–	–	–
		Calculated value	2.25	2.06	1.58	0.75

Based on the analysis of the monitoring data of the underground powerhouse, this paper inverses the rock mass parameters: when the class II surrounding rock parameters are Em=15 GPa, Poisson's ratio 0.25, cohesion c=1.0, and friction angle 48°, the numerical analysis results are in good agreement with the field measured results.

In general, using the in-situ stress characteristics analyzed above and the rock mass parameters estimated empirically, the calculated surrounding rock excavation response characteristics of the underground powerhouse cavern group are consistent with the surrounding rock displacement and stability characteristics revealed by the monitoring data. The rock mass parameters obtained by inversion are also close to the empirical estimation results, indicating that the empirical prediction is consistent with the actual excavation exposure.

REFERENCES

Deng H., Li J., Wang L. (2008). Study on mechanic parameters of bolted fractured rock mass under unloading. *Rock and Soil Mechanics*, 29 (4): 1027–1030.

Jiang Q., Feng X., Xiang T. (2011). Numerical simulation study on stability analysis and intelligent dynamic optimization design of large caverns. *Journal of Rock Mechanics and Engineering*, 30 (3): 524–539.

Marinos P., Hoek E. (2000). GSI: A Geologically Friendly Tool for Rock Mass Strength Estimation. *Melbourne: GeoEng 2000 Conference*.

Zhang C., Zhou H., Feng X. (2007). Evaluation of geotechnical engineering stability based on failure proximity. *Journal of Rock Mechanics and Engineering*, 28 (5): 888–894.

Zhu H., Zhang Q., Zhang L. (2013). Review on research progress and application of Hoek Brown strength criterion. *Journal of Rock Mechanics and Engineering*, 32 (10): 1945–1963.

Advances in Civil Engineering and Environmental Engineering – bin Yahya & Bin Daud (Eds)

Preparation and application study of aerogel thermal insulation mortar material

Zhihang Li, Liwen Tan, Guodong Zhang, Meijun Liu, Yujia Zhang, Yuexinghong Yin, Fang Deng, Mingli Wang, Zhongjun Tang & Ting Hu*
Chongqing College of Architecture and Technology, Chongqing, China

ABSTRACT: Through the aerogel insulation mortar test, the mix ratio and the preparation method are adjusted, and the prepared aerogel insulation mortar takes into account the advantages of the insulation mortar material system. It can be applied to a particularly complex base surface, good thermal insulation performance, and can easily meet the energy-saving design standard of 65% in hot summer and cold winter areas, especially for housing buildings that design sub-household insulation in the northern region and conventional thermal insulation mortar cannot meet energy-saving requirements. It has the advantages of good fire retardant safety, good temperature, chemical stability and green environmental protection, good thermal insulation performance, simple construction, and good economy.

1 INTRODUCTION

Enclosure structure is one of the important sources of building energy consumption. Improving the working performance of envelope insulation materials has become an important way to improve building energy efficiency. Traditional inorganic thermal insulation materials in the current development requirements of energy conservation and emission reduction development, gradually can not meet the requirements (Xia & Yang 2021). In 2020, General Secretary Xi Jinping proposed that "China's carbon dioxide emissions will peak by 2030, and strive to achieve carbon neutrality by 2060". To achieve this goal, the construction industry, which accounts for more than 50% of the national carbon emissions, is urgent to develop new efficient and energy-saving building materials, and the research and development of new wall insulation materials is an effective way to reduce building energy consumption (Zhao et al. 2021). Li Sha et al. (2021) With glass beads as the light aggregate, cement and fly ash as the compelling materials, adding glue powder and cellulose ether and other additives, using hydrophilic and hydrophobic aerogels to replace glass beads in the insulation mortar, the related properties of the insulation mortar are analyzed and characterized. Sun Yuqing et al. (2021) because of the high cost and low strength of traditional aerogel insulation materials, cheap expansion perlite and SiO2 aerogel are used to prepare aerogel expansion perlite composite insulation materials, and test the preparation process and performance of it. Based on the current situation of domestic and foreign research and the best mix ratio, the influence of related factors on thermal conductivity (Wang et al. 2019). Yuan Xinzhu (Yuan 2019). The influence of the different raw materials of inorganic insulation mortar is discussed. Liu Jing et al. (2018). PTemp software was used to simulate the heat transfer performance of three materials in summer and winter temperature distribution in Beijing and compare the extreme temperature curve in summer. The more the number of glass beads, the smaller the dry density, thermal conductivity, compressive strength, and water absorption of insulation mortar; the incorporation of silicone can reduce the water absorption

*Corresponding Author: 4920049510@qq.com

 DOI 10.1201/9781003349563-31

under the other properties of insulation mortar; the incorporation of gas phase silica can reduce the thermal conductivity under the other properties of insulation mortar (Wu et al. 2018).

2 THERMAL INSULATION PRINCIPLE OF AEROGEL THERMAL INSULATION MORTAR

The 21st century is a new century of sustainable development. With the improvement of the requirements of energy conservation and consumption reduction and the increasingly prominent energy problems, more and more countries put building energy conservation in the first place. In recent years, China has put forward a series of policies such as the "double-carbon" plan, and the development of energy-saving building materials is the only way. Aerogel is a kind of light nano-solid material, its solid phase, and the void structure are nano-scale, with ultra-light, adiabatic, insulation, and other advantages, it also has an excellent performance in acoustics, catalysis, optics, and electricity. With the deepening of research and extensive application, the application of aerogel in the construction field will improve the energy-saving effect by order of magnitude, and the advantages and prospects of aerogel are increasingly shown. Its product form mainly includes aerogel glass, aerogel thermal insulation coating, aerogel fiber composite material, aerogel thermal insulation felt, aerogel vacuum thermal insulation board, aerogel mortar, and concrete, etc.

Aerogel insulation mortar refers to a building material made of barrier insulation material and mortar material, used for the insulation layer of structures and building surfaces. Insulation is to prevent heat from the interior through the envelope, in which the wall part of the building heat reaches 30%.

From the heat insulation point of view of heat insulation, heat is blocked outside, heat transfer mainly has the following four parts: ① gas molecule heat conduction; ② gas heat transfer; ③ solid material heat conduction; ④ infrared radiation heat transfer. The insulation aggregate nanopore silicon aerogel and ordinary insulation aggregate have an essential improvement in these four aspects.

The heat transfer process of thermal insulation materials is performed in the solid phase, and a situation occurs when the thermal conductivity process occurs between the aggregates. That is because the surface of aggregate particles is rough and uneven, the contact area between aggregate and aggregate is very small, and small pores will form between aggregate and aggregate, so the area of vertical and heat transfer path will become very small, and heat transfer is difficult. When these voids are filled with air, the thermal conductivity of the air is much less than the thermal conductivity of the aggregate.

Reduce the heat conduction of solid materials: To minimize the heat conduction of solid materials, the solid thin walls used as a gas barrier should be as thin as possible. According to the diameter of the 60 nm thin wall sphere as a theoretical calculation model, if the percentage of solid volume is controlled at 5%, then the thickness of the spherical thin wall should be controlled at about 2nm, and our aerogel is close to this theoretical model. And the silicon aerogel solids themselves also have a very low thermal conductivity.

Reduce infrared radiation heat transfer: heat radiation is also an important kind of heat transfer mode, when the temperature difference between hot and cold surfaces is more than 100°C, infrared radiation heat transfer will occupy the dominant position of heat transfer. And as the temperature increases, this trend will be more pronounced (according to Boltzmann's law, the heat transfer of infrared radiation increases four times the temperature). The products we tested are used at room temperature and can not reach the temperature of infrared radiation as the main heat transfer mode. Thermal insulation materials, because of their loose and porous properties, virtually cause a lot of barriers to the path of thermal radiation, so the effect of radiation heat transfer becomes worse.

To sum up, the unique structure and properties of nanopore silicon adiabatic materials reduce the thermal conductivity of the four parts of the heat transfer mode to a very low level. With the good adiabatic performance of glass beads, the total high adiabatic performance of mortar is finally realized. The final model of composite insulation mortar is shown in Figure 1.

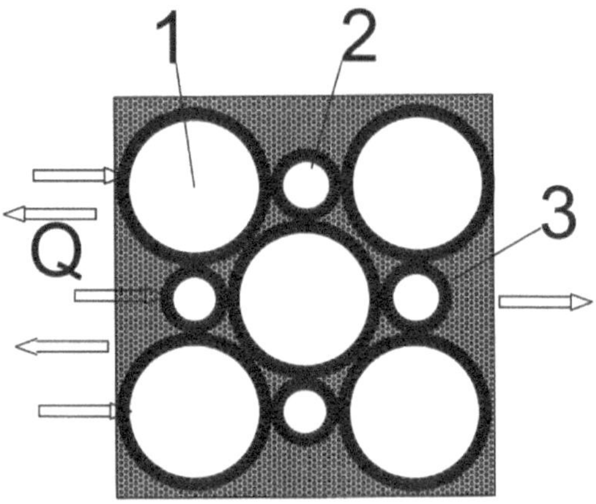

1-Glass beads; 2-silicon aerogel; 3-cement and other gelatinate components

Figure 1. Model of thermal insulation mortar.

3 PREPARATION PROCESS OF AEROGEL THERMAL INSULATION MORTAR

3.1 *The mass mix of the mixture is calculated and the mixture is determined in Table 1*

Table 1. Mix ratio of aerogel thermal insulation mortar.

raw material	cement /g	flyash /g	Silicon aerogel paste /g	Glass microbeads /g	Redisperse adhesive powder / g	wood fibre /g	polypropylene fibre /g	cellulose ether /g
quality	1.3m	0.2m	1.5m	0.67m	1.8%	0.8% m	0.5% m	0.8% m

Because aerogel paste and ordinary silica gas gel particles are different, different construction methods will affect the insulation mortar sink and thermal conductivity, so the study of different construction methods on the thermal conductivity test, summarizes the most suitable mortar preparation method, can maximize the aerogl cream filling effect, reduce the influence of moisture in aerogel cream. The preparation process is shown in Figures 2~4.

Figure 2. Scale out of aerogel thermal insulation mortar mixture.

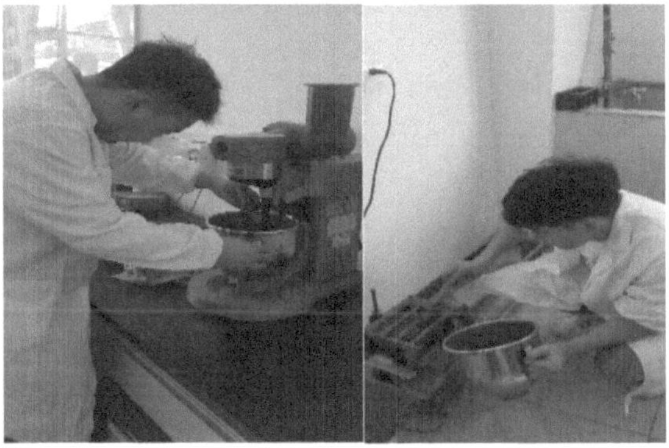

Figure 3. Mixing and molding of aerogel thermal insulation mortar mixture.

Figure 4. Test piece forming and curing.

The technical indicators of the aerogel thermal insulation mortar developed by us are tested, as shown in Figures 5 and 6. It can be concluded that its dry apparent density is 241.4 kg/m^3; The thermal conductivity is as low as 0.053 W / (m·K), lower than that of similar products on the market; the compressive strength is 0.52 MPa; the linear shrinkage rate is 0.157%, and the pressure shear bond strength is 80 Kpa.

Figure 5. Cube test piece of aerogel thermal insulation mortar.

Figure 6. Thermal conductivity test piece of aerogel thermal insulation mortar.

4 SUMMARY

4.1 *Good fire prevention and flame retardant safety*

Compared with organic insulation materials, all components of silica aerogel insulation mortar are inorganic materials, so it has the characteristics of a non-flame retardant. And expansion polystyrene board, extruded polystyrene board, and polystyrene particles these organic insulation materials although their thermal insulation performance is very good, because it is an organic material, belongs to flammable insulation materials, once a fire, these organic insulation materials are easy to burn, causing safety accidents. Moreover, through the research, it is found that the thermal conductivity of inorganic insulation mortar changes with the change in temperature, and the thermal conductivity of inorganic insulation mortar will also increase with the increase of temperature. However, when the insulation layer n decreases, there will be no chemical change, and the change of its structure is not easy to fall off, so it is safe to use.

4.2 *Good temperature, chemical stability and green environmental protection*

Silica dioxide aerogel insulation mortar is composed of pure inorganic material, each component is inert, and performance does not change with temperature. Silica dioxide aerogel insulation mortar can be achieved in acid and alkali resistance, corrosion resistance, adapt to various atmospheric environments, and can be the same life as the building exterior wall, there is no maintenance problem. The production process of its raw materials will not pollute the air and water. Silica dioxide aerogel insulation mortar insulation system is non-toxic, tasteless, no radioactive pollution, the use of engineering will not discharge harmful gases, and is a green material. But the life of organic insulation materials is short, when reaching its service life, it is difficult to replace, and also produces a large amount of waste causing environmental pollution.

4.3 *Good thermal insulation and heat insulation performance*

SiO_2 aerogel has ultra-low thermal conductivity and is the best thermal insulation material ever found. Through experimental research, we found that adjusting the proportion of various components of silica aerogel insulation mortar and optimizing the mix ratio can reduce the thermal conductivity of mortar to 0.053W / (m K), with good energy saving effect, reduce carbon emissions and save heating and cooling costs.

4.4 *Simple construction and good economy*

To minimize the heat conduction of solid materials, the solid thin walls acting as a gas barrier should be as thin as possible. And through the adjustment of the performance of the insulation mortar, with the requirements of mechanical properties and the actual use of the functional requirements, can be completely applied in different building structures, to adapt to a variety of base walls, close bonding, and no gap. During site construction, stir evenly according to the optimized mix ratio, and then apply it directly to the wall mechanically or manually. The construction process is simple, the short construction period, and the quality is easy to control.

REFERENCES

Li Sha, Fang Meng, Wang Dan, Wang Yiheng. Preparation and characterization of mixed affinity and hydrophobic aerogel glass microbead insulation mortar [J]. *Concrete World*. 2021, (06):58–60.

Liu Jing, Wang Jianming, Cheng Kai, Gaoming Yong. Numerical simulation and comparative analysis of the thermal insulation performance of aerogel thermal insulation mortar [J]. *Municipal Technology*. 2018, 36(01):196–198+202.

Sun Yuqing, Hui Lili, Zhao Na, Wang Qiaoyun. Preparation of aerogel perlite and its application in cement base [J]. *Cohere*. 2021, 46(05):18-22.

Wang Xiu, Cheng Gong, Zhang Wenjing. Study on thermal conductivity of aerogel insulation mortar and building energy consumption [J]. *Concrete and Cement Products*. 2019, (12): 70–73.

Wu Hongyan, Fei Haihua, Xia Lu. Study on the preparation method and influencing factors of thermal insulation mortar [J]. *Low-temperature Building Technology*. 2018, 40(12):10–11.

Xia Tian and Yang Wenwei. Research and development of aerogel thermal insulation mortar material for building outer envelope structure [J]. *Building Energy Conservation*. 2021, 49(07)':106–110.

Yuan Xinzhu. Study on the preparation method and influencing factors of thermal insulation mortar [J]. *Fujian Building Materials*. 2019, (05): 14–15.

Zhao Jianwei, Shang Yang, Cui Jie. Application of aerogel in wall insulation materials [J]. *New Materials Industry*. 2021, (02): 57–60.

© 2023 The Author(s), ISBN: 978-1-032-39397-1

Static mechanical properties and strengthening mechanism of nano carbon fiber modified concrete

Biao Ren
College of Aeronautical Engineering, Air Force Engineering University, Xi'an, China

Jinyu Xu*
College of Aeronautical Engineering, Air Force Engineering University, Xi'an, China
Northwestern Polytechnical University, School of Mechanics, Civil Engineering and Architect, Xi'an, China

Zhihang Wang, Yan Lv, & Liangxue Nie
College of Aeronautical Engineering, Air Force Engineering University, Xi'an, China

ABSTRACT: As a widely used inorganic composite material, the brittleness of concrete is an important reason to limit the improvement of concrete performance. To improve the performance of concrete and study the modification effect of carbon nanofibers (CNFs) on concrete, this paper poured CNFs reinforced concrete specimens with different contents, and studied the static compressive strength, splitting tensile strength, and flexural strength of each group of specimens. The influence of CNFs on the micropore structure of concrete was analyzed by Mercury intrusion porosimetry (MIP). The results show that the addition of the appropriate amount of CNFs can enhance the static mechanical properties of concrete to varying degrees, and the optimal content of CNFs is 0.3%. CNFs can effectively reduce the content of macropores in concrete and optimize the pore structure of concrete. When the content is 0.3%, harmful holes and multi-harmful holes are reduced by 34.7% and 45.6% respectively.

1 INTRODUCTION

As the largest amount of inorganic materials in the world, concrete has low prices and high strength. However, ordinary concrete is a brittle material with low tensile strength. During the curing process, irregular micro-cracks are often formed due to drying shrinkage, resulting in low impermeability, frost resistance, and carbonation resistance of concrete (Miao & Mu 2022; Wang et al. 2020, 2019). In the process of use, people often add reinforcing materials to the concrete to improve its performance of the concrete. Common reinforced concrete includes nanomaterial reinforced concrete, polymer reinforced concrete, and fiber reinforced concrete (Ferdous et al. 2020; Yu et al. 2019; Yang et al. 2021). Among them, fiber-reinforced concrete has attracted the attention of many researchers because of its excellent performance.

CNFs reinforced materials are composite materials that have been gradually developed in recent 10 years. Early research works mainly focused on reinforced polymer-based (Chen & Wang 2005; Liu et al. 2015) and metal-based (Li et al. 2016; Xu et al. 2007) materials, and the research on CNFs reinforced cement-based materials is very rare. CNFs have the dual properties of common carbon fiber and nanomaterials. The research on their effects on the mechanical properties of concrete is very important for the development of concrete materials. Abu al rub R k et al. (2011) found that untreated CNFs can improve the mechanical properties of cement-based materials to a certain extent, while acidified CNFs (sulfuric acid and nitric acid) can weaken the mechanical properties of cutting materials. Gdoutos E E (2016) used a two-parameter fracture model to determine the

*Corresponding Author: zhh17713821307@163.com

DOI 10.1201/9781003349563-32

fracture parameters of CNFs reinforced cement mortar to characterize its fracture mechanical properties. The results show that CNFs can effectively improve the microstructure of mortar matrix, resulting in significant improvement of all fracture and mechanical properties. The research results of Ludvig p (2022) show that CNFs have no obvious effect on the setting time of cement mortar, but can significantly improve its compressive and flexural tensile strength. Meng w (2018) studied the effects of CNFs on the rheological properties, hydration dynamic properties, self-shrinkage, and pore structure of ultra-high performance concrete. At the same time, to ensure the uniform dispersion of nanomaterials in ultra-high performance concrete, an optimized dispersion method is proposed. Cui Lei (Trey 2018) studied the dynamic modulus, tensile strength, and deformation performance of CNFs modified asphalt concrete.

At present, there are few studies on the relationship between the static mechanical properties and micro strengthening mechanism of CNFs modified concrete. Therefore, it is necessary to carry out a comprehensive static mechanical test of CNFs modified concrete and analyze the reinforcement mechanism of CNFs in combination with the micro test.

Based on this, in this paper, fiber reinforced cement concrete specimens with 0%, 0.1%, 0.2%, 0.3%, and 0.5% content are configured, and the electro-hydraulic servo material test system is used as a technical means to study the compressive strength, splitting tensile strength and flexural strength of CNFs reinforced concrete under static load, and to analyze the reinforcement mechanism and micro pore structure characteristics of nanocarbon fiber modified concrete by MIP.

2 MATERIALS AND METHODS

2.1 *Test materials*

The raw materials of CNFs reinforced concrete mainly include cement, water, aggregate, nano carbon fiber, and concrete admixture.

Cement: P.O42.5 ordinary portland cement produced by Shaanxi Qinling Cement Co., Ltd., with a fineness of 1.6%, initial setting time of 3 H and final setting time of 12 h. Water is domestic water with pH value greater than 4. The fine aggregate is well-graded clean Bahe medium sand, with a fineness modulus of 2.8 and silt content less than 0.9%. The coarse aggregate is limestone-graded crushed stone with particle size of 5~20 mm, the density is 2.71 g/cm^3, the bulk density is 1.63 g/cm^3, and the silt content is less than 0.2%. CNFs are made of nano carbon fiber produced by *Beijing Deke Daojin Technology Co., Ltd.*, with purity of 99%, length diameter ratio of 70, monofilament diameter of 100~200mm, resistivity less than 0.012 Ω·cm, and density of 0.18 g·cm^{-3}.

Concrete admixtures mainly include defoaming agents and water reducing agents. The function of a defoamer is to reduce the bubbles caused by various active substances and mixing during concrete preparation. The defoaming agent used in this test is NOPCO NXZ type metal soap defoaming agent produced by Nippon NOPCO. The appearance is light yellowish brown turbid liquid. The specific gravity is 0.89~0.95 at 20 ° and the pH value is 5.1. The main function of water reducing agent is to improve the workability of concrete and increase its strength of concrete. It is purchased from *Hebei Shengtong Building Materials Technology Co., Ltd.* It is a light yellow transparent liquid with a density of 1.19 g/cm^3 and a pH value of 6.

2.2 *Specimen preparation and test*

The mix proportion of test pieces is designed according to *The Code for Mix Proportion Design of Ordinary Concrete* (2011).

The test pieces used in this paper are poured based on the "*Sand Wrapping Method*", and strictly comply with the relevant requirements of *The Standard for Test Methods of Physical and Mechanical Properties of Concrete* (GB/T 50081-2019).

The poured concrete specimens shall be removed after standing for 1d, and then cured under standard curing conditions for 28 d. After 28 d, the static mechanical properties of each group of specimens shall be tested. The mechanical tests carried out in this paper are mainly carried out by

HYY electro-hydraulic servo material test system. The bending test and splitting tensile test need to be carried out with the corresponding auxiliary equipment. The bending test needs the help of four-point loading fixture, and the splitting tensile test needs to be completed with the help of a steel cushion block, steel cushion strip, and steel support.

3 RESULTS & DISCUSSION

3.1 *Static mechanical test results*

The static mechanical test results are also calculated according to the relevant provisions in the standard for test methods of the physical and mechanical properties of concrete.

In consideration of the discreteness of concrete test, and following the provisions of the *Standard for Test Methods of Physical and Mechanical Properties of Concrete*, each data has been tested by three groups of test pieces.

During the flexural test of concrete, the brittle failure of the specimen is particularly obvious, there is no residual strength, and the curve descent section is very steep. Therefore, for the convenience of research, only the rising section is taken as the bending curve.

3.2 *Analysis of test results*

3.2.1 *Analysis on compressive properties of concrete*

Figure 1. Typical compressive failure of specimen.

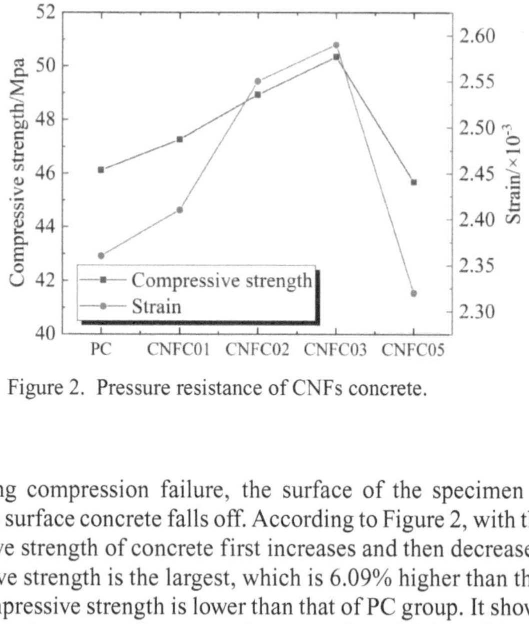

Figure 2. Pressure resistance of CNFs concrete.

It can be seen from Figure 1 that during compression failure, the surface of the specimen is dominated by longitudinal cracks, and the surface concrete falls off. According to Figure 2, with the increase of CNFs content, the compressive strength of concrete first increases and then decreases. When the content is 0.3%, the compressive strength is the largest, which is 6.09% higher than that of PC. When the content is 0.5%, the compressive strength is lower than that of PC group. It shows that CNFs can improve the compressive performance of concrete, but excessive addition of CNFs may reduce the strength of concrete. The variation trend of compressive strain with the amount of admixture is roughly the same as that of compressive strength. With the increase of CNFs content, the failure strain of concrete specimens first increases and then decreases, and cnfc03 increases by 11.6% compared with PC.

3.2.2 *Analysis of tensile properties of concrete*

The splitting tensile strength of concrete is an important index to evaluate its tensile performance. Figure 3 shows the typical splitting failure characteristics of the specimen. The relationship between the splitting tensile strength of the specimen and the displacement of the loading point is shown in Figure 4.

Figure 3. Typical splitting tensile failure.

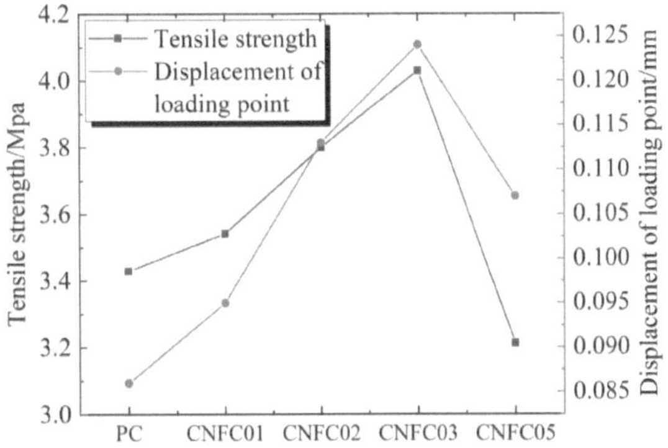

Figure 4. Splitting tensile mechanical properties of concrete.

It can be seen from Figure 4 that the influence of CNFs on the splitting tensile properties of concrete also tends to increase at first and then decrease, and the optimal increase in the content is still 0.3%. In this case, the splitting tensile strength increased by 17.50% and the displacement of the loading point increased by 0.69 mm. In addition, as the fiber content continues to increase, the splitting and tensile properties of the specimens gradually decrease. When the content of 0.5%, the splitting tensile strength of the specimen decreased by 20.28% compared with that of 0.3%.

3.2.3 *Analysis on flexural properties of concrete*

The bending test is to add four point bending test auxiliary equipment based on the compression test device to test the bending performance of the specimen under four-point load. The bending test process and typical failure characteristics of the specimen are shown in Figure 5. Due to the limited test conditions, the maximum displacement of the fracture surface cannot be measured. Therefore, the bending deformation resistance of the test piece is characterized by the rising displacement of the lower end of the auxiliary equipment. The laws of CNFs content, flexural strength, and failure displacement are drawn as shown in Figure 6.

Figure 5. Bending failure of test piece.

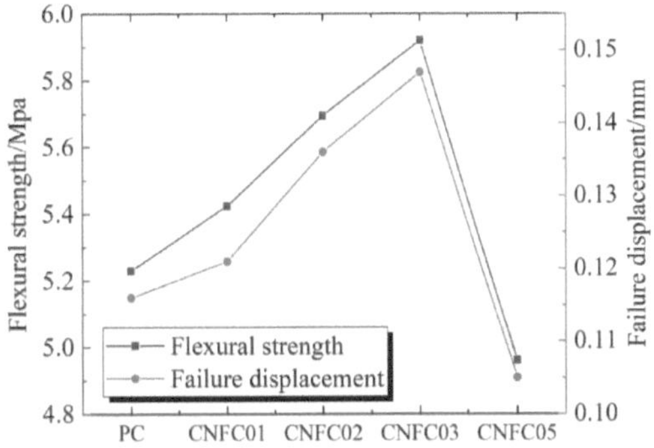

Figure 6. Flexural mechanical properties

It can be seen from Figure 6 that the flexural mechanical properties of concrete reach the optimal state when the content of CNFs is 0.3%. The maximum flexural strength increased by 0.69 MPa, with an increase of 13.19%, and the failure displacement increased by 26.72%. At the same time, with the further increase of fiber content, the flexural properties weaken significantly. Compared with PC group, the flexural strength and failure displacement were weakened by 5.16% and 9.48% respectively at 0.5% content.

4 MECHANISM ANALYSIS

4.1 *Macro modification mechanism*

The macro improvement mechanism of CNFs on concrete performance can be analyzed from the following aspects:

(1) Crack inhibition

The brittleness of concrete is an important factor limiting its performance of concrete, and the reason for the high brittleness of concrete is the poor tensile performance of cement paste and interface transition zone. When the concrete is under pressure, it will generate both compressive stress and tensile stress, and the tensile strength of the concrete is far lower than the compressive strength. Under the action of tensile stress, the micro-cracks in the concrete begin to develop, and finally form through cracks, which is macroscopically manifested as the disintegration of the concrete.

After adding CNFs into the concrete, CNFs will be evenly dispersed into concrete. When the crack develops through the fiber, the fiber will restrict the development of the crack until the fiber is pulled off or pulled out due to the large aspect ratio of the fiber and the strong bonding with the concrete. In this way, the limit stress of internal failure of concrete can be increased from the tensile strength of concrete to the tensile strength of fiber or the bond strength between fiber and concrete. In addition, the addition of CNFs in concrete can also limit the initial damage of concrete caused by drying shrinkage and water loss during curing.

(2) Enhancing effect

The monofilament diameter of CNFs is about 100~200 mm, and 100~200 mm holes are a kind of hole that pose a great threat to the internal structure of concrete. Adding CNFs into the concrete can fill the internal pores of the concrete, reduce the number of harmful pores, refine the pores and reduce the porosity, to improve the compactness and performance of the concrete.

(3) Small size effect

CNFs monofilament has a very small diameter and belongs to nanomaterials. Adding CNFs into concrete can exert its small size effect and improve the performance of concrete. On the one hand, CNFs can improve the internal particle gradation of concrete and optimize the concrete structure. On the other hand, CNFs have a large surface area and strong surface activity, which can promote cement hydration and increase the amount of hydration products. At the same time, CNFs can also provide adsorption sites for hydrated calcium silicate, making the hydrated calcium silicate form a network structure with CNFs as the core during the growth process. It can improve the structure of cement stone and the performance of concrete.

4.2 *Micro pore structure analysis*

The PoreMaster33 automatic mercury intrusion instrument produced by *Cantar instrument company of the United States* is used in this test. The mercury porosimeter is equipped with two low-pressure analysis stations, the test aperture range is 6.4 nm ~650 μm, and the measurement error of mercury inlet and outlet is less than 0.1 μL. Before the test, set the mercury contact angle to 140° and the mercury inlet pressure range to 20~30000 psi.

The samples are taken from the concrete fragments of the compression test. Random sampling shall be conducted at each part of the test piece. Then it shall be broken into small pieces of appropriate size and put into the drying oven for complete dehydration after cleaning the dust on the surface. Finally, a mercury injection test shall be conducted and the test data shall be recorded.

According to the pore classification method of academician Wuzhongwei, the pore distribution of each group of specimens is divided into multiple harmful pores (> 200 nm), harmful pores (50~200 nm), less harmful pores (20~50 nm), and harmless pores (< 20 nm). The pore size distribution is shown in Figure 7.

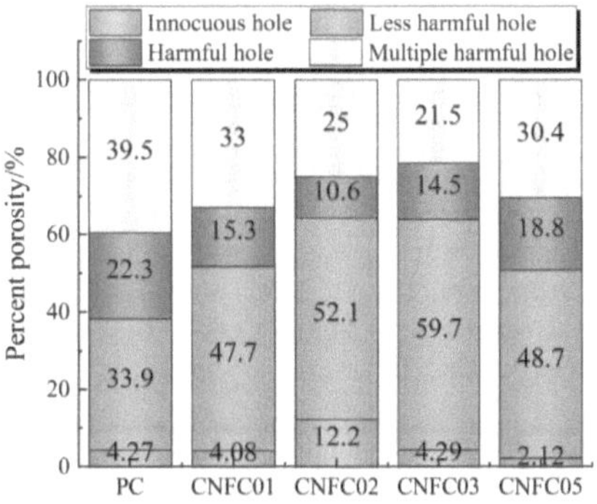

Figure 7. Aperture distribution.

It can be seen from Figure 7 that adding graphene oxide can significantly improve the pore structure of concrete. The main performance is that the percentage of harmful holes and multi-harmful holes is reduced, and the aperture is refined. With the increase of CNFs content, the number of harmful holes and holes in concrete decreased significantly, and the effect was the most obvious when the content was 0.3%, which was 34.7% and 45.6% less than that of PC harmful holes and multi harmful holes, respectively. It shows that CNFs can refine the internal pore diameter of concrete and improve its performance of concrete.

5 CONCLUSION

1. An appropriate amount of CNFs can enhance various static mechanical properties of concrete to varying degrees, but too much (0.5%) will weaken its performance.
2. The macro modification mechanism of CNFs on concrete mainly includes three aspects: crack resistance, reinforcement, and small-size effect. The performance of concrete is improved from three aspects: inhibiting the development of concrete cracks, refining concrete pores, and optimizing the structure of cement paste.
3. CNFs can effectively reduce the average pore diameter of concrete and greatly reduce the content of macropores in concrete. When the content is 0.3%, harmful pores and multi-harmful pores are reduced by 34.7% and 45.6% respectively, but the content of small pores will increase slightly.
4. Through the research, it is found that the mechanical properties of concrete can be significantly improved by adding an appropriate amount of CNFs. In the follow-up, supplemented by durability and engineering technology research, CNFs reinforced concrete technology can be developed and improved.

ACKNOWLEDGMENTS

This article is supported by the National Natural Science Foundation of China. The project approval number is 51908548.

REFERENCES

Abu Al-Rub R K, Tyson B M, et al. (2011) Mechanical properties of nanocomposite cement incorporating surface-treated and untreated carbon nanotubes and carbon nanofibers [J]. *Journal of Nanomechanics and Micromechanics*, 2(1): 1–6.

Chen H.Y., Wang J.H. (2005) Carbon nanofibers and their applications in polymers [J]. *FRP/Composite*, 2005 (3): 41–43.

Gdoutos E E, Danoglidis P A. (2016) Portland cement mortar nanocomposites at low carbon nanotube and carbon nanofiber content: A fracture mechanics experimental study [J]. *Cement and Concrete Composites*, 70: 110–118.

Li F., Kang W.M., Cheng B.W. et al. (2016) Preparation and electrochemical properties of silver loaded hollow carbon nanofibers [J]. *Materials Engineering*, 44 (11): 56–60.

Liu Y, Hu Y, Liu T, et al. (2015) Mechanical behavior of high density polyethylene and its carbon nanocomposites under quasi-static and dynamic compressive and tensile loadings [J]. *Polymer Testing*, 41: 106-116.

Ludvig P, Calixto J M F, Ladeira L O, et al. (2022) Analysis of cementitious composites prepared with carbon nanotubes and nanofibers synthesized directly on clinker and silica fume [J]. *Journal of Materials in Civil Engineering*, 29(6): 06017001.

Meng W, Khayat K H. (2018) Effect of graphite nanoplatelets and carbon nanofibers on rheology, hydration, shrinkage, mechanical properties, and microstructure of UHPC [J]. *Cement and Concrete Research*, 105: 64–71.

Miao C.W., Mu S. (2022) Development and prospect of concrete technology [J]. *Silicate Bulletin*, 39 (01): 1–11.

Trey. (2018) Experimental study on nano carbon fiber modified asphalt concrete [J]. *SME Management and Technology*, (22): 139–140.

W. Ferdous et al. (2020) Optimal design for epoxy polymer concrete based on mechanical properties and durability aspects[J]. *Construction and Building Materials*, 232(C): 117229–117229.

Wang A.G., Zheng Y., Zhang Z.H., Liu K.W., Ma R., Sun D.S. (2019) Research progress in improving the durability of concrete by modifying geopolymer cementitious materials [J] *Materials Reports*, 33 (15): 2552–2560.

Wang C.X., Liu L., Cao F.B., Lu L. (2020) Experimental study on mechanical properties of recycled concrete after freeze-thaw cycle [J]. *Journal of Building Structure*, 41 (12): 193–202.

Xu S.J., Zhang M., Fu S.Y. (2007) Preparation of nano carbon fiber/copper matrix composites by laminated die casting [J]. *Materials Science and Technology*, 15 (4): 503–506.

Yang J. and Chen B.C. and Nuti C. (2021) Influence of steel fiber on compressive properties of ultra-high performance fiber-reinforced concrete [J]. *Construction and Building Materials*, 302.

Yu W, Long Q, Zhao Y. C, et al. (2019) Physical properties of concrete containing graphene oxide nanosheets [J]. *Materials*, 40(11): 1417–1427.

Advances in Civil Engineering and Environmental Engineering – bin Yahya & Bin Daud (Eds)

A study on interfacial bonding properties between UHPC and ordinary concrete

Songji Xu
Shandong Hi-Speed Hubei Development Co., Ltd., Wuhan, China

Xiang Li & Wenxiang Fu
Hubei Wuma Expressway Co., Ltd., Wuhan, China

Fansheng Kong, Qiuyang Xiong & Fang Xu*
Faculty of Engineering, China University of Geosciences, Wuhan, China

ABSTRACT: The mechanical properties of the bonding interface between ultra-high-performance concrete (UHPC) and ordinary concrete (NC) are compared and analyzed, and the influencing factors of the UHPC bonding interface are reviewed. The changes in bonding properties of UHPC with different sulphoaluminate content were studied. It was found that the addition of steel fiber significantly improved the bonding strength of UHPC, which was briefly discussed. The mechanical test methods and test results of the bonding surface are discussed, and the research results on the mechanical properties and bonding strength of UHPC are described. The results show that the distribution of UHPC fiber has a significant effect on the interfacial bonding properties; The mechanical properties and time performance of the UHPC bonding interface are obviously better than that of the NC interface, and the time performance is better.

1 INTRODUCTION

Ultra-high-performance concrete (UHPC) has ultra-high strength and super durability, which is in line with the development trend of high strength, low carbon and environmental protection of concrete structures. UHPC is outstanding in the field of bridge engineering, has been successfully applied to more than 400 new bridges, and has also performed well in the field of bridge maintenance and repair. There are research circles on UHPC repairing bridge decks, expansion joints, piers, etc (Zhou et al. 2020). The superior performance of UHPC makes the reinforced bridge structure work normally under high-strength loads and harsh environments avoid defects such as aging and corrosion of repair materials, reduces later maintenance costs, and reduces energy consumption and carbon emission The gap between old and new materials is the weak link in repairing the structure. The bridge deck of a bridge in Switzerland has 30%–35% of the joint surface peeling after 2 years of concrete repair, and even more serious, the joint surface peeling completely occurs only 8 months after repair. Therefore, the good, efficient and lasting gap is the key to successful reinforcement, which determines the safety and durability of bridge reinforcement engineering Therefore, the research on the interface gap between UHPC and normal concrete (NC) has gradually become a hot spot. Taking the UHPC-NC bonding interface as the research object, this paper combs and analyzes the current research status and existing problems from two aspects of mechanical properties and influencing factors, and discusses the contents to be further studied in this field.

*Corresponding Author: xufang@cug.edu.cn

DOI 10.1201/9781003349563-33

2 TEST

2.1 *Materials*

The cement used in the experiment is P.O 52.5 ordinary portland cement and R.SAC 42.5 fast hardening high strength sulphoaluminate cement. Its basic composition and related properties are shown in Table 1 below.

Table 1. Chemical composition of P.O 52.5 ordinary portland cement (%).

Composition	Fe_2O_3	Al_2O_3	CaO	SiO_2	MgO
Content	4.52	5.75	64.5	23.2	1.3

2.2 *Test ratio design*

This experiment mainly studies the influence of the content of sulphoaluminate cement on the bonding performance of UHPC. A unified matrix mix proportion is adopted. The matrix mix proportion is obtained through a large number of mix proportion tests in the early stage and has the best mechanical properties. The specific mix proportion of the test is shown in Table 2. The concrete flexural test block made in this test is a 40×40×160mm cube standard test block that shall be cured in the curing box, and the bonding properties of test blocks at different ages shall be measured.

Table 2. The mixing ratio of the research experiment on the preparation method.

Test group	Portland cement	Sulphoaluminate cement	Silica fume	Fly ash	Steel fiber	Quartz sand	Water reducing agent	water-binder ratio
A1	750	0	170	180	120	990	35	0.18
A2	675	75	170	180	120	990	35	0.18
A3	600	150	170	180	120	990	35	0.18

2.3 *Test process*

(1) Weigh the raw materials required for the test according to the designed mix proportion.
(2) First wet the mixing pot, then pour the weighed materials into the mixing pot, dry mix for 2 min, then add the weighed water and stir slowly for 5 min. Finally, evenly distribute the steel fiber and stir quickly for 5 min, and the mixing is completed.
(3) First, conduct a fluidity test on the mixed UHPC. The first part of the test steps is to install 40 mm × 40 mm × 160 mm. It is formed in a 160 mm steel mold, placed at room temperature, covered and cured with plastic film, and its compressive and flexural strength are tested at the age of 1d, 3d, 7d and 28d. The second part of the test is then carried out, first at 40 mm × 40 mm × 160 mm ordinary concrete specimens with cross sections are placed in the 160 mm steel formwork, then poured into the mixed materials for molding, placed at room temperature, covered and cured with plastic film, and the bonding strength is tested at the ages of 1d, 3d, 7d, 14d and 28d.

3 ANALYSIS OF MECHANICAL STRENGTH AND WORKING PERFORMANCE OF UHPC

3.1 *Flow performance*

It can be seen in Figure 1 that when the content of sulphoaluminate cement gradually increases, the fluidity of fresh UHPC slurry first increases and then decreases. It can be seen from the graph change that when the content of sulphoaluminate cement is below 20%, the fluidity increases obviously, and when the content of sulphoaluminate cement is above 20%, the fluidity decreases slowly; When the content of sulphoaluminate cement is 20%, the fluidity of slurry is 163 mm, and the fluidity of slurry is good.

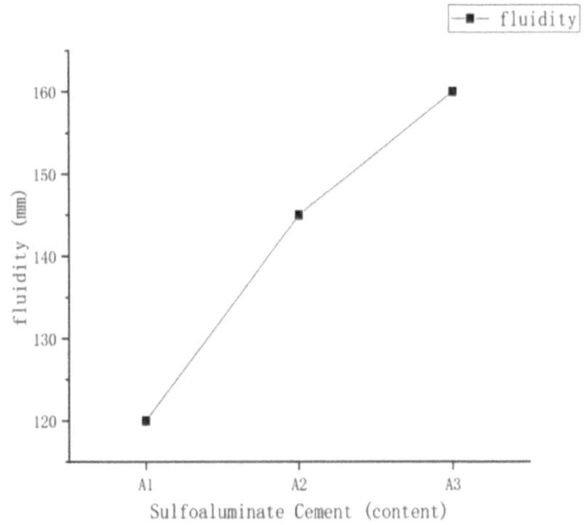

Figure 1. Effect of different sulphoaluminate content on the fluidity of UHPC slurry.

3.2 *Effect of sulphoaluminate cement content on mechanical properties*

Table 3 shows that sulphoaluminate cement content can enhance the early strength of UHPC. On the whole, the strength of UHPC shows a downward trend with the increase of sulphoaluminate cement content. When the age is 1d, the UHPC strength first increases and then decreases with the content of sulphoaluminate cement, and reaches the maximum when the content of sulphoaluminate cement is 20%. The flexural strength at 3d,7d and 28d reached the maximum when the content of sulphoaluminate cement was 10%. When the content of sulphoaluminate cement is more than 10%, it shows a downward trend, and the anti-shrinkage phenomenon appears. Therefore, the best content of sulphoaluminate cement is about 10%.

Table 3. UHPC compressive strength and flexural strength of different sulphoaluminate cement contents (MPa).

Test group	Age	A1	A2	A3
Compressive strength	1d	23.55	26.97	20.6
	3d	55.2	78.12	44.53
	7d	90.11	96.27	76.46
	28d	120.2	122.13	88.6
Flexural strength	1d	4.52	6.4	6.55
	3d	8.95	13.55	11.2
	7d	13.23	14.3	12.75
	28d	16.2	18.13	15.1

3.3 *Strengthening effect of steel fiber on the bonding strength of UHPC*

In the test, it is found that adding steel fiber material to fresh UHPC concrete can also improve the interfacial bonding performance between new and old concrete. The tensile elastic modulus of steel fiber is very high, which can significantly improve the tensile compression ratio of the concrete substrate. Chen, P. W. et al. (1995), Xie, H. C. et al. (2003) added carbon fiber into new concrete and found that the interfacial stiffness of new and old concrete increased by 80%–120%. The principle of strengthening the bonding strength of the bonding interface is shown in Figure 2, which mainly includes the following three aspects: ① the shrinkage of UHPC decreases after adding fiber, which can effectively reduce the shrinkage stress and the risk of shrinkage crack; ② The disordered fibers improve the roughness of UHPC interface, and the fibers at the interface extend into the pores of the ordinary concrete interface to bring higher mechanical bite force; ③ In the process of interfacial stress, the bridging effect between fiber and fiber can effectively inhibit the development of microcracks and change the failure mode of stiffening surface.

3.4 *Effect of sulphoaluminate cement content on UHPC-NC interface bonding strength*

It can be seen from Figure 2 that the content of sulphoaluminate cement can enhance the interfacial bonding strength of UHPC-NC. On the whole, with the increase of the content of sulphoaluminate cement, the interfacial bonding strength increases first and then decreases. The binding capacity of UHPC-NC with different sulphoaluminate cement content is obtained by comparing the interfacial bonding strength with the measured flexural strength. When the age is 1d, the UHPC strength first increases and then decreases with the content of sulphoaluminate cement, and reaches the maximum when the content of sulphoaluminate cement is 10%. The bond strength at 3d, 7d and 14d reached the maximum when the content of sulphoaluminate cement was 10%. When the content of sulphoaluminate cement is more than 10%, it shows a downward trend, and the bonding strength shows a backward shrinkage phenomenon, so the best content of sulphoaluminate cement is about 10%.

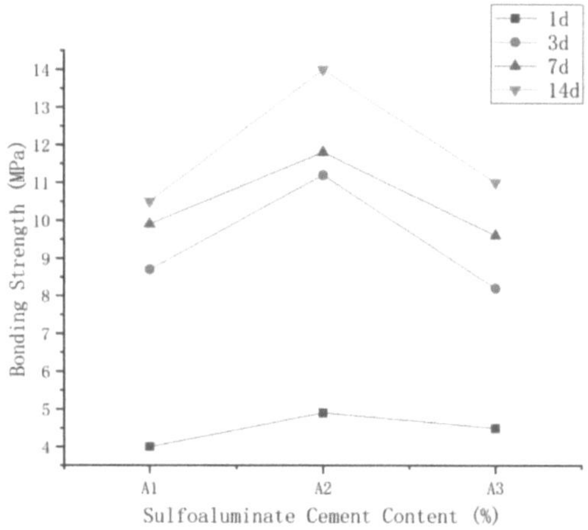

Figure 2. Effect of different sulphoaluminate cement content on bonding strength of UHPC-NC interface.

Figure 3 shows the failure interface of the bonding test. The failure surfaces occur on one side of the old concrete, and the slurry attached to the failure surface is the old slurry from the old concrete specimen. Therefore, it can be seen that UHPC material has excellent performance and good bonding performance. Due to the low bonding strength, when a fracture occurs, the interface is often the fracture surface. Therefore, in the flexural test, whether the slurry is attached to the failure surface at the interface, whether there is a crack and whether the crack propagation passes through the interface is to judge the bonding ability of UHPC-NC interface with different sulphoaluminate cement content.

Figure 3. The bond failure test and specimen failure interface.

4 CONCLUSION

Based on the results and discussions presented above, the conclusions are obtained as below:

(1) The fluidity of composite cementitious system concrete reaches the maximum when the content of sulphoaluminate cement is 30%. Combined with the development trend of mechanical strength, the optimal content of sulphoaluminate cement should be controlled between 10% ~ 20%. This content has a good molding effect, good fluidity and high early strength.
(2) Sulphoaluminate cement can enhance the interfacial bonding performance of UHPC-NC to a certain extent. When the content is 10%, the interfacial bonding performance of UHPC-NC is the best. The flexural strength at 14d reaches 14 MPa, and the bonding strength at 1d, 3d, 7d and 14d reaches the maximum when the content of sulphoaluminate cement is 10%. When the content of sulphoaluminate cement is greater than 10%, it decreases, and the bonding strength shrinks elephant.
(3) It is found that the fiber distribution and fiber direction at the interface will affect the transverse displacement of the failure surface. The steel fiber at the interface extends into the pores of the ordinary concrete interface to bring higher mechanical bite force and enhance the bonding strength.

ACKNOWLEDGEMENT

This work was financially supported by the Science and technology project of Shandong Hi-speed Group Co., Ltd (HSB2020119), Enterprise Technology Innovation Project of Shandong Province (202060100083) and the Science and Technology Project of Shandong Expressway Bridge Maintenance Co., Ltd (2020–07).

REFERENCES

Chen, P.W., Fu, X.L., Chung, D.D.L. (1995) Improving the bonding between old and new concrete by adding carbon fibers to the new concrete. *Cement. Concrete. Res.*, 25 (3).

China concrete & cement-based products association-UHPC. (2020) Technology and application development report of China UHPC in 2019. *Concrete. World.*, (02): 30–43.
Li, J.K., Li, Y., Li, Y.J. (2016) SWMM-based evaluation of the effect of rain gardens on urbanized areas. *Environ. Earth. Sci.*, (75): 17–31.
Lin, Z.Z., Yang, H., Chen, H.M. (2019) Influence of fillers on the removal of rainwater runoff pollutants by a permeable brick system with a frame structure base. *Water. Sci. Technol.*, 80(11): 2131–2140.
Shi, C.J, Wu, Z.M., Xiao, J.F., Wang, D.H., Huang, Z.Y., Fang, Z. (2015) A review on ultra-high-performance concrete: Part I. Raw materials and mixture design. *Constr. Build. Mater.*, 101:741–751.
Xie, H.C., Shen Y.B. (2003) Improvement of bonding performance of new and old concrete by carbon fiber reinforced concrete. *J. Civ Eng.*, (10): 15–18.
Yan, Y.W. (2020) Preparation and application of ultra-high-performance concrete (UHPC). *Build. Technol. Dev.*, 47 (07): 145–146.
Yin, X.X., Lai, J.Z., Zhou, J.H., Kang, N.Y., Du, L.Y. (2021) Numerical simulation of ultra-high speed penetration resistance of ultra-high performance concrete gradient materials. *J. Chin. Ceram. Soc.*, 49 (11): 2435–2449.
Zhang, X.H., Zhang, S., Luo, Y.M., Wang, L. (2020) Effects of interface orientations on bond strength between old conventional concrete and new self-consolidating concrete. *ACI. Struct. J.*, 117 (5): 23–29.
Zhou, J.T., Zhou, L., Yang, J., Wang, Z.S. (2020) Summary of research on interfacial adhesion between UHPC and ordinary concrete. *J. Jiangsu. U: Nat Sci Ed.*, 41 (04): 373–381.

Advances in Civil Engineering and Environmental Engineering – bin Yahya & Bin Daud (Eds)
© 2023 The Author(s), ISBN: 978-1-032-39397-1

Research on shrinkage rate of UHPC composite cement-based material for bridge expansion joints

Lei Li, Xianbing Ma & Jianghai Hu
Shandong Hi-speed Hubei Development Co., Ltd., Wuhan, China

Fansheng Kong, Qiuyang Xiong & Fang Xu*
Faculty of Engineering, China University of Geosciences, Wuhan, China

ABSTRACT: Because of the excellent properties of ultra-high-performance concrete (UHPC), it has potential in the use of bridge expansion joints (BEJ). And crack resistance is one of the important indicators of concrete applied to BEJ. The dry shrinkage rate is one of the indicators to characterize the crack resistance of concrete. In this paper, combined with the dry shrinkage rate and mechanical properties of the new UHPC composite cement-based material for BEJ, its performance research is taken out. It would provide practical advice for the practical application of the new UHPC composite cement-based material for BEJ. The results show that sulphoaluminate cement can obviously reduce the shrinkage of the new bridge expansion joint UHPC and reduce the risk of cracking. When the content of sulphoaluminate cement is 5%, the dry shrinkage rate of the new UHPC for BEJ basically keeps stable after 14 days.

1 INTRODUCTION

According to the Statistical Bulletin on the Development of Transportation Industry in 2020, the total mileage and density of roads in China have been increasing by 2020. And the mileage of road maintenance has reached 99.0%. By the end of 2020, the length of national highway Bridges has also increased. Freight vehicles and tonnage maintained an increasing trend compared with last year. By 2020, nearly 70% of the commercial passenger and cargo volume in the whole society will be transported by road. At the same time as highway bridge construction, with the situation of overloading, bridge expansion joints (BEJ) have an important influence.

Ultra-High-Performance Concrete (UHPC) has good properties of compressive, flexural, impact and other mechanical properties. UHPC can be applied to BEJ. Crack resistance is one of the important indexes of concrete applied to bridge expansion and contraction rate is one of the indexes of concrete crack resistance. UHPC is characterized by the low water-binder ratio (0.14–0.19), high content of cementation material, high content of active mineral admixture and low content of coarse aggregate (Huang & Sun 2017), accompanied by large self-shrinkage during condensation hardening and post-hardening service, and high risk of early cracking (Zhang et al. 2014). The shrinkage of UHPC is usually better than that of ordinary concrete, and the shrinkage development speed is fast. It is greatly affected by curing conditions. The self-shrinkage of UHPC accounted for a larger proportion than the drying shrinkage of the total shrinkage, and the shrinkage range was 80–170 $\mu\varepsilon$ (Vande Voort et al. 2008). UHPC is mainly self-contraction, accounting for 79%~90% of the total contraction (Delplace et al. 2012; Fehling et al. 2005; Li et al. 2016; Maeder et al. 2004). UHPC shrinkage mainly presents the characteristics of fast shrinkage deformation rate in the early stage, large deformation and small shrinkage deformation in the late stage (Graybeal, 2005;

*Corresponding Author: xufang@cug.edu.cn

DOI 10.1201/9781003349563-34

Soliman, 2011; Zhao et al. 2013). After 24h of standard curing in an environment of 20°C and 50% relative humidity, the test results of shrinkage are also different at different initial measurement times (Shink & Vacher 2008). Shink et al. (Shink & Vacher 2008) and Habel et al. (2006) found that UHPC basically stopped in 90 days with little growth.

In this paper, by testing the drying shrinkage rate of the new UHPC composite cement-based material for BEJ, the influence of the drying shrinkage and mechanical development of the new UHPC composite cement-based material for BEJ is studied. It has certain practical application significance.

2 TEST

2.1 *Materials*

P.O 52.5 ordinary Portland cement (PC) and R. SAC 42.5 fast hard sulphoaluminate cement (SAC) were used. Fly ash (FA) is grade I fly ash. Silica fume (SF) is with an average particle size of 89 nm, with a specific surface area of 1.85×104 m^2/kg. Superplasticizer (SP) is with a water reduction rate of more than 30%; Fine aggregate (A) was selected to use 20–40 mesh, 40–80 mesh, 80–120 mesh quartz sand; The steel fiber (F) is with the length of 13mm, with the diameter of 0.3mm. The chemical composition of cementitious materials is shown in Table 1.

Table 1. Chemical composition of cementitious material (%).

Composition	Fe_2O_3	Al_2O_3	SO_3	CaO	SiO_2	MgO
PC	4.51	5.76	/	64.5	23.1	1.4
SAC	1.43	5.85	1.78	48.9	5.85	1.78
FA	9.35	31.62	1.21	4.57	52.52	0.73
SF	0.14	0.25	0.58	0.36	95.4	0.47

2.2 *Test ratio design*

The experimental design mix ratio group was group A1~A5. The water-binder ratio is 0.18. By changing the content of SAC, the influence of the content of SAC on the dry shrinkage of UHPC composite cement-based material was studied. The specific match of the test is shown in Table 2.

Table 2. Test mix ratio.

Test group	PC	SAC	SF	FA	F	A	SP
A1	0.682	/	0.154	0.164	0.110	0.900	0.032
A2	0.648	0.034	0.154	0.164	0.110	0.900	0.032
A3	0.614	0.068	0.154	0.164	0.110	0.900	0.032
A4	0.545	0.137	0.154	0.164	0.110	0.900	0.032
A5	/	0.682	0.154	0.164	0.110	0.900	0.032

2.3 *Test procedure*

(1) The raw materials were weighed according to the design mix ratio.
(2) Wetting the mixing pot, then pouring the materials into the mixing pot, drying the mix for 2 min, then adding the weighed water, and stirring slowly for 5 min. Finally, put in the steel fibers and stir for 5 min.
(3) The mixed UHPC was first tested for fluidity. The steel die was formed in 40 mm × 40 mm × 160 mm, and covered with plastic film at 25°C for curing. The dry shrinkage rate and

compressive strength and flexural strength were tested at 1 day, 3 days, 7 days and 28 days. The XRD analysis was tested in 28 days.

3 TEST RESULTS AND DISCUSSIONS

3.1 *The early performance*

As shown in Figure 1, SAC has a significant impact on the early performance of the new UHPC composite cement-based material for BEJ. The fluidity, compressive strength (1 day) and flexural strength (1 day) of the new UHPC composite cement-based material for BEJ all increase first and then decrease with the content of SAC. It can be seen that when the content of SAC is 0.068, the early performance of the new UHPC composite cement-based material for BEJ is the best.

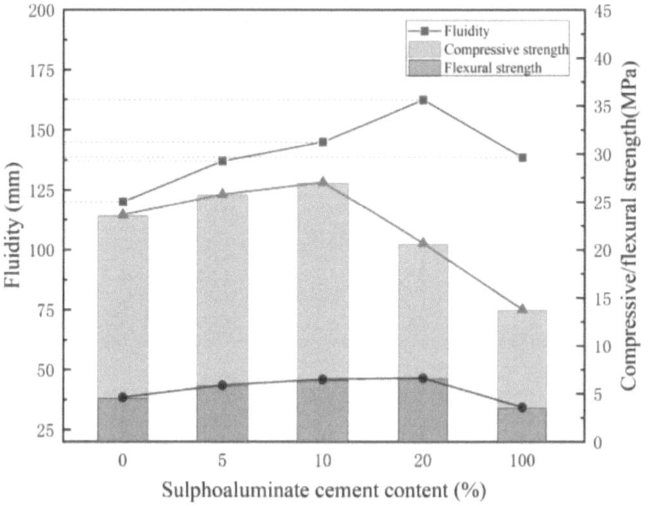

Figure 1. Early performance with SAC.

3.2 *Drying shrinkage rate*

As shown in Figure 2, when the content of SAC increases, the drying shrinkage rate of UHPC in 1 day, 3 days, 7 days and 28 days shows a trend of first increasing and then stabilizing. When the SAC content reached 0.034, there is expansion at the beginning of 3 days. However, the addition of SAC has an effect on the dry shrinkage rate of UHPC, and with the increase of SAC, the dry shrinkage rate of UHPC will gradually decrease. The dry shrinkage rate of the new UHPC for BEJ with the dosage of 5% SAC basically keeps stable after 14 days.

3.3 *Compressive strength*

As shown in Figure 3, when the content of SAC increases, the compressive strength of UHPC shows a trend of first increasing and then decreasing. When the content of SAC reaches 0.068, its compressive strength reaches the maximum value. However, the addition of SAC has a negative effect on the 28 days' compressive strength of UHPC, and with the increase of SAC content, the 28 days' compressive strength of UHPC decreases rapidly compared with other ages.

3.4 *Flexural strength*

Figure 4 shows that when the content of SAC increases, the flexural strength of UHPC at 1 day, 3 days, 7 days and 28 days shows the tendency of first increasing and then decreasing. When the ratio

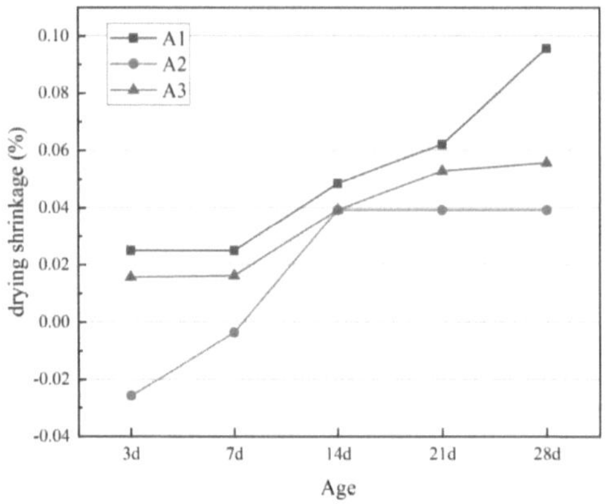

Figure 2. Effects of SAC on the drying shrinkage rate: d (day).

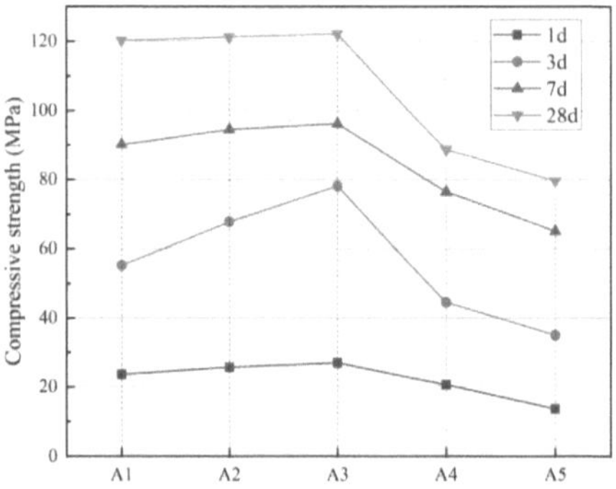

Figure 3. Effects of SAC on the compressive strength.

of SAC reaches 0.068, its flexural strength reaches the maximum value. However, the addition of SAC has a negative effect on the 28 days' flexural strength of UHPC, and the 28 days' flexural strength of UHPC decreases with the increase of the addition of SAC.

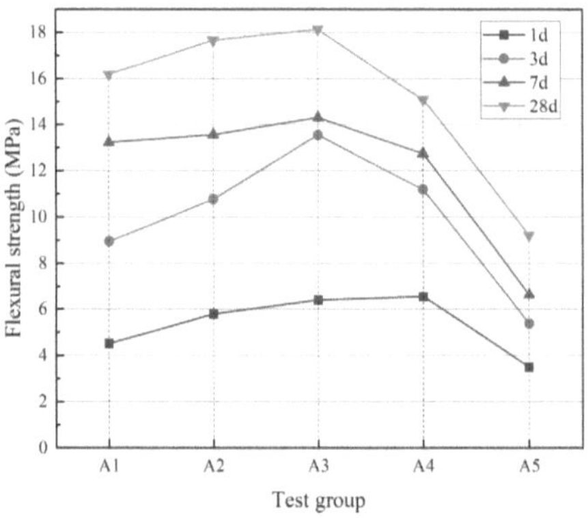

Figure 4. Effects of SAC on the flexural strength.

3.5 *XRD analysis*

Figure 5 shows that with the increase in the proportion of SAC, the diffraction peak of C_3S gradually weakens, and the diffraction peak of $C\bar{S}$ and AFt gradually increases. This indicates that SAC can effectively promote hydration and produce C-S-H gel in UHPC composite cement-based materials. SAC and PC can promote the mineral composition hydration reaction and produce more AFt. AFt provides certain shrinkage resistance.

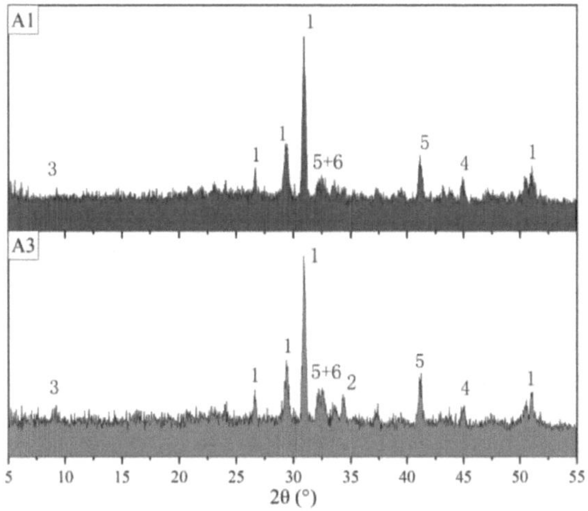

Figure 5. XRD patterns in 28 days: 1-C-S-H, 2-C-A-H, 3-AFt, 4-$C\bar{S}$, 5-C_2S, 6-C_3S.

4 CONCLUSION

Based on the results and discussions presented above, the conclusions are obtained as below:

(1) Due to its excellent properties, low porosity and high compactness, UHPC has good impermeability. UHPC can be applied to bridge expansion for maintenance. Crack resistance is one of the important indexes of concrete applied to bridge expansion and contraction rate is one of the indexes of concrete crack resistance.
(2) When the SAC is added, the early strength of the new UHPC composite cement-based material for BEJ is enhanced, and the recommended dosage is between 5 and 10%. However, with the increase of SAC, the 28 days strength of UHPC will be adversely affected.
(3) When the SAC is mixed, the early drying shrinkage rate of the new UHPC composite cement-based material for BEJ will be affected, and the recommended dosage is 5%. The dry shrinkage rate of the new UHPC for BEJ with the dosage of 5% SAC basically keeps stable after 14 days. At 1 day, 3 days, 7 days and 28 days, the drying shrinkage rate of UHPC firstly increases and then stabilizes. With the increase of SAC, the drying shrinkage rate of UHPC will decrease.

ACKNOWLEDGEMENT

This work was financially supported by the Science and technology project of Shandong Hi-speed Group Co., Ltd (HSB2020119), Enterprise Technology Innovation Project of Shandong Province (202060100083) and the Science and Technology Project of Shandong Expressway Bridge Maintenance Co., Ltd (2020-07).

REFERENCES

Delplace, G., Hajar, Z., Simon, A. (2012) *Precast Thin Shells made of UHPFRC for a Large Roof in a Wastewater Treatment Plant Near Paris*. In: Schmidt, M., Fehling, E., Glotzbach, C., Fröhlich, S., Piotrowski, S. *3rd International Symposium on UHPC and Nanotechnology for High Performance Construction Materials*. Kassel University Press, Kassel. pp. 1011–1018.

Fehling, E., Schmidt, M., Teichmann, T., Bunje, K., Bornemann, R., Middendorf, B. (2005) *Entwicklung, dauerhaftigkeit und berechnung ultra-hochfester betone* (UHPC). Kassel University Press, Kassel.

Graybeal, B.A. (2005) *Characterization of the behavior of ultra-high-performance concrete*. University of Maryland, College Park.

Habel, K., Charron, J.P., Denarie, E., Bruehwiler, E. (2006) Autogenous deformations and viscoelasticity of UHPFRC in structures. Part I: Experimental results. *Mag. Concrete. Res.*, 58(3): 135–145.

Huang, W., Sun W. (2017) Effects of limestone addition on hydration development of ultra-high-performance concrete. *J. Southeast. U: Nat. Sci. Ed.*, 47(4): 751–759.

Li, C., Luo, X., Chen, B.C., Wei, J.G. (2016) *Experimental on Shrinkage Performance of Ultra-High-Performance Concrete under Sealed Condition*. In: Shi, C.J., Wang, D.H. 1st *International Conference on UHPC Materials and Structures*. RILEM Publications, New York. pp. 420–429.

Maeder, U., Lallemant Gamboa, I., Chaigmon, J., Lombard, J.P. (2004) *Ceracem, a New High-Performance Concrete: Characterizations and Applications*. In: Schmidt, M., Fehling, E., Geisenhanslüke, C. (2004) *1st International Symposium on Ultra-High-Performance Concrete*. Kassel University Press, Kassel. pp. 59–68.

Shink, M., Vacher, J.P. (2008) EFFIX design®—properties and applications of a creative concrete. In: Fehling, E., Schmidt, M., Stürwald, S. (2008) *2nd International Symposium on Ultra-High-Performance Concrete*. Kassel University Press, Kassel. pp. 847–854.

Soliman, A.M. (2011) Early-age shrinkage of ultra-high-performance concrete: Mitigation and compensating mechanisms. *The University of Western Ontario*, London.

Vande Voort, T.L., Suleiman, M.T., Sritharans. (2008) Design and performance verification of ultra-high performance concrete piles for deep foundations. *Iowa State University*, Ames.

Zhang, Y.S., Zhang, G.R., Li, S.C. (2014) Study on early autogenous shrinkage of ultra-high performance cementitious composite. *J. Build. Mater.*, 17(1): 19–23.

Zhao, J., Lian, H.Z., Jin, J.C. (2013) A new model of steel-concrete composite-ultra-high-performance concrete (UHPC/UHPFRC): shrinkage and cracking, high-temperature-performance, permeability and durability and design guide. *China Concrete*, 12: 60–71.

Advances in Civil Engineering and Environmental Engineering – bin Yahya & Bin Daud (Eds)
© 2023 The Author(s), ISBN: 978-1-032-39397-1

Research on flying gangue movement character under the different angle

Ming Liu, Kangfei Xiao & Lin Sun*
School of Environment and Safety Engineering, Liaoning Petrochemical University, Fushun, China

ABSTRACT: Aiming at the problems of dynamics behavior is not perfect between the flying gangue and the bottom plate of working face in steeply dipping seam, The goal of this study is to understand the working face of the flying gangue of longwall mining in steeply dipping seam, and based on distinct element method (DEM), the motion model of flying gangue is established. By using numerical simulation, the motion character for flying gangue of longwall mining working face in steeply dipping seam is systematically studied. The accurate calculation for motion of flying gangue will provide a reliable theory basis for improving control technology of flying gangue and ensuring safe mining in a steeply dipping seam.

1 INTRODUCTION

Many random factors exist due to the environmental complexity in steeply dipping seam mining; moreover, the flying gangue is easy to slide. In reality, the motion of flying gangue is very complicated, so it is difficult to have an accurate calculation for flying gangue movement character (Liu et al. 2016; Wu et al. 2020, 2013; Wang & Jiao 2016; Wu et al. 2019; Wu et al. 2014; Yun et al. 2017). Therefore, the flying gangue is used as an analytical model in this paper The purpose is to develop a practical and useful method for determining input parameter values for distinct element method (DEM) simulation of three-dimensional flying gangue motion on slopes. DEM provides a powerful analytical tool for modeling flying gangue behaviors, such as trajectories, velocities, and energies of flying gangue. Distinct element analysis of flying gangue requires known values of five parameters, as shown in Figure 1. These parameters are normal spring constant K_n tangential spring constant K_s normal damping ratio η_n, tangential damping ratio η_s, and friction coefficient μ.

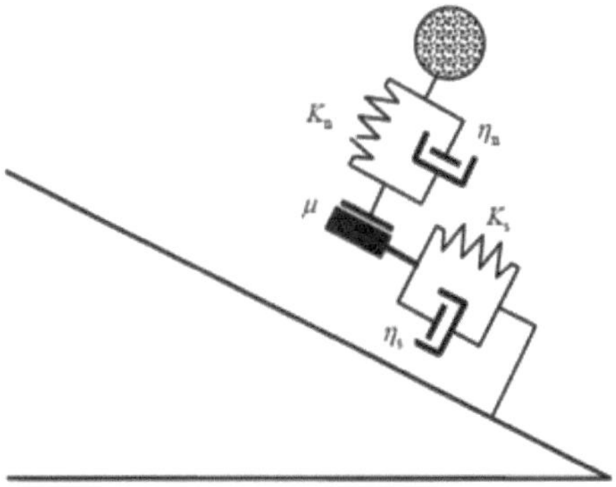

Figure 1. DEM contact model between slope and flying gangue.

*Corresponding Author: sunlin0626321@163.com

The normal spring constant and the tangential spring constant can be obtained according to the measuring method of the elastic wave velocity of rock strata, by using the following formulas:

$$K_n = \frac{1}{4}\pi\rho V_n^2, \quad K_s = \frac{1}{4}\pi\rho V_s^2 \tag{1}$$

where V_n is normal elastic wave velocity, and V_s is tangential elastic wave velocity.

Damping ratio can be determined by the following formulas:

$$\eta = 2\xi\sqrt{Km} \tag{2}$$

where ξ is damping coefficient. The relationship between the recovery coefficient and damping coefficient can be expressed as:

$$\xi = \frac{\ln e}{\sqrt{\ln^2 e + \pi^2}} \tag{3}$$

Although DEM numerical simulation of the flying gangue behavior has not been widely carried out in practice, an effective and efficient method has not been established as yet for determining the input parameters which govern flying gangue motion and may differ greatly from one slope to another due to the difference in geometry and rock properties involved in the slope. Assessment of the dynamic DEM parameters is rather difficult because flying gangue motion depends on many factors including the geometry of flying gangue and slope, deformation and strength characteristics of slope materials, and bumpiness. Another difficulty associated with the determination of the above-mentioned parameters is that they cannot be obtained directly from flying gangue experiments. A flying gangue experiment can produce a trajectory of the used flying gangue and its velocity distribution, but it cannot directly provide DEM parameters. In particular, it is difficult to assess the magnitude of these parameters considering the effects of deformation of the slope when the flying gangue collides against the slope and other barriers on the slope.

In this paper, a distinct element method is presented between the flying gangue and a given slope. The contact area between the flying gangue and the slope can be treated as the boundary. At the same time, the displacement and rotation are allowed off the discrete block and even completely leave, and in the process of computing, the new contact surface must be automatic identification to ensure the calculation run smoothly, the contact models of two rigid blocks and deformed blocks are shown in Figures 2 and 3.

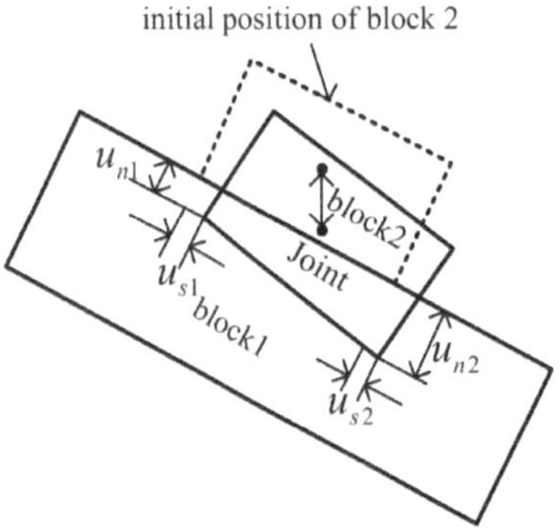

Figure 2. Contact between two rigid blocks.

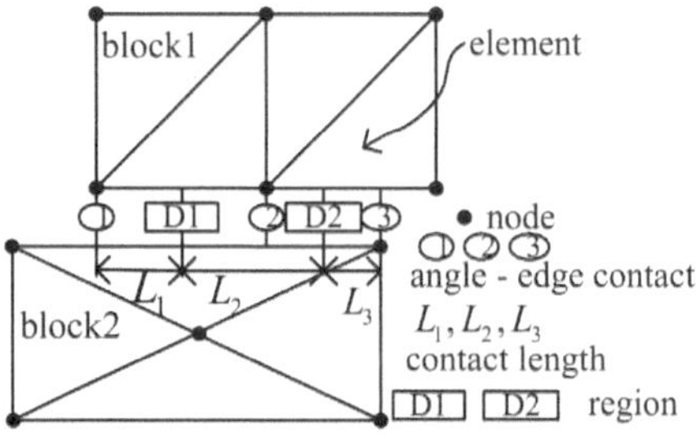

Figure 3. Contact between the two deformation blocks.

2 THE MOVEMENT CHARACTERISTICS OF FLYING GANGUE UNDER THE DIFFERENT MINING HEIGHT

The experience of numerical simulation has shown that whether the results of numerical simulation are closer to the actual result depends on the accuracy of parameter selection. This paper adopts the Mohr-Coulomb yield criterion to determine the destruction of the rock mass in order to do simulation, which is closer to the engineering practice.

(1) The 35° and 45° calculation models of large mining height was set up in steeply dipping seam respectively, the model of the constitutive properties and material parameter were set, fixing boundary, giving a initial condition, calculating initial equilibrium;
(2) The working face length is 110m, the mining height is 4.5m, and the strike advance is 4m at a time.
(3) Output the results of typical images and data, comparing and analyzing the movement characteristics of flying gangue under different mining heights.

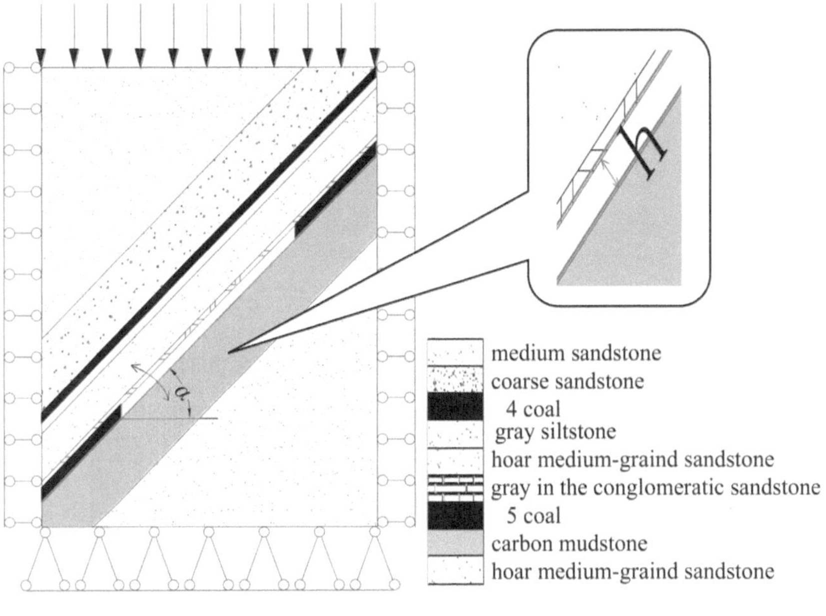

Figure 4. A model for the numerical calculation with different mining height.

3 FLYING GANGUE MOVEMENT CHARACTER UNDER THE DIFFERENT DIP ANGLE

The motor processes for flying gangue derived by wall caving in the upper working face under the 35° dip angle of the coal seam as shown in Figure 5.

Figure 5 shows the wall caving, free fall, hit bottom, bounce, the second fall, the skid and stop after the second hit bottom plate occurred in turn, the flying gangue hit the bottom of the working face two times, bounced one time, the movement distance of 15.0 m. The process is as follows: the coal wall entered a free fall after wall caving under its own gravity, the movement speed had an ever-increasing; it collided with the bottom of working face for the first time, then bounced the reaction plate left bottom bounce. Its maximum height is 0.23 m from the floor. It rose to a certain height, fell again, and hit a second collision plate. The distance between the two collision plates is 0.8 m. That is, the waste material flies 1.2 m before it is cleared away. After the second impact with

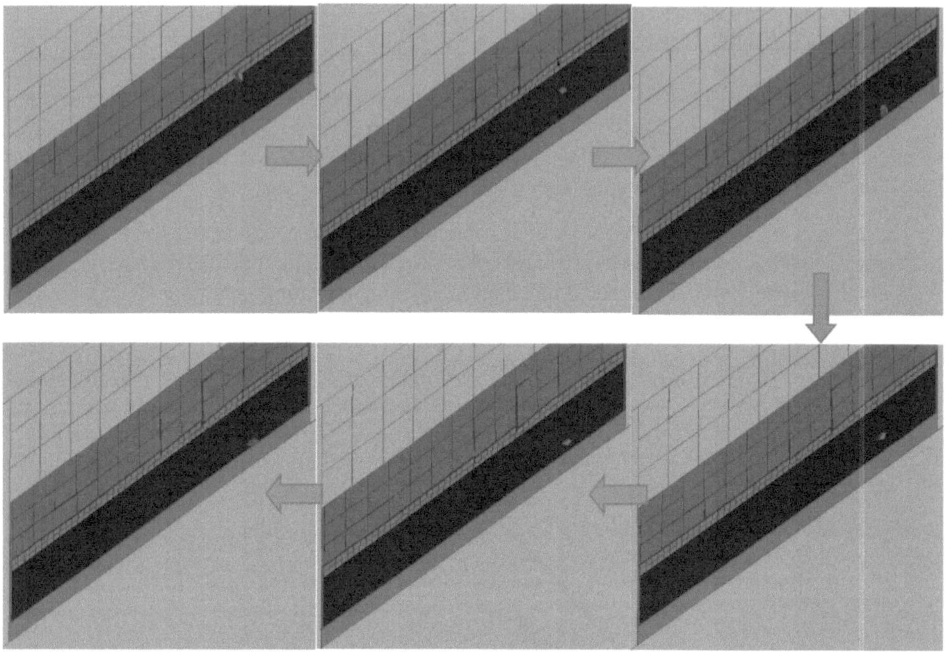

Figure 5. Process of flying gangue movement in 35° coal seam.

the floor, because the energy of the fly ash decreases obviously, it does not leave the floor again, but along the slope face smoothly rolls to a stop due to bottom friction and other resistances of the coal waste body. The 35° seam fly ash relationship between displacement and time is numerically simulated with a time step of 0.002 s and a time of 40 cm, as shown in Figure 6.

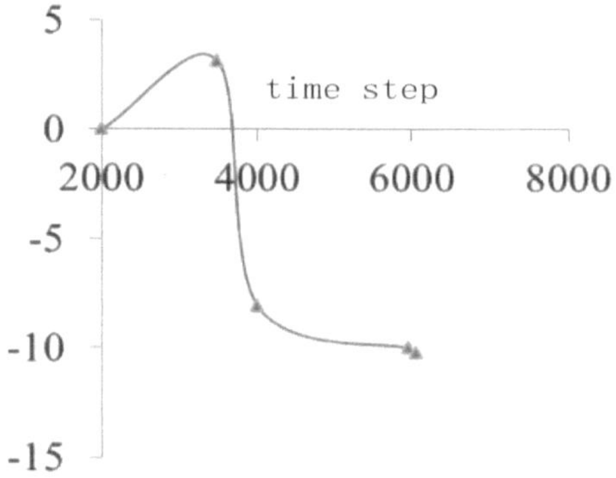

Figure 6. Relationship between time and the x direction displacement of flying gangue.

Figure 7 shows the process of flying gangue movement in a 45° coal seam. From left to right, the flying gangue is in free fall, the first collision between the flying gangue and the bottom plate inclines to the working face and bounces up, the second bouncing is the maximum distance from the bottom plate and stops rolling along the working face. The first time is 0.6m. It moves along the inclination of the working face for 8m and then falls to the bottom plate, colliding with the bottom plate. After bouncing up again, it moves along the inclination direction for a distance and then falls to the bottom plate and continues sliding.

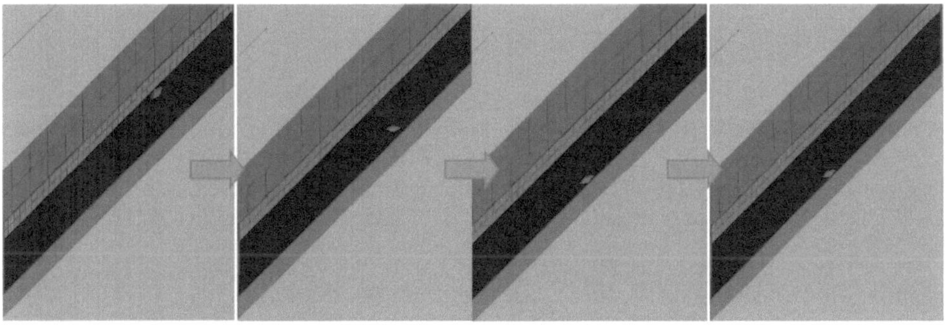

Figure 7. Process of flying gangue movement in 45° coal seam.

As shown in Figure 8, the displacement curve of the flying gangue with a 45° angle shows that the flying gangue moves in free fall to the bottom plate of the first collision, and the displacement and velocity values keep increasing. After the collision, part of the energy is dissipated, and the velocity value decreases. Meanwhile, the displacement increases slowly. When the flying gangue moves downward from the maximum position away from the bottom plate of working face and under the condition of work done by its own gravity, the velocity value increases. Colliding with the bottom plate for the second time, the velocity value decreases. Then the flying gangue rolls along the working face tendency and stops at a distance of 0.34m. In the whole process, the flying gangue tilts 24.7m along the working face.

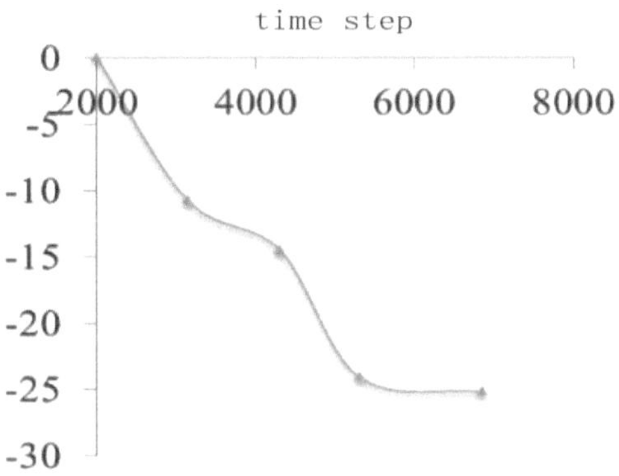

Figure 8. Relationship between time and the x direction displacement of flying gangue.

4 CONCLUSION

The analysis of motion characteristics of flying gangue at different dip angles shows that the dip angle plays a decisive role in the formation of flying gangue, and the large gangue and lump coal are difficult to self-stabilize under the action of dip angle after the coal wall spalling of the inclined longwall working face of steeply dipping seam. With larger the dip angle of coal seam, the gravity work of flying gangue increases, the friction effect of bottom plate decreases, and the more times the flying gangue moves along the working face and intermittently leaves the bottom plate. In the process of leaving the bottom plate intermittently, part of the energy is dissipated in the internal fissure expansion, and shape changes every time the flying gangue collides with the bottom plate, and the flying gangue splits and disperses into multiple pieces. When the flying gangue has energy and cannot make itself leave the bottom plate intermittently (bouncing up), it starts to roll and slide along the bottom plate and stops moving after a certain distance.

ACKNOWLEDGEMENTS

This work was financially supported by the FuShun Revitalization Talents Program (FSYC202107004), and the National Natural Science Foundation of China (51604213).

REFERENCES

Liu M, Wu Y P, Xu G. 2016. Research on rule of flying gangue in steeply dipping seam mining. *Coal Technology*, vol. 35, pp. 17–18.

Wang J A, Jiao J L. 2016. Criteria of support stability in mining of steeply inclined thick coal seam. *International Journal of Rock Mechanics and Mining Sciences*, vol. 82, pp. 22–35.

Wu Y P, Hu B S, Xie P S. 2019. A new experimental system for quantifying the multidimensional loads on an on-site hydraulic support in steeply dipping seam mining. *Experimental Techniques*, pp. 1–15.

Wu Y P, Liu K Z, Yun D F, Xie P S, Wang H W. 2014. Research progress on the safe and efficient mining technology of steeply dipping seam. *Journal of China Coal Society*, vol. 39, pp. 1611–1618.

Wu Y P, Yun D F, Xie P S, Wang H W, Lang D and Hu B S. 2020. Progress, practice and scientific issues in steeply dipping coal seams fully-mechanized mining. *Journal of China Coal Society*, vol. 45, pp. 24–34.

Wu Y, Xie P, Wang H. 2013. Theory and practices of fully mechanized longwall mining in steeply dipping coal seam. *Mining Engineering*, vol. 65, pp. 35–41.

Yun D F, Liu Z, Cheng W D, Fan Z D, Wang D F, Zhang Y H. 2017. Monitoring strata behavior due to multi-slicing top coal caving longwall mining in steeply dipping extra thick coal seam. *International Journal of Mining Science and Technology*, vol. 27, pp. 179–184.

Engineering survey technology and construction risk assessment

Advances in Civil Engineering and Environmental Engineering – bin Yahya & Bin Daud (Eds)
© 2023 The Author(s), ISBN: 978-1-032-39397-1

Research on the implementation method of big data management engineering archives

Yuling Huang*, Chaoqun Leng & Liping Zhan
Nanchang Institute of Technology, Nanchang, Jiangxi, China

ABSTRACT: Along with the rapid development of information technology, big data is playing an important role of support technology in the field of modern modeling. Especially, since the current social actual application of all sectors of big data has become more and more frequent, the rise of the actual application of big data leads people to research big data from all walks of life, and also lead to an increase in application research. As an important part of social development, archives management can protect the basic rights and interests of the people to a certain extent. As the center of archives management, archives management personnel need to take archives management as the central point, and pay attention to improving the overall quality of archives management, to provide support for archives management work.

1 INTRODUCTION

The continuous innovation of technology in the era of big data not only promotes the continuous improvement of the construction engineering archives management level but also greatly promotes the information construction of construction engineering archives management. The information construction of construction engineering archives management is the necessary stage of the development process of construction engineering archives management. It is of great practical significance for construction engineering archives management (Zhang 2010).

2 ARCHIVES MANAGEMENT

Archives management refers to the process of the project research, the design of the project, the implementation of the project, and the results of the materials formed by the sorting, ghosting, and refining, so as to record, standardize, and guide the management behavior of the project research. Archives management is also known as archival work (Gao 2015). It is the general term of the business work of archives (offices) that directly manages archives entities and archives information and provides the use of services, and it is also the most basic part of the national archives cause. Archives management is mainly within the filing of documents produced by administrative organs for management and application operations. Its general process includes file generation, file classification, file catalog, and file sorting of several processes. These works are only relatively stable rather than absolute, and because modern archives management has become a complex system, there are also multilevel methods. The first level is divided into archival entity management and archival information development, and each subsystem is divided into several levels of small systems. Archives entity management is divided into the collection, sorting, identification, storage, statistics, and other work links. Archives information development is divided into information

*Corresponding Author: qgjxzbt@163.com

processing and information output. Information processing consists of catalog compilation, editing literature compilation, and compiling reference materials, and information output is composed of multiple service activities such as reading, copying, consulting, correspondence, lending, and publishing and exhibition. The entire archive management system and its subsystems form a feedback mechanism in operation. The development of archives management modernization will also have a new impact on the structure of archives management work. The ultimate purpose of archives management is to provide archival information for social practice services, and the structure of the archives management system is set according to this purpose. Each of these works is essential and has certain procedures. They form an organic whole and play their own role in realizing the overall function of the archives management system, but also associate and restrict each other. For example, the value appraisal work is sometimes carried out in combination with the collection and sorting work, and even the preliminary appraisal when the documents are filed.

The object of archives management is the archives and the service object is the user of archives. The basic contradiction that need to be solved is the contradiction between the scattered, chaotic, miscellaneous, large quantity, solitary situation and the requirements of social utilization, systematic, high quality, special reference, and extensive use of archives. The degree of social satisfaction for archives depends on the continuous improvement of the archives management level, while the level of archives management should adapt to the growing social needs. The two are in the process of constant contradictions from adapting to adapting, so as to promote the development of archives management work forward. From the history of thousands of years, archives management develops from independent systems to independent systems, from simple to complex management, from experience to scientific management, from manual to computer management, and from closed to open systems (Gao 2015).

3 THE NECESSITY OF FILE MANAGEMENT IN THE ERA OF BIG DATA

Archives work from a narrow sense is the archives business work, is a series of business work around the archives around the temple, and is the use of scientific principles and methods to manage archives, for the work of social services. The necessity is mainly included in the following aspects.

3.1 *With all the needs for social development*

With the continuous development of the social economy, the construction of a digital city has become the basic requirement of urban development, and such social development puts forward higher requirements for archives management (Wang 2021). Therefore, the relevant departments in society need to take archives management as the focus of the work, carry out effective quality management and optimization according to the actual requirements, try their best to improve the management quality of archival information, and realize the effective preservation of archives management materials. In the resource optimization of archives management, if the obtained data resources cannot be comprehensively analyzed to a certain extent, the efficiency of the management of archives management will be reduced and is not conducive to the stable and effective development of the society (Liu 2021). Therefore, archives management needs to conduct real-time docking with big data, optimize and integrate with cloud resources, clarify the core goals of data archives management, and comprehensively follow up on the development of society.

3.2 *Data resource management*

In the good situation of big data operation and development, data management is also a big problem. In the face of countless data, in order to effectively avoid data chaos, we can use platforms such as cloud storage to conduct orderly resource integration. Through the sorting of paper manuscripts, we can give full play to the core mechanism of document resource management, so as to improve the utilization efficiency of electronic file information and realize the centralized management of massive files, through the voluntary inquiry and improvement of relevant procedures, thus providing a good guarantee for the data optimization of file resources (Dong 2021).

3.3 *Convenient the storage and management of archives*

After extensive data integration and optimization, it is necessary to store and back up archives information resources. With the increase in electronic data and files, if information resources are not integrated and managed promptly, data confusion and loss may occur, or certain losses may be caused (Shen 2000). At this time, the storage and management of archives became particularly important. Through the application of big data resources, the data resources of archives management can be improved, and the information storage and complete analysis are effectively realized on this basis, which improves the service quality of electronic archives and promotes the continuous improvement of the quality of electronic archives service (Li 2021).

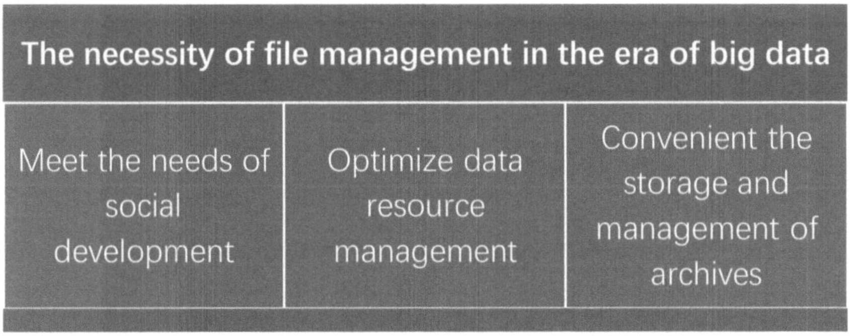

Figure 1. The necessity of file management in the era of big data.

4 PROBLEMS FACED BY BIG DATA MANAGEMENT ENGINEERING ARCHIVES

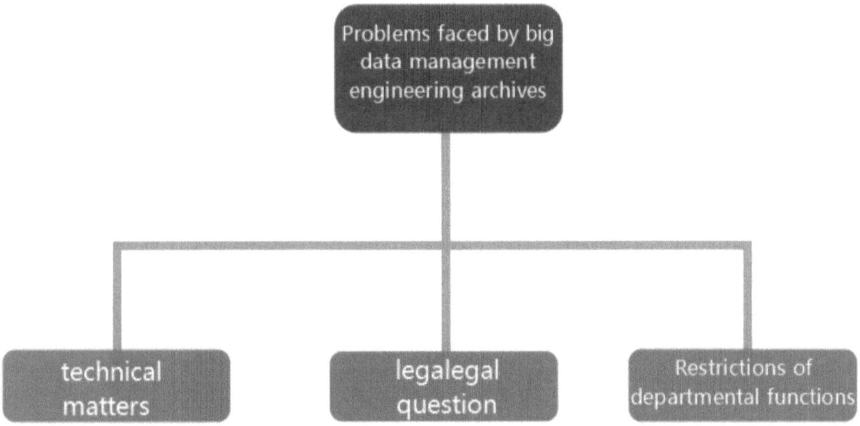

Figure 2. Problems faced by the big data management engineering files.

4.1 *Technical matters*

Huge data are filled with computers, and it is impossible to complete this degree of record volume only relying on the human brain (Lu 2017). In today's era, for the file management problem under big data, according to the passage of time and the increase of data volume, the system should also be improved to a certain extent. In the process of technical analysis and optimization, if technical management cannot conform to the development of the times, it will fall on others on the basis, thus causing certain constraints on the integration of archival information. In view of this situation, at the same time as the old and new technology reform, the imperfect traditional technology should also be abandoned by the times, leaving where it is beneficial to the modern management archives system, taking its essence from its dross, so as to clear a part of the obstacles for archives management (Zhang 1997).

4.2 *Legal question*

In the process of technology optimization and information integration, the management of big data also faces certain challenges, which will cause certain constraints to voluntary information innovation and the need for file management. Moreover, China is relatively short on network information management, and anyone can release any information on the Internet, whether true or false. This leads to the phenomenon of wrong information that people receive. Meanwhile, the lack of timely judgment and integration of information will also lead to further false dissemination of information. The emergence of this phenomenon will also affect the management of archival resources.

4.3 *Restrictions of departmental functions*

In the optimization of file management and the optimization of difficult management functions, the perfect information management mechanism is more difficult for now. In the process of optimizing and integrating information, it is also necessary to improve the basic functions of the department and strive to achieve the stable integration of the value of information management. At the same time, under the background of big data development, the design of departmental functions has certain restrictions to some extent, and the emergence of this phenomenon will bring certain constraints to the optimization of departmental functions (Li 2017).

5 IMPLEMENT THE MEASURES FOR THE INFORMATION CONSTRUCTION OF PROJECT ARCHIVES MANAGEMENT

5.1 *Improve the awareness of archives management personnel on the construction of project management information construction*

From the perspective of the information construction of the engineering archives management, the understanding of the engineering archives management has a very important impact on the work efficiency of the management of its archives. In the process of information construction of project archives management, its staff is the main force of information construction (Zhao 1997). If they do not achieve a correct understanding of the importance of engineering archives information construction, then it is difficult to develop on the road to promote the construction of project archives. Therefore, in order to ensure the information construction of engineering archives management, it is necessary to carry out a comprehensive, systematic, and complete training in information construction for archives management personnel, so as to improve the cognitive level of archives management personnel for information construction. To ensure that the archives management personnel can fully realize the importance of project archives management from the consciousness level, so as to play a strong role in promoting the process of information construction of project archives management (Liu 2021).

5.2 *Take the initiative to promote the information process of project archives management*

In order to smoothly promote the information process of project management, first, we should broaden the idea of information management, in the information construction of advanced information technology as the basis, and then on this basis of the project archives management mode of reform and innovation (Wang 2021). At the same time, it is necessary to combine the actual situation of archives materials to improve the secondary utilization rate of archives materials, to ensure that the essence of the project archives management can be fully excavated and effectively utilized. Second, it is necessary to constantly improve the informatization level of engineering archives management, store the data in the corresponding electronic files, under the management and control of all kinds of archives management software, and integrate the management of all kinds of electronic files, so as to improve the work efficiency of archives management. Finally, to ensure the safety of engineering archives management data and to further improve the safety of archives management system, all kinds of engineering archives data can be stored.

5.3 *Strengthen the system construction of the information work of project archives management*

In order to promote the information work of project archives management, it is necessary to establish a file management information system in line with the actual work system. China has set up relevant specific implementation units for enterprises in the field of engineering management, so it is necessary to combine the actual situation of engineering archives management and innovate in the management system, so as to ensure to better adapt to the archives level of engineering management, and then save the information construction of engineering archives management under the background of the era of big data (Li 2017).

6 CONCLUSION

Under the current information technology optimization stage and the background of big data integration, it is an opportunity yet a challenge for the information construction of project management. The construction of the engineering archives management system has a long way to go. The managers of project management archives must carry out theoretical innovation, learn to use big data technology, change the way of thinking about traditional archives, create the mode of project archives management, and strive to build a good project archives management system. This paper gives a comprehensive elaboration and systematic analysis of the background of big data, which has some reference significance to the staff engaged in engineering archives management (Gao 2015).

REFERENCES

Dan Zhang. On the legal management of engineering archives [N]. *Yellow River News*, 2010-12-09 (003).

Hua Gao. Research on the implementation methods of big data management engineering archives [J]. *Lauder World*, 2015 (S3): 48–49.

Jia Dong. Research on electronic management under the background of big data [J]. *Small and Medium Enterprises Management and Technology* (MS), 2021 (10): 37–39.

Jiangli Lu. Implementation methods for big data management project archives [J]. *Global Market Information Guide*, 2017 (17): 107.

Jicheng Zhang, Changren Chen. Data management based on PDM [J]. *Archives and Construction*, 2017 (02): 84–87.

Kaihua Zhao. Archives of engineering design drawing with computer technology [J]. *Forestry Science and Technology Information*, 1997 (04): 28–29.

Qiangdong Shen. Implement the electronic archives management project to promote the modernization of enterprise archives management [J]. *Engineering Construction and Design*, 2000 (01): 35–36.

Shujuan Wang. Application of innovation consciousness in the comprehensive file management of public institutions [J]. *Heilongjiang Archives*, 2021 (04): 136–137.
Wei Li, Zhao Weimin. Engineering conception and preliminary exploration of archives management [J]. *Urban Construction Archives*, 2017 (04): 90–91.
Wei Wang. The application of scientific management and modern technology in archives management [J]. *Modern Enterprise*, 2021 (09): 56–57.
Wenjuan Li. The information development trend of archives management [J]. *Office Automation*, 2021,26 (18): 56–57 + 22.
Yong Liu. Implementation strategy of strengthening file objective management [J]. *Heilongjiang Archives*, 2021 (04): 156–157.

Advances in Civil Engineering and Environmental Engineering – bin Yahya & Bin Daud (Eds)
© 2023 The Author(s), ISBN: 978-1-032-39397-1

Risk assessment and identification of wellbore instability before deepwater drilling

Yang Du
Cnooc (China) Co., LTD hainan Branch, Hainan, China

Kejin Chen*
China University of Petroleum, Beijing, Beijing, China

Yuguang Liu & Yi Huang
Cnooc (China) Co., LTD hainan Branch, Hainan, China

Ting Sun & Shu Jia
China University of Petroleum, Beijing, Beijing, China

ABSTRACT: Drilling is one of the most direct risk sources for deepwater oil and gas exploration. Without a clear understanding of formation pressure, it is straightforward to cause drilling accidents like kick, blowout, lost circulation, and sticking. According to the Worldwide Offshore Accident Databank (WOAD) statistics, accidents during drilling accounted for 95.22% of all deepwater platform operations from 1975 to 2014. Therefore, this study introduces the formation pressure prediction model commonly used in drilling projects. It indicates that the uncertain ground stress and geomechanical parameters used in the model are the crucial cause of the inaccurate prediction results. Then, we use the Monte-Carlo simulation method to analyze the probability distribution of the parameters. Based on the uncertain distribution of formation pressure, the study can finally obtain a safe drilling fluid density window with confidence and derive the quantitative risk assessment of wellbore instability at a certain depth. The case analysis of an X oil field block in the South China Sea shows that the recommended safe drilling fluid density window is consistent with the actual field application. The accuracy of the possible wellbore instability accident prediction results at a certain depth before drilling reaches 93.1%. It proves that the study can provide accurate quantitative risk assessment and early warning of possible wellbore instability accidents. The study helps to ensure safe and efficient drilling operations.

1 INTRODUCTION

The petroleum industry generally considers water depth from 500 meters to 1500 meters as deep water and water depth greater than 1500 meters as ultra-deep water. Wellbore instability is the most challenging problem in deepwater drilling engineering. Unjustified drilling fluid density will cause a series of downhole complex accidents such as wellbore collapse, constriction, leakage, sticking, blowout, and reservoir pollution (Aboozar 2018). However, due to the confidential and harsh underwater environment and the errors in logging interpretation of geological information, wellbore stability analysis exists significant uncertainties before drilling (Aboozar 2017). The hot research issues in the field of deepwater drilling at home and abroad focus on improving the accuracy of safe drilling fluid density windows, assessing the wellbore instability risk, and identifying possible accidents while drilling.

*Corresponding Author: 18853919728@163.com

DOI 10.1201/9781003349563-37

2 ANALYSIS OF WELLBORE INSTABILITY CAUSES

By the end of 2016, a total of 17 wellbore instability accidents, such as lost circulation, kicking, and sticking, had occurred in the South China Sea. The detailed statistics of the accidents are shown in Table 1. The causes of accidents mainly include geological disasters, marine environment, narrow drilling fluid density window, and improper operation of personnel. The proportion of accident causes is shown in Figure 1.

Table 1. Drilling accidents resulting from wellbore instability in the South China Sea.

Accidents Type	Counts
Lost Circulation	8
Kicking	6
Sticking	3
Amounts	17

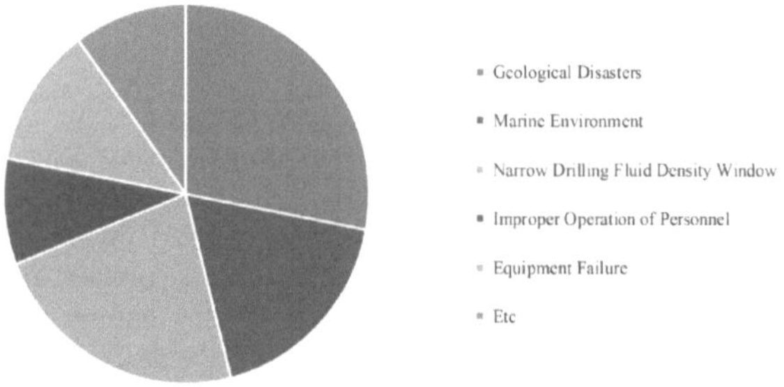

Figure 1. Causes of drilling accidents in the South China Sea.

It can be seen from the historical operation engineering practice. One of the causes of wellbore instability in deepwater drilling is the misjudgment of drilling fluid density windows. The direct cause of this misjudgment is lacking a clear understanding of geomechanical input parameters. On the other hand, the logging data of adjacent wells are often selected for formation pressure prediction in the same block. The assumption based on the consistency of regional geomechanical parameters aggravates the misjudgment degree of safe drilling fluid density window.

Thus, kicking will occur when the predicted drilling fluid density is lower than the pore pressure. When the predicted drilling fluid density is higher than the fracture pressure, it will induce lost circulation. The wellbore collapse will occur when the predicted drilling fluid density is higher than the collapse pressure (Asadi 2016). Therefore, kicking, lost circulation, and wellbore collapse are the main types of wellbore instability accidents in deepwater drilling.

In this study, we choose the Monte-Carlo simulation to obtain the expected distribution characteristics of logging interpretation geomechanical input parameters. It helps to build a safe operation window with reliability. Then, comparing the safe operation window with the probability distribution of formation pressure, it's able to identify wellbore instability accidents in different depths.

3 QUANTITATIVE EVALUATION MODEL OF WELLBORE INSTABILITY RISK BEFORE DRILLING

3.1 *Prediction model of safe drilling fluid density window*

While drilling, the rock support starts to be replaced by the fluid column pressure of the drilling fluid, thus changing the stress distribution of the original formation. As shown in Figure 2, the horizontal maximum in-situ stress is σ_H, and the minimum in-situ stress is σ_h. The overlying formation pressure in the vertical direction is σ_V, and the column pressure of the drilling fluid in the borehole is p_i.

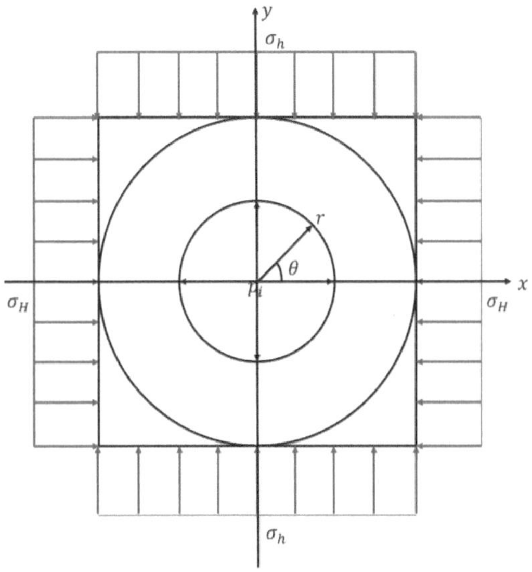

Figure 2. Stress model of borehole.

When calculating the stress distribution at the wellbore, model radius r in Figure 2 equals a. Adding the nonlinear correction and the original pore pressure p_p at the wellbore, the stress distribution at the wellbore is shown in Eq. (1), Eq. (2) and Eq. (3).

$$\sigma_r = p_i - \alpha p_p \tag{1}$$

$$\sigma_\theta = \eta[(1 - 2\cos 2\theta)\sigma_H + (1 + 2\cos 2\theta)\sigma_h - p_i] - \alpha p_p \tag{2}$$

$$\sigma_r = p_i - \alpha p_p \tag{3}$$

Where η is the nonlinear correction coefficient of wellbore stress, generally takes 0.95, and α is the effective stress coefficient.

From the mechanical point of view, the main reason for the wellbore collapse is the low pressure of the liquid column in the borehole, which makes the stress of the rock around the wellbore exceed the shear strength itself. Finally, it will result in shear failure. When it comes to the wellbore fracture, the main reason is that the drilling fluid column pressure in the borehole is too high. It makes the rock stress around the borehole exceed the tensile strength itself. Tensile damage will occur soon (Chen 2008).

In this model, the Mohr-Coulomb strength criterion is selected as the criterion of wellbore collapse. It is also applied in the collapse pressure calculation. According to this criterion, when the shear stress τ on a plane of borehole rock is greater than the shear strength limit τ_0, the borehole rock will fail.

According to the envelope line of the Mohr-Coulomb criterion in Figure 3, the liquid column pressure of drilling fluid should be controlled to ensure that the stress circle of the borehole is always under the rock failure zone, so it should satisfy Eq. (4).

$$\eta\left(3\sigma_H - \sigma_h - p_i\right) - \alpha p_p \leq \left(p_i - \alpha p_p\right)\tan^2\left(45^\circ + \frac{\varphi}{2}\right) + 2c\tan\left(45^\circ + \frac{\varphi}{2}\right) \tag{4}$$

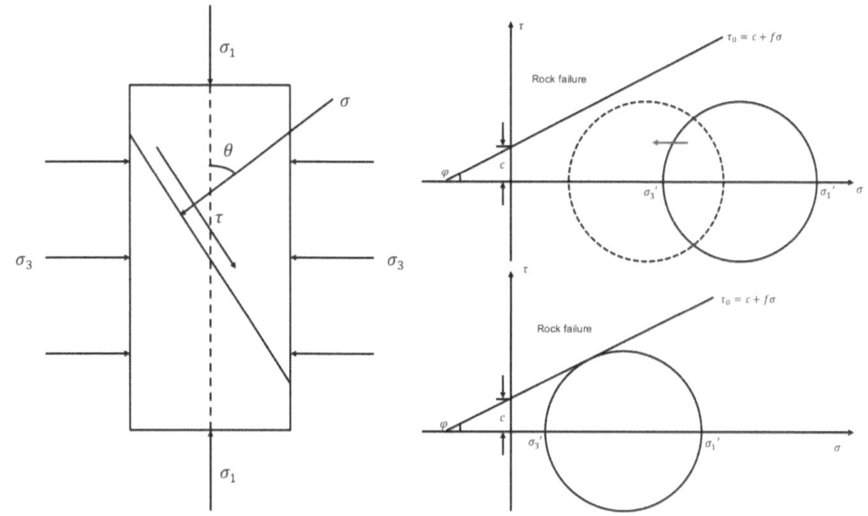

Figure 3. Envelope line of Mohr-Coulomb criterion.

Thus, the wellbore collapse pressure p_c is shown in Eq. (5).

$$p_c = \frac{\eta\left(3\sigma_H - \sigma_h\right) - 2c\tan\left(45^\circ + \frac{\varphi}{2}\right) + \alpha p_p\left[\tan^2\left(45^\circ + \frac{\varphi}{2}\right) - 1\right]}{\eta + \tan^2\left(45^\circ + \frac{\varphi}{2}\right)}. \tag{5}$$

The maximum tensile stress intensity criterion is selected as the criterion of wellbore fracture. Assuming the tensile strength of the rock around the borehole is S_t, the density of drilling fluid should be guaranteed to prevent the wellbore from fracturing, which is shown in Eq. (6).

$$p_i - S_t \leq 3\sigma_h - \sigma_H - \alpha p_p \tag{6}$$

Thus, the wellbore fracture pressure p_f is shown in Eq. (7).

$$p_f = 3\sigma_h - \sigma_H - \alpha p_p + S_t \tag{7}$$

In order to ensure the wellbore stability during drilling, the equivalent density of formation pore pressure, collapse pressure, and fracture pressure is used to form the safe drilling fluid density window.

$$\min\{P_p, P_c, P_f\} \leq Pi \leq \max\{P_p, P_c, P_f\} \tag{8}$$

3.2 *Uncertainty analysis method of ground stress and geomechanical parameters*

Combining Eq. (5) and Eq. (7), the accuracy of the collapse pressure and fracture pressure prediction results is affected by ground stress and geomechanical parameters, such as in-situ stress, pore pressure, cohesive force, internal friction angle, and tensile strength. Thus, the Monte-Carlo simulation method can be used to analyze the uncertainty characteristics of the above-mentioned in-situ stress and rock mechanics parameters.

The specific processing steps are as follows. Based on the in-situ stress and geomechanical input parameters converted from the logging data, we can obtain the mean value and standard deviation of the above parameters. Then, according to each input parameter's maximum and minimum values, the number and the interval of packets are determined. Finally, the counts and frequency of the scattered in-situ stress and geomechanical input parameters in the same sedimentary environment are obtained.

3.3 *Reliable quantitative characterization of safe drilling fluid density window*

Input the in-situ stress and geomechanical parameters into Eq. (5), Eq. (7), and Eq. (8). The cumulative probability distribution of formation pore pressure, collapse pressure, and fracture pressure can be determined statistically, as shown in Figure 4. The existence of the interference zone means that there is no safe drilling fluid density window with a probability of 100%. The intersection point shows that the kicking and collapse probability is equal. Any cumulative probability density under the intersection point indicates the reliability of wellbore stability. The intersection of a certain degree of reliability with pore and collapse pressure constitutes a corresponding drilling fluid density operation window.

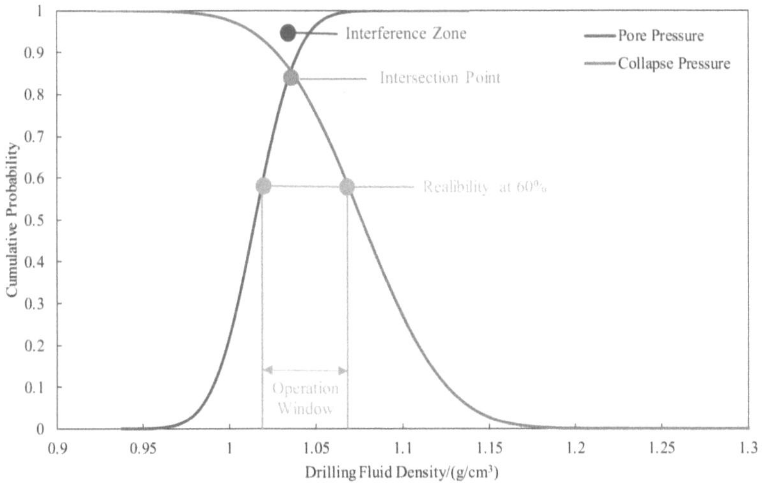

Figure 4. Safe drilling fluid density window with reality.

4 CASE ANALYSIS

This thesis chooses an actual X oil field block in the South China Sea. The water depth of the sea is about 520 meters, and the reservoir is buried deep at about 1900 meters. The logging materials of the same deposition environment in the range of 956 meters to 1520 meters are selected. The upper portion is in large grey mudstone, light grey thin to medium powder sandstone, and calcium powder sandstone; the lower portion is mainly in bulk grey mudstone and trace powder sandstone.

4.1 *Prediction of formation three pressure profile*

Logging interpretation of in-situ stress and geomechanical data are input into the safe drilling fluid window prediction model to obtain formation pore pressure, collapse pressure, and fracture pressure within the depth of the well, as shown in Figure 5.

It can be seen that the density of the safe drilling fluid should be controlled within the limited range of pore pressure and collapse pressure. The operation window is extremely narrow, with some areas as low as 0.001g/cm^3. The formation's pore pressure and collapse pressure at the depths of 1237.79 meters, 1343.71 meters, and 1379.52 meters overlap. It indicates that the area is prone to occur wellbore instability, resulting in lost circulation and kicking accidents.

4.2 *Uncertainty analysis of geostress and geomechanical parameters*

The statistical result shows that the uniaxial compressive strength is between 5.68MPa and 20.37MPa. The tensile strength is between 0.47MPa and 1.69Mpa. The minimum horizontal ground

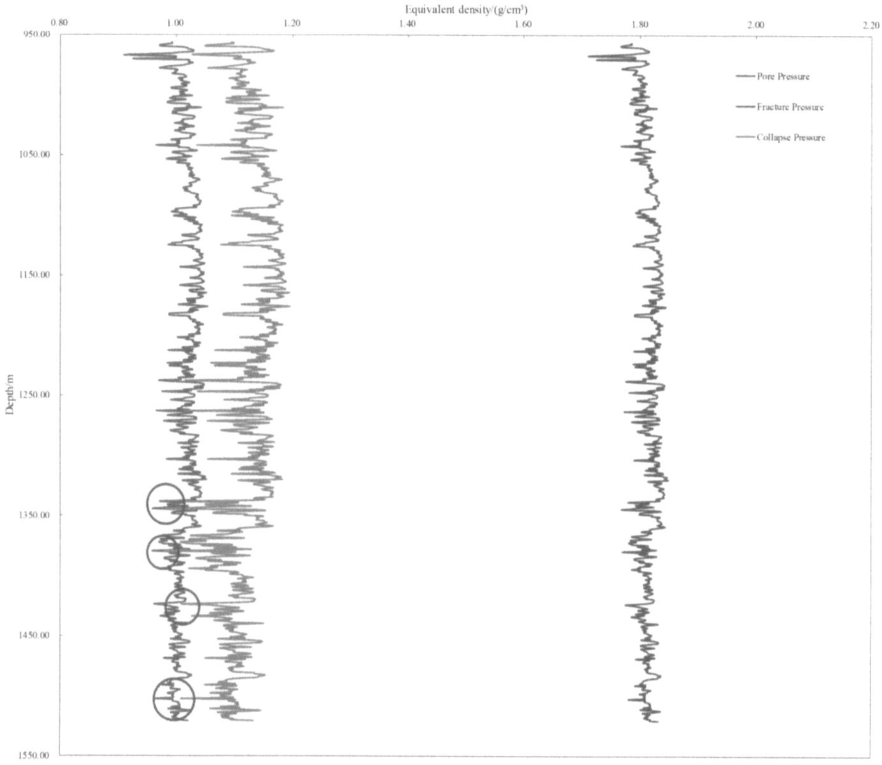

Figure 5. Pore pressure, collapse pressure and fracture pressure profile.

stress is between 1.41MPa and 1.57MPa, while the maximum horizontal ground stress is between 1.71MPa and 1.85MPa. The cohesive force is between 0.55MPa and 3.28MPa. The internal friction Angle is between 24.57°C and 24.67°C. The quantitative characterization of uncertainty is shown in Table 2 and Figure 6.

Table 2. Statistical characteristics of ground stress and geomechanical parameters.

			Cumulative Probability	
Parameters	Average	Standard Deviation	5%	95%
σ_v	10.71	3.32	10.78	14.99
S_t	0.89	0.27	0.90	1.25
σ_3	1.55	0.02	1.55	1.58
σ_1	1.83	0.02	1.83	1.85
c	1.34	0.59	1.36	2.11
φ	24.64	0.02	24.65	34.67

4.3 *Quantitative evaluation of wellbore instability risk and identification of possible accident types*

The cumulative probability distribution of pore pressure, collapse pressure, and fracture pressure in the depth range of 956 meters to 1520 meters is calculated, as shown in Figure 7 and Figure 8.

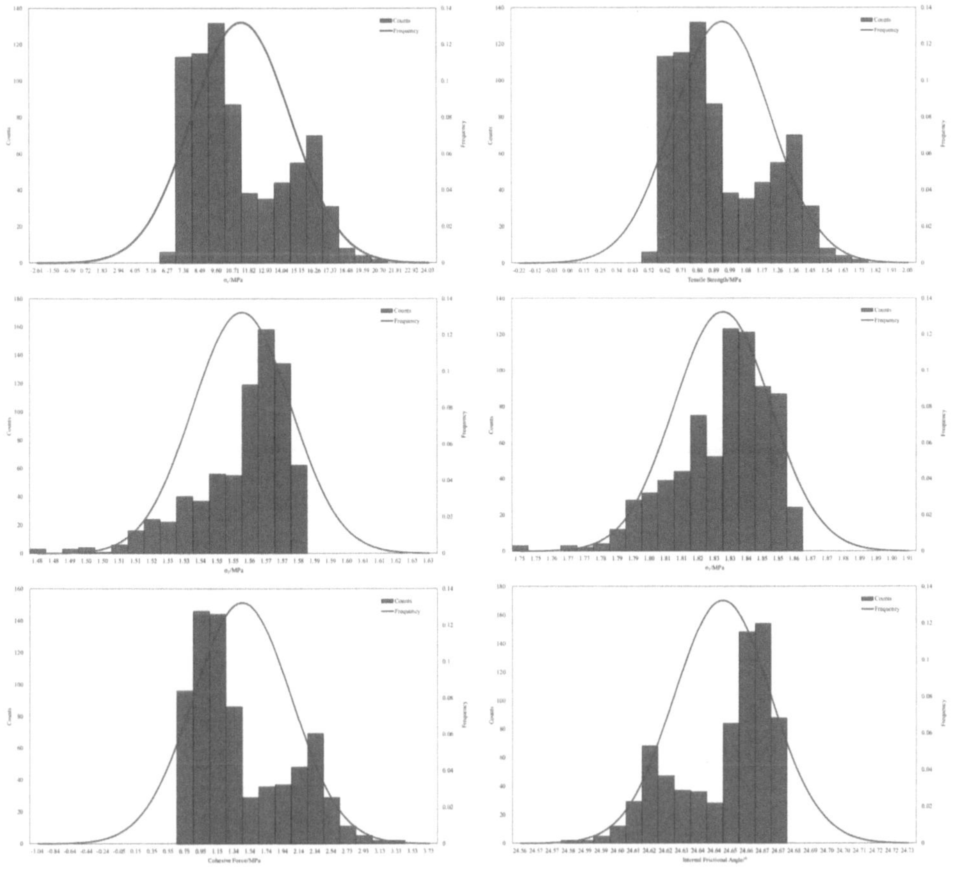

Figure 6. Normal distribution characteristics of geostress and geomechanical input parameters.

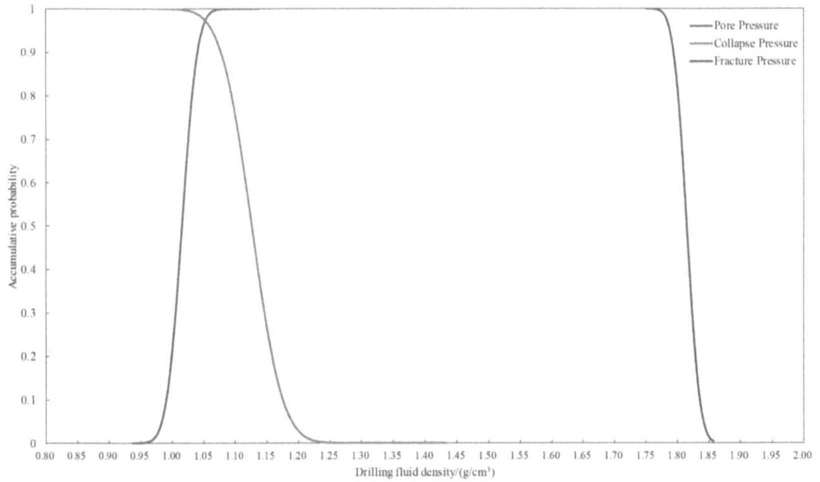

Figure 7. The cumulative probability distribution of pore pressure, collapse pressure and fracture pressure.

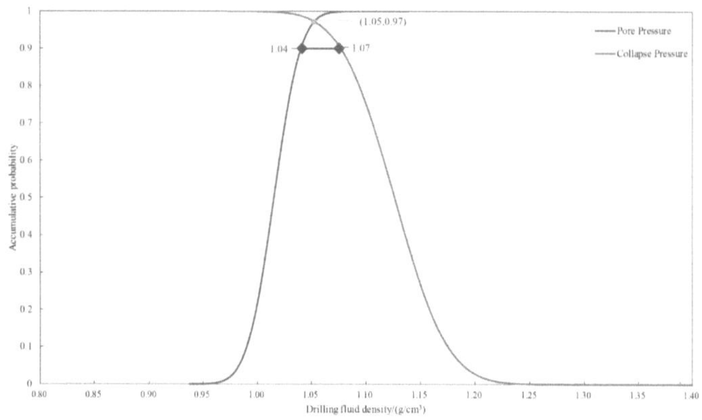

Figure 8. Distribution of pore pressure interferes with collapse pressure.

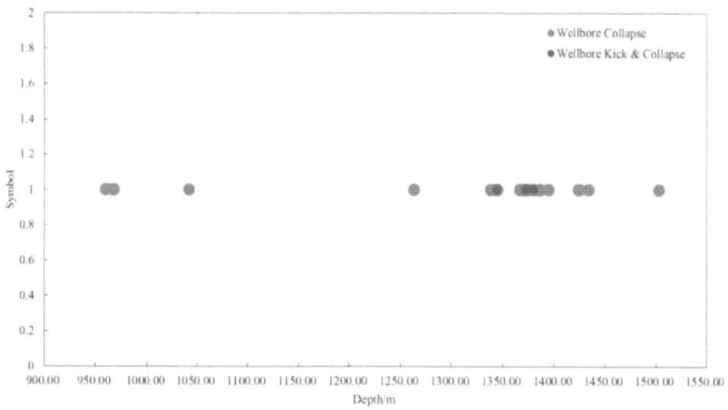

Figure 9. Prediction of possible drilling accidents types in which depth under the density window between 1.04 g/cm^3 and 1.07g/cm^3.

It can be concluded that the accumulative probability distribution of formation pore pressure interferes with the accumulative probability distribution of collapse pressure. On the one hand, it means that there is no safe drilling fluid density window with a probability of 100%. No matter the value of drilling fluid density selected, the risk of wellbore instability will occur. On the other hand, it means that the loss and kick are likely to occur in the same zone during drilling. The corresponding drilling fluid density at the intersection point is 1.05g/cm^3, with an equal probability of wellbore collapse and kick. In order to successfully drill through the formation with a relatively safe probability, engineering needs to control the probability of reliability above 90%, which corresponds to the recommended safe drilling fluid density window between 1.04 g/cm^3 and 1.07g/cm^3. Under such an operation window, the predicted accident types in which depth are shown in Figure 9. It reaches 93.1% accuracy compared with the actual drilling situation.

The recommended density window is only 0.03g/cm^3, which is recommended to be widened to 0.15g/cm^3 during the well structure design. At the same time, dynamic kill technology should be used to flexibly control drilling fluid density to maintain safe and efficient drilling.

5 CONCLUSION

1) The outburst reason for the wellbore instability in deepwater drilling is the safe drilling fluid density window misjudgment. The direct cause is lacking a clear understanding of the geomechanical input parameters.

2) This thesis proposes a quantitative evaluation model of wellbore instability risk before drilling. It can form a safe drilling fluid window with reliability and evaluate the risk of wellbore instability. Comparing the safe operation window with the probability distribution of formation pressure, it's able to identify wellbore instability accidents in different depths. In the case analysis, it reaches 93.1% accuracy compared with the actual drilling situation.
3) Narrow pressure window is a common problem in deepwater drilling. The local pressure window is as narrow as 0.001g/cm^3, which can easily induce wellbore instability. It is suggested that, based on the prediction of a safe drilling fluid density window under, the well structure optimization design should be carried out to broaden the pressure window. and the model inversion prediction should be carried out. Then, the model presented in this thesis can be used for inverse detection of the optimized design results.

ACKNOWLEDGMENT

The authors gratefully acknowledge the support from the fund items: The research on tube-soil interaction mechanism and drilling parameter control method in the process of pipe jetting in deepwater Drilling, No.2020M670581; Analysis and control method of the time-varying Bearing capacity of deepwater subsea Wellhead coupled with dynamic Action, No.2021T140736; Study on the mechanism of gas hydrate formation decomposition and weakening during drilling process (2462021YJRC006).

REFERENCES

Aboozar, G., Podgornov, V.M., Rebetsky, Y.L., Ghasemi, M., and Shaybakov, A.L. Comprehensive wellbore stability analysis using elastic and porothermoelastic models (In Russian) [J]. *Neftyanoe Khozyaistvo – Oil Industry*, 2018, vol: 8; pp: 14–18.

Alsubaih, A.A., H.H. Dahm, and I.A. Alsubaih . "History matching, sensitivity and uncertainty analysis to improve wellbore stability model predictions." *Arma*, 2017.

Asadi, M.S., Khaksar, A., Ring, M.J. and Yin, K.Y. Comprehensive geomechanical modeling and wellbore stability analysis for infill drilling of high-angled wells in a mature oil field. *SPE Asia Pacific Oil & Gas Conference and Exhibition*, 2016.

Chen Mian & Jin Yan, Zhang Guangqing. *Petroleum-related rock mechanics* [M]. Beijing: Science Press, 2008.

Daniel, Moos, Pavel, Peska, Thomas, Finkbeiner, Mark and Zoback. Comprehensive wellbore stability analysis utilizing quantitative risk assessment [J]. *Journal of Petroleum Science & Engineering*, 2003, vol: 38, pp: 97–109.

Todd, B., M. Reese, and S. Brown. "Operationalizing upstream process safety for drilling operations by adopting an integrated risk assessment approach." *SPE/IATMI Asia Pacific Oil & Gas Conference and Exhibition*, 2019.

Wei Kai, Guan Zhichuan, Liao Hualin, Shi Yucai and Liu Yongwang. Assessment method of borehole instability risk [J]. *Journal of China University of Petroleum* (Edition of Natural Science), 2013, vol: 37; pp: 62–66.

Wei Kai, Ma Jinshan, Guan Zhichuan, Qi Jintao & Zhang Xiaochun. Determination of wellbore instability area and analysis of influencing factors[J]. *Mechanics in Engineering*, 2014, vol: 36, pp: 54–58.

Yi Huang, Yanan Shen, Zhichuan Guang, etc. Quantitative risk assessment of wellbore stability of Yingqiong Basin [J]. *Fault Block Reservior*, 2019, vol: 26, pp: 380–384.

Yingjie Chen, Chuanguang Deng, Tianshou Ma. A risk assessment method of wellbore instability based on the reliability theory [J]. *Drilling Engineering*, 2019, vol: 39, pp:97–104.

Zhang Lisong, Yan Xiangzhen, Tian Zhonglan, Yang Henglin and Yang Xiujuan. A collapse pressure analysis in reservoir drilling based on the reliability method [J]. *Acta Petrolei Sinica*, 2012, vol: 33; pp: 477–482.

Zhang Lulu, Zhang Jie, Xu Yao & Li Xu. Reliability theory in geotechnical engineering [M]. *Shanghai: Tongji University Press*, 2011.

Advances in Civil Engineering and Environmental Engineering – bin Yahya & Bin Daud (Eds)
© 2023 The Author(s), ISBN: 978-1-032-39397-1

Research on social stability risk assessment of major highway projects: Take Quanzhou-Nanning expressway (G72) Guangxi Guilin to Liuzhou section reconstruction and expansion project as an example

Zhenli Ou
Nanning Normal University; Guangxi Transportation Research & Consulting Co., Ltd., Nanning, Guangxi, China

Hongri Zhang* & Hongming Li
Guangxi Transportation Research & Consulting Co., Ltd., Nanning, Guangxi, China

Gengfei Li
Guangxi Communications Investment Group Finance Co., Ltd., Nanning, Guangxi, China

Weidang Luo
Guangxi Transportation Research & Consulting Co., Ltd., Nanning, Guangxi, China

ABSTRACT: Major highway engineering projects have the characteristics of the large-scale, long construction period, and huge numbers of people involved. As social development faces more and more factors that affect social stability, its social stability risk assessment has also attracted more and more attention. This article is aimed at major highway engineering projects, based on the relevant regulations of national risk management and the basic framework of social stability risk assessment generally implemented at home and abroad. Take risk investigation, risk identification, risk estimation, risk prevention and mitigation measures, the expected risk level after the measures are implemented, and the conclusion of risk analysis as the research ideas. Take the Quanzhou-Nanning Expressway (G72) Guangxi Guilin-Liuzhou Section Reconstruction and Expansion Project as an example to conduct a social stability risk analysis. Finally, 10 main risk factors were identified. The comprehensive evaluation index of social stability risk was changed from 0.338 to 0.152 through preventive and mitigating measures. The impact of risk factors is minimized. The research content can provide a basis for corresponding strategies, and at the same time provide a reference to the social stability risks assessment of major highway projects.

1 INTRODUCTION

The practice of social stability risk assessment in my country began with the exploration of Suining City, Sichuan Province in 2006. After being implemented by the National Development and Reform Commission in major fixed-asset investment projects in 2012, the scope of application has continued to expand, gradually expanding to major projects, major decisions, major reforms, and other major government issues (Yang 2021; Zhu 2016; Zou et al. 2019). However, the current domestic social stability risk assessment for major construction projects is still in the exploratory stage. There is a lack of systematic and pertinent research, and there are few analytical methods of social stability risks, and there are limited reference cases, and there are even fewer studies

*Corresponding Author: 253541461@qq.com

 DOI 10.1201/9781003349563-38

specifically aimed at social stability risk assessment of major highway projects (Dong et al. 2018; Tian et al. 2018; Zhou et al. 2015).

Major highway projects have the characteristics of a large-scale, long construction period, and many people are involved. Social stability risk assessment research of major highway projects has profound significance (Xu et al. 2020; Zhang et al. 2012). The Quanzhou-Nanning Expressway (G72) Guangxi Guilin-Liuzhou Section Reconstruction and Expansion Project (Hereinafter referred to as "G-L section") has a total length of 100.168km, passing through 3 counties, 8 townships, and 27 villages, with a total investment of approximately 13.659 billion yuan. The project will have a significant impact on the lives and economic lives of the people along the route. Therefore, this article analyzes the social stability risks of the "G-L section" to provide a basis for countermeasures and provide a reference to the analysis of similar social stability risks.

2 MATERIALS AND METHODS

This article takes the "G-L section" as an example and uses a combination of qualitative and quantitative methods to conduct social stability risk assessment and analysis. It mainly uses qualitative analysis methods such as interview method, questionnaire survey method, literature method, risk comparison method, expert scoring method, and quantitative analysis methods such as risk probability-impact matrix, analytic hierarchy process, and comprehensive risk index method. Here is a detailed description of the quantitative analysis method.

2.1 *Risk probability-impact matrix*

The risk probability-impact matrix is established for risk estimation. The risk probability-impact matrix takes the probability of the occurrence of a risk factor as the abscissa, and the impact on the risk factor of the project after the occurrence of the risk factor is the ordinate. The specific evaluation criteria are as follows.

a. Risk probability. According to the probability of the occurrence of risk factors, the risk probability is divided into five grades: very high, high, medium, low, and very low, and are assigned scores respectively, and the probability of occurrence frequency and risk accidents are described. As shown in Table 1.

Table 1. Risk probability grade table.

Probability level	Very high	High	Medium	Low	Very low
Scoring value	5	4	3	2	1
Accident description	Very frequently	More frequent	Generally	Occasionally	Rarely happens
Probability	81%~100%	61%~80%	41%~60%	21%~40%	0~20%

b. Degree of influence. According to the impact on the project after the risk occurs, the impact level is divided into five impact levels: severe, large, medium, small, and negligible for qualitative judgment, as shown in Table 2.

Table 2. The degree of influence grade table.

Severity	Severe	Large	Medium	Small	Negligible
Scoring value	5	4	3	2	1
Quantitative Judgment Standard	81%~100%	61%~80%	41%~60%	21%~40%	0‘20%

c. Degree of risk. Use the risk probability-impact degree matrix (Figure 1) to determine the degree of risk. Generally, the degree of risk is divided into three levels: high-risk, medium-risk, and low-risk, among which low-risk can be further subdivided into tolerable, self-sustainable, and negligible, as shown in Table 3.

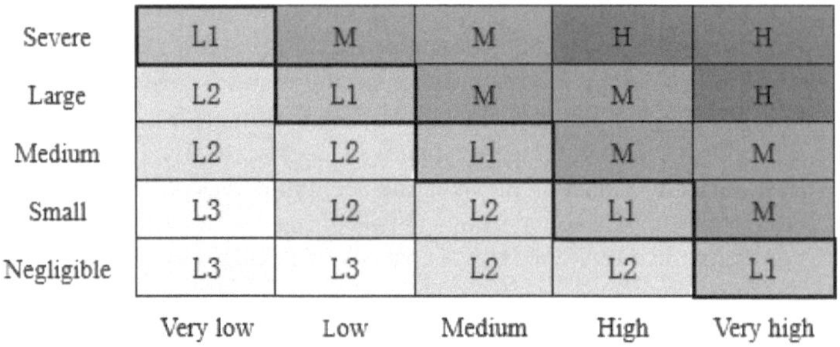

Figure 1. Risk probability-impact matrix.

Table 3. Risk level table.

			Low risk		
Degree of risk	High risk	Medium risk	Tolerable	Self-Sustainable	Negligible
Corresponding to the matrix region	H	M	L1	L2	L3
Index pessimistic value	1	0.64	0.36	0.16	0.04

2.2 *AHP*

The analytic hierarchy process, abbreviated as AHP, is a decision-making analysis method of hierarchical weights. AHP uses less quantitative information to mathematicize the thinking process of decision-making. It is especially suitable for problems that are difficult to fully quantitatively analyze. This paper uses the AHP to determine the weights of the importance of the main risk factors relative to the occurrence of risks.

2.3 *The comprehensive risk index method*

The comprehensive risk index method is one of the commonly used risk grade evaluation methods. The general steps for using this method to determine the risk grade are: (1) Determine the weight. According to the actual situation of the project, an expert group was formed to sort the single risk factors according to their importance, assign values of the sorted risk factors, and calculate the weight W in a unified manner. (2) Assign a value to the risk degree R of each risk factor. (3) Calculate the comprehensive risk index. Establish a comprehensive risk evaluation index table, and obtain a comprehensive risk index S according to Formula 1. The higher the comprehensive risk index value, the higher the risk of the proposed project. (4) Rating risk. Analyze the degree of single-factor risk, comprehensive risk index, survey results, and risk events to determine the level. The reference standard of risk grades evaluation is shown in Table 4. If one of the standards is reached, it can be judged as the corresponding grade.

$$S = \sum_{i=1}^{n} R_i W_i \tag{1}$$

Table 4. Reference standards for judging the social stability risk level of the proposed project.

Risk level	High (Significant negative impact)	Middle (Large negative impact)	Low (General negative impact)
Overall Judging Standard	Most people have opinions and reactions to the implementation of the project, which may trigger large-scale mass incidents.	Some people have opinions and strong reactions to the implementation of the project, which may cause conflicts.	Most people understand and support it, but a few people have opinions on the implementation of the project.
The criteria for judging risk events	Such as attacking and besieging party and government agencies, key departments, and key areas and places; incidents of beating, smashing, looting, burning, and other collective weapons, gatherings to cause disturbances, casualties, illegal assemblies, demonstrations, strikes, etc.	Such as collective petitions, extreme personal incidents, confinement of construction sites, blocking off traffic, negative public opinion in the media, etc.	Such as personal non-normal petitions, meditation, pulling banners, shouting slogans, distributing promotional materials, disseminating harmful information, etc.
Judgment criteria for the number of participants in risk events	More than 200 people	20 people~200 people	20 people or less
Single-factor risk assessment standard	2 or more single factors are high risk or 5 or more single factors are medium risk	One single factor is high risk or 2~4 single factors are medium risk	One single factor is the medium risk or both are single factor low risk
Comprehensive risk index evaluation criteria	>0.64	0.36~0.64	<0.36

3 RESULTS AND DISCUSSION

3.1 *Risk investigation*

According to the actual situation of the project, the survey scope is determined to be 500m on both sides of the project centerline (covering the scope of possible negative impacts during highway construction and operation). The representative research objects are identified as villagers in 27 administrative villages in 8 towns and towns involved in land acquisition and demolition, Huangmian State-owned Forest Farm, and other villagers that are not involved in land acquisition and demolition but are affected by the environment. It mainly adopts the combination of sampling survey and case investigation, comprehensively adopts methods such as observation method, interview method, and literature method, and conducts risk investigation into the legality, rationality, feasibility, and controllability of the implementation of the project construction.

Table 5. Identification table of main risk factors.

Number	Type of risk	Occurrence stage	Risk factors	Possible consequences
1	Land acquisition, demolition, and compensation	After implementation	Employment and life of land-expropriated farmers	Mob
2		After implementation	Quantity and quality of resettlement housing	Mob
3		Implementation phase	Compensation standards for land and house acquisition	Protest
4		Implementation phase	Compensation procedures and methods for land and house acquisition	Protest, mob
5	Technical economy	Before implementation	Engineering scheme	Protest
6	Ecological environmental impact	In the implementation phase, after implementation	Water environmental impact	Protest
7		Implementation phase	Other influences (cultural relics)	Destruction of cultural relics, Work stoppage
8	Project management	Implementation phase	Civilized construction and quality management	Protest, mob, Work stoppage
9	Economic and social impact	Implementation phase	Impact on surrounding traffic	Traffic congestion
10	Compatibility with society	Implementation phase	Social contradictions leftover from history	Protest, mob, Work stoppage

3.2 *Risk factor identification*

This article mainly adopts the comparison table method and refers to the identification results of the expert survey method, interview method, and field observation method to identify the social stability risk. First, according to the analysis of the types, causes, distribution, impact performance, impact degree, and probability of occurrence of various risk factors of the social stability of the project, the risk factors are sorted out. According to the project stage where the risk may occur, combined with the mutual adaptability of the local economy, society, and the project, a total of 51 risk factors in nine categories are sorted out. (The nine types of risk factors are policy planning approval procedures, land acquisition and demolition compensation, technical economy, ecological environmental impact, project management, economic and social impact, safety and health, social compatibility, and business risk.) Finally, 10 major single risk factors were screened and identified from 51 risk factors, as shown in Table 5.

3.3 *Risk assessment*

Risk estimation is the process of comprehensively analyzing and estimating the overall risk level of the project based on estimating the degree of risk of a single risk factor. The project risks estimation process is mainly as follows. First, a combination of expert experience survey and evaluation methods is adopted to estimate the risk probability and influence degree range of the main factors, and the risk degree level of the single factor is judged according to the risk probability-influence matrix. Then, take the most pessimistic value according to the single-factor risk level. Determine the weight of each factor according to the analytic hierarchy process. Refer to Table 3 and use the comprehensive risk index method to weigh and stack to obtain the comprehensive

Table 6. Comparison table of risk changes of various factors before and after the measures.

Number	Risk factors	Risk probability		Influence level		Degree of risk	
		Before measures	After the measures	Before measures	After the measures	Before measures	After the measures
1	Employment and life of land-expropriated farmers	Low	Very low	Small	Negligible	Low (Self-Sustainable)	Low (Negligible)
2	Quantity and quality of resettlement housing	Low	Low	Large	Medium	Low (Tolerable)	Low (Self-Sustainable)
3	Compensation standards for land and house acquisition	Medium	Medium	Medium	Medium	Low (Tolerable)	Low (Tolerable)
4	Compensation procedures and methods for land and house acquisition	Medium	Low	Medium	Small	Low (Tolerable)	Low (Self-Sustainable)
5	Engineering scheme	High	Low	Large	Medium	Medium risk	Low (Self-Sustainable)
6	Water environmental impact	Low	Very low	Medium	Small	Low (Self-Sustainable)	Low (Negligible)
7	Other influence	Low	Very low	Small	Small	Low (Self-Sustainable)	Low (Negligible)
8	Civilized construction and quality management	Medium	Low	Small	Negligible	Low (Self-Sustainable)	Low (Negligible)
9	Impact on surrounding traffic	Medium	Medium	Medium	Small	Low (Tolerable)	Low (Self-Sustainable)
10	Social contradictions leftover from history	Low	Very low	Small	Negligible	Low (Self-Sustainable)	Low (Negligible)

evaluation index of the project. Finally, according to the risk grade evaluation standard (Table 4), the risk grade result is given.

3.3.1 *Initial risk assessment*

The initial risks pessimistic value and weight value of the main risk factors of the project are based on the foregoing calculations as shown in Table 7. According to the calculation formula for the comprehensive evaluation index of social stability risk, the initial comprehensive evaluation index Si of the "G-L section" social stability risk is 0.338. At the same time, the project risks survey showed that 99% of the 326 people surveyed supported the construction of the project. No people chose to petition to express their personal opinions, and 66% of the people chose to report to the relevant government departments. According to the initial comprehensive evaluation index and risk investigation situation, combined with Table 4, the initial projected risk level can be judged as low risk of a qualitative and semi-quantitative comprehensive evaluation.

3.3.2 *Risk prevention measures*

In terms of risk prevention and mitigation measures, the social stability risks prevention and mitigation measures adopted by the project are generally based on risk control, supplemented by risk self-support. The goal of the project is to ensure and promote the smooth implementation of the project and the normal operation of the project after construction, considering the legitimate and reasonable interests of all parties, maintaining social stability, and promoting harmonious development. It mainly proposes a series of preventive and resolving measures from various aspects such as policy planning approval procedures, land acquisition, and demolition compensation, technology, and economy, ecological environmental impact, project management, economic and social impact, safety and health, media public opinion, etc. This is not elaborate.

After the measures are implemented, the paper uses the expert scoring method again to analyze and evaluate the project's main risk probability and impact degree. The risk degree is determined according to the risk probability-influence matrix in Figure 1. The analysis of the risk changes of the main factors before and after the measures are shown in Table 6. It can be seen from Table 6 that after the implementation of the project measures, all risk factors are low risks. After the prevention and mitigation measures are implemented, the risk probability or impact degree of the project's main risks will be reduced to varying degrees. Therefore, the degree of risk is correspondingly lower than before implementation.

3.3.3 *Risk assessment after implementing risk prevention measures*

Same as above, refer to Table 3 and use the comprehensive risk index method to obtain the expected comprehensive risk index after the project has implemented the measures. The pessimistic value and weight value of the main risk factors after the implementation of the project measures are shown in Table 7. According to the calculation formula for the comprehensive evaluation index of social stability risk, the expected comprehensive risk index of the project after the implementation of the measures is 0.152. Combining the risk investigation and project reality, according to Table 4, the project risk level after the qualitative and semi-quantitative comprehensive evaluation measures is low risk. This shows that prevention and mitigation measures can effectively reduce the degree of influence of social stability risk factors.

Table 7. The pessimistic value and weight value table of the main risk factors before and after the implementation of the project.

	Initial		After the measures	
Risk factors and comprehensive risk index	R	W	R	W
Employment and life of land-expropriated farmers	0.16	0.024	0.04	0.024
Compensation standards for land and house acquisition	0.36	0.151	0.36	0.151
Quantity and quality of resettlement housing	0.36	0.132	0.16	0.132
Compensation procedures and methods for land and house expropriation	0.36	0.132	0.16	0.132
Engineering scheme	0.64	0.151	0.16	0.151
Water pollution impact	0.16	0.092	0.04	0.092
Other influence	0.16	0.057	0.04	0.057
Civilized construction and quality management	0.16	0.125	0.04	0.125
Impact on surrounding traffic	0.36	0.112	0.16	0.112
Social contradictions leftover from history	0.16	0.024	0.04	0.024
Comprehensive Risk Index *S*	**0.338**		**0.152**	

4 CONCLUSIONS

At present, the theoretical system of social stability risk assessment for major highway projects in my country is not perfect. Aiming at major highway engineering projects, this paper adopts a combination of qualitative and quantitative methods, taking the reconstruction and expansion project of the Quanzhou-Nanning Expressway (G72) Guangxi Guilin-Liuzhou section as an example to conduct social stability risk assessment and analysis. Based on this risk assessment model, 10 main risk factors were identified: employment and life of land-expropriated farmers, quantity and quality of resettlement housing, compensation standards for land and house expropriation, compensation procedures and methods for land and house expropriation, engineering scheme, water environment impact, other impacts (cultural relics), civilized construction and quality management, impact on surrounding traffic, social conflicts left over from history, etc. The comprehensive evaluation index of social stability risk changed from 0.338 to 0.152 after prevention and mitigation measures. The impact of risk factors is minimized. The results show that the method adopted is reasonable and effective and can provide a basis for corresponding strategies. This assessment model can provide a reference to the social stability risk assessment of major highway projects. However, the applicability of this model to social stability risk assessment in other fields remains to be studied. In the future, more extensive research will be conducted.

ACKNOWLEDGMENTS

This article is financially supported by the project plan of Guangxi Zhuang Autonomous Region Transportation Department and Guangxi Zhuang Autonomous Region Quality and Technical Supervision Bureau. According to the "Notice of the Transportation Department of Guangxi Zhuang Autonomous Region on Issuing the Guangxi Transportation Standardization Project Plan in 2019" (Guangxi Transportation Research & Consulting Co., Ltd. [2019] 136) and the "Guangxi Zhuang Autonomous Region Market Supervision Bureau on issuing the second batch of Guangxi local standards in 2020" Notification of Project Plan" (Guangxi Zhuang Autonomous Region Market Supervision Bureau Letter [2020] 1832), finally completed.

REFERENCES

Dong Z, Wang H, Dong X, Liu S. (2018) Road project social stability risk assessment method based on entropy weight extension matter-element model [J]. *China Journal of Highway and Transport*, 31(09): 191–198.

Tian Z, Yang X, Zhou S. (2018) Social stability risk assessment of road engineering based on the analytic hierarchy process: Taking a national road reconstruction project in Beijing as an example [J]. *Highway Engineering*, 43(02): 265–269.

Xu Z, Wu J, Cao R. (2020) Research hotspot and trend analysis of social stability risk based on knowledge graph [J]. *Journal of Northwest University* (Philosophy and Social Sciences Edition), 50(03): 53–63.

Yang J. (2021) An empirical study on key risk factors in leadership decision-making of major events [J]. *Leadership Science*, (10): 112–115.

Zhang C, Huang D, U.Lall, Hua J. (2012) Research on the social system stability of major water conservancy projects based on self-organization theory [J]. *China Population, Resources, and Environment*, 22(11): 109–115.

Zhou S, Zhang X, Tian Z, Yang X. (2015) Research on the pre-assessment method of social stability risk in the early stage of road engineering based on the two-stage method [J]. *Highway Engineering*, 40(01): 9–13.

Zhu D. (2016) Research on social stability risks assessment of major decision-making matters [M]. *Science Press, Beijing.*

Zou D, Chen C, Chen S. (2019) Rights guidance dilemma and mechanism design of social stability risk assessment for major events [J]. *Leadership Science*, (06): 74–77.

Advances in Civil Engineering and Environmental Engineering – bin Yahya & Bin Daud (Eds)
 ISBN: 978-1-032-39397-1

Risk assessment of pilotage process in Guangzhou port based on unascertained measurement model

Fagen Wang*
Merchant Marine College of Shanghai Maritime University. Shanghai, China

Hao Chen
Guangzhou Pilot Station, Guangdong, China

Shenping Hu* & Shaoyong Xuan*
Merchant Marine College of Shanghai Maritime University. Shanghai, China

ABSTRACT: Guangzhou harbor is one of the busiest waters for marine traffic in China. In order to ensure the safe pilotage of large ships in Guangzhou port, it is necessary to identify high-risk segments in the port waters when large ships enter the port. In this paper, through the analysis of factors affecting pilotage environment risk, an unascertained measurement mathematical model is introduced, and the copula algorithm is used to measure the weight relationship between risk factors, and a risk assessment model based on the unascertained measurement of the Guangzhou port waters pilotage process is established. Combining multiple scenarios of actual entry of large ships into the port, a quantitative risk assessment was carried out for important pilot segments in the waters of Guangzhou port. The case verification shows that: in the pilotage of large ships, the overall risk of the ship in the post-pilotation section is relatively high. Tide velocity is one of the key factors affecting the berthing safety of ships.

1 INTRODUCTION

As an important port in the Pearl River Delta Economic Zone and the departure port of the Maritime Silk Road, Guangzhou Port is a golden port supporting regional economic development. With the development of China's economy, the flow of ships in Guangzhou Port is increasing. At the same time, with the development of large-scale ships, new requirements have been put forward for the safety of ship piloting in Guangzhou Port. The pilotage risk assessment of Guangzhou Port has important practical significance to ensure the safety of ships and ports.

The risk assessment of Guangzhou Port waters has received widespread attention from researchers in recent years, and great progress has been made. Xuan et al. (2013) proposed an improved risk assessment method (PRA) based on the hybrid of Bayesian theory and navigation flow calculation, analyzing the risk characteristics and development in Guangzhou Port; Lu (2021) introduced the model of entropy weight theory conducts an objective risk assessment on the navigation risk of the port area. Taking the navigation risk of Guangzhou Port as an example, the risk level of Guangzhou Port is relatively safe, which is in line with reality; Zhang et al. (2021) resorted to the Monte Carlo model (MC) obtained the characteristics of multiple random samples by a simulation experiment and analyzed the navigation risk distribution characteristics of Guangzhou

*Corresponding Authors: 202030110213@stu.shmtu.edu.cn; smu_mra<smu_mra@163.com and syxuan@stu.shmtu.edu.cn

DOI 10.1201/9781003349563-39

Port under the adverse conditions of visibility 0-12km with small sample data. At the same time, the risk assessment of navigable waters is also affected by the navigable environment. Wang et al. (2016) established a pilotage environment risk model based on the entropy weight TOPSIS method, and evaluated each evaluation object based on the risk closeness, so as to quantitatively evaluate the navigation environment of the waterway; Gan et al. (2018) aimed at the randomness and ambiguity of the pilotage environment risk, established pilot environment risk assessment model based on the standard normal cloud matter element , and determined the risk level of the pilotage environment by calculating the correlation between the matter element and the standard normal cloud-matter element; Chen et al. (2019) proposed a dynamic weight quantification model of the secondary subjective weight adjustment of the subjective risk perception at the level of the navigation environment impact factor and the objective weight adjustment based on the characteristics of the accident based on maritime accident samples, and further clarified the risk mechanism of shipping accidents. Chen et al. (2020) introduced a navigational environment assessment model based on the catastrophe progression method improved by the CRITIC method, which effectively reduced the subjective problem of decision-making in the evaluation process. In comparison with other mainstream models, the evaluation results are consistent, which proves the effectiveness and practicality of the model are improved.

Indeed, the risk is the emergence of a multi-factor coupled system, which is characterized by nonlinearity and high ambiguity. The introduction of the mathematical model of unascertained measurement is of great help to the risk assessment of the navigation environment. Li et al. (2009) took Taizhou Port and Jiaxing Port as examples established a mathematical model of unascertained measurement of port water area risks, and obtained the risk grading sequence of each segment of the port water area. Comparison with actual investigations. The effectiveness of the safety assessment of pilot waters. Zhao et al. used the unascertained measurement theory to deal with the randomness and uncertainty of the evaluation factors in the risk assessment of passenger ship night sailing. Unascertained degree models are widely used, for example, in mining development (Yang et al. 2021), aviation (Han & Yang 2020), tunnel construction (Di et al. 2021), natural disaster prevention and control (Wang & Wang 2017), etc. These studies have greatly promoted the development of the application of the unascertained degree model.

In summary, in order to more accurately and objectively analyze the risk of the pilotage process of Guangzhou Port, a mathematical model of the unascertained measure-copula function was established to analyze the risk of each important pilotage segment, and combined with actual scene conditions, put forward the risk assessment of the piloting process of each important pilotage segment.

2 RISK ASSESSMENT METHOD

2.1 *Risk of ship pilotage process*

Risk is a collection of the possibility and consequences of accidents. Due to the instability of external environmental factors and the uncertainty of the potential evolution of internal factors, the possibility and consequences of risk are dynamic over time. The functional relationship is shown in formula (1):

$$R_{(t)} = f\left(F_{(t)}, C_{(t)}\right) \tag{1}$$

(1): $R_{(t)}$ Show the value of risk at the moment; $F_{(t)}$ Indicates how often the risk occurs at the moment; $C_{(t)}$ Indicates the consequences of a risk event at the moment.

In the piloting process of a ship, from the starting position of piloting to the completion of berthing, the collection of risk events in time constitutes the risk of the ship's piloting.

2.2 *Risk index system of ship pilotage process*

Based on the risk system theory, the composition of ship pilotage risk is composed of human, machine, environment and management. This article selects environmental factors as the evaluation index to evaluate the risk of each pilot segment during the piloting process. Through visits and surveys of pilots on the front line, eight main assessment indicators that affect pilotage safety have been identified, they are tide height, tidal current, waterway width, waterway depth, turning range, VTS control, traffic density and waterway conditions. Through literature review and a questionnaire survey of senior pilots, the 8 evaluation indicators are divided into 5 levels as shown in Table 1.

Table 1. Classification standard of ship pilotage risk assessment indicators.

Indicators	C_1 Lower Risk (LR)	C_2 Lower Risk (L)	C_3 Moderate Risk (M)	C_4 High Risk (H)	C_5 Higher Risk (HR)
tide height F_1	1	2	3	4	5
tide current F_2	1	2	3	4	5
waterway width F_3	$(1/\infty, 1/800)$	$[1/800, 1/500)$	$[1/500, 1/300)$	$[1/300, 1/100)$	$[1/100, 1/0)$
waterway depth F_4	$(1/\infty, 1/16]$	$(1/16, 1/12]$	$(1/12, 1/8]$	$[1/8, 1/4)$	$[1/4, 1/0)$
turning angle F_5	1	2	3	4	5
VTS control F_6	1	2	3	4	5
traffic density F_7	1	2	3	4	5
waterway conditions F_8	1	2	3	4	5
value	<0.2	[0.2-0.4]	(0.4-0.6]	(0.6-0.8]	>0.8

Note: The number of 1/U in the table, U is the actual value.

3 THE UNASCERTAINED MEASURE RISK MODEL

Unascertained mathematics is proposed according to the needs of construction engineering theory research. This opens a new way to study the mathematical expression and processing methods of unascertained information and solves the problem of the expression and processing of "incomplete information". Generally, there are continuous unascertained numbers and discrete unascertained numbers. Because risk has nonlinear and discrete characteristics, the discrete unascertained number function is the main application of risk assessment.

3.1 *Classification standard matrix*

Suppose the sample R has n regions in its evaluation space, and each region corresponds to m evaluation index F. r_i represents the i_{th} evaluation sample; F_j represents the j_{th} evaluation index; the evaluation value F_j of r_i is recorded as r_{ij}, therefore, the evaluation sample R_i can be expressed as:

$$R_i = \{r_{im}\}, (1 \le i < n) \tag{2}$$

Assuming that the sample R has k_{th} evaluation levels, the evaluation space

$$U = \{C_1, C_2, \cdots, C_k\} \tag{3}$$

of the sample R, and the evaluation level vector $C_1, C_2, \cdots, C_k$ is linearly increasing or decreasing. Assuming the risk variable

$$C_k > C_{k-1} \tag{4}$$

there is

$$C_1 < C_2 < \cdots < C_{k-2} < C_{k-1} < C_k \tag{5}$$

In summary, the classification standard matrix can be written as:

$$\begin{cases} \{C_1 \cdots C_k\} \times \begin{Bmatrix} F_1 \\ \vdots \\ F_m \end{Bmatrix} = \begin{bmatrix} a_{11} \cdots a_{1k} \\ \vdots \\ a_{m1} \cdots a_{mk} \end{bmatrix} \\ a_{m1} < a_{m2} < \cdots < a_{mk}, or, a_{m1} > a_{m2} > \cdots > a_{mk} \end{cases} \quad (6)$$

3.2 *Single index unascertained measure*

Calculate the unascertained measure

$$\chi_{ijk} = \chi(r_{ij} \in C_k) \quad (7)$$

that the j_{th} indicator of the i_{th} evaluation sample belongs to the class C_k, satisfying:

$$\begin{cases} \chi(r_{ij} \in U) = 1, 0 \leq \chi(r_{ij} \in C_p) \leq 1 \\ \qquad\qquad\qquad i = 1, 2, \cdots, n; \\ \dfrac{\sum_{l=1}^{p} \chi(x_i \in C_l)}{\left| r_{ij} \in \bigcup_{l=1}^{p} \right|} = \chi, j = 1, 2, \cdots, m; \\ \qquad\qquad\qquad p = 1, 2, \cdots, k \end{cases} \quad (8)$$

If the χ pair of evaluation space U satisfies normalization and additivity, then the χ satisfies the above formula is called an unascertained measure.

Assuming

$$a_{j1} < a_{j2} < \cdots < a_{jk} \quad (9)$$

then:

$$\begin{cases} \chi_{ij1} = 1 & k \neq 1; r_{ij} \leq a_{j1} \\ \chi_{ij1} \times \left(a_{jl+1} - a_{jl}\right) = a_{jl+1} - r_{ij} & a_{jl} \leq r_{ij} \leq a_{jl+1} \\ \chi_{ijl+1} \times \left(a_{jl+1} - a_{jl}\right) = r_{ij} - a_{jl} & a_{jl} \leq r_{ij} \leq a_{jl+1} \\ \chi_{ijk} = 0 & k < h \| k > h + 1 \end{cases} \quad (10)$$

3.3 *Weight of each indicator*

The index weight w_i is the main method to measure the importance of the index. The method of determining the weight includes subjective and objective, combination weighting method combines subjective and objective. The indicator weight is important to meet the following conditions:

$$w_1 + w_2 + \cdots + w_i = 1; \quad i \in [0, 1] \quad (11)$$

This article uses Copula's method to obtain.

3.4 *Sample identification and sorting*

The evaluation grade $\{C_1, C_2, \cdots, C_k\}$ has a continuous increasing or decreasing relationship, and the confidence identification criterion can be used. If it is satisfying with

$$C_i > C_{i+1} (i = 1, 2, \cdots, k - 1) \quad (12)$$

the confidence

$$\sigma \in (0.6, 1] \quad (13)$$

$$\sigma_i = \min\left\{\sigma : \sum_{l=1}^{p} \chi_{il}(C_l) \geq \sigma, 1 \leq p \leq k\right\} \quad (14)$$

The result is determined: the sample r_i belongs to the class C_{ki}.

In order to be able to identify the risk of the evaluation samples in the same category C_{ki}, and to make a reasonable sorting, based on the classification of Table 1,

$$\sigma = 5, C_1 < C_2 < \cdots < C_k \quad (15)$$

you can calculate according to the value pair of the arithmetic sequence Ψ_{ri} with a tolerance of 0.25 and compare and sort according to the obtained results.

$$\Psi_{ri} = \sum_{p=1}^{k} n_p \chi_{ri}\{0, 0.25, 0.5, 0.75, 1\} \quad (16)$$

4 EXAMPLE APPLICATION

4.1 *Scene description*

In order to research the risk of the pilotage segment in detail, the pilotage process is divided into 9 segments through the collection and analysis of the actual ship pilotage data in different periods, and the pilotage risk level selected for each segment in the four scenarios is determined, see figure 1, T/S means Tide Station.

Scenario 1: Take the LNG carrier "LERIC" as an example. length 217 m, width 34 m, and draft 11 m. At 0700 hours on July 21, 2021, under full load conditions, the pilot will board the ship from Zhen Xiangwei pilot station and pilot the ship to Jiufeng No. 1 Petrochemical Terminal of Dong Guang, completing the berth at 1300.

Scenario 2: Take the oil tanker "SALUTE LEGEND" as an example. length 183 m, width 32 m, and draft 11.5 m. At 0800 hours on July 21, 2021, under full load conditions, the pilot boarded the ship from the Zhen Xiangwei pilot station and pilot the ship to Jiufeng No. 1 Petrochemical Terminal of Dong Guang, completing the berth at 1400.

Scenario 3: Take the oil tanker "CHANG HANG CHENG GONG" as an example. length 185 m, width 32 m, and draft 11.7 m. At 0900 hours on July 21, 2021, under full load conditions, the pilot boarded the ship from the Zhen Xiangwei pilot station and pilot the ship to Jiufeng No. 2 Petrochemical Terminal of Dong Guang, completing the berth at 1500.

Scenario 4: Take the oil tanker "CHANG HANG HU PO" as an example. length 185 m, width 32 m, and draft 11.7 m. At 1000 hours on July 21, 2021, under full load conditions, the pilot boarded the ship from Zhen Xiangwei pilot station and pilot the ship to Jiufeng No. 2 Petrochemical Terminal of Dong Guang, completing the berth at 1600.

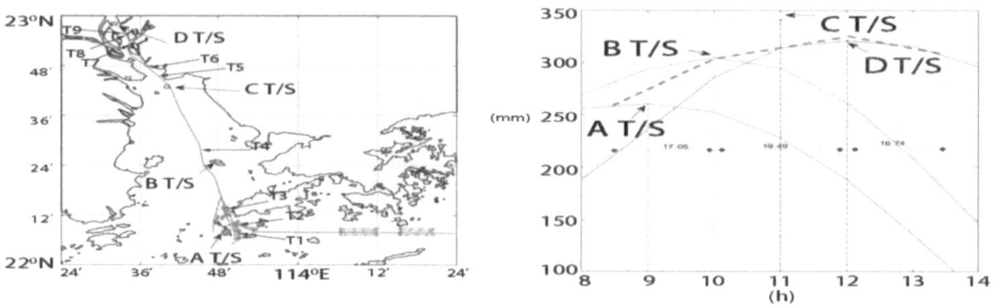

Figure 1. General situation of Guangzhou port and tide station information.

4.2 *Data acquisition and processing*

Taking scenario 1 as an example, the evaluation index values of each pilot segment are obtained after analysis, as shown in Table 2.

Table 2. Evaluation index value of each segment.

Area/Index	F_1	F_2	F_3	F_4	F_5	F_6	F_7	F_8
T1	2	3	1/500	1/17	1	3	2	1
T2	2	3	1/500	1/17	2	3	2	1
T3	2	3	1/385	1/17	1	3	2	1
T4	2	3	1/385	1/17	3	1	3	2
T5	1	3	1/385	1/17	2	1	3	3
T6	1	2	1/180	1/13	2	1	3	3
T7	1	1	1/180	1/13	2	1	4	3
T8	1	1	1/180	1/13	4	1	4	2
T9	1	1	1/500	1/12	2	1	3	2

Use the above mathematical model in conjunction with Table 2 to process the data, and see Table 3 for the comprehensive evaluation vector of the risk of each pilot segment.

Choose confidence 0.6 to get the risk of each pilot segment. Use formula (16) to process the data in Table 3 to get the relative risk value of each pilot segment. It can be seen from the data that there are 3 medium-risk pilot segments and 6 low-risk pilot segments, as shown in Table 3.

Table 3. Comprehensive risk measure evaluation vector for each segment.

Area/Risk	C_1	C_2	C_3	C_4	C_5	value	result
T_1	0.3872	0.2653	0.2282	0	0.1194	0.2998	L
T_2	0.2637	0.3888	0.2282	0	0.1194	0.3307	L
T_3	0.3872	0.2653	0.2282	0	0.1194	0.2998	L
T_4	0.2487	0.2519	0.3800	0	0.1194	0.3723	L
T_5	0.3781	0.1235	0.3789	0	0.1194	0.3397	L
T_6	0.2370	0.2442	0.2583	0	0.2606	0.4508	M
T_7	0.3576	0.1235	0.1225	0.1358	0.2606	0.4546	M
T_8	0.3576	0.1225	0	0.2594	0.2606	0.4858	M
T_9	0.3576	0.2460	0.1358	0	0.2606	0.39	L

4.3 *Result and analysis*

According to the application process of 4.2, scenario 1-4 is applied and calculated, and the results can be obtained as shown in Figure 2.

a) During the entire piloting process, the risk of the rear section is relatively high. That is because the ship is approaching the berth and the speed is reduced. At this time, it is not conducive to the ship's control, and the ship's maneuverability is easily affected by environmental factors such as tidal currents, especially for large ships. The law of the "bathtub curve" also confirmed this point by Guo et al. (2021) Therefore, in the daily training of pilots, it is necessary to focus on improving pilots' handling skills when the ship in low speed, as well as to reduce the impact of tidal currents on ships berthing to prevent accidents happen.

b) The risk of berthing operations during high and low tides is significantly lower than in other adjacent periods. In the latter part of the pilotage, based on the data from the tide station D, shown in figure 1, the pilotage risk of the ship shows a gradual upward trend with high flat tide-first

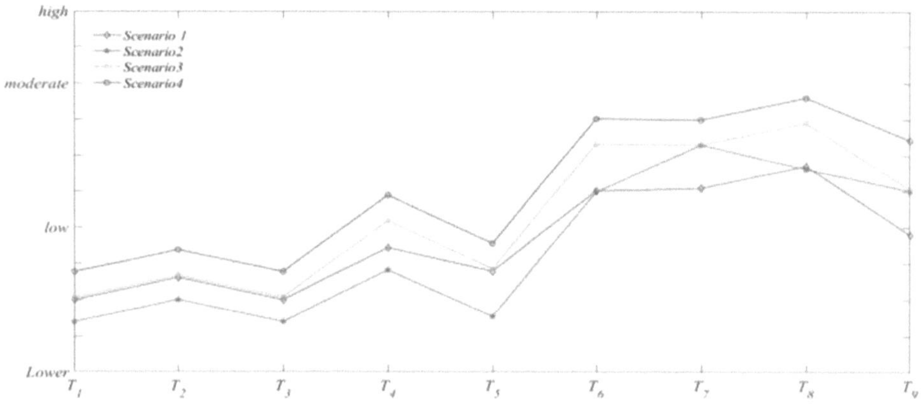

Figure 2. Pilotage of ships in Guangzhou port.

fall-rapid fall. It is in line with the conclusion that the risk of ship berthing is low during the “high and flat tide” window (Guo et al. 2021).

c) In the T_8 segment, which is towing stage, the risk is relatively higher. This is because the ship’s speed is decelerating quickly, and the towing operation is carried out. At this moment, the ship is still in the waterway, which is limited by the width of the waterway, affected by the tide, and the turning angle. The large impact makes the risk level of this pilot segment higher. The waterway management department should pay attention to this segment and make emergency arrangements, such as dispatching additional emergency tugs, delineating emergency anchorages, and traffic control measures to strengthen the safety management of the waterway.

d) Compared with the T_5 pilot segment, the risk level of the T_6 pilot segment has been significantly increased. Compared with the value of various indicators, the decrease in the water depth is an important factor in the increase of the risk of this pilot segment. Therefore, when the water depth changes, safety speed should be taken in time. Regarding changes in water depths of different pilot segments, formulating and strictly implementing safe speeds is a necessary measure to ensure the safety of ships.

e) The pilotage risk of Guangzhou port is generally at a low level. (Xuan et al. 2013) With the full commissioning of the Guangzhou port deep-water waterway widening project and reasonable port planning and construction in recent years, the accident probability of Guangzhou Port has been at a low level. However, the burden of ship traffic has increased based on the increasing throughput of Guangzhou Port, and relevant departments must actively ensure the navigation safety of Guangzhou Port.

4.4 *Discussion*

Based on the risk of the piloting process in four scenarios, combined with the time change of each index, as shown in Figure 3, the role of environmental risk factors in Guangzhou Port is discussed, and the results are as follows:

a) In the T_3 segment, the waterway is relatively straight and the risk of turning angles is reduced, but the overall risk level increases with the smaller waterway width. Therefore, during port navigation, the width of the waterway is one of the necessary conditions to ensure the safe navigation of ships. In a waterway, a large turning angle can be appropriately widened to ensure the safe navigation of ships.

b) In the T_4 segment, under the conditions that the risk of turning angles, traffic density, and waterway conditions have increased, the overall risk of this segment has not shown a significant increase with the strengthening of VTS control. Therefore, the rise of the control level of VTS in the port area plays an important role in the safety management of the waterway. VTS plays an irreplaceable role in information provision, emergency support, and action coordination. In order to improve the management of the waterway, the scope and level of VTS control can be appropriately expanded and upgraded to ensure the safety of ship pilotage in the port area.

c) In the T_5 segment, the emergency anchorages and the wealthy waters on both sides are reduced, which increases the risk of waterway conditions, but the overall risk level of the waterway is reduced with relatively straight in this segment, Therefore, when designing and planning artificial waterways, minimizing their turning points, and turning angles are an important means to ensure the safety of the waterways.

d) In the T_7 segment, although the flow rate is stable, the overall risk increases as the traffic density increases. Therefore, at waterway intersections and ships encountering water, it is necessary to strengthen waterway traffic control, and pilots should be vigilant and use safe speeds.

e) In the T_9 segment, the ship is in the stage from near berth to complete berth. With the assistance of the tugboat and its own limited maneuverability, the ship slowly approaches the berth. At this time, environmental factors such as water depth and waterway width have little effect, but the berthing risk increases with the increase of tide speed, it confirms that the tide is one of the key factors affecting ship berthing safety once again. In terms of waterway management, breakwaters should be built for piers affected by strong currents; for different ship types, berthing standards at safe tide speeds should be verified to ensure the berthing safety of ships.

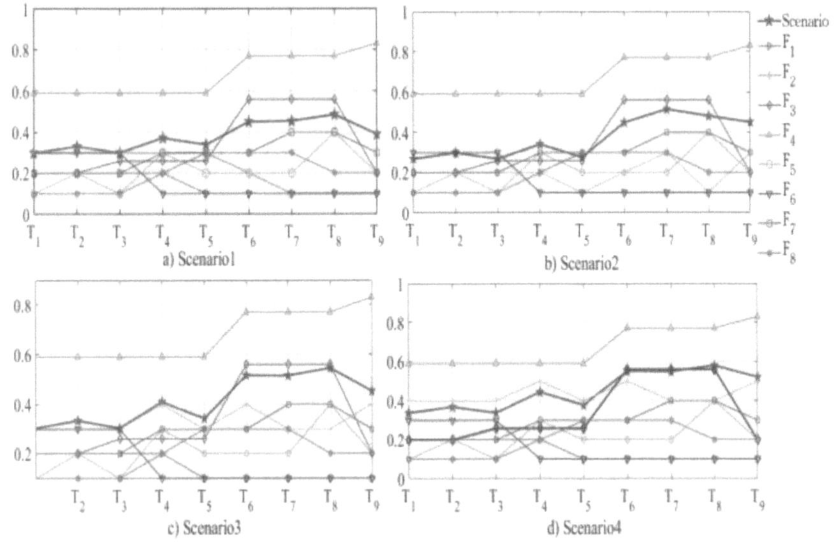

Figure 3. Influence of environmental factors on pilotage process.

5 CONCLUSIONS

In this paper, through the application of an unascertained measurement mathematical model, combined with actual scenarios, a quantitative and objective analysis of the safety of ship pilotage in Guangzhou Port is carried out, avoiding the blindness of only subjective and qualitative analysis.

Based on the division of 9 pilot segments, the reasons for the differences in risk levels of each pilot segment are analyzed in detail, as well as the corresponding intervention measures, for reference by the relevant authorities of the port.

Since risk is a complex system with multi-factors coupling and it is necessary to comprehensively and systematically consider the formation mechanism of its dynamic evolution. Therefore, it is necessary to further study the relationship between the factors, so as to better improve the safety management of the port.

ACKNOWLEDGMENT

Fund project: Funded by the Key Research and Development Program of China (2019YFB1600602).

REFERENCES

Chen, S.J., Sun, J, Zhao C.Y., et al. (2019) Dynamic weighting quantification model of navigation environment risk of maritime accident. *Navigation of China*. vol. 42(04):18–23.

Chen, W.J., Zhang, P.F., Jiang, S.Q., et al. (2020) Navigation environment risk assessment of the waterway based on improved catastrophe progression method. *Journal of Safety and Environment*. vol. 20(05):1617–1623.

Di, Q, Gu, W.H., Zhao, H.Y. (2021) Risk assessment of gas disaster in tunnel construction based on unascertained measurement theory. *Journal of Railway Science and Engineering*. vol. 18(03):803–812.

Gan, L.X., Cheng, X.D., Wang, C. (2018) On the evaluation for the pilotage navigation environment risks based on cloud matter element theory. *Journal of Safety and Environment*. vol. 18(02):418–422.

Guo, Y.L., Zhang, T, Hu, S.P, et al. (2021) Risk simulation model of ship piloting process in multi-period windows at tidal port. *Journal of Safety and Environment*. vol. 21(01):49–55.

Han, Y.J., Yang, Y.Q. (2020) Health status evaluation of aero-engines based on combination weighting method and unascertained measure model. *Operations research and Management Science*. Vol. 29(04):204–211.

Li, W, Hu, S.P., You, Q.H., et al. (2009) Unascertained measure model for port traffic risks assessment. *Journal of Shanghai Maritime University*. vol. 30(03):13–17+21.

Lu, J.H. (2021) Safety risk assessment of guangzhou port's navigation based on entropy catastrophe theory. *Journal of Guangzhou Maritime University*. vol. 29(02):9–13.

Wang, C, Jiang, F.C., Ma, Q.D. (2016) Pilotage environmental risk assessment of waterways based on entropy weight and TOPSIS model. *Journal of Safety and Environment*. vol. 16(03):33–37.

Wang, Y.Z., Wang, X.F. (2017) evaluation of rainstorm disaster prevention and mitigation capacity in qinling mountains huayang scenic area based on the unascertained measure model. *Journal of Natural Disasters*. vol. 26(01):67–79.

Xuan, S.Y., Li, F.M., Hu, S.P., et al. (2013) Risk assessment of Ship Navigation in the Waters of Guangzhou Harbor. *Navigation of China*, vol. 36(02): 96–99.

Yang, Y.P., Huang, P.S., Chen, G.G. (2021) Safety evaluation of tailing pond based on FIM-optimization unascertained measure. *Journal of Safety and Environment*. vol. 21(03):996–1004.

Zhang, J, Zhang, Z.J., Zhou, X.Y. (2021) Navigation risk assessment of Guangzhou port based on monte carlo method. *Journal of Meteorology and Environment*. vol. 37(03):132–138.

Zhao, H.D., Liu, J.X., Liu, Y, et al. Evaluation of navigation risk for Inland-river passenger ships at night-time through improved uncertainty measure model. *Journal of Transport Information and Safety*. vol. 36(01):28–34.

Advances in Civil Engineering and Environmental Engineering – bin Yahya & Bin Daud (Eds)

Study on risk assessment of Wupai tunnel engineering of Jingheshan-Nanning Expressway in Liuzhou, Guangxi

Hongming Li, Hongri Zhang*, Jianhai Wei & Yihong Pang
Guangxi Transportation Research & Consulting Co., Ltd.Nanning, Guangxi, China

ABSTRACT: Based on a detailed analysis of engineering geology, hydrogeological characteristics, and adverse geological effects of the Wupai Tunnel, the LEC method is used to identify and evaluate the risks of rock mass engineering geology, classification of tunnel surrounding rock, stability of tunnel site area, stability of entrance and exit slope, stability of tunnel surrounding rock, rock burst and unsymmetrical pressure. Meanwhile, the safety hazards of this tunnel are put forward, and special treatment measures are formulated for the evaluation results. The study results provide reliable data support and risk control for the overall improvement of tunnel construction quality and construction level.

1 INTRODUCTION

Highway tunnels in China are characterized by wide distribution, complex environments, and geological conditions. Tunneling has many operation steps and has the characteristics of stage dynamics, so it is necessary to actively carry out risk management and control work. Scholars at home and abroad have done a lot of research on tunnel risk assessment in theoretical methods, mathematical models, program design and development, and achieved some results. However, at present, there is still a lack of quantitative methods and applied research on tunnel risk, which are logical, systematic, and complex. In this paper, the Wupai Tunnel Engineering of Jingheshan - Nanning Expressway in Liuzhou, Guangxi (hereinafter referred to as Wupai Tunnel Engineering) is taken as the research object. Based on a detailed analysis of the basic geological characteristics, the LEC method is used to identify and assess the risks, and corresponding measures are put forward, to provide a basis for improving the construction quality and level of the tunnel project.

2 MATERIALS AND METHODS

Tunnel risk assessment is mainly to analyze and assess the construction sections of tunnel projects with an overall risk assessment level of Class III (high risk) or above, identify the risk sources according to their operational risks and similar projects, and semi-quantitatively estimate the major risk sources, and propose corresponding countermeasures.

In this risk assessment, firstly, the LEC method is used for risk estimation, and the product of three indicators related to the system risk rate is used to evaluate the risk of casualties in the system: L is the possibility of accidents; E is the frequency of human exposure to this dangerous environment; C is the loss consequence caused by accident. Risk score $D = L \times E \times C$. The larger the value of D, the greater the danger of the system, which means it needs to be taken measures to change the possibility of accidents or reduce the frequency of human exposure to a dangerous environment, or reduce accident losses, to ensure safety.

*Corresponding Author: 253541461@qq.com

DOI 10.1201/9781003349563-40

3 RESULTS AND DISCUSSIONS

3.1 *Engineering geological analysis of Wupai Tunnel*

3.1.1 *Geologic characteristics*

Wupai Tunnel is located in Wupai Village, Gukang Village, Guosui Township, Xincheng County, Laibin City. The tunnel site area is a karst peak cluster depression, with large topographic relief, mostly bare solitary peaks, with a ground elevation of 281.5 ~ 8 m and a relative height difference of about 193 m. The exposed strata are mainly the quaternary eluvial layer (Q^{el+dl}), and the lower Permian Maokou stage (P_1m) containing limestone and siliceous dolomite. The geotectonic structure of the tunnel site is located in northwest of the horseshoe shield of the "mountain-shaped" structure in Guangxi (Cui 2018; Jiang 2017). The main structural features in the region are Guwan fault (F4), Jialu syncline (G1), and Eshan anticline (G2), as shown in Figure 1. The tunnel site area and its surrounding structures are relatively developed, and most of them occurred in the Mesozoic era. Now, all tectonic movements tend to be stable. No signs of neotectonic activity are found in the investigation, and the regional stability is good. The regional ground motion peak acceleration is 0.05 g, the characteristic period of the ground motion response spectrum is 0.35 s, and the basic earthquake intensity is Degree VI. The overall engineering geology is complex.

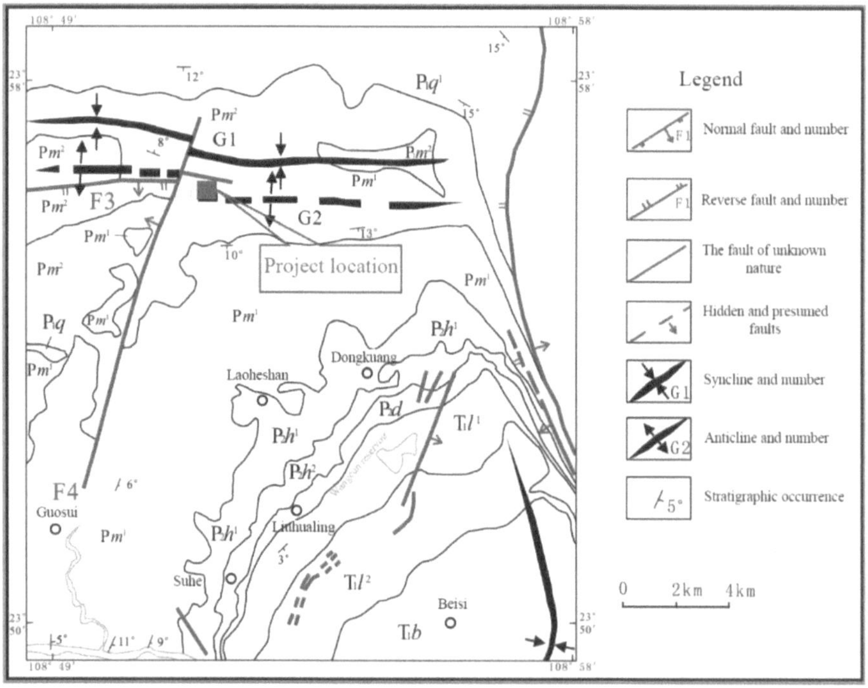

Figure 1. Outline of geological structure.

3.1.2 *Hydrogeological characteristics of tunnel engineering.*

a) Engineering geological characteristics of tunnel rock mass

The engineering geological soil mass is mainly divided into Quaternary eluvial breccia (Qel+dl-1), which is brown, slightly wet and slightly dense, distributed at the bottom of the depression, with a layered thickness of 1-10 m and a seismic elastic longitudinal wave velocity of VP = 500 m/s; Silty clay (Qel+dl-2) is yellowish-brown and slightly wet, which is mainly distributed in the depression bottom, gentle slope surface and carbonate dissolution trench in the tunnel site area. The thickness exposed by drilling varies from 5.0 m to 11.8 m, the number of standard penetration blows is 16 ~ 22, and the seismic elastic longitudinal wave velocity is VP = 500 m/s; Silty clay containing gravels (Qcol+dl-3) is yellowish-brown and gray, with liquid limit index < 0-0.30 and an average value of 0.14, and the seismic elastic longitudinal wave velocity is VP = 500 ~ 800 m/s. Tunnel mileage ZK73 + 420 ~ 660 m and YK73 + 430 ~ 660 m are located in this stratum.

The engineering geological rock mass is mainly composed of highly weathered siliceous dolomite in the lower Permian Maokou stage (P_1m), which is light gray-light yellow, and mainly distributed in the south of the F2 fault in the tunnel site area. Only CK2 and SK3 boreholes are exposed, and the seismic elastic longitudinal wave velocity is $V_P = 1350 \sim 1600$ m/s. Moderately weathered siliceous dolomite, light gray, is mainly distributed in the south of the F2 fault in the tunnel site area, and the seismic elastic longitudinal wave velocity is $V_P = 4600$ m/s. Medium weathered limestone, light gray-gray, medium-thick layered structure, relatively broken and complete rock mass, and the seismic elastic longitudinal wave velocity is $V_P = 3000 \sim 4600$ m/s. The saturated uniaxial compressive strength of the rock samples taken from the borehole is 29.729.738.3 ~ 66.8 MPa, with an average value of 53.79 MPa, which belongs to hard rock. The rock mass is relatively broken (Xing & Wang 2019; Xia et al. 2018), which is the main rock stratum in the tunnel site area, distributed throughout the tunnel site area, in which the tunnel bodies of mileage ZK72 + 060 ~ ZK73 + 420 m and YK73 + 040 ~ YK72 + 430 m are located.

b) Characteristics of karst underground river

The tunnel site area of Wupai Tunnel is located in the hydrogeological unit of the karst underground river, which is distributed in peak cluster valley and depression landform area, structurally developed in Maokouji limestone along the north-south direction and discharged into Hongshui River near Nengrengtun in Suhe Village. The overall hydraulic slope of the underground river is 11‰, and the upstream dry season discharge is 63 L/s; There are many exposed water points downstream, which flow out of the surface in the local rainy season to form seasonal surface runoff. A 50-meter-long open-flow section appeared in the underground river near Hongshui River (Li et al. 2018; 2017; Zhou & Qin 2018). The underground river has a double-deck exit, the lower exit is located in the riverbed below the low water level of Hongshui River, and the upper exit is located on the floodplain with an elevation of 74 m. The mainstream of the underground river develops to the southeast after passing through the water-dissipating tunnel S06 in Jiafutun, enters the tunnel site area near the water-dissipating tunnel S07, develops along with the depression, joins another tributary in the northwest direction near the water-dissipating tunnel S09, and then develops to the south and flows out of the tunnel site area. In the tunnel site area, it is estimated that the development elevation of the underground river pipeline is about 200 m, which is lower than the design elevation of the tunnel. See Figure 2 for the schematic profile.

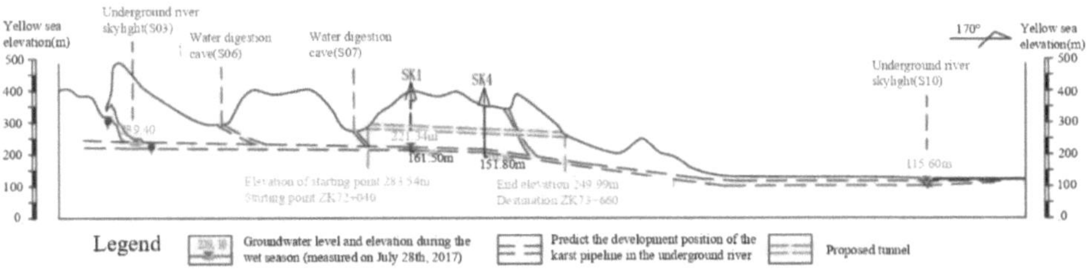

Figure 2. The relationship between the proposed Wupai Tunnel and the underground river karst pipeline.

The designed tunnel body of the Wupai Tunnel is located above the underground river karst pipeline and the stable groundwater level, which is not greatly affected by the underground river and karst pipeline. However, the surface depression and the underground river pipeline often form vertical crevice dissolution channels, which are the main water guide zone for surface water to seep into groundwater through the water-dissipating tunnel. When the tunnel passes through these crevice vertical dissolution channels, coupled with a large amount of rainwater injection during the rainy season, this will pose a great threat to the tunnel construction, which should not be ignored.

c) Characteristics of tunnel water inflow

According to the engineering geological and hydrogeological conditions of the tunnel, combined with the division of water-bearing rock groups, combined with the geological characteristics such as topography and geomorphology, the infiltration coefficient method, runoff modulus method and groundwater dynamics method are used to roughly calculate the water inflow of the tunnel.

Table 1. Calculation results of tunnel water inflow (m^3/d).

	Dry Season	Rainy Season
Infiltration Coefficient Method	1718.29	13715.28
Runoff Modulus Method	2798.66	20989.96
Groundwater Dynamics Method	10714.6	21429.21

It can be seen from the calculation results (Table 1) that the normal water inflow during tunnel excavation in the dry season is small, the water inflow is mainly wet and drip-like, and the local water inflow is linear. After heavy rainfall in the rainy season, there may be transient rain or water gushing in strands. Especially, it is speculated that there may be slit caves with a certain width above the underground river, which is the channel for rainwater to inject water into the deep underground river. When the tunnel passes through these slit caves, the large amount of rainwater injected poses a certain threat to the construction. However, as long as the karst waterway channel is not blocked and the waterway is kept open during excavation, no water disaster will be formed (He 2018; Yuan et al. 2016)

d) Tunnel surrounding rock classification.

According to the geological characteristics and engineering characteristics, the surrounding rocks of Wupai Tunnel mileage ZK72 + 060-ZK73 + 420 m and YK72 + 040-YK73 + 430 m can be divided into rocky surrounding rocks, and the lithology is moderately weathered limestone containing flint. The surrounding rocks of the mileage ZK73 + 420-660 m and YK73 + 430-660 m are classified as the soil surrounding rocks, and the lithology is silty clay containing gravel. According to groundwater development of surrounding rock, hardness and integrity of rock, moisture degree of soil surrounding rock, basic quality index BQ of surrounding rock and correction value [BQ], the detailed classification of surrounding rock is determined by comprehensive evaluation combined with qualitative characteristics of the rock mass.

See Table 2 for calculation results of basic quality index BQ and modified value [BQ] of the surrounding rock.

Table 2. Basic quality index BQ of surrounding rock and revised value [BQ].

Rock Character		R_C	K_V	BQ	K_1	K_2	K_3	[BQ]
Moderately weathered limestone containing flint.	Weak karst development	60	0.51	397.5	0.1	0.3	0	357.5
	Moderately developed karst	60	0.51	397.5	0.3	0.3	0	337.5
	Strong karst development	46.99	0.43	338.5	0.6	0.3	0	248.5
Fault Breccia	Strong karst development	31.23	0.66	348.7	0.8	0.3	0	238.7

According to the engineering geological and hydrogeological characteristics of the surrounding rock of the tunnel, the detailed level of the designed surrounding rock can be comprehensively divided into III ~ V levels, of which V level can be further divided into two sub-levels according to the lithology of the surrounding rock (Meng et al. 2017), namely V_1 of the rock surrounding rock and V_2 of the soil surrounding rock.

3.1.3 *Adverse geological processes*

a) Karst

There are many types of karst individual forms in the tunnel site and its surroundings, and the surface and underground forms are well developed (Li et al. 2016; Jiang 2016). Karst depression is the most widely distributed karst form in the tunnel site, mostly round and oval, with a depth of 50 ~ 200 m generally, and deep depression can reach a depth of 200 ~ 300 m. Water dissipation tunnels are generally developed at the bottom, which is the channel where surface water is transferred to the ground (Chen 2016; Zhou et al. 2008). Karst caves at the entrance and exit of the tunnel and the tunnel body are rare, mainly of shallowly buried underground type, such as beaded karst caves exposed by drilling holes SK2 and SK3, where water-dissipating tunnels or skylights are formed in some areas, in which water-dissipating tunnels S07 are located at the eastern edge of depressions on the road section ZK71 + 850 ~ ZK72 + 080, with an elevation of about 279 m and a water flow of about 0.895 L/s; The water dissipation tunnel S09 is located in the southeast corner of the depression on the road section ZK73 + 000 ~ ZK73 + 160, and the inlet opening is not obvious, with an elevation of about 340 m.

On the plane, karst is strongly developed in the tunnel site area, with chainages ZK72+040 ~ ZK72+200 and ZK72+550 ~ ZK73+500 on the left line and chainages YK72+060 ~ ZK73+450 on the right line. The karst is mainly developed in dissolved fissures and caves, with an elevation of 120-310 m and a deep karst development. The cave-encountering rate is 40%, and the linear karst rate is 8.55%. See Table 3 for the karst caves and soil caves exposed by drilling.

Table 3. Situation of karst cave and soil cave exposed by drilling.

Hole Number		SZZK72415	SZZK73365	SK2		SK3			CK3	
Type of Encounter Hole		Karst cave	Karst cave	Karst cave		Karst cave			Earth cave	
Roof of Tunnel (m)	Top depth	71.8	15.5	57.6	96	29.5	90.2	116	8.7	52
	Elevation of Floor Top	332.09	319.89	331.2	292.8	316.58	255.88	230.08	298.47	255.17
Tunnel Floor (m)	Top Depth	72.4	17.3	79.3	105.6	34	102.2	130	13.1	53.3
	Elevation of Floor Top	331.49	318.09	309.5	283.2	312.08	243.88	216.08	294.07	253.87
Height of Cave (m)		0.6	1.8	21.7	9.6	4.5	7.6	14	4.4	1.3
Filling Situation		Cavity	Semi-filled plastic clay	Filled with soft plastic clay and a small number of weathered fragments of original rock			Semi-filled soft plastic clay	Fully filled soft plastic clay	Cavity	Cavity

In the vertical direction, karst in the tunnel site area develops between 100 and 350 m. Drilling reveals that karst in hole SK2 and hole SK3 mainly develops in three elevation sections of 280 ~ 330 m, 220 ~ 250 m, and 100 ~ 200 m, and forms an underground karst channel, through which groundwater flows from north to south. In the 1980s, the karst ground collapse group was induced in the valley on the right side of tunnel mileage YK72+765 ~ 835 m due to dam construction and mountain pond water storage. A water dissipation tunnel (S09) is developed at the bottom of the depression on the left side of the mileage ZK73+010 ~ 140 on the right line of the tunnel, which is the main drainage channel for surface water in the depression. All the above strongly prove that there is a strong karst development zone in this section.

b) Sloping soil

The exit of the tunnel is located on the colluvial soil of the tongue-shaped slope inclined to the south, and the gullies on the east and west sides meet at ZK73+800 m in front of the tunnel site area. The lithology of the colluvial soil is composed of eluvial silty clay and silty clay containing gravels, with a thickness of 30-85m. Transverse soil cracks are developed. Due to the long-term rain erosion on the surface, deep and steep gullies are gradually formed on the east and west sides of the tongue-shaped accumulation body where the tunnel is located, and meet in front of the accumulation, thus forming a free surface around the accumulation body.

3.2 *Analysis and identification of engineering geological risks*

3.2.1 *Evaluation and prediction of regional stability of tunnel site*

The regional geological structures are relatively developed, and most of them occurred in the Mesozoic era. Now, all tectonic movements tend to be stable, and no signs of neotectonic activities are found. The three faults developed in the tunnel area are all non-brand new active faults, and the ground surface is mostly covered by Quaternary, so the regional stability is good (Zhang et al. 2017).

3.2.2 *Stability evaluation of tunnel entrance slope and back slope*

The design elevation of the entrance of the left and right lines of the proposed tunnel in Liuzhou is 280 m, which is located at the transitional position between the angle of the karst peak slope and the lower depression, with a natural slope of 12 ~ 40°C. The slope is mainly composed of moderately weathered limestone, and the shallow karst is well developed. the surface layer is covered with thin silty clay, and the surrounding rock of the entrance section is classified in detail as Grade *V*. There are no bad geological phenomena such as landslides and collapses, and the natural slopes are stable. The entrance section is a shallow buried section, the rock mass is broken and the stability is poor, and the block falling phenomenon may occur in the developed section of local dissolution fissures. The elevation slope can be graded according to the slope rate of 1RU 0.75, and the active protective net should be used in some areas (Chen 2018). The design elevation of the exit of the left and right line at the end of Nanning is about 250 m. The slope is located on the tongue-shaped collapse slope accumulation below the karst peak, the east and west sides are cut by gullies, and the two gullies flow from north to south, which intersects about 140 m in front of the tunnel exit. The ground slope is about 15°C, and the left exit is located at the accumulation body and the edge of the gully, and the ground slope is more than 45°C.

The accumulation body of the collapsed slope is mainly composed of gravel-bearing silty clay from the Quaternary residual slope with a large thickness, and the surrounding rock of the exit section is classified as grade V in detail. no adverse geological phenomena such as landslides and collapse have been found in the investigation of the present situation. According to the calculation of Lizheng geotechnical slope, under the condition of the natural slope, the stability system of the slope is 1.62, which is in a stable state, and the stability coefficient is 1.26 under rainstorm conditions, but because most of the gullies are V-shaped valleys, the slopes on both sides are steep. when it is eroded by Rainwater for a long time, it develops to the depth of slope collapse, thus reducing the ability of self-stability, when the tunnel is excavated or under the condition of a rainstorm. Destruction of the original stress state, resulting in slope instability and the formation of landslides, collapses are more likely, and the impact is greater.

3.2.3 *Stability analysis of tunnel surrounding rock*

Tunnel surrounding rock is mainly composed of moderately weathered medium-thick limestone with strong weathering resistance, complete rock mass and well-developed local dissolution fractures. The exit section is soil surrounding rock. Statistically, the tunnel is dominated by Class V surrounding rock, accounting for 50.00% of the total length, followed by Class IV, accounting for 30.12% of the total length, and Class III, accounting for 19.88% of the total length. See Table 4 for details.

Table 4. Classification statistics of tunnel surrounding rock.

Tunnel	The total length of the tunnel (m)	The rock surrounding rock grade						Grade of soil surrounding rock	
		Class III (m)	Proportion (%)	Class IV (m)	Proportion (%)	Class V (m)	Proportion (%)	Class V (m)	Proportion (%)
Left line	1620	310	19.14	500	30.86	570	35.19	240	14.81
Right line	1600	330	20.63	470	29.38	570	35.63	230	14.38
Add up to	3220	640	19.88	970	30.12	1140	35.40	470	14.60

The stability of tunnel surrounding rock is not only related to its lithologic characteristics and strength but also related to the complexity of geological structure, groundwater activities and other conditions. In the joint, fissure development zone and structural fracture zone, the surrounding rock mass is severely broken, which is prone to geological disasters such as mud outbursts and water inrush and has the functions of collecting and guiding water during the wet season.

3.2.4 *Rockburst and bias*

According to the preliminary discrimination formula of ground stress:

Extreme stress discriminant:

$$R_C/\sigma_{max} = R_C/\lambda(\gamma H) < 2 \tag{1}$$

High-stress discriminant:

$$2 < R_C/\sigma_{max} = R_C/\lambda(\gamma H) \leq 4 \tag{2}$$

In which: R_C – saturated uniaxial compressive strength of rock (MPa), taken as 84.5MPa.

σ_{max} – is the maximum initial stress value (MPa) perpendicular to the axial direction of the hole.

λ_0 – is the proportional coefficient, which is 1.2 according to a large number of measured data at home and abroad.

γ – the average weight of rock and soil mass (kN/m^3), taking 25kN/m^3.

H – buried depth of tunnel (m).

According to the discriminant inverse calculation H, the buried depth of the extremely high-stress area is H≥1408 m, and the burial depth of the high-stress area is 1408 m> H≥70 m. The maximum buried depth of the tunnel is 148 m, so there is no extremely high stress or high stress.

According to the lateral slope of the tunnel roof and the tunnel buried depth, it is judged that the soil surrounding rock of the tunnel exit section ZK73+420 ~ ZK73+660 m and YK73+430 ~ YK73+660 m is of Grade V, and the ground lateral slope ratio is less than 1: 0.75. The tunnel is shallow buried and has the phenomenon of partial pressure, so it is easy to have the possibility of large deformation of the surrounding rock, collapse, or roof fall during tunnel excavation. Therefore, short footage should be adopted during tunnel excavation, and the support should follow the excavation.

3.2.5 *Tunnel construction scheme*

The new Austrian tunneling method (NATM) is adopted for the construction of the hidden tunnel. For the V-grade surrounding rock section of the main tunnel, advance support should be done first, and three steps and seven steps should be adopted for the construction of the reinforced section of the tunnel entrance, and soil and gravel tunnel entrance section. The shallow and deep buried sections of Class V surrounding rock tunnel are excavated by the method of retaining core soil in the arc pilot tunnel. The distance between the upper and lower sections of the shallow buried section is controlled at 5 ~ 10 m, and that between the upper and lower sections of the deep-buried section is controlled at 8 ~ 10 m. The low side of the shallow buried unsymmetrical pressure section should be excavated first. For the V-grade surrounding rock section of the main tunnel, advance support should be made during construction excavation, and the excavation should adopt the method of ascending and descending steps; The construction time of the secondary lining shall be determined according to the monitoring and measurement results, and the distance between the secondary lining of Class V surrounding rock and the working face shall be controlled within the range of 3 ~ 5 times of the hole diameter, and the possibility of hazardous accidents caused by unreasonable construction scheme is medium and the degree of harm is large.

4 CONCLUSIONS AND SUGGESTIONS

(1) The tunnel area has good stability. the peak acceleration of basic ground motion in this area is 0.05 g, the corresponding basic earthquake intensity is VI degree, and the characteristic period of basic ground motion response spectrum is 0.35 s, so local geological data should be collected strictly.

(2) The surrounding rock of the tunnel is mainly moderately weathered limestone containing flint, which is divided into three grades: III, IV, and V. The tunnel is mainly of Grade V, accounting for 50.00% of the total length, followed by Grade IV, accounting for 30.12% of the total length, and Grade III accounting for 19.88% of the total length. The entrance slope and back slope of the tunnel have good stability. The slopes on both sides can be graded at 1: 0.5, and the back slope can be graded at 1: 0.75. The stability of the exit slope and the back slope of the tunnel is relatively poor, so it is suggested that the slope should be placed according to the slope ratio of 1: 1.5, and reasonable and effective supporting measures should be taken. In addition, the right side of the tunnel exit is located in the gully, so the drainage of surface water and the blocking measures of loose rock and soil in the gully should be taken.

(3) Deep-developed dissolved gaps, joints and underground karst pipelines have a large water inflow, and geological disasters such as strong water inrush or surface water pouring into the tunnel may occur after tunnel excavation. Special construction schemes must be prepared according to the standard specifications and design requirements, to ensure that they are organized and implemented according to the schemes. It is strictly forbidden to change construction methods without authorization.

(4) The soil surrounding rocks in the tunnel exit sections ZK73+420 ~ ZK73+660 m and YK73+430 ~ YK73+660 m are of Grade V, and there is a certain bias, so attention should be paid to the bias design. Drainage ditches and intercepting ditches should be set on the top surface of the entrance slope and the back slope and around them, and the arrangement should be comprehensively considered with the subgrade drainage system.

(5) Dissolved fissures, small caves, and gullies (grooves) may be encountered in the cave body, which can be treated by measures such as burying and crossing; Local karst is strongly developed, and there is the possibility of meeting large hall-type karst caves and large karst pipelines. When tunneling, there is the possibility of collapse, roof fall, water inrush, and mud outburst, so emergency measures should be taken. It is necessary to implement the provisions of advance hydrogeological exploration and prediction, and immediately stop work and evacuate people if the monitored data (exploration) exceeds the standard, and it is strictly forbidden to take risks in construction operations; The number of field workers must be strictly controlled. The mechanized operation should be carried out on the heading face. It is strictly forbidden to organize construction work by overstaffing.

REFERENCES

Chen N. 2018. Preliminary study on route selection strategy in mountainous areas with severe terrain hazards [D]. *Southwest Jiaotong University*.

Chen X. 2016. Main types and analysis and evaluation of engineering geological disasters in loess tunnel [D]. *Chang'an University*.
Cui M. 2018. Suitability evaluation and engineering design of geological environment protection and land reclamation of oil and gas mines in Maowusu sandy land [J]. *Mineral Exploration*, (5):1037–1041.
He D. 2018. Analysis of engineering geological properties and bearing capacity characteristics of silt in Zhengdong New District of Henan Province[J]. *Mineral Exploration*, (10):2031–2037.
Jiang F. 2017. Evaluation of hydrodynamic conditions for soil cave development in Wuxuan, Guangxi [J]. *Hydrogeology and Engineering Geology*, (2): 90–95.
Jiang Y, Wang R, Zhu J. 2016. Experimental study on a 3D geomechanical model of tunnel anchor of Wujiagang Bridge [J]. *Chinese Journal of Rock Mechanics and Engineering* (S2):4103–4113.
Li S, Tang S, Liu H, et al. 2016. Landslide risk assessment of Heishui River area in Songxin, Sichuan Province based on ArcGIS [J]. *Mineral Exploration*, (5):864–869.
Li S, Wang K, Li L, et al. 2017. Formation mechanism and development trend of water inrush disaster in karst tunnel [J]. *Journal of Mechanics*, (1):22–30.
Li S, Xu Z, Huang X, et al. 2018. Structural classification, geological identification, disaster-prone model and typical case analysis caused by water and mud inrush in tunnels [J]. *Chinese Journal of Rock Mechanics and Engineering*, (5):1041–1069.
Meng Z, Peng J, Huang Q, et al. 2017. Study on deformation response of subway tunnel caused by fault dislocation [J]. *Engineering Geology*, (6):1624–1632.
Xia Y, Dong X, Xiong Z. 2018. Study on the safe thickness of water-resisting strata on the face of deep and long tunnels based on shear failure [J]. *Hydrogeology and Engineering Geology*, (2):57–66.
Xing J, Wang J. 2019.In-situ stress characteristics and rockburst prediction of Guigala expressway tunnel [J]. *Hydrogeology and Engineering Geology*, (2):170–178.
Yuan J, Deng G, Xu F, et al. 2016. Multivariate statistical analysis of hydrochemical characteristics and influencing factors of karst groundwater in northern Bijie City [J]. *Geology of China*, (4):1446–1456.
Zhang P, Sun Z, Wang Q. 2017. In-situ stress measurement and stability analysis of surrounding rock in the northern section of Muzhailing deep-buried tunnel [J]. *Journal of Geomechanics*, (6):893–903.
Zhou L, Huang X, Cao X. 2008. Engineering geological conditions of Jiulingshan deep-buried extra-long tunnel [J]. *Journal of Guilin University of Technology*, (1):54–58.
Zhou Z, Qin Z. 2018. Comparative study on permeability coefficient calculated by single hole steady flow pumping test [J]. *Mineral Exploration*, (4):706–713.

Advances in Civil Engineering and Environmental Engineering – bin Yahya & Bin Daud (Eds)
© 2023 The Author(s), ISBN: 978-1-032-39397-1

Study on comprehensive risk evaluation of subway station

Cui Ding*
School of Safety Engineering, China University of Labor Relations, Beijing, China

ABSTRACT: In order to evaluate the comprehensive risk of the subway station, the safety risk evaluation index system of the subway station was established by analyzing the safety risk factors of the subway station, and four first-class indexes and twenty-six second-class indexes were obtained. Based on the analytic hierarchy process (AHP) of triangular fuzzy number and fuzzy comprehensive evaluation method, the comprehensive risk assessment model and risk classification standard of subway stations were established. Taking Beijing Huayuanqiao subway station as an example, a comprehensive risk assessment was carried out, and the risk grades of the evaluation objects and the importance ranking of the evaluation indexes were obtained, and the weak links of the subway station were determined. The risk control measures of the subway station were put forward from four aspects including management factors, personnel factors, equipment and facilities factors, and environmental factors to guide the safe operation of subway station.

1 INTRODUCTION

The subway is an important part of the urban rail transit system. The subway plays a significant guiding role in effectively improving the efficiency of urban transportation. It has become an important policy measure to effectively solve the problems of current urban traffic and large automobile exhaust emissions. However, with the vigorous development of China's subway industry, safety accidents are constantly occurring. There have been many research achievements on risk assessment of subway at home and abroad, mainly focusing on fire risk assessment of subway operation (Mi et al. 2015; Wang et al. 2017), risk assessment model of crowded stampede accident (Huo et al. 2016; Hashash et al. 2001; Li et al. 2016), construction of risk assessment system of subway operation (Wang et al. 2018), the theoretical model of cause mechanism of subway accident (Xia 2019), but there is little research on overall risk assessment of subway station. In this paper, the safety risk of subway stations will be evaluated and analyzed, and the safety evaluation system model will be established from four aspects: human, machine, environment and management, and the analytic hierarchy process of triangular fuzzy number and fuzzy comprehensive evaluation method will be used to evaluate the overall safety risk of the subway station.

2 SAFETY RISK EVALUATION INDEX SYSTEM OF THE SUBWAY STATION

2.1 *Safety risk evaluation index system of the subway station*

The subway traffic accidents are not only related to the system units in the subway station but also closely related to external environmental factors. From the view of system safety theory, the risk factors of subway stations can be roughly divided into four categories: human factor, equipment factor, environmental factor, and management factor.

*Corresponding Author: cui_ding@163.com

DOI 10.1201/9781003349563-41

(1) Human factor

Human subjective behavior has a great influence on the development of events. The accidents frequently happen in subway stations as their high-density personnel and uneven personnel quality. On the one hand, the main reason is that the professional skills of station staff are not professional and daily risk investigation and risk prediction are not enough. On the other hand, passengers have low safety awareness and uncivilized behaviors, such as seizing seats, crowding, and blocking doors, which will reduce latent safety accidents.

(2) Equipment factor

As an important place in the subway service system, the subway stations include a variety of automatic and non-automatic facilities and equipment, such as power supply and water supply equipment and facilities, communication, passage, fire and other safety protection facilities, detection, supervision, and control system, etc. So, the normal operation of subway stations and the safety of passengers will be greatly threatened if any kind of equipment can't work correctly.

(3) Environmental factor

Environmental factors include not only natural environmental factors but also the social environment, such as the operating environment of equipment, the working environment of employees, etc. Since some equipment can only maintain normal working conditions under the given environment, such as temperature, humidity, etc.

(4) Management factor

Management factor is also an important factor affecting the safe operation of subway stations. At present, there are still many problems in the management of subway stations. For example, personnel safety education and skill training to deal with safety accidents are not enough. The pre-job training of new employees is ignored.

Based on the above analysis, the human factor, equipment factor, environmental factor, and management factor were taken as the first-class indexes of safety risk assessment of subway stations. According to the characteristics of each first-class indicator and on-site investigation, the second-class indexes were determined. The safety risk index system of the subway station is shown in Table 1.

2.2 *The calculation of the evaluation index weight*

The weight calculation methods mainly include the analytic hierarchy process, expert evaluation method, Delphi method, triangular fuzzy number analytic hierarchy process, and so on. The triangular fuzzy number analytic hierarchy process is widely used for its accuracy and practicability (Huang 2020). In this paper, the triangular fuzzy number analytic hierarchy process is used to calculate the weight of evaluation objectives.

According to the pairwise comparison of each element by each expert, the triangular fuzzy number judgment matrix of first-class indexes and second-class indexes can be obtained. In this matrix, each index has a corresponding fuzzy evaluation interval [l, u], which is used to reflect the confidence of experts in the evaluation results. The larger the (u-l), the smaller the confidence is. The smaller the (u-l), the greater the confidence is. The value of each element of the fuzzy judgment matrix corresponds to the importance of one factor over another factor, using the scale method of 1~9 and its reciprocal, as shown in Table 2.

The calculating process of index weight based on the triangular fuzzy number is as follows:

(1) According to the results of expert evaluation, the fuzzy judgment matrix is established: $A=(a_{ij})_{n*n}$, where the element $a_{ij} = [l_{ij}, m_{ij}, u_{ij}]$ is a closed interval with m_{ij} as the median.

(2) Consistency test of the median matrix. The median matrix $M=(m_{ij})_{n*n}$ is calculated, and its maximum eigenvalue M_{max} is calculated. according to the traditional AHP method, consistency is checked.

(3) Establish the fuzzy evaluation factor matrix $E=(e_{ij})_{n*n}$.

$$e_{ij} = \begin{cases} 1, i=j \\ 1-s_{ij}, i \neq j \end{cases} \quad (1)$$

Table 1. Safety risk assessment indexes of the subway station.

	First-class indexes	Second-class indexes	
Safety risk evaluation index system of subway station U	Human factor U1	Weather the security check is strict	U11
		Misoperation of driver	U12
		Misoperation of passenger	U13
		Fatigue operation of personnel in the monitoring room	U14
		Violate regulations of maintainer	U15
		Dereliction of duty by station staff	U16
		The dispatcher misdirected	U17
		Passengers throw cigarette butts or arson	U18
	Equipment factor U2	Door or screen door system failure	U21
		Aging or failure of elevator components	U22
		Whether the firefighting facilities are complete	U23
		Short circuit of transformer or distribution room line	U24
		Improper protection of substation box	U25
		Aging of electromechanical equipment such as substation box or distribution room	U26
		Train connection parts are aged or screws are loose	U27
	Environmental factor U3	Extreme weather	U31
		Foreign bodies in the tunnel	U32
		Chaotic subway lines	U33
		The exposed power cord on the wall	U34
		Flammable materials in the subway station	U35
		The unreasonable layout of facilities in a subway station	U36
	Management factor U4	Whether the safety management rules and regulations are sound	U41
		Whether safety education and training are conducted frequently	U42
		Whether emergency drills are conducted frequently	U43
		Whether the accident responsibility be implemented for individuals	U44
		Whether the management personnel have strong safety awareness	U45

Table 2. 1-9 Quantity scale.

Scale (ij)	Meaning	Scale (ij)	Meaning
1	It means that the latter factor is extremely important compared with the former factor.	6	It means that the former factor is slightly more important than the latter factor.
2	It means that the latter factor is strongly more important than the former factor.	7	It means that the former factor is obviously more important than the latter factor.
3	It means that the latter factor is obviously more important than the former.	8	It means that the former factor is strongly more important than the latter factor.
4	It means that the latter factor is slightly more important than the former.	9	It means that the former factor is extremely important compared with the latter factor.
5	Compared to the two factors, they are equally important.	reciprocal	The value of factor i compared with factor j is equal to the reciprocal of the value of factor j compared with factor i.

$$s_{ij} = \frac{u_{ij} - l_{ij}}{2m_{ij}} \tag{2}$$

Equation (2) is the standard deviation rate, which is directly proportional to the degree of ambiguity and inversely proportional to the reliability.

(4) Calculate the adjustment matrix Q.

$$Q = M \times E = \begin{bmatrix} 1 & \cdots & m_{1n} \\ \vdots & \ddots & \vdots \\ m_{n1} & \cdots & 1 \end{bmatrix} \times E \tag{3}$$

Where M is the median matrix of triangular fuzzy numbers.

(5) The matrix Q is transformed into a judgment matrix Q′ with diagonal 1, then the weight W of each index is calculated.

2.3 *The comprehensive evaluation method of subway station risk*

The fuzzy comprehensive evaluation method (Cui 2021) is used to evaluate the overall risk level of the subway station. The evaluation process is as follows:

(1) Determine the evaluation weight, and establish the weight matrix $W = (W_1, W_2, W_3 \ldots, W_n)$ of the criterion layer and index layer.

(2) The survival of the fittest of the evaluation objects is usually expressed by grades, and the set composed of grades is called comment set V = (V1, V2, V3, V4, …Vm), which indicates the safety degree of each factor in the index layer.

(3) Calculate the membership degree of each factor in the index layer. In this paper, the equal proportion method will be used to determine the membership degree of a single factor, and experts will directly evaluate and score the factors. The frequency of the number of people in each level will be taken as the membership degree of this factor, and the evaluation matrix R of the index layer will be obtained.

(4) Calculate the fuzzy evaluation result to obtain a comprehensive membership degree vector, then the evaluation matrix is respectively operated according to Equation (4) to obtain a comprehensive membership degree vector K_U.

$$K_U = W \cdot R \tag{4}$$

(5) Calculate the comprehensive score and grade. In view of the actual situation of the evaluation object, the safety level V of each factor in the index layer is assigned, and then multiplied by the corresponding parameters in the evaluation weight matrix K_U, and then summed, so that the comprehensive score f of the subway station can be obtained. According to Table 3, the safety risk level of the evaluation object is obtained.

Table 3. Evaluation of safety grade.

[8, 10]	[6, 8)	[4, 6)	[2, 4)	[0, 2)
Safe	Safer	General safety	Danger	Very dangerous

3 CASE APPLICATION

3.1 *Comprehensive risk assessment*

In this paper, the safety risk of Huayuanqiao Subway Station of Beijing Metro Line 6 is comprehensively evaluated. Huayuanqiao Subway Station is a three-story underground station, which adopts an island platform design, in which the first floor is the equipment floor, the second floor is the station hall floor, and the third floor is the platform floor, with 6 entrances and exits. Each evaluation factor of the index is divided into five grade standards V= (V1, V2, V3, V4, V5), in which V1, V2, V3, V4, and V5 respectively represent excellent, good, fair, poor, and extremely poor. The environmental safety of Huayuanqiao Station was scored, including excellent 10 points, good 8 points, general 6 points, poor 4 points and extremely poor 2 points. The risk factors of the Huayuanqiao subway station were evaluated by authoritative experts in the subway field, and the evaluation index weight and factor record degree were obtained, as shown in Table 4.

According to Formula (4), the index record degree K_U of the criterion layer is obtained:

$K_U = W \cdot R = (0.13, 0.45, 0.3, 0.1, 0)$

The overall score of the laboratory is $p = 7.1$ after calculating, and according to Table 3, the safety degree of Beijing Huayuanqiao Subway Station is "relatively safe".

3.2 *Suggestions*

According to the above evaluation results, the importance of human, machine, environment, and management is in the order of management factor > human factor > equipment and facilities factor > environmental factor, and suggestions and measures are put forward from these four aspects respectively.

(a) Management factors

The weight value of management factors is 0.4, which is the highest among humans, machines, the environment, and management. In the secondary indicators of management factors, "emergency drill", "responsibility implementation" and "safety awareness" all scored "poor". Therefore, the following suggestions are given:

(1) Formulate reasonable rules and regulations for subway stations, and make clear the division of labor and tasks of staff.

(2) Improve the emergency response capability, improve the emergency plan, and organize the station staff to conduct emergency drills regularly.

(3) Establish a responsibility implementation system, so that once an accident occurs, it can be investigated and implemented by individuals.

(4) Carry out safety education and training regularly, make safety publicity slogans and post them in eye-catching positions in subway stations.

(B) Human factors

The weight value of human factors is 0.28, which is the second highest factor among the four factors. Among the eight second-level indicators with personnel factors, "whether the security check is strict" accounts for the largest proportion, and the score is the highest in the "poor" rating. So, the security check should be strengthened.

(C) Equipment factors

The weight value of the facility factor is 0.2, and the weight value of elevator components aging, door or screen door system failure and whether firefighting facilities are complete are the highest among the secondary indicators. According to the above analysis, the following suggestions are given:

(1) For vehicle systems, screen door systems, and other equipment with a high wear rate in daily use, they shall be regularly overhauled or maintained.

(2) For electromechanical systems and power supply systems, insulation measures should be taken to prevent electric shock or fire accidents.

(d) Environmental factors

The weight value of environmental factors is 0.12, which is the lowest among the four factors. According to the scoring results, it is suggested that the cleaning staff in the station should clean up the garbage regularly and in time to ensure that there is no combustible garbage accumulation.

Table 4. Index weight and factor record degree of the subway station.

First-class indexes	weight	Secondary indexes	Fuzzy evaluation matrix	A	B	C	D	E	weight
Human factor U1	0.28	Weather the security check is strict	R1	0	0.50	0.22	0.28	0	0.23
		Misoperation of driver		0	0.28	0.58	0.14	0	0.11
		Misoperation of passenger		0.22	0.64	0.14	0	0	0.17
		Fatigue operation of personnel in the monitoring room		0	0.5	0.28	0.22	0	0.10
		Violate regulations of maintainer		0.14	0.44	0.42	0	0	0.08
		Dereliction of duty by station staff		0	0.36	0.42	0.22	0	0.11
		The dispatcher misdirected		0	0.28	0.58	0.14	0	0.13
		Passengers throw cigarette butts or arson		0	0.64	0.22	0.14	0	0.07
Equipment factor U2	0.2	Door or screen door system failure	R2	0.22	0.22	0.56	0	0	0.21
		Aging or failure of elevator components		0.28	0.22	0.50	0	0	0.25
		Whether the firefighting facilities are completed		0.22	0.64	0.14	0	0	0.17
		Short circuit of transformer or distribution room line		0.42	0.44	0.14	0	0	0.13
		Improper protection of substation box		0.14	0.50	0.36	0	0	0.10
		Aging of electro-mechanical equipment such as substation box or distribution room		0.64	0.14	0.22	0	0	0.08
		Train connection parts are aged or screws are loose		0.14	0.64	0.22	0	0	0.06
Environmental factor U3	0.12	Extreme weather	R3	0	0.50	0.50	0	0	0.10
		Foreign bodies in the tunnel		0.22	0.42	0.36	0	0	0.26
		Chaotic subway lines		0.64	0.22	0.14	0	0	0.32
		The exposed power cord on the wall		0	0.50	0.36	0.14	0	0.06
		Flammable materials in the subway station		0	0.42	0.36	0.22	0	0.10
		The unreasonable layout of facilities in a subway station		0.22	0.14	0.28	0.36	0	0.16
Management factor U4	0.4	Whether the safety management rules and regulations are sound	R4	0.28	0.22	0.50	0	0	0.10
		Whether safety education and training are conducted frequently		0.36	0.22	0.42	0	0	0.15
		Whether emergency drills are conducted frequently		0.14	0.44	0.28	0.14	0	0.18
		Whether the accident responsibility be implemented for individuals		0	0.28	0.50	0.22	0	0.35
		Whether the management personnel have strong safety awareness		0	0.22	0.56	0.22	0	0.22

4 CONCLUSION

By analyzing the risk characteristics of the subway station, the risk evaluation index system of subway station was established from four aspects: human, machine, environment, management, and the comprehensive risk evaluation model of subway station were established based on a triangular fuzzy analytic hierarchy process and comprehensive fuzzy evaluation method, which had been applied in Huayuanqiao subway station.

(1) The safety degree of Beijing Huayuanqiao subway station is "relatively safe".
(2) The order of importance of humans, machines, environment, and management is management factor > human factor > equipment factor > environmental factor, so the risk control of management should be strengthened in subway stations.
(3) The risk control measures of subway stations are put forward from management factors, human factors, equipment factors, and environmental factors to guide their safe operation.

ACKNOWLEDGMENT

This work was supported by the Fundamental Research Funds for the Central Universities of China (No.21ZYJS012).

REFERENCES

Cui Ding. Risk assessment of physical and chemical analysis laboratory based on the laboratory risk assessment model [C]. *2nd International Conference on Urban Engineering and Management Science*, 2021, 293–296.

Hashash Y Ma, Hook Jj, Schmidt B, et al. Seismic design and analysis of underground structures [J]. *Tunneling and Underground Space Technology*, 2001, 16(4): 247–293.

Huang Xiangwan, Zhu Wanhong, Xu Wenting. Risk evaluation model of underground protective engineering construction based on AHP of triangular fuzzy number [J]. *Protective Engineering*, 2020, 42(04): 41–44.

Huo Yumang, Song Shouxin, GU Yibo. Risk assessment of crowd crushing and trampling accidents in subway stations based on combined weights [J]. *Safety and Environmental Engineering*, 2016, 23(5): 139–143.

Li Wen-Hang, Gong Jian-Hua, YU Ping, et al. Modeling, simulation and analysis of group trampling risks during escalator transfers [J]. *Physica A: Statistical Mechanics and its Applications*, 2016, 444: 970–984.

Mi Hong-Fu, Xiao Guo-Qing, Wang Lin-Yuan, et al. On the fire risk evaluation for the subway station based on the probability fuzzy theory [J]. *Journal of Safety and Environment*, 2015, 15(5):16–20.

Wang Jianbo, Peng Longbiao, Li Na et al. Fire Risk Evaluation of subway station based on PCA-RBF neural network [J]. *Industrial Safety and Environmental Protection*, 2017, 43(9): 67–70.

Wang Yi-Min, Luo Yue, Yu Heng, et al. Evaluation method of security risk on crowded metro station [J]. *Journal of Traffic and Transportation Engineering*, 2018, (20)5: 198–207.

Xia Runhe. Analysis of metro tunnel collapse accident based on accident mechanism and responsibility and its prevention strategy [J]. *Tunnel Construction*, 2019, (39)10: 1602–1609.

Advances in Civil Engineering and Environmental Engineering – bin Yahya & Bin Daud (Eds)
© 2023 The Author(s), ISBN: 978-1-032-39397-1

Research on deformation control technology of tunnel argillaceous surrounding rock

Jian Xiang
CCCC Third Harbor Consultants Co. Ltd, Shanghai, China

Xinyu Yan* & Zhuqing Tan
CCCC Second Harbor Engineering Co. Ltd, Wuhan, China

ABSTRACT: Based on the large deformation problem easily caused by the excavation of tunnels in cement-rich stratum, the engineering geological characteristics of mudstone and the combined support system are summarized and analyzed, and the reinforcement effect is analyzed according to the field monitoring and measuring data using the numerical simulation of the auxiliary support measures such as locking anchor pipe and radial grouting with the 3DEC. The results show that the settlement of the vault and the displacement around it tend to be stable, less than 1mm/d, about 20 days after excavation. Taking active reinforcement measures can control the deformation of surrounding rock, and the research results can provide a reference for similar mudstone tunnel design and construction.

1 INTRODUCTION

In the process of tunnel construction, muddy surrounding rock is a kind of common rock mass. Because of its unique physical chemistry property and environmental sensitivity, it is usually classified as a soft rock mass. At present, many scholars and engineers have studied the engineering characteristics and deformation control of tunnel argillaceous surrounding rock and have obtained rich research results. In view of the lining cracking problem of the tunnel crossing the tertiary mudstone stratum, Liang Qingguo et al. carried out the corresponding indoor test and field tests and analyzed the cause and mechanism of the lining deformation and cracking. Yuan Tie et al. analyzed the correlation between the preferred joint dip angle and the range of plastic zone of the multi-jointed mudstone rock mass of the Menghua railway tunnel and studied the influence of the joints on the stability of the surrounding rock and the lining support structure. Huang Mingqi optimizes the construction method and puts forward some control measures to solve the problem of large deformation in the initial support of the tunnel passing through the carbonaceous mudstone stratum. Liang Huiru studied the softening effect of carbonaceous mudstone by water expansion and gave the judging condition of the stability of tunnel anchor pipe-steel arch frame. Wuyazun analyzed the stability coefficient of three-step temporary invert excavation utilizing the strength reduction method and FLAC3D software. He Changguo studied the deformation law of red bed sand-mudstone soft stratum encountered by most tunnels on the Chengdu-Guizhou Railway and suggested that the deformation amount should be reasonably reserved, and the initial support stiffness should be optimized.

Many research results show that the deformation mechanism of surrounding rock, the time-space effect of tunnel excavation, and the stress-strain relationship of the support structure are very complex. Taking the sandy mudstone stratum of highway tunnels as the research object, this paper

*Corresponding Author: 347832087@qq.com

DOI 10.1201/9781003349563-42

summarizes and analyzes the engineering geological characteristics of mudstone. Based on the three-step temporary invert method, the auxiliary supporting measures such as anchor pipe with lock foot and radial grouting are simulated.

2 CONTROL MEASURE OF MUDSTONE FORMATION DEFORMATION

2.1 *Project profile*

In cement-rich surrounding rock stratum, the rock mass shows some creep characteristics, such as large deformation, fast deformation rate, long duration, slow convergence, etc. Taking the two-way two-lane expressway tunnel excavated by the three-step temporary invert method as an example, in the mudstone stratum with rich groundwater, when the surrounding rock creep develops continuously, the supporting structure usually bears a larger force, in order to control the deformation, taking the common strong weathering silty mudstone as the research object, the related physical parameters of the tunnel mudstone are as follows: the free expansion rate is 27%~95%, the softening coefficient is 0.28, the natural compressive strength is 0.67~4.8 Mpa and the natural bulk density is 19.8 ~ 22 kN/m^3. The level of surrounding rock is Vb.

Figure 1. Three steps temporary invert method of tunnel.

2.2 *Numerical model*

In this paper, 3DEC software is used to carry out simulation analysis without considering the plastic failure of natural rock and soil in the course of the natural deposition. After the initial in-situ stress field is generated, the multi-mode simulation calculation is started. Assuming that the excavation section size of two-way two-lane expressway tunnel is 12.25 m × 9.6 m (width × height), a three-dimensional calculation model is established according to the design section of the tunnel, and the corresponding parameters of rock and soil are assigned to it, the boundary conditions of the model impose normal constraints around and at the bottom. The Rock and soil mass adopts the Mohr-coulomb elastoplastic constitutive model. The overall calculation model is shown in Figure 2 below. Two working conditions are set for analysis: 1 locking anchor pipe working condition (Φ42 and Φ108, length 4 m), and 2 radial grouting working conditions.

The physical and mechanical properties of the surrounding rock and the supporting structure of the tunnel are determined by an indoor geotechnical test, as detailed in Table 1.

2.3 *Analysis of calculation results*

(1) Working condition of anchor pipe with lock foot (Φ42 and Φ108, length 4 m)

The relevant parameters of anchor pipe support with lock foot are shown in Table 2.

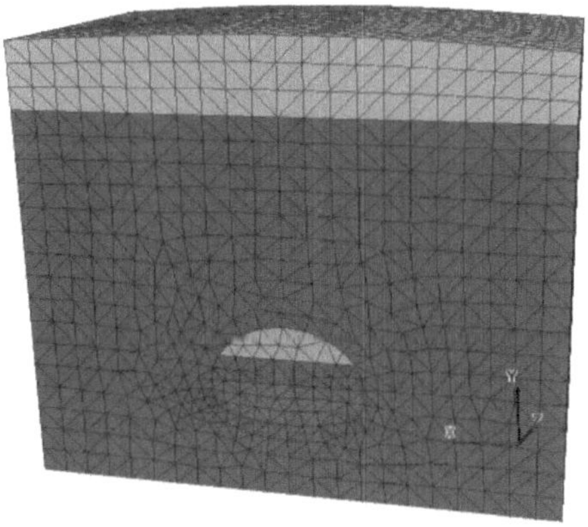

Figure 2. Mesh generation of the computational model.

Table 1. Physical and mechanical parameters of materials.

Materials	Heavy kN/m^3	Cohesion kPa	Angle of internal friction	Deformation modulus MPa	Poisson ratio
Fully weathered mudstone	17.9	5	11	1.5	0.30
Strongly weathered mudstone	19.8	27	16	4.8	0.30
Grouting layer	24	100	30	13000	0.20
Initial support	23	–	–	23000	0.20

Table 2. Mechanical parameter values of anchor pipe support.

Type of anchor pipe	Elastic modulus E(Pa)	Tensile yield strength/N	Area/m^2	Grouting stiffness/Pa.	Grouting intensity N/m
φ42	2e11	2.46e5	1.81e-3	1.35e11	2.76e7
Φ42	2E11	2.46e5	1.81e-3	1.35 E11	2.76 e7
φ108	2e11	2.46e5	9.16e-3	1.35e11	2.76e7
Φ108	2E11	2.46e5	9.16 E-3	1.35 E11	2.76 e7

In the model, the cable bolt structure unit built in 3DEC software is used to simulate the lock-foot anchor pipe. The cable unit assumes that the anchor bolt is fully combined with the rock mass. When the surrounding rock deforms, the deformation of the surrounding rock is resisted by the interaction of the grouting material between the bolt and the rock mass, which mainly depends on the axial action of the bolt and the shear action of the mortar surface, by using the shear resistance distributed on the interface between the anchor and grouting as a damper, the simulation of the anchor is realized. Liner element built in 3DEC software is used to simulate the initial support of tunnel excavation, which can resist shear and moment load and can also bear the tension and compression stress in the main direction.

The effect of anchor pipe with lock foot on the vertical displacement of the tunnel is analyzed. The cloud figures of vertical displacement during tunnel excavation under the condition of anchor pipe reinforcement are shown in Figures 3 and 4.

According to the vertical displacement cloud diagram after tunnel excavation and support, it can be seen that under the condition of Φ42 lock-foot anchor pipe reinforcement, the settlement of the tunnel vault is about 11.31 cm, and under the condition of Φ108 lock-foot anchor pipe

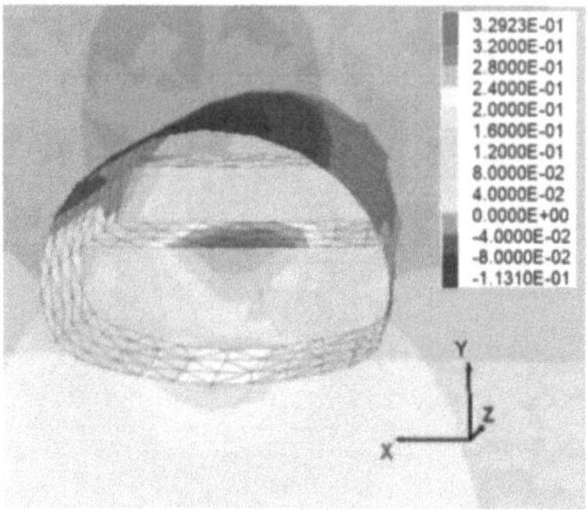

Figure 3. Vertical displacement of the tunnel with Φ42 anchor pipe reinforcement.

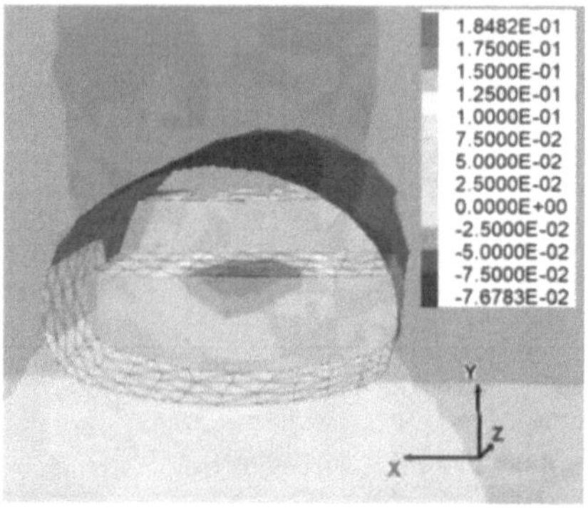

Figure 4. Vertical displacement of the tunnel with Φ108 anchor pipe.

reinforcement, the maximum settlement of the tunnel arch is about 7.67 cm. It can be seen that under the condition of poor geology, the large diameter anchor pipe is used to reinforce the tunnel, and the effect of controlling the tunnel excavation deformation is good.

In general, Φ42 steel pipe is used to replace anchor rods in soft rock stratum, and the length of the anchor pipe should be lengthened appropriately when the tunnel passes through a cement-rich stratum. The use of large-diameter steel pipe and good pin grouting can slow down the subsidence of the surrounding rock.

(2) Radial grouting

Based on the Φ108 anchor rod, radial grouting is carried out at the same time, and the reinforcement areas of both sides of the tunnel are taken at 5 meters each. The effect of radial grouting on the horizontal displacement of the tunnel is analyzed. Through simulation calculation and analysis, the horizontal displacement cloud map of tunnel excavation after grouting reinforcement is shown in Figures 5 and 6 below.

As can be seen from the cloud diagram of the horizontal displacement, after the tunnel excavation and support are completed, the horizontal displacement of the tunnel without the radial grouting reinforcement is about 7.51 cm, and that with the radial grouting reinforcement is about 7.51 cm. The maximum horizontal displacement of the tunnel is about 3.88 cm, which shows that the radial grouting is more effective in controlling the deformation of the tunnel. Therefore, under the premise of anchor pipe reinforcement in cement-rich stratum, radial grouting can be added for reinforcement, which can better control the deformation of surrounding rock.

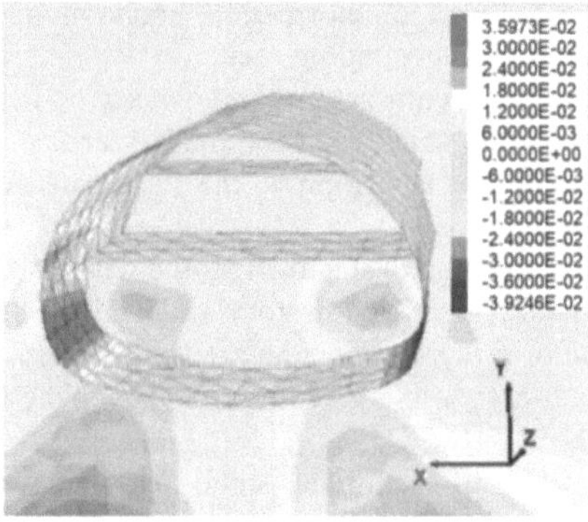

Figure 5. Horizontal displacement of the tunnel without radial grouting reinforcement.

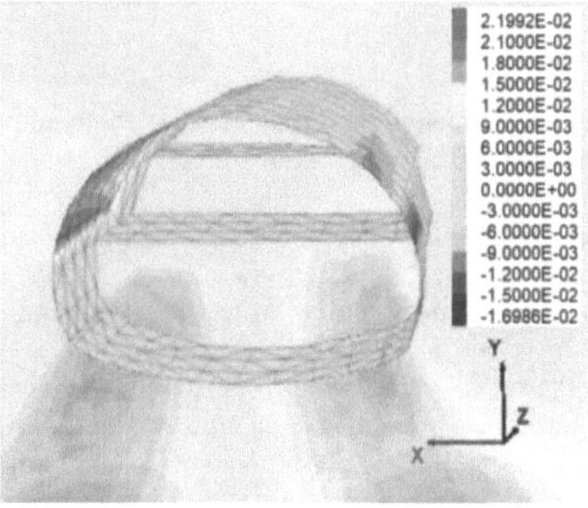

Figure 6. Horizontal displacement of the tunnel strengthened by radial grouting.

3 MONITORING DATA ANALYSIS

In order to verify the reliability of numerical analysis and analyze the combined reinforcement effect of large diameter anchor pipe with locking foot and radial grouting, the zK32 + 503 section was selected to carry out 26-day monitoring and measurement of the crown settlement and peripheral displacement, the measurement frequency once a day. The deformation speed of vault settlement and peripheral displacement is faster after tunnel excavation, and the variation trend is generally the same. About 12 days after tunnel excavation, 74.7% of the total convergent deformation can be achieved, and about 17 days after excavation, 87.5% of the total convergent deformation can be achieved. About 20 days after excavation, the settlement of the vault and the displacement around it tend to be stable, less than 1mm/d. It can be concluded that the adoption of large diameter anchor pipe and radial grouting reinforcement has an obvious effect on the stability of tunnel surrounding rock and the safety of construction.

4 CONCLUSION

(1) The water stability of mudstone formation is usually poor, so the foundation of choosing active combined support measures is to master its engineering geological characteristics.
(2) The excavation and support system of argillaceous surrounding rock should be selected according to the comprehensive comparison of the stratum engineering geological conditions,

hydrogeological conditions, predicted deformation, economic factors, etc., adopting one or more measures to strengthen the surrounding rock.

(3) Locking anchor pipe and radial grouting are important measures in the initial support system. In the rich cement rock stratum, they should be combined as far as possible and take the initiative to strengthen the support concept and improve the supported grade, and it can control the deformation of the surrounding rock.

(4) The field monitoring and measuring data show that the deformation of the surrounding rock may last for a long time in the cement-rich stratum, and it still needs at least 17 days to become stable under the condition of using large diameter anchor pipe and radial grouting, in actual construction, continuous observation should be carried out on the cement-rich strata to prevent potential safety hazards.

(5) Considering the simplification of the model, some load parameters in the calculation may be different from the actual ones. The results are only used as a reference, and the actual deformation is mainly monitored in the field.

REFERENCES

HE Changguo (2018). Analysis of reasonable reserved deformation amount of large-span tunnel in soft surrounding rock and preliminary optimization of initial support stiffness. J. *Tunnel Construction*, 38(z2): 227–231.

Huang Mingqi (2020). Deformation controlling techniques of carbon mudstone tunnel with high terrestrial stress. J. *Railway Construction Technology*, 44(4): 35–44.

LIANG Huiru (2020). Research on mechanical stability of feet-lock pipe and steel arch in soft rock tunnel. J. *Journal of Beijing Jiaotong University*, 44(4): 35–44.

WU Yazun (2018). Analysis of the tunnel surrounding rock stability of weak mudstone based on strength reduction method. J. *The Chinese Journal of Geological Hazard and Control*, 29(5): 66–70.

YUAN Tie (2020). Effect of joints on stability of mudstone tunnel. *J. Highway*, (2): 288–293.

Zhang Qinpeng (2016). Test study on the mechanical characteristics of mudstone tunnel on Baoji-Lanzhou PDL.J. *Journal of Railway Science and Engineering*, 13(11): 2134–2143.

Advances in Civil Engineering and Environmental Engineering – bin Yahya & Bin Daud (Eds)

Research on automatic segment arrangement technology of shield machine

Fulin Geng, Ji Zhao & Qingang Wu
Beijing Municipal No.4 Construction Engineering Co., LTD, Beijing, China

Weiwei Sun*
Beijing Information Science &Technology University, Beijing, China

ABSTRACT: 3D rapid modeling technology of segment and tunnel is the foundation of tunnel construction visualization and Information management. With the in-depth application of BIM (Building Information Modeling) technology in tunnel construction, segment and tunnel 3D rapid Modeling technology have important application value for tunnel construction Information management. With the gradual in-depth application of BIM (Building Information Modeling) technology in the field of domestic architecture, this technology has important application value. In this paper, parameterized modeling of shield tunnel was studied by Dynamo platform, and 3d tunnel modeling program is constructed by visual programming language, which can be used for rapid modeling of the tunnel. The tunnel model can be generated by directly inputting parameters, and parameterized modeling of the tunnel can be realized. The operability and practicability of the system are verified by the engineering practice of a sample section of the Beijing subway.

1 GENERAL INSTRUCTIONS

Shield construction is widely used in China's transportation, water conservancy, electric power, underground space development, and other fields of tunnel engineering because of its high degree of automation, low labor intensity, fast construction schedule, and small impact on the surrounding environment. It plays a pivotal role in China's infrastructure construction. The shield segment is the most important assembly member in shield construction. And as the primary lining of the shield tunnel, the shield segment is also the permanent lining of the shield tunnel. Prefabricated segment splicing technology is commonly used in shield tunnel construction. Due to the different shapes and structures of segments and the complex and changeable assembling forms, designers and constructors must have strong spatial imagination if they only use plane drawings to express segments.

BIM is a data-based tool that uses digital technology to express geometric, physical, and functional information about construction projects and is applied in engineering design and construction management. BIM technology is not only modeling, but also information sharing. In recent years, BIM technology has gradually become a research hotspot in tunnel engineering construction and management. Internal and external scholars conducted many studies on the modeling and application of BIM technology in urban rail transit construction. Compared with domestic, foreign scholars have conducted more detailed research on BIM in tunnels. Christian Koch et al. proposed a tunnel information modeling framework that integrates four interrelated subdomain models and linked project performance data. Frodl Stephan et al. discussed how to establish a 3D BIM design model which can be derived from construction drawings and proposed establishing 3D BIM geometric

*Corresponding Author: 332915161@qq.com

DOI 10.1201/9781003349563-43

model. Hyunjoo Kim et al. proposed an intelligent three-dimensional model for tunnel construction, which can electronically represent a large amount of information in the project. Zhang Liang et al. studied how to apply BIM technology to parametric modeling of shield segments and to deal with the problems encountered in the construction process. Tang Yanmei et al. determined a segment sequencing plan by designing a standard segment and turning loop segment, and established a three-dimensional model of a shield tunnel through BIM technology. Xu Litong et al. studied the use of Civil3D+Dyanmo+Revit interaction to carry out refined modeling of tunnels and realize the closure of engineering data. However, once the design scheme is changed, it needs to be modified and integrated with different software, lacking a linkage mechanism. The above studies mainly focus on the modeling process and segment layout of shield tunnels based on BIM technology, and some of them carry out parametric modeling research, but the secondary development technology of the parametric automatic generation model is still immature.

According to the structural characteristics of tunnel segments, this paper puts forward the modeling idea of tunnel segments, and develops the automatic assembly technology for the whole process of interval segments, so as to realize the rapid modeling of segment and tunnel, and provide strong technical support for visualization and information management of tunnel construction.

2 OVERALL PROCESS OF PARAMETERIZATION MODELING

In the work of automatic segment arrangement, the segment model should be created first, and then the center line of the tunnel should be processed. After reading the center line information through Dynamo, an automatic segment arrangement will be carried out. The basic flow is shown in Figure 1.

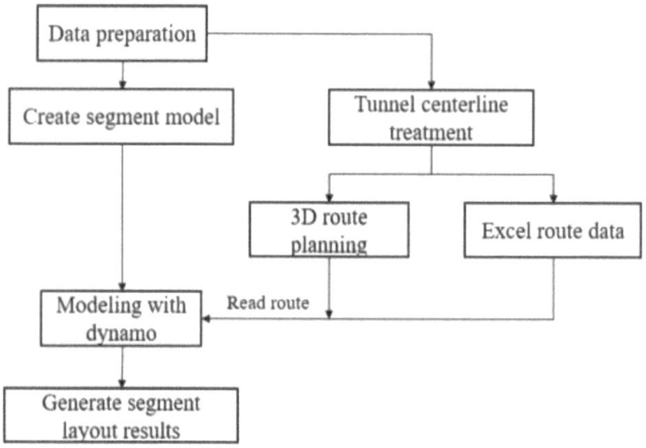

Figure 1. The basic flow of automatic segment arrangement.

2.1 *Shield segment modeling*

In this paper, a standard double-sided wedge-shaped universal ring is adopted. Each ring segment is assembled from six precast reinforced concrete segments, including a capping block, two connecting blocks, and three standard blocks. The outer diameter of the lining ring is 3.2m, the inner diameter is 2.9m, the thickness is 0.3m, and the width is 1.2m, as shown in Figure 2.

The 3d model of the segment is achieved by establishing an adaptive family with adaptive points. Since the segment is a flat segment, eight edges and corners of the segment need to be controlled by eight adaptive points, and four arcs of the segment need to be controlled by four adaptive points. The adaptive family established here has 12 adaptive points. At the same time, the accuracy of the model can be improved by creating 16 or more adaptive points according to specific requirements. The segment adaptive family is shown in Figure 3.

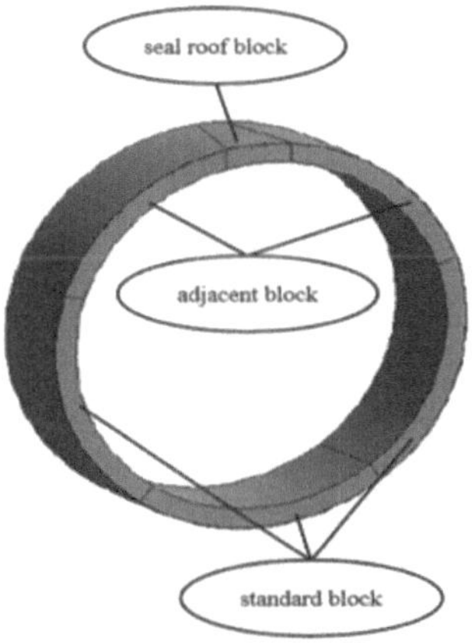

Figure 2. Pipe ring composition.

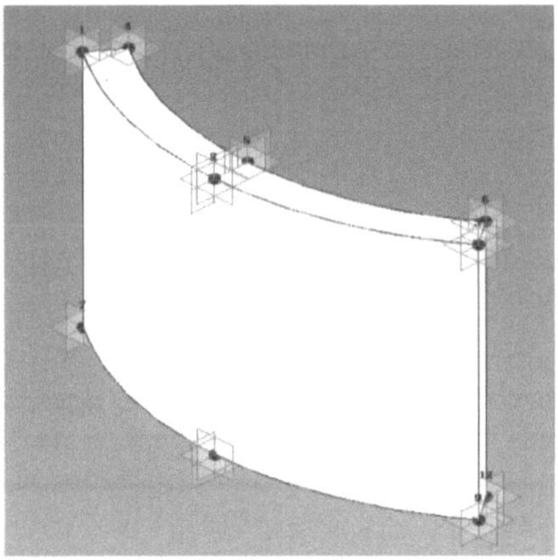

Figure 3. Segment adaptive family.

Section parameter settings are shown in Figure 4, with an outer diameter of 3.2m, a thickness of 0.3m, and a width of 1.2m. Standard block and adjacent block models can be generated respectively through parameter control, as shown in Figure 5. The input parameter of the example model is standard block Angle 67.05°. Adjacent block parameters (front): the outer arc length is 3.6699, and the inner arc length is 3.3165. Segment parameters (back): the outer arc length is 3.8699, and the inner arc length is 3.5165. The radian value is converted into the Angle value by formula.

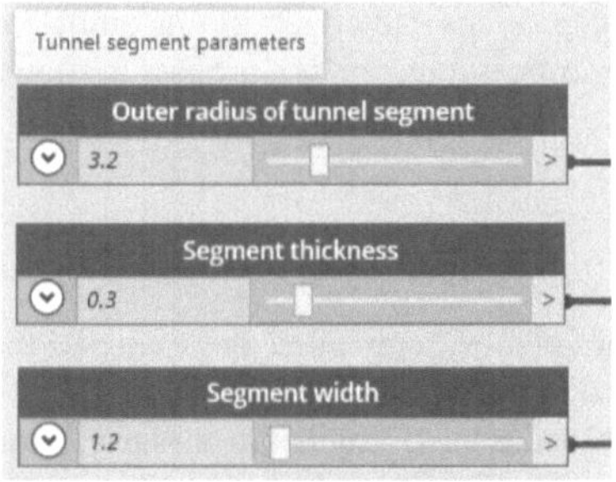

Figure 4. The segment parameters.

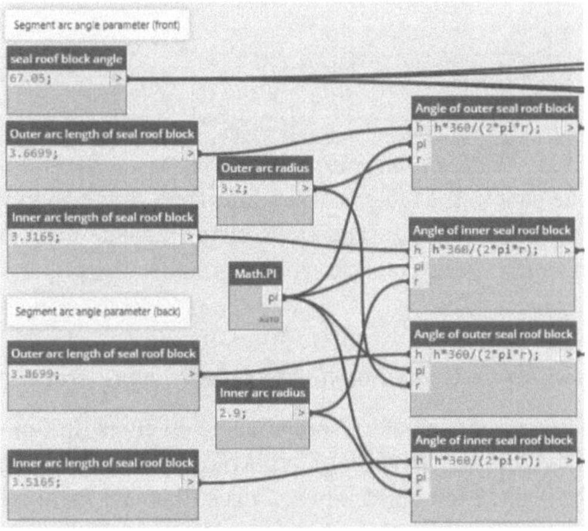

Figure 5. Standard block and adjacent block parameters.

The seal roof block model is automatically calculated and generated according to the Angle between the input standard block and adjacent block, as shown in Figure 6.

2.2 *Processing of tunnel center line*

The center line of the tunnel is a three-dimensional curve, which is the reference line of the circular point alignment of the pipe ring structure. The tunnel segments are spliced along the center line of the tunnel. In actual construction, although there is a book of shield machine driving circuits, it is in the process of shield tunnel on both sides of the water and soil pressure is inconsistent,

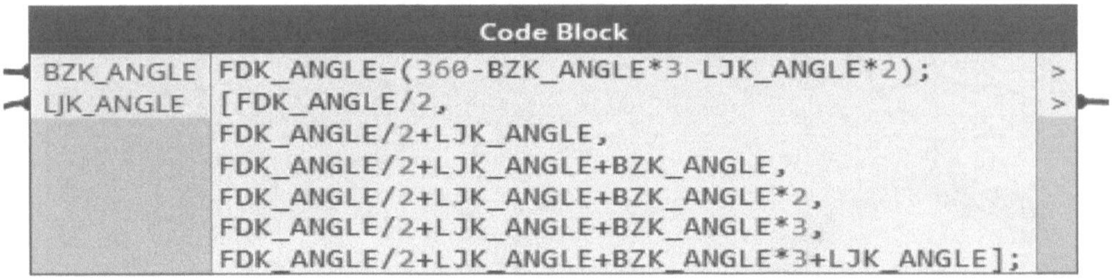

Figure 6. The seal roof block.

synchronous grouting pressure, cutter wear, and so on factors, slightly offset with the designed route in advance, and make the tunnel centerline eventually became an irregular space curve.

2.2.1 *Existing model processing*

Read point coordinates in Dynamo and use PolyCurve.ByPoints node to generate spline curves, as shown in Figure 7.

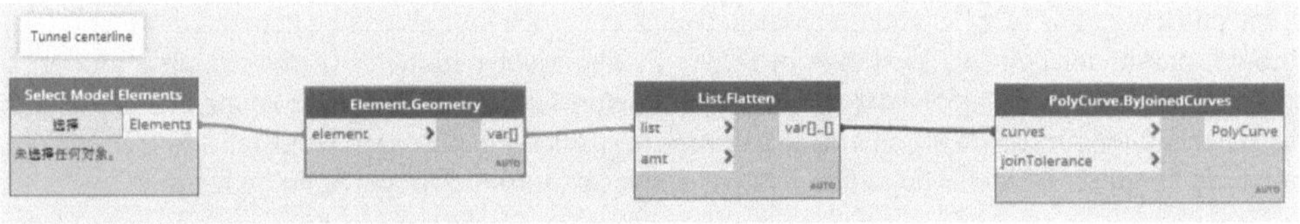

Figure 7. Processing flow.

2.2.2 *Table data reading mode*

The data in the table shall include three-dimensional coordinate values and offsets of pile points, as shown in Figure 8. The Geometry. Translate node is used for processing, and the process is shown in Figure 9. In practice, you can choose the processing mode according to your needs.

2.3 *Tunnel visual modeling programming process*

After importing the center line, the center line of the tunnel can be segmented according to the fixed arc Length through code blocks such as Curve. Points At Chord Length From Point to obtain the segmented Point. Then, Curve.Plane At Parameter code block is used to generate a Plane on the center line of the tunnel obtained in front. The position of the Plane is determined by the key nodes segmented in front, and the normal line of the Plane is aligned with the tangent line of the center line of the tunnel. Circle. By Plane Radius code block is then used to create concentric circles on the generated Plane with the segmented points as the center of the Circle. The specific process is shown in Figure 10.

2.4 *Ring the number statistics*

According to the layout result, the straight line segment, right turn segment, and left turn segment can be distinguished by different colors, and the number of loops used by different segments can be counted, as shown in Figure 11.

Station	Northing (Y/m)	Easting (X/m)	altitude(m)	describe	deviation(m)	remarks
12+139.050	498287.088	301000.312	19.781	P6	0.000	ZX
12+140.000	498287.238	300999.374	19.783	P6	0.000	
12+141.000	498287.395	300998.387	19.785	P6	0.000	
12+142.000	498287.552	300997.399	19.788	P6	0.000	
12+143.000	498287.709	300996.412	19.790	P6	0.000	
12+144.000	498287.867	300995.424	19.792	P6	0.000	
12+145.000	498288.024	300994.436	19.794	P6	0.000	
12+146.000	498288.181	300993.449	19.796	P6	0.000	
12+147.000	498288.338	300992.461	19.798	P6	0.000	
12+148.000	498288.495	300991.474	19.801	P6	0.000	
12+149.000	498288.653	300990.486	19.803	P6	0.000	
12+150.000	498288.810	300989.499	19.805	P6	0.000	
12+151.000	498288.967	300988.511	19.807	P6	0.000	
12+152.000	498289.124	300987.523	19.809	P6	0.000	
12+153.000	498289.282	300986.536	19.811	P6	0.000	
12+154.000	498289.439	300985.548	19.814	P6	0.000	
12+155.000	498289.596	300984.561	19.816	P6	0.000	
12+156.000	498289.753	300983.573	19.818	P6	0.000	

Figure 8. Tabular data.

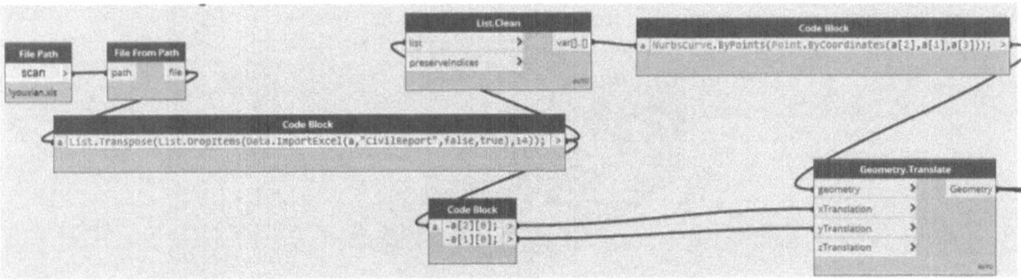

Figure 9. Processing flow.

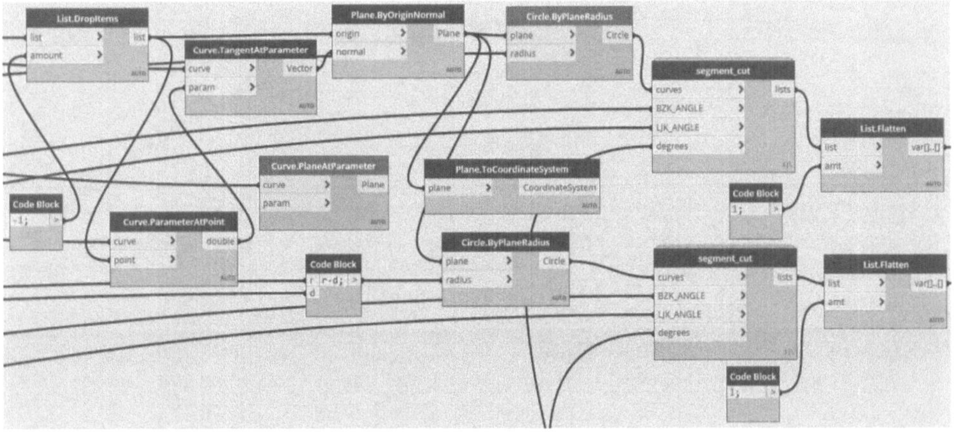

Figure 10. Processing flow.

3 APPLICATION INSTANCE

Combined with a subway tunnel in Beijing, the modeling program is applied to engineering modeling. The basic parameters of segment construction are given in the previous paper.

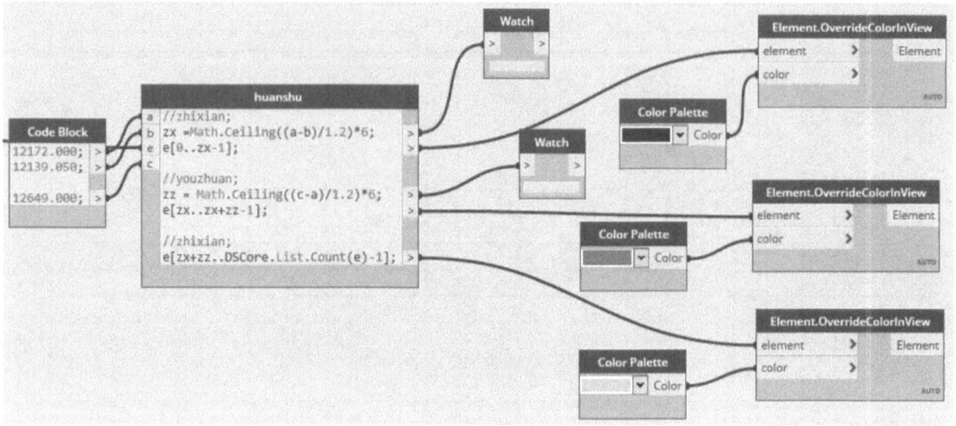

Figure 11. Ring the number statistics.

The total length of the tunnel is 620m, which is assembled with staggered joints. After defining the above parameters of the segment, the program compiled previously can be used to input relevant parameters to realize the model creation from segment to ring and from ring to the tunnel. The final tunnel model rendering is shown in Figure 12.

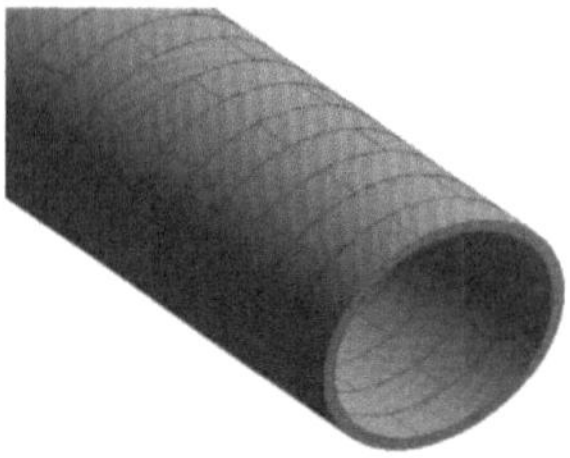

a) Local model

b) Overall tunnel model

Figure 12. Tunnel model renderings generated.

4 CONCLUSIONS

In this paper, parameterized modeling of shield tunnel is studied by Dynamo platform, and 3d tunnel modeling program is constructed by visual programming language, which can be used for rapid modeling of the tunnel. The tunnel model can be generated by directly inputting parameters, and parameterized modeling of the tunnel can be realized. Through this research, the efficiency and convenience of tunnel modeling can be improved, and fast and efficient tunnel modeling can be realized, which provides a reference for the application of BIM technology in underground engineering.

REFERENCES

Frodl S. Dipl.-Ing., Mayer P.-M. Dr.Ing. BIM modeling in mined tunneling discussions and recommendation for the structural geometry to be considered in 3D models. *Geomechanics and Tunnelling*, 2018, 11(4):357–365.

Kim H., Shen Z., Moon H., et al. Developing a 3D intelligent object model for the application of construction planning/simulation in a highway project. *KSCE Journal of Civil Engineering*, 2016, 20(2):538–548.

Koch C., Vonthron A., König M. A tunnel information modeling framework to support management, simulations, and visualizations in mechanized tunneling projects. *Automation in Construction*, 2017, 83: 78–90.

Tang Y.M., Zhang C.C., Wang X.L. Application of BIM technology in tunnel segments. *Soil Eng. and Foundation*, 2019, 33(3):290–294.

Xu L.T., Yuan X.C., Cui Y.J., et al. Application of BIM to mountain tunnel based on Autodesk platform. *Tunnel Construction*, 2020, 40(2):267–275.

Zhang L, Liu J. Application research of BIM Technology in construction of shield segments. *Construction Technology*, 2019, 48(12):16–18+27.

Advances in Civil Engineering and Environmental Engineering – bin Yahya & Bin Daud (Eds)
© 2023 The Author(s), ISBN: 978-1-032-39397-1

An experimental study on stress wave detection of damaged wood based on the stickup sensors

Chengwei Peng*
Guangdong Provincial Academy of Building Research Group Co., Ltd., Guangzhou, Guangdong Province, China
State Key Laboratory of Subtropical Building Science, South China University of Technology, Guangzhou, Guangdong Province, China

Wenlong Lv
Guangdong Provincial Academy of Building Research Group Co., Ltd., Guangzhou, Guangdong Province, China

ABSTRACT: Damage is a common occurrence in ancient wooden structures, and its impact on the mechanical properties of the structures cannot be ignored. A stress wave is a typical method to detect the internal damage to wood, and the needle-type sensor is usually used, which will cause slight damage to the outer surface of the wood. The protection of ancient wooden structures needs strict requirements, and the use of stickup sensors can avoid micro damage on the outer surface of wooden components, which is more appropriate from this perspective. In this paper, 17 different wood damage detection states are designed with the shape, area, location, and type of the damaged area and the number of measuring points as the running parameters. Based on the stickup sensors, a stress wave detector (FAKOPP 3D Acoustic Tomograph) is used for testing, and Arbor Sonic 3D produced by FAKOPP is used for graphic processing. The testing results show that the data gotten by the stickup sensors can obtain the stress wave detection images showing the hollows and the decayed areas. The stress wave detection image cannot effectively distinguish the shape of the damaged area, but can roughly display its position. The uneven side surface and the wood shrinkage crack impact the stress wave detection images to a certain extent. The test damage degree is higher than the actual damage degree on average, and the test damage degree deviation of the specimen with the damage type of hollow is larger. The increase in the number of measuring points can make the shape of the damaged area in the stress wave detection image and the test damage degree closer to the actual situation.

1 INTRODUCTION

Chinese historical architecture is an important carrier of Chinese civilization and a valuable heritage in the world's architectural art. In Guangdong Province, for example, 4,084 historical buildings have been registered as of May 2022. Among them, the structural types are mainly brick wood structures and wood structures, accounting for about 62% of the total. For one thing, the heat or humidity of the external environment will cause the breeding of borers. In addition, the ancient wooden structure has been used for a long time, which is easy to produce different types and degrees of damage. For another, some historical buildings have been activated and utilized to form new functions, which may change the original stress state of the structure to a certain extent and aggravate the damage. Although the damages in ancient wooden structures often develop slowly, it may cause quantitative

*Corresponding Author: 874541364@qq.com

 DOI 10.1201/9781003349563-44

changes (Wang 2006). Therefore, for the safety of ancient wooden structures, internal damage detection research is an important part.

At present, scholars have studied various methods such as the micro drilling resistance method (Winistorfer 1995, C 2001), the ultrasonic method (Kandemir-Yucel 2007), and the stress wave method (An 2013, Wang 2015, Yang 2004) for wood internal damage detection, and there are also reports on the combined application of multiple detection methods (Zhu 2011, Zhu 2015). Among them, the stress wave method has the characteristics of visualization and quantification. However, it usually uses needle-type sensors, which will cause slight damage to the outer surface of the wood during detection (Li 2015). For one thing, the needle-type sensor is difficult to penetrate and stabilize some dense ancient architectural wood (such as Merbau); for another, the protection of ancient wooden structures needs strict requirements, and the use of stickup sensors can avoid micro damage on the outer surface of wooden components, which is more appropriate from this perspective. There are few reports on stress wave detection using stickup sensors in the existing literature. Therefore, 17 stress wave detection tests of damaged wood based on stickup sensors are carried out and discussed the impacts of the area, shape, location, and type of the damaged area, and the number of measuring points on the detection results.

2 TEST DESIGN

2.1 *Specimen design and material characteristics*

In this paper, five specimens are made, all of which are pie-shaped discs with a diameter of 40 cm and a thickness of 10 cm. The material used is Merbau. Each specimen is subjected to three times artificial simulated damage processing. With the increase in processing times, the area of damage area gradually increases. The three damaged processing states of each specimen form a series, which is represented by different letters. For Series C, R, and T, the damaged areas are round, square, and regular triangular, respectively, and all are located in the center of the specimens; the damaged types are all hollow. For Series D, the damaged area is circular and is located at the edge of the specimen; the damage type is hollow. For Series H, the damaged area is circular and is located in the center of the specimen; the damage type is decay (dense drilling in the damaged area is used to simulate the decaying state of loosely connected wood fiber tissue).

In this paper, 17 stress wave detection tests are carried out, and the parameters of each detection state are shown in Table 1. The test number consists of a series of letters and the number of simulated damage processing. For example, R2 refers to the detection state of the specimen of Series R after two times of artificial simulated damage processing. C2 (8) and C2 (6) indicate that C2 is detected by eight sensors and six sensors, respectively. Ten sensors are used for other detection states.

Referring to references (Xie 2014, Xie 2015), the damage degree is defined as the ratio of the area of the damaged area to the cross-sectional area of the specimen. There are two types of damage degree, the actual damage degree η_{act}, and the test damage degree η_{test}, which are obtained by calculating according to the size of the actual specimen and the data of the stress wave detection diagram, respectively.

2.2 *Test method*

In this paper, the FAKOPP 3D acoustic tomography stress wave detector is used for testing. Figure 1 (a) shows the fixing method of the stickup sensor. First, a small iron sheet of 1 cm $\times$ 1 cm is pasted at the measuring point, and then the sensor is attached to it (the sensor has built-in magnetic materials). After the sensors are uniformly arranged along the outer side of the specimen, the testing can be started (Figure 1 (b)). Knock each sensor in turn with a small hammer, and then the average stress wave velocity between different sensors can be acquired through Arbor Sonic 3D. After graphic processing, the color legend of the stress wave detection image is shown in Figure 1 (c).

Table 1. Details of the specimens and test results.

Test number	Shape of the damaged area	Area of the damaged area (cm^2)	Number of measuring points	Damage type	Actual damaged degree η_{act} (%)	Test damaged degree η_{test}(%)
C1	round	78.54	10	hollow	6.25	15.06
C2	round	240.53	10	hollow	19.14	39.32
C3	round	490.87	10	hollow	39.06	61.80
C2(8)	round	240.53	8	hollow	19.14	42.13
C2(6)	round	240.53	6	hollow	19.14	71.76
R1	square	100.00	10	hollow	7.96	12.87
R2	square	200.00	10	hollow	15.92	18.82
R3	square	312.50	10	hollow	24.87	34.21
T1	regular triangle	32.48	10	hollow	2.58	33.22
T2	regular triangle	99.46	10	hollow	7.91	43.22
T3	regular triangle	202.97	10	hollow	16.15	54.04
D1	round	78.54	10	hollow	6.25	12.54
D2	round	157.08	10	hollow	12.50	32.31
D3	round	235.62	10	hollow	18.75	56.48
H1	round	78.54	10	decay	6.25	11.05
H2	round	240.53	10	decay	19.14	22.40
H3	round	490.87	10	decay	39.06	37.96

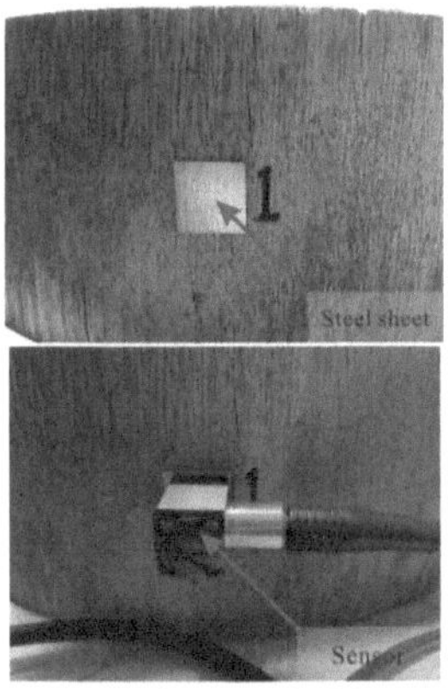

(a) Schematic diagram of stickup sensor

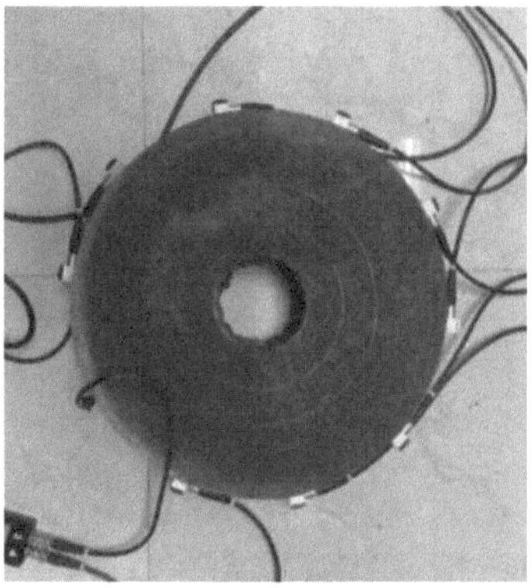

(b) Test diagram

(c) Color legend of stress wave detection images

Figure 1. Testing device.

3 TEST RESULTS AND ANALYSIS

3.1 *Shape and area of the damaged area*

Figure 2 shows the specimens and stress wave detection images of Series C, R, and T. The following conclusions can be drawn from Figure 2 and the test data in Table 1:

(1) The stress wave detection image showing the hollow area can be obtained from the data gotten by the stickup sensors. For Series C, R and T, the test damage degree is higher than the actual damage degree.

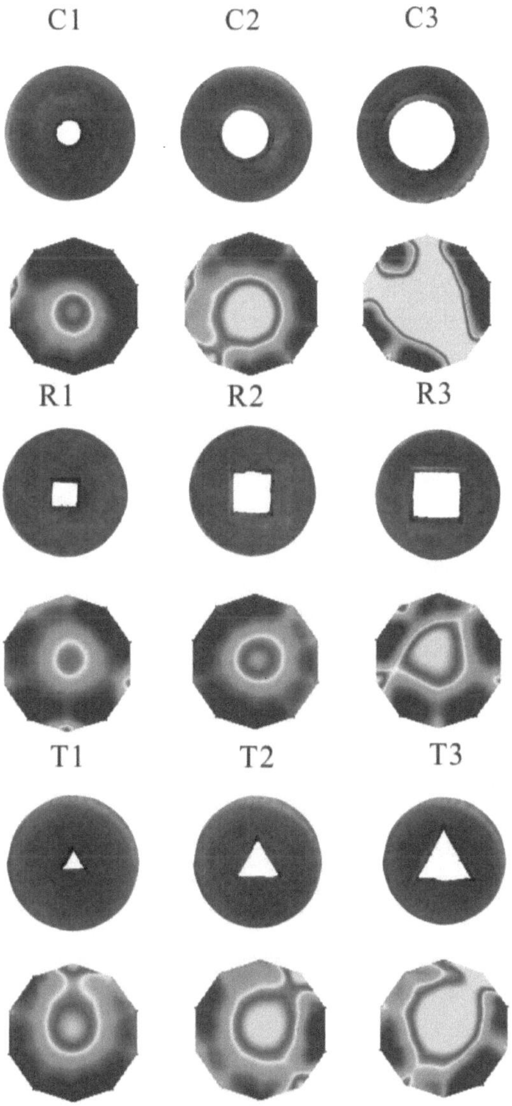

Figure 2. Comparison of shape and area of the damaged area.

(2) The stress wave detection image is not sensitive to the shape of the damaged area. When the hollow area is small (η_{act} is less than 7. 96%), the hollows of different shapes are oval in the stress wave detection images. With the gradual increase of the hole area, the shape of the hole displayed in the image also changes constantly, and some edges and corners appear, but there is a large deviation from the actual situation. This is consistent with the conclusion obtained by using the needle-type sensors (Li 2015).
(3) The hollows with different shapes have a great influence on the deviation of the test results. When the damaged area is a square, round and regular triangle, the average ratio of the test damage degree to the actual damage degree is 1.39, 2.02, and 7.02, respectively.
(4) The uneven side surface of the specimen will cause small gaps between the iron sheet and the wood at the measuring point. In addition, the wood itself will also produce cracks due to

drying shrinkage. These two situations will affect the stress wave detection image to a certain extent. This phenomenon is also common in other series of test results (Figure 3).

3.2 *Location of the damaged area*

Figure 4 shows the specimens diagram and stress wave detection images of Series D. The following conclusions can be drawn from Figure 4 and the test data in Table 1:

(1) The stress wave detection image can roughly represent the position of the damaged area and cannot accurately reflect its connectivity. When there is a crack between two adjacent non-connected hollows, the detected image may display a connected hollow.
(2) The test damage degree is higher than the actual damage degree. On average, the ratio between the two is 2.53. With the increase of the actual damage degree, the test damage degree deviation increases. When the actual damage degree is 6.25%, 12.50%, and 18.75%, the test damage degree is 2.01, 2.58, and 3.01 times the actual damage degree, respectively.

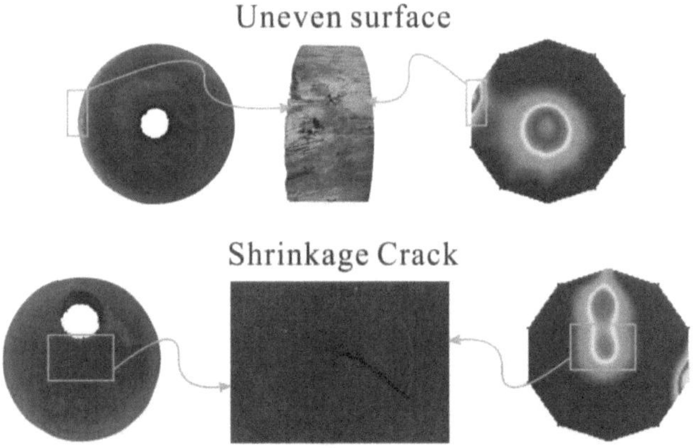

Figure 3. Influence of uneven surface and wood shrinkage crack.

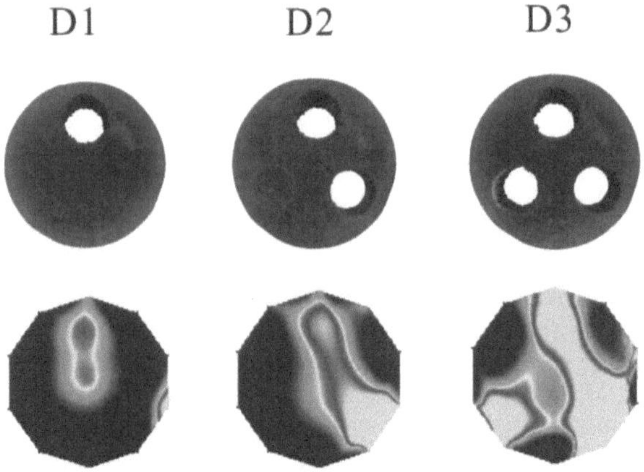

Figure 4. Comparison of location of the damaged areas.

3.3 *Damage type*

Figure 5 shows the specimens and stress wave detection images of Series H. The following conclusions can be drawn from Figure 5 and the test data in Table 1:

(1) The stress wave detection image showing the decayed area can be obtained from the data gotten by the stickup sensors, and the shape of the decayed area in the image is consistent with the actual situation to some extent.
(2) The average ratio of the test damage degree to the actual damage degree is 1.30. With the increase of the actual damage degree, the test damage degree deviation decreases. When the actual damage degree is 6.25%, 19.14%, and 39.06%, the test damage degree is 1.77, 1.17, and 0.97 times the actual damage degree, respectively.

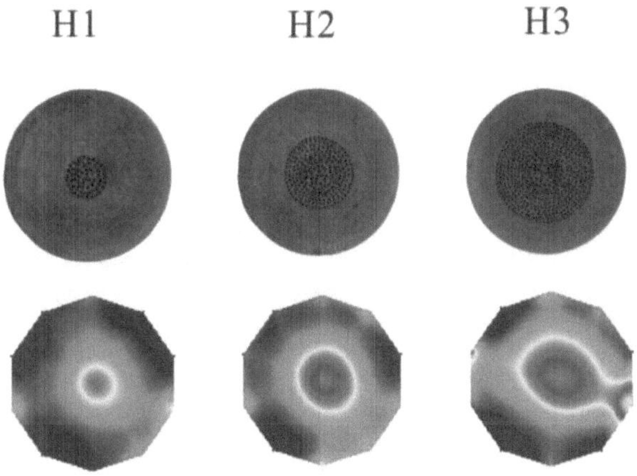

Figure 5. Artificial simulated decayed specimens and detection images.

3.4 *Number of measuring points*

Figure 6 shows stress wave detection images of C2, C2 (8), and C2(6). The following conclusions can be drawn from Figure 6 and the test data in Table 1:

(1) The number of measuring points has a great influence on the stress wave detection images. When the number of measuring points is six, the shape of the damaged area is distorted. With the increase in the number of measuring points, the shape of the damaged area is closer to the actual situation.

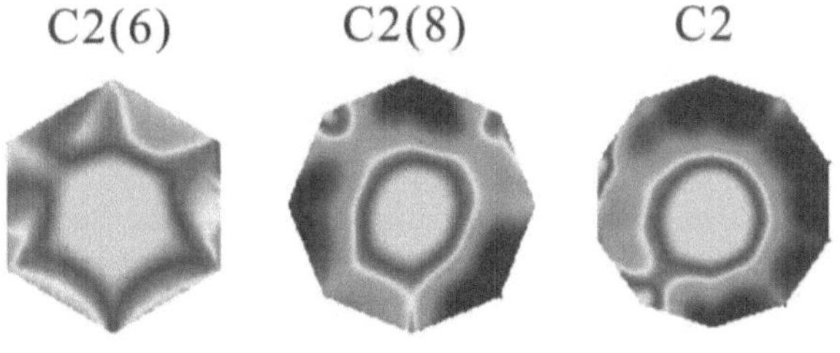

Figure 6. Comparison of the number of measuring points.

(2) With the increase in the number of measuring points, the test damage degree deviation decreases. When the number of measuring points is 6, 8, and 10, the test damage degree is 3.75, 2.25, and 2.05 times the actual damage degree, respectively.

4 CONCLUSIONS

According to the test results of this paper, the following conclusions can be drawn:

(1) The stress wave detection images showing hollows and decayed areas can be obtained from the data gotten by the stickup sensors.
(2) The stress wave detection image cannot effectively distinguish the shape of the damaged area, but can roughly show its position. The uneven side surface and the wood shrinkage crack have an impact on the stress wave detection image to a certain extent.
(3) On average, the test damage degree is higher than the actual damage degree, and the test damage degree deviation of the specimen with the damage type of hollow is greater.
(4) The increase in the number of measuring points can make the shape of the damaged area in the stress wave detection image and the test damage degree closer to the actual situation.

REFERENCES

An, Y. (2013). Two-dimensional imaging technique of wood defects based on stress wave. D. *Chinese Academy of Forestry*. (In Chinese)

C, Ceraldi. V. et al. (2001). Resistographic inspection of ancient timber structures for the evaluation of mechanical characteristics. *J. Materials & Structures*, 34(1), 59–64.

Kandemir-Yucel, A. & Tavukcuoglu. A. & Caner-Saltik. E. N. (2007). In situ assessment of structural timber elements of a historic building by infrared thermography and ultrasonic velocity. *J. Infrared Physics & Technology*, 49(3), 243–248.

Li, X. (2015). Key technology research on material performance and damage detection for wooden components of ancient Chinese buildings. D. *Beijing University of Technology*. (In Chinese)

Wang, X. H. (2006). Study on the variety and the NDE of aged wood physical and mechanical properties of ancient architecture. D. *Inner Mongolia Agricultural University*. (In Chinese)

Wang, Z. C. & G. H. Li. & H. L. Feng. et al. (2015). A method of wood defect identification and classification based on stress wave and SVM. J. *Journal of Nanjing Forestry University (Natural Sciences Edition)*, 39(3), 130–136. (In Chinese)

Winistorfer, P. & W. Xu. & R. Wimmer. (1995). Application of a drill resistance technique for density profile measurement in wood composite panels. *J. Forest Products Journal*, 45(6), 90–93.

Xie, Q. F. & B. Du. & S. Li. et al. (2015). Tests for aseismic behaviors of damaged dovetail mortise-tenon joints of ancient timber buildings. J. *Journal of Vibration and Shock*. 34(4), 165–170. (In Chinese)

Xie, Q. F. & P. J. Zheng. & W. Xiang. et al. (2014). Experimental study on seismic behavior of damaged straight mortise-tenon joints of ancient timber buildings. J. *Journal of Building Structures*. 35(011), 143–150. (In Chinese)

Yang, X. C. (2004). Testing theories and experiments of log inner decay based on stress-wave technology. D. *Northeast Forestry University*. (In Chinese)

Zhu, L. & H. J. Zhang. & Y. L. Sun. et al. (2011). Determination of mechanical properties of ancient architectural timber based on stress wave and micro-drilling. J. *Journal of Northeast Forestry University*, 39(10), 81–83. (In Chinese)

Zhu, L. & H. J. Zhang. & Y. L. Sun. et al. (2015). Mechanical properties non-destructive testing of wooden components of Korean pine based on stress wave and micro-drilling resistance. J. *Journal of Nanjing Forestry University (Natural Sciences Edition)*, 37(2), 156–158. (In Chinese)

Advances in Civil Engineering and Environmental Engineering – bin Yahya & Bin Daud (Eds)

Research on entity relation extraction of the BERT-BiLSTM-CRF model: The perspective of NIMBY conflict in China

Qing Yang & Mengting Chen*
School of Safety Science and Emergency Management, Wuhan University of Technology, Hubei, Wuhan, China

Hui Zhou & Tianyu Wan
School of Management, Wuhan University of Technology, Hubei, Wuhan, China

ABSTRACT: Over the last decades, waste disposal has brought many troubles to the world's environment and human health. NIMBY (Not-In-My-Back-Yard) conflicts caused by waste have been one of the major issues globally. Moreover, current measurements usually have limited effect. Positioning the critical public issues accurately could facilitate the construction of a waste treatment system. However, the materials on politically sensitive issues often have intense obscurity, which is hard to recognize. Aiming to explore the potential knowledge of the NIMBY events, this work utilizes NLP (natural language processing) and some typical cases of the NIMBY problem in China from 2007 to 2020 to recognize the key factors and conflict structure. A BERT-BiLSTM-CRF model is employed to extract triples of entity relations from unstructured texts, leading to a case database of NIMBY incidents and autonomous recognition of related text. The NIMBY event in Shanghai J district in 2018 is used to check the feasibility of sensitive text mining. The results show that the application of BERT-BiLSTM-CRF with specific text could help discover and position NIMBY entity relation extraction, with the highest F value reaching 0.890. This model also has detectably prominent performance on the validation set, with all three values above 0.8, which provides a new insight for conflict decision-making. This work can improve the content of NIMBY's crisis management system and can be extended to general social and environmental issues.

1 INTRODUCTION

Waste incineration plants are an urban public infrastructure. However, while enhancing the city's comprehensive capacity and helping the city escape from the waste siege dilemma, they often lead to NIMBY(Not-In-My-Back-Yard) conflicts due to the psychological barriers of surrounding residents. "NIMBY" was introduced into the academic world by O'Hare (O'Hare 1977) in 1977 to describe the phenomenon in which public perception of a facility that may cause environmental pollution or affect human health is higher than expected, resulting in social conflicts. With the rapid development of China's economy, many kinds of projects have been established consecutively and have more or less polluted the surrounding environment, causing dissatisfaction and fear among the affected population. Under the circumstances, positioning the NIMBY issues precisely could facilitate the construction of a waste treatment system (Feng 2017). It is deemed that positioning the NIMBY conflicts is ultimately a task of semantic analysis. Advances in machine learning provide a good solution to this problem. The BERT model proposed by Google in 2018 has performed surprisingly well in Squad1.1, the top-level test of machine reading comprehension (Devlin al. 2018), and can be well utilized in QA(Questioning-Answer) and reading comprehension (Adeyinka

*Corresponding Author: molly0226@whut.edu.cn

DOI 10.1201/9781003349563-45

al. 2021), recommendation system (Wang al. 2021), text classification (Singh al. 2021), sequence labeling (Sun al. 2021) and other tasks. Information extraction is to convert unstructured or semi-structured natural language text into structured data. Information extraction is designed to extract structured information from large-scale unstructured or semi-structured natural language texts (Liu al. 2021). And name entity recognition (NER) and relation extraction (RE) are two paramount subtasks of information extraction (Qiao al. 2021),, which are key components of many scientific literature mining tasks (Gao al. 2021). Entity and relation extraction is used to identify and assign a label to each unit of a given input sequence (Lin al. 2019), and sequence labeling plays an important role by automatically analyzing a sequence to assign category labels to each part (Lin al. 2020).

Due to its advantageous performance, the BERT language model is used to replace the traditional Word2Vec language for semantic representation and combined with the improved bidirectional Long Short-Term Memory (Bi-LSTM) model based on the self-attention mechanism to extract contextual features adaptively (Li al. 2021). For this reason, the Conditional Random Fields (CRF) are added on the basis of this model, which considers the semantic information in the sequence and the distance dependence problem, and can automatically extract features and achieve better performance (Liu al. 2021). Aiming at extracting entity relations from non-structure sensitive texts and reducing the interference caused by manual annotation errors on the data set, an improvement and application of the BERT-BiLSTM-CRF knowledge extraction model is conducted in this work. The results of the extraction process provide a scientific reference for the decision-making of the NIMBY conflict.

2 METHODOLOGY

BERT is an unsupervised natural language processing and training model (Vaswani 2017) and a multi-layer bidirectional transformer encoder. The structure of the primary model is the stacking of transformers. The transformer typically uses language modeling targets to pre-train extensive collections of documents, and the resulting model can then be transferred to target tasks with relatively little supervised training data, resulting in state-of-the-art performance in many NLP tasks. Each module comprises multi-head self-attention, standardization, residual connection, and a feedforward neural network. The specific implementation process is mainly divided into two stages, model pre-training and model fine-tuning. In the pre-training stage, due to the large order of magnitude of model parameters, abundant data training is needed. However, owing to the unsupervised model, it only needs to crawl or use open-source data sets. In the fine-tuning stage, the model needs to be fine-tuned for specific tasks to achieve better results. BERT only needs an additional output layer to fine-tune the pre-trained BERT, and it can meet various tasks without modifying the model for specific tasks.

Graves A. et al. proposed a BiLSTM model, in which sequence data are processed forward and backward through two separate hidden layers, namely forward LSTM and backward LSTM, to capture the bidirectional semantic dependence and merge the outputs at the same time (Hartigan 2021). Hence it implements forward and backward transfer of sequential data and models data dependencies in both directions, identifying the time nuances in the corpus (Pogiatzis 2020). The structure of the model is shown in Figure 1, where x_t, y_t is the input sequence and the output sequence at the t moment, and two LSTM layers form a bidirectional LSTM layer. Whereas one-way LSTM only focuses on the meaning of past sentences, while BiLSTM provides remote context access for comprehensive learning of specific problems in terms of input direction and results (Khan al. 2021). The benefit of this approach is that BiLSTM can recognize time nuances in the corpus and can be used to better model dependencies in both directions (Cai al. 2021). Therefore, the BiLSTM model is adopted for the contextual sequence labeling task in this work.

The Conditional Random Field is a typical discriminant model proposed by Lafferty et al. in 2001 (Lafferty 2001). It is a discriminant model for labeling and separating sequence data. It tries to build a conditional probability distribution model on a label sequence given an observed sequence (Vaswani 2017). Therefore, the CRF model is introduced here to capture the link relation between

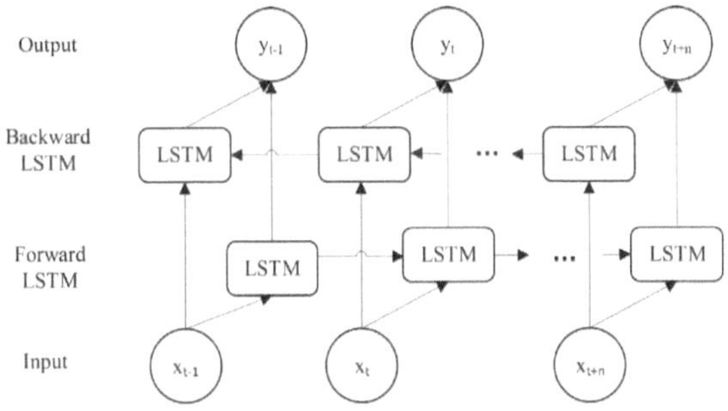

Figure 1. The structure of the Bi-LSTM model. With the backward and forward LSTM, it can better capture the two-way semantic dependence.

the two tags and constrain unreasonable tags. The CRF layer can mark the global optimal sequence and substitute the hidden layer state sequence with the optimal one.

The input of the CRF module is the word vector trained by the BERT and BiLSTM layers, represented as $x=\{x_1,x_2,\ldots,x_n\}$, and the optimal sentence-level sequence is obtained, represented as $y=\{y_1,y_2,\ldots,y_n\}$.

In accordance with the CRF calculation principal (Gridach 2018), the score of a certain label sequence l of sentence s is calculated first by the feature function of CRF, which can be represented as follows:

$$score\,(l|s)=\sum_{j=1}^{m}\sum_{i=1}^{n}\lambda_j f_j(s,i,l_i,l_{i-1}) \tag{1}$$

Where, f_j and λ_j denote the feature value and the weight of the word vector output by the BERT and BiLSTM layers, respectively.

By indexing and standardizing the score through the Softmax function(Qin 2021), the probability value $p(l|s)$ of the labeling sequence l can be obtained, which is represented in Equation 2. Among all labeling sequences, the l with the highest probability value is considered to be the final labeling sequence.

$$p\,(l|s)=\frac{\exp[score(l|s)]}{\sum_{l'}\exp[score(l'|s)]}=\frac{\exp[\sum_{j=1}^{m}\sum_{i=1}^{n}\lambda_j f_j(s,i,l_i,l_{i-1})]}{\sum_{l'}\exp[\sum_{j=1}^{m}\sum_{i=1}^{n}\lambda_j f_j(s,i,l'_i,l'_{i-1})]} \tag{2}$$

The model proposed in this paper is composed of three modules, BERT, BiLSTM, and CRF. Its framework is shown in Figure 2. Each sentence described in the case is used as an input into the

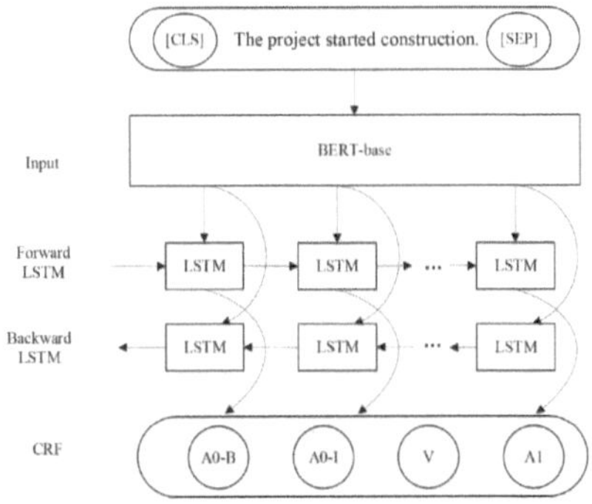

Figure 2. The structure of the BERT-BiLSTM-CRF model. The model consists of a BERT layer, forward LSTM layer, backward LSTM layer, and CRF layer.

BERT module layer for pre-training, and then the output word vector sequence is semantically coded in the LSTM layer, the score of the tag is calculated and then sent to the CRF layer, after that the optimized tag sequence is calculated and the marking results are output. The BERT-BiLSTM-CRF model requires a lot of corpus learning research, and it can grasp the grammatical and semantic features and better understand the context.

3 EXPERIMENT AND RESULTS

3.1 *Dataset*

In this work, 13 articles on the urban waste crisis are selected as a training set, and 41 NIMBY cases of national waste incineration from 2007 to 2020 are selected as a test set. Then, the triples of the entity relation are extracted.

We search 463 scientific papers with "waste" as the theme and "NIMBY" as the keyword on CNKI (Yang al. 2021). According to the NIMBY entity relation extraction rules, 13 critical articles are selected, and each semantic unit and semantic role are manually judged and labeled according to the semantic annotation template. We first formulate the labeling rules, train the labeling personnel, spend one week carrying out multiple rounds of labeling for the articles, and verifies the labeling results. Finally, several random tests are performed, and inaccurate information is noted. The results of the manual annotation of 13 articles are used as the training set to train the model's parameters and optimize its performance.

The 41 domestic NIMBY incidents from 2007 to 2020 are also involved in this paper. The description of the cases comes from the relevant news reports and some literature excerpts. The case's development process is described as four stages: the nascent stage, the development stage, the climax stage, and the mitigation stage. The two crucial subjects of the NIMBY event, namely the government and the public, their behavior and the results, are described in detail. These cases are stored and constructed in the NIMBY crisis database of urban waste.

3.2 *Standards of labeling and evaluation*

The Jieba algorithm is adopted to segment the case, and the BERT-BiLSTM-CRF model is employed for labeling. For a single-text sentence task, two unique tags [CLS] and [SEP] are added to indicate the beginning and end of each sentence, respectively. These two output markers are necessary for the CRF, which needs to model the dependence between the labels. In the process of segmentation, all the words in the text are marked by their part of speech, and the marking rules are shown in Table 1. BIO (Begin-Inside-Outside) labeling method is used in name entity recognition sequence labeling. B, I, and O represent the beginning part of the entity, the middle part of the entity, and the nonentity part, respectively. In addition, marking the triples utilizes the A0-V-A1 method. A0 denotes the first entity, which is the issuer of the action, V denotes the action relation, and A1 denotes the second entity, which is the undertaker of the action.

Table 1. The notation of the part-of-speech tagging.

Notation	Part of speech
/n	noun
/v	verb
/p	preposition
/m	numeral
/c	conjunction
/a	adjective
/ad	adverb

Accuracy rate, recall rate, and F score, which are commonly used in NLP classification tasks, serve as evaluation criteria to compare the effects of different models (Wang 2021). The formulas of the three standards are defined as follows:

$$\text{Recall} = \frac{n_c}{n_c + n_m} \times 100\% \tag{3}$$

$$\text{Precision} = \frac{n_c}{n_c + n_f} \times 100\% \tag{4}$$

$$F = \frac{2 \times \text{Precision} \times \text{Recall}}{\text{Precision} + \text{Recall}} \times 100\% \tag{5}$$

Where, n_c represents the total number of scenes detected correctly, n_m represents the total number of undetected scenes, n_f represents the total number of falsely detected scenes. F1 measure believes that the precision rate and the recall rate have the same weight, but in some scenarios, we may think that the precision rate will be more significant.

3.3 *Experimental analysis*

This paper extracts entity relation triples from the case of NIMBY events in two cases, namely the BERT-BiLSTM-CRF model and BERT-CRF, to evaluate the performance of the models and calculates the extraction accuracy rate, the recall rate, and the F1 value, as shown in Table 2. It also combines two methods of supervised learning and unsupervised learning. In the unsupervised phase, BERT is utilized to perform unsupervised feature learning on the labeled training set. In the supervised phase, the BiLSTM neural network is superior to the one-way network in terms of time-related feature extraction and is faster and more accurate than other standard methods based on LSTM (Pogiatzis 2020). Using the CRF model to mark and slice serialized data, the effect of sequential labeling is better than the traditional machine learning model (Qin 2021). According to whether the BiLSTM-CRF layer and the BERT body are split or not, the learning rates under the two conditions are calculated. At the same time, a full connection layer is added as a classifier, and the output dimension of the full connection layer is determined without using BiLSTM, which is directly submitted to the CRF layer for processing to compare the extraction results of the two models.

Table 2. The results of the test set.

		Split learning			Non-split learning		
		Precision	Recall	F1	Precision	Recall	F1
BERT-BiLSTM-CRF	A0	0.863	0.919	0.890	0.862	0.879	0.870
	V	0.847	0.897	0.871	0.833	0.909	0.869
	A1	0.865	0.892	0.878	0.849	0.904	0.875
BERT-CRF	A0	0.851	0.873	0.861	0.834	0.851	0.842
	V	0.847	0.892	0.868	0.839	0.896	0.866
	A1	0.836	0.898	0.865	0.865	0.884	0.874

As shown in Figure 3, the accuracy rate, recall rate, and F1 value of the model are all above 0.8 under the influence of BERT regardless of the learning condition, which means that the BERT model achieves an excellent result in the process of extracting the triples of sensitive text entity relation.

Compared with non-split learning, split learning has superior performance. The average F1 under the split learning mode's influence is 0.71% higher than that under non-split learning. The average F1 of the model with and without the BiLSTM layer is 0.875 and 0.862, respectively. The

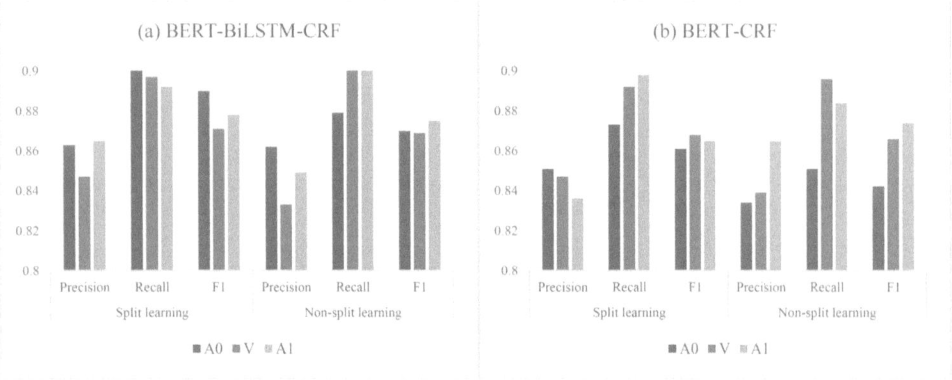

Figure 3. Model performance comparison. (a): The extraction results of the BERT-BiLSTM-CRF model. Under the influence of the BERT-BiLSTM-CRF model, the precision rate, recall rate, and F1 value are greater than 0.8. (b)The extraction results of the BERT-CRF model. Without the BiLSTM layer, the precision rate and the F1 value have decreased distinctly.

addition of the BiLSTM layer increases the model performance by 1.48%, and the F1 values on A0, V, and A1 are increased by 3.34%, 0.35%, and 0.81%, respectively. The model has the highest accuracy in extracting A0. A1 is the second most accurate three times in the four experiments, and V is the second most accurate in one, indicating that the model performs better in entity recognition than in relation recognition. In accordance with the results above of entity relation triple extraction, it can be seen that the pre-trained BERT model on large-scale unlabeled corpus proposed in this paper has a beneficial effect on the entity relation information extraction, and it provides scientific support and a realistic basis for the actual test.

4 COMPARISON VALIDATION

After extracting entity relation triples, applying comparison validation to analyze the feasibility of text mining is expected. The paper intends to use the extracted entity relations as the original data set, after removing the stop words, count the top 35 high-frequency words from the entity relation, and then try to utilize high-frequency words to restate the NIMBY conflict caused by the waste transfer station in J district in Shanghai in 2018, to judge the accuracy and effectiveness of text mining.

First, the triples of entity relations extracted from the BERT-BiLSTM-CRF model under the condition of split learning are taken as original data, stop words are removed, and the top 35 high-frequency entity relation words are counted. The terms and frequencies of high-frequency words are shown in Table 3.

It can be seen from the table that high-frequency entities mainly include government, villagers, residents, organizations, people, waste incineration, projects, events, etc. Among them, waste incineration and projects are the main body of NIMBY incidents, that is, the direct cause of NIMBY conflicts. Villagers, residents, organizations, and the public are the participants of NIMBY incidents, while the government is the undertaker of NIMBY incidents. High-frequency relational words can be roughly divided into two categories. The first category is the construction progress words of waste incineration projects, such as construction, site selection, stop, start, announcement, construction, relocation, put into use, environmental assessment, planning, construction, etc. The second category is public sentiment words, such as opposition, protest, petition, boycott, application, administrative reconsideration, questioning, etc. By combing and analyzing the high-frequency words of NIMBY cases, the scope and characteristics of NIMBY from the event can

Table 3. High-frequency vocabulary of NIMBY cases.

No.	word	Freq.	No.	word	Freq.	No.	word	Freq.
1	project	34	2	waste incineration	34	3	oppose	34
4	construct	34	5	waste	34	6	resident	33
7	protest	30	8	villagers	28	9	government	28
10	location	27	11	organize	27	12	environmental assessment	27
13	cease	26	14	populace	26	15	release	24
16	group	24	17	operation	23	18	opinion	23
19	process	23	20	event	22	21	intense	22
22	power plant	21	23	gather	21	24	build	20
25	apply	20	26	Administrative Reconsideration	20	27	petition	18
28	move	18	29	put into use	18	30	plan	18
31	finish	17	32	resist	17	33	construction	17
34	publish	16	35	suspect	14			

be gradually understood. Cases are manifested mainly as individuals or communities opposed to the construction of certain NIMBY facilities or land use because the benefits of such construction are shared by all of the society, but the adverse consequences are mostly borne by the local residents, so the local public strongly opposes the construction. The main approaches people resist NIMBY projects include mainly petitions, gathering behaviors affecting social order, and applying for administrative reconsideration to suspend the construction of the project.

In 2018, the NIMBY conflict at the waste transfer station in District J of Shanghai lasted about a year. Due to the concern of the surrounding residents about the environmental pollution caused by the waste transfer station and the impact on their health, there have been unpleasant phenomena such as protests and intentional gatherings affecting the traffic. The evolution of the NIMBY conflict in the J district of Shanghai can be divided into four stages, namely the embryonic stage, the development stage, the climax stage, and the mitigation stage. The description of the case is drawn as a timeline diagram, and both the high-frequency words of the case and the extraction results of the triple entity relation of the BERT-BiLSTM-CRF model are shown in Figure 4, corresponding to Figure 4 (a), 4 (b), and 4 (c).

After sorting out the NIMBY incident at the waste transfer station in the J District of Shanghai, it can be seen that the NIMBY conflict mainly revolves around the government and residents. After the government announced the project, it caused dissatisfaction among the people, so a series of irrational activities began to form, such as raising banners to protest, petitioning for a response, gathering to expand influence, etc. The asymmetry of information and the inaction of relevant departments intensified the group contradiction, which affected social stability and people's normal life. According to the retelling of the NIMBY case, the BERT-BiLSTM-CRF model is employed to extract the entity relation triples, and the top 35 high-frequency words are counted, as shown in Figure 4(b).

This work conducts experiments on the case under the condition of split learning on account of its excellent performance. The calculated accuracy rate, recall rate, and F1 value are all greater than 0.8, and the best F1 value reaches 0.879, which indicates the extraordinary performance in the entity relation extraction of the Shanghai NIMBY conflict. The same as before, the extracted entities and relations also include two types of subjects involved in the NIMBY incident: the public and the government. The conflict between the two due to the garbage transfer station has caused the public to take some irrational behaviors. The government was required to respond positively, but the inaction of the relevant departments contributed to the dissatisfaction of the residents, and eventually, the project ended up with nothing definitive. This is also more consistent with China's current trend in the construction of NIMBY facilities, with more suspended construction and less success remaining. It is also found that, among the 35 high-frequency words in the case, 13 belong

(a) The development of the NIMBY incident.

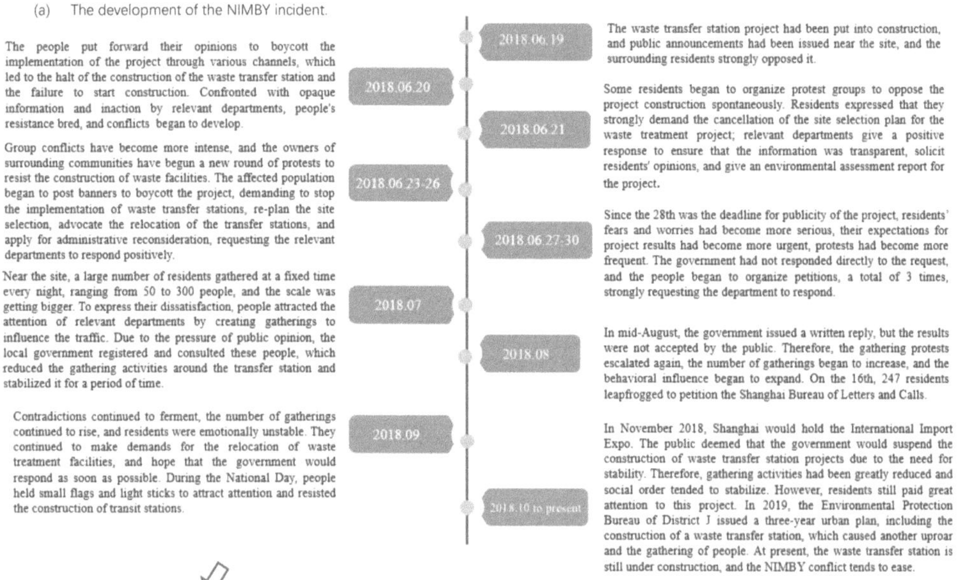

(b) High-frequency words in this case.

No.	word	Freq.	No.	word	Freq.	No.	word	Freq.
1	resident	11	13	surrounding	4	25	nearby	2
2	transfer station	10	14	resist	4	26	issue	2
3	project	10	15	government	4	27	publish	2
4	begin	9	16	activity	4	28	oppose	2
5	waste	8	17	opinion	3	29	through	2
6	behavior	7	18	directly	3	30	propose	2
7	gather	7	19	demand	3	31	implement	2
8	construct	6	20	response	3	32	cease	2
9	public	6	21	number	3	33	information	2
10	department	5	22	continue	3	34	transparent	2
11	protest	5	23	bud	2	35	relevant	2
12	more	5	24	site	2			

(c) Performance of model implementation.

	Precision	Recall	F1
A0	0.903	0.812	0.855
V	0.908	0.831	0.868
A1	0.892	0.866	0.879

Figure 4. The results of comparison validation. (a) The timeline of the 2018 NIMBY conflict in Shanghai, (b) High-frequency words in this case, (c) Precision rate, recall rate and F1 value of the BERT-BiLSTM-CRF model.

to the high-frequency words of the entity relation triples extracted from the BERT-BiLSTM-CRF model, and 10 of the first 20 words overlap. These overlapping words include several essential entities of the NIMBY conflict, such as transit stations, projects, residents, governments, and departments. There are also words that reflect the public's attitudes and behavior in the face of NIMBY problems, such as boycotts, opposition, gathering, protest, etc. It also involves the situation of NIMBY facilities, such as construction, implementation, and suspension. In general, similar to the high-frequency words extracted from the model, the high-frequency words, in this case, basically reflect the crucial information about the NIMBY event in the J District of Shanghai.

The comparison validation fully illustrates that the entity relation triple extraction model for NIMBY events proposed in this work performs quite well in actual case analysis, reflecting the feasibility of sensitive text mining.

5 DISCUSSION

In today's society, people's demand for public services is increasing, and environmental awareness is growing more assertive at the same time. When the social services that the public needs will sacrifice the interests of some surrounding residents, these people tend to have a lot of negative emotions and even irrational behaviors, leading to the appearance of NIMBY conflict. Due to the political particularity of the NIMBY problem, public opinion on the Internet and political documents are less publicized, and much of the empirical evidence is drawn from Western countries (Sun,2016).

Therefore, the paper starts with the literature and a few reports for sparse text mining, selects 13 relevant academic articles as the training set, 41 NIMBY cases as the test set, and extracts entity relations based on the BERT-BiLSTM-CRF model, finally measures the real efficiency. The experimental results demonstrate that the BERT model favors the entity relation triples extraction task, with the highest F value reaching 0.890. Finally, taking the waste transfer station event as the validation set, it is found that the BERT-BiLSTM-CRF model offers an effective way of knowledge extraction. Meanwhile, the high-frequency words of the Shanghai case and the high-frequency words extracted from the NIMBY case database overlap, which indicates that the BERT-BiLSTM-CRF model proposed in this paper has universal applicability and efficiency in the extraction of entity relations in the field of NIMBY events.

Entity relation extraction is widely adopted in the general field, but narrowly applied in the NIMBY problem. In the data collection process, research usually uses open-source text data sets (Ma,2021), most of which are web texts. Due to the political sensitivity of the NIMBY issue, there are few relevant materials on the Internet. Therefore, sparse text mining is employed to collect information on the NIMBY cases through relevant papers and a few reports. Regarding the entity relation extraction model, the existing researches involve the end-to-end joint entity and relation extraction method based on convolutional neural network and recurrent neural network, the extraction method based on the combination of the BERT model and domain knowledge, the multi-level hierarchical graph attention network method, etc. The commonly used BERT-BiLSTM-CRF model is adopted in this paper, which has proved to achieve excellent results. BERT simplifies feature extraction in preprocessing. LSTM focuses on long-distance text information to solve the problem of gradient explosion or gradient disappearance generated during recurrent neural network training. The CRF considers the relation between neighboring tags and can obtain an optimal prediction sequence. In addition, the present model provides a possible way to position the NIMBY conflict, which bridges the gap between the NIMBY crisis theory and authentic cases.

6 CONCLUSION

This work focuses on the urban waste NIMBY conflict problem to explore the knowledge of the NIMBY problem from the relevant literature and examples. For this purpose, a BERT-BiLSTM-CRF model based on neural networks and deep learning is proposed to extract the entity relation of NIMBY events. The 13 authoritative papers concerning the NIMBY problems are used as training sets to train various neural network parameters. The 41 NIMBY conflict cases of urban waste from 2007 to 2020 are taken as the test set to measure the model's performance. And the case in the J district of Shanghai in 2018 serves as the comparison validation set. It explains the feasibility of sensitive text mining and analysis. The experimental results of the test set reveal that the extraction effect of split learning is better than that of non-split learning. The average accuracy rate for the extraction of entities from the BERT-BiLSTM-CRF model is 0.858, the average recall rate is 0.902, and the average value of F1 is 0.880 under the influence of split learning. The model also has detectably prominent performance on the validation set, with all three values above 0.8. The top 35 high-frequency words in the test set and the validation set are counted, respectively, and it is found that the coincidence rate of the two is 37%. Moreover, the coincidence rate of the first 20 high-frequency words in the verification set reached 50%. The experimental results show that the results of the entity relation extracted by the model have a beneficial effect on the NIMBY problem, mirroring the corresponding practical significance. Texts related to NIMBY conflicts are characterized by sensitivity and sparsity. After data processing of these rare and unstructured texts, the sequence labeling task is completed, and entity relation triples are extracted, which can clearly reflect the connection among the subjects involved. By counting the frequency of each word, the main subjects surrounding the development of the matter and the content and the manifestation of their contradictions can be intuitively reflected, to accurately locate the NIMBY problem, find a breakthrough to resolve the conflict and apply it downstream tasks. Furthermore, the extraction results of the test set can be used to restore the general process of the general NIMBY event, which has universal applicability.

In future research, the experimental results will be applied in downstream tasks, such as constructing knowledge graphs and researching complex networks. At the same time, more text data will be introduced to train the model to improve efficiency and accuracy, hence exploring more potential knowledge of NIMBY problems, which could provide a new insight for conflict decision-making.

6.1 *Funding Statement*

This research was funded by the Major Project of the National Social Science Foundation of China, grant number 16ZDA045.

6.2 *Data Availability*

The data used to support the findings (including 13 authoritative papers and NIMBY event data) of this work are available from the corresponding author upon request.

6.3 *Acknowledgments*

The authors would like to thank Hui Zhou and Tianyu Wan for their helpful suggestions and technology support. The authors are very grateful to the editors and reviewers for their valuable comments and suggestions.

6.4 *Declaration of competing interest*

The authors declare that they have no known competing financial interests or personal relations that could have appeared to influence the work reported in this paper.

REFERENCES

Adeyinka, A., Serestina, V., Oluwaseun, O., Mario, V. (2021). Deep learning approach for medical image analysis. *Computational Intelligence and Neuroscience*. vol.2021.

Cai, R., Qin, B., Chen Y., et al. (2020). Sentiment analysis about investors and consumers in energy market based on BERT-BiLSTM. *IEEE Access*, vol.8, pp.171408–171415.

Devlin, J., Chang, M., Lee, K., Toutanova, K. (2018). BERT: Pretraining of deep bidirectional transformers for language understanding. https://arxiv.org/abs/arXiv:1810.04805.

Feng, Y., Zhang, H., Hao, W., Chen, G., Voulodimos, A. (2017). Joint extraction of entities and relations using reinforcement learning and deep learning. *Computational Intelligence and Neuroscience*. vol.2017, 2017.

Gao, S., Kotevska O., Sorokine, A., Christian, J.B. (2021). A pre-training and self-training approach for biomedical named entity recognition. *PLOS ONE*, vol.16, e0246310.

Gridach, M., Haddad, H. (2018). Arabic named entity recognition: A bidirectional GRU-CRF approach. *Transactions on Computational Science* XXXVIII. vol.10761, pp.264–275, 2018.

Hartigan, J.A. (2021). Estimation of a convex density contour in two dimensions. *Journal of the American Statistical Association*, vol.82, no.397, pp.267–270.

Khan, M., Wang, H., Riaz, A., Elfatyany, A., Karim, S. (2021). Bidirectional LSTM-RNN-based hybrid deep learning frameworks for univariate time series classification. *The Journal of Supercomputing*.https://doi:10.1007/s11227-020-03560-z.

Lafferty, J., Mccallum, A., Pereira, F. (2001), Conditional random fields: Probabilistic models for segmenting and labeling sequence data. *Proc. 18th International Conf. on Machine Learning*, vol.951, pp.282–289.

Li, S., Liu, Y. (2021). News video title extraction algorithm based on deep learning. *IEEE Access*, vol.9, pp.12143–12157.

Lin, C. et al. (2019). A Bi-LSTM mention hypergraph model with encoding schema for mention extraction. *Engineering Applications of Artificial Intelligence*. Oct. (2019):175–181.

Lin, C. et al. (2021). ASRNN: A recurrent neural network with an attention model for sequence labeling. *Knowledge-Based Systems*. Volume 212, 2021.

Liu, J., Duan, X., Zhang, R., Sun, Y., Guan, L., Lin, B. (2021). Relation classification via BERT with piecewise convolution and focal loss. *PLOS ONE*, vol.16, e0257092.

Ma, P., Jiang, B., Lu, Z., Li, N., Jiang, Z. (2021). Cybersecurity named entity recognition using bidirectional long short-term memory with conditional random fields. *Tsinghua Science and Technology*. vol.26, pp.259–265.

O'Hare, M. (1977). Not on my block you don't: Facility siting and the strategic importance of compensation. *Public Policy*. vol.25, no.4, pp.407–458.

Pogiatzis, A., Samakovitis, G. (2020). Using BiLSTM networks for context-aware deep sensitivity labeling on conversational data. *Applied Sciences*, vol.10, pp.8924.

Qiao, B., Zou, Z., Huang, Y. et al. (2021). A joint model for entity and relation extraction based on BERT. *Neural Computing and Applications*, https:// doi:10.1007/s00521-021-05815-z.

Qin, Q., Zhao, S., Liu, C. (2021). A BERT-BiGRU-CRF model for entity recognition of chinese electronic medical records. *Complexity*. vol. 2021, pp.1–11.

Singh, M., Jakhar, A.K., Pandey, S. (2021). Sentiment analysis on the impact of coronavirus in social life using the BERT model. *Social Network Analysis and Mining*, vol.11, no.1.

Sun, C., Yang, Z., Wang, L. et al.(2021). Biomedical named entity recognition using BERT in the machine reading comprehension framework. *Journal of Biomedical Informatics*, vol.118, pp.103799.

Sun, L., Zhu, D., Chan, E.H.W. (2016). Public participation impact on environment NIMBY conflict and environmental conflict management: Comparative analysis in Shanghai and Hong Kong. *Land Use Policy*. vol.58, pp.208–217.

Vaswani, A., Shazeer, N., Parmar, N. et al. (2017). *Attention is all you need*. https://arXiv:1706.03762v5.

Wang, J., Zhang, G., Wang, W., Zhang, K., Sheng, Y. (2021). Cloud-based intelligent self-diagnosis and department recommendation service using Chinese medical BERT. *Journal of Cloud Computing*. vol.10, no.1.

Wang, J., Zhang, G., Wang, W., Zhang, K., Sheng, Y. (2021). Cloud-based intelligent self-diagnosis and department recommendation service using Chinese medical BERT. *Journal of Cloud Computing*. vol.10, no.1, pp.1–12.

Yang, Q., Zhou, H., Liu, X. et al. (2021). Reconstruction of ER Network from Specific Academic Texts for the Governance of MSW-NIMBY Crisis in China. *Complexity*, 2021,2021.

Advances in Civil Engineering and Environmental Engineering – bin Yahya & Bin Daud (Eds)
© 2023 The Author(s), ISBN: 978-1-032-39397-1

Evaluation study of olfactory landscape in Hangzhou grand canal Asian Games Park

TianMei Du* & TianLong Cao
Zhejiang University of Technology, Zhejiang, China

ABSTRACT: The interrelationship among human, olfactory landscape, and spatial environment has become the focus of olfactory landscape research, and the study of the olfactory landscape for urban parks is beneficial in providing design methods for urban park environment construction. Taking Hangzhou Grand Canal Asian Games Park as an example, this paper constructs the olfactory landscape evaluation model of Grand Canal Asian Games Park by spatial syntax and analytic hierarchy process, builds the evaluation model from plant elements, animal elements, natural abiotic elements, and artificial elements, determines the index weights and derives the judgment matrix, so as to obtain the evaluation results of the olfactory landscape of Grand Canal Asian Games Park, and analyzes the model for the construction of the olfactory landscape environment of the city park and make design suggestions.

1 INTRODUCTION

China's urban construction has transitioned from the stage of quantitative growth of rough urbanization to the stage of pursuing the improvement of urban quality and livability, and according to the China Agricultural and Rural Development Report 2020, it is pointed out that: China's urbanization rate is expected to reach 65.5% by 2025. The problems of insufficient supply of ecological products, degradation of ecological environment, and degradation of ecological service quality of green areas brought by rapid urbanization can no longer meet the residents' good demand for the quantity, type, and cultural characteristics of green areas, and the residents' urban diseases cannot be effectively alleviated. Therefore, constructing an urban park environment is particularly important, and the optimization and harmony of the relationship between "people – park – city" is an important element in creating a better life. This paper constructs an evaluation model between "human-olfactory landscape-spatial environment" to create a practical basis for the construction of urban parks.

2 OLFACTORY LANDSCAPE

Olfactory landscape refers to the olfactory environment perceived by a person in a place. The main purpose of the concept of olfactory landscape is to explore the relationship between people, odor, and the environment. Tuan (1977) argues that odor can give distinctive characteristics to objects and places. Rodaway (2002) argues that each place has its unique olfactory environment, which is closely related to the local geography and the lifestyle of its inhabitants. Xiao (2021) believes that olfactory landscape uses odor as a medium and olfactory perception to create landscape imagery.

*Corresponding Author: 2112015059@zjut.edu.cn

 DOI 10.1201/9781003349563-46

2.1 *Status of domestic research*

Liu's (1956) study of traditional garden "fragrance landscape" revealed that there are different forms of planting in different spaces according to the characteristic of "flower fragrance." Gao (2011) verified the hypotensive health function of the scenting area, mainly composed of plants. Zhang (2019) argued that the right scent could relax people and achieve better therapeutic effects. Therefore, the main domestic research on olfactory landscape includes classical garden plants, plant healing, health, etc. The research methods are mostly subjective, and there is a lack of necessary theoretical guidance on the pollution management and practice of the olfactory environment.

2.2 *Status of foreign research*

Henshaw (2014) explored the odors people perceive in different spaces and their perceptions of odors, examining the forms of built environments and the components that influence the experience of odors. Diaconu (2011) organized, explored, and recorded the tactile sensations and odors at public transportation nodes while investigating local residents' memories of odors and the various meanings of odors. Bouchard (2013) reveals the "temporal pattern" triggered by smell, exploring how people's memory of smell affects the spatial-temporal perception of cities. Badach (2018) presents an "olfactory landscape plan" for odor pollution in a waste disposal area in Gdańsk, Poland. The main research contents in foreign countries are olfactory perception and evaluation, odor characteristics, odor concentration, odor distribution, etc. The research methods are mostly quantitative, such as making "odor walks," "odor maps," "spatial distribution map and diffusion range map," quantitative evaluation method, and physiological evaluation methods such as autonomic nervous system index, central nervous system index, and endocrine system index. These methods are based on objective conditions, which are more in line with the laws of nature and can help to obtain more accurate evaluation results in line with human physiology, but at the same time, they ignore the role of human subjectivity and psychology in the relationship of "human-olfactory landscape-spatial environment."

3 METHODS

3.1 *Overview of the study area*

The Grand Canal Asian Games Park is located in Shenhua Block, Gongshu District, Hangzhou, covering an area of 701 mu, and is the largest new Asian Games venue in the main urban area of Hangzhou, as well as the first comprehensive urban sports park in Zhejiang Province that integrates a stadium, a park, a sports field, and commercial facilities. The main feature of the Grand Canal Asian Games Park is its ecological, participatory, and technological nature. The "Flower Festival" and the culture of the God of Flowers are mainly inherited and carried forward from both hard and soft aspects.

3.2 *Data sources*

The data on the research area of the thesis come from field research and the official website of the park; the data of the importance assigned to the index system of the thesis come from the scoring by experts in ecology, architecture, landscape architecture and design; the data of the questionnaire survey come from field research.

3.3 *Research methods*

3.3.1 *Spatial syntax*

The limit distance a person can smell is 3 m, and the maximum distance at which a person is most willing to walk is 500 m. Therefore, the radius was divided into three classes: 3 m, 500 m, and global. The locations of crowd gathering and dispersion points in Hangzhou Grand Canal Asian Games Park were obtained by the integration degree, average step depth, selection degree, and fit degree of spatial syntax.

3.3.2 *Analytic hierarchy process*

AHP is a calculation method to determine the weight of indexes, and the index factors are compared two by two through expert scoring to determine their relative importance, after which a judgment matrix is established, and the weight value of each factor is determined through a series of calculations.

The discrete points of 3 meters and 500 meters that people are most willing to reach on foot are derived by the spatial syntax and spatially compared with the main olfactory nodes derived from the AHP to discuss whether the distribution points of the olfactory landscape of Hangzhou Grand Canal Asian Games Park overlap with the paths walked by people, thus verifying the rationality of the spatial distribution of the Asian Games Park itself.

4 CONSTRUCTION OF OLFACTORY LANDSCAPE EVALUATION MODEL OF HANGZHOU GRAND CANAL ASIAN GAMES PARK

4.1 *Spatial syntax to calculate crowd gathering and dispersion points*

In the analysis of the integration degree, the global integration degree better indicates that there are no clear crowd gathering and dispersion points in the Asian Games Park. Under the condition of a 500 m walking radius, among which the largest spatial access potential is the walkway flyover. The whole space of Hangzhou Grand Canal Asian Games Park spreads from the middle to both sides in a cross shape. The core of the walking space with an integration degree of 500 meters is in the "Hualing Twelve Square Fashion District" (as shown in Table 1).

Table 1. Selectivity, integration of R3, R500, Rn.

Radius	integration	Selectivity
R3		
R500		
Rn		

The analysis of the degree of choice focuses on the degree of choice of 3 meters and 500 meters. In the range of 3 meters, people's walking space cores are mainly in the north and south gates of Hangzhou Umbrella Stadium, Huayuangang Station of Metro Line 10, and Starry Sky Camp. In the range of 500 meters, people are more willing to choose the west trail of Hangzhou Umbrella Stadium, the connection between the two trails, and the runway in the southwest for access. The global analysis, on the other hand, shows that urban roads are more favored (as in Table 1).

In summary, the main node locations of Hangzhou Grand Canal Asian Games Park are at the north and south gates of Hangzhou Umbrella Stadium, the Huayuangang Station of Metro Line 10, the staging area, the Starry Sky Camp, Hangzhou Sports Lottery – National Fitness Court, and the West 1 entrance and exit (Figure 1).

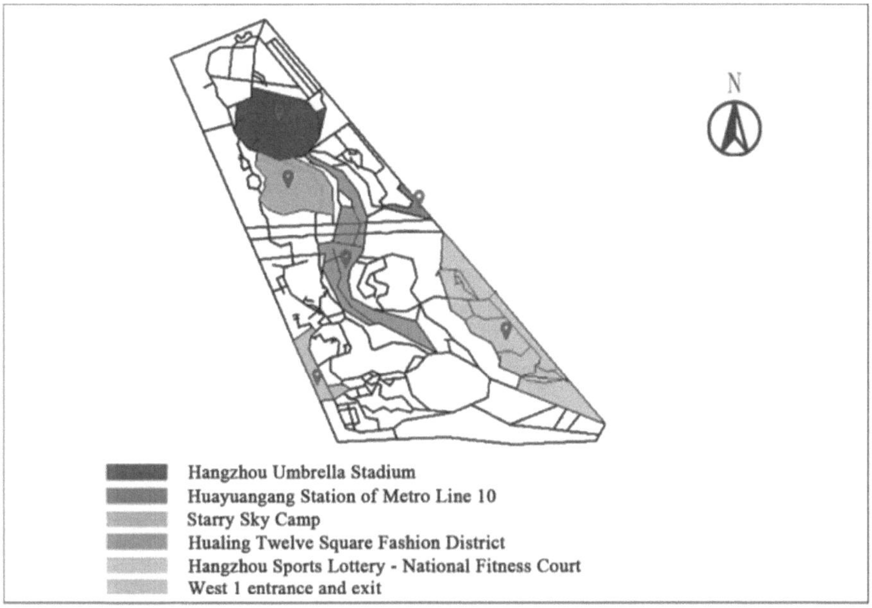

Figure 1. Crowd gathering and dispersion points of R3 in Hangzhou Grand Canal Asian Games Park.

4.2 *Construction of olfactory landscape level evaluation index system*

The olfactory landscape in the Grand Canal Asian Games Park can bring a sensory experience to visitors, and different elements bring different feelings. Therefore, it is necessary to distinguish and screen the smells in the park and find the landscapes that have positive and negative effects on visitors. The specific screening conditions are: (1) to have comprehensive coverage of the olfactory landscape of the Grand Canal Asian Games Park; (2) the indicator elements should be able to bring the olfactory sensory experience to the visitors; (3) to be universally representative of the majority of olfactory landscapes in the Grand Canal Asian Games Park. As a whole, the olfactory landscape in the Grand Canal Asian Games Park consists of natural elements and artificial elements, and the natural elements are divided into three parts: plant elements, animal elements, and natural abiotic elements. Plant elements are one of the main landscape elements in the Grand Canal Asian Games Park. The odors emitted from the branches, leaves, flowers, fruits, seeds, and other organs of plants can bring different olfactory sensations to visitors, and different types of plants such as trees, shrubs, lawns, and aquatic wetland plants bring different olfactory sensations to visitors. The scent of animal elements often attracts the attention of visitors, and the scent emitted by different types of animals can bring visitors a feeling of enjoyment or detachment. The water resources in

Table 2. Selectivity, integration, and fit of R3, R500, and Rn.

Radius	R2	
In-R500&in-Rn	0.242944	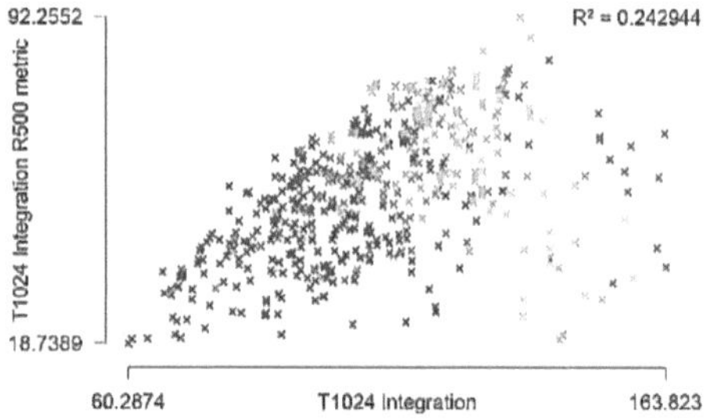
Ch-R3&in-Rn	0.175887	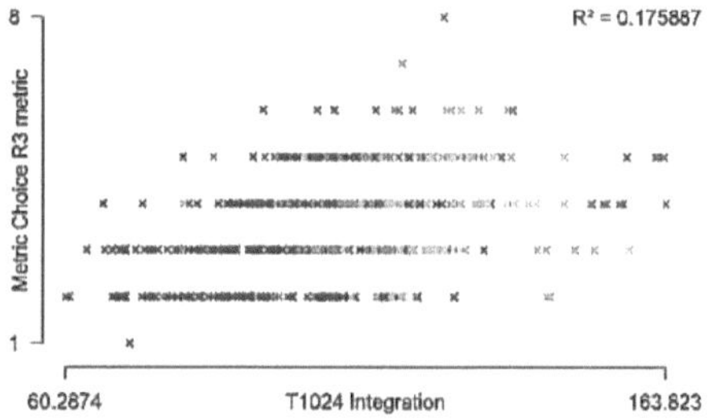
Ch-R500&in-Rn	0.0814373	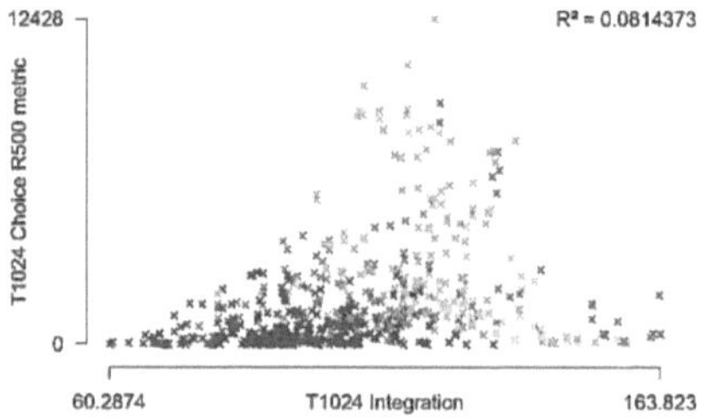
Ch-R3&ch-R500	0.0489718	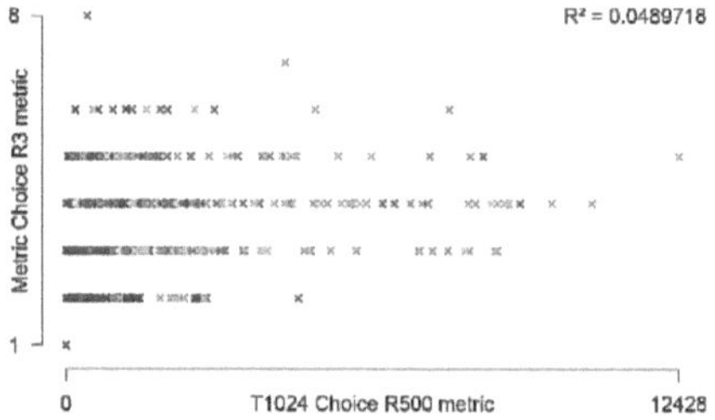

the park can attract many visitors, and the smell of buildings, plastic materials, rocks, soil, and air pollution can bring different sensory and emotional experiences to visitors. Based on the above factors, the criterion layer of the Grand Canal Asian Games Park evaluation system consists of plant elements, animal elements, natural abiotic elements, and artificial elements, and the specific index system is shown in Table 3.

Table 3. Olfactory landscape evaluation index system of Hangzhou Grand Canal Asian Games Park.

Target layer A	Criterion layer B	Indicator layer C
Evaluation of Olfactory Landscape in Hangzhou Grand Canal Asian Games Park A_1	Plant elements B_1	Odor of trees C_1 Odor of shrubs C_2 Odor of Lawns C_3 Odor of aquatic wetland plants C_4
	Animal elements B_2	Odorof terrestrial animals C_5 Odor of aquatic animals C_6 Odor of amphibians C_7
	Natural Abiotic Elements B_3	Odor of rivers C_8 Odor of artificial lakes C_9 Odor of surface flow wetlands C_{10} Odor of soils C_{11}
	Artificial elements B_4	Odor of rockery stackings C_{12} Odor of construction materials C_{13} Odor of sports trails C_{14} Odor of air pollution C_{15} Odor of garbage C_{16}

4.3 *Determining index weights*

Different index factors have different degrees of influence on the olfactory landscape of the Grand Canal Asian Games Park, so the importance degree of each evaluation factor should be judged. In this paper, we use AHP to evaluate the olfactory landscape of the Grand Canal Asian Games Park. Based on the scoring of the experts, the relative importance of the index factors is expressed by the proportional quantified value of the 1~9 points system, and then the corresponding $n \times n$ order judgment matrix is established, and the weights of each index factor are determined by solving the eigenvectors.

The relative importance of the two factors is quantified by expert scores, and the proportional scale of 1~9 points was used to measure the judgment matrix quantitative values, and their meanings are shown in Table 4.

Table 4. Importance of quantification value and meaning.

Quantified values	Factor i compared to factor j
1	Compared to the two elements, both are equally important
3	Compared to the two elements, the former is slightly more important than the latter
5	Compared to the two elements, the former is more important than the latter
7	Compared to the two elements, the former is strongly more important than the latter
9	Compared to the two elements, the former is more extremely important than the latter
2 4 6 8	Intermediate value of two adjacent judgments
aij=aji	Countdown

Normative column averaging calculates the sum of each element mi of each row of the judgment matrix a; the sum of the elements of each row of a is normalized; the vector is the requested weight vector.

5 RESULTS AND ANALYSIS

5.1 *Evaluation index system weights determination*

When the result of the consistency ratio CR=CI/RI is less than 0.1, the criterion of the consistency test is met. The CR values shown in Tables 5–9 are all less than 0.1, indicating that the consistency test results pass and the data are referable.

Table 5. Indicator weights of the criterion layer.

Criterion layer B	Eigenvector	Weighting value	Maximum characteristic root	CI	CI
Plant elements B_1	3.0801	0.5379			
Animal elements B_2	0.3102	0.0542			
Natural Abiotic Elements B_3	0.6043	0.1055	4.0080	0.0027	0.003
Artificial elements B_4	1.7321	0.3025			

Table 6. Indicator weights of the plant elements.

Indicator layer C	Eigenvector	Weighting value	Maximum characteristic root	CI	CI
Odor of trees C_1	2.7108	0.5023			
Odor of shrubs C_2	0.8409	0.1558			
Odor of Lawns C_3	1.5651	0.29	4.0104	0.0035	0.004
Odor of aquatic wetland plants C_4	0.2803	0.0519			

Table 7. Indicator weights of the animal elements.

Indicator layer C	Eigenvector	Weighting value	Maximum characteristic root	CI	CI
Odorof terrestrial animals C_5	2.5198	0.6154			
Odor of aquatic animals C_6	1.2599	0.3077	3.0000	0	0
Odor of amphibians C_7	0.315	0.0769			

Table 8. Indicator weights of the natural abiotic elements.

Indicator layer C	Eigenvector	Weighting value	Maximum characteristic root	CI	CI
Odor of rivers C_8	0.6687	0.1174			
Odor of artificial lakes C_9	1.2779	0.2242			
Odor of surface flow wetlands C_{10}	0.3433	0.0602	4.0080	0.0027	0.003
Odor of soils C_{11}	3.4087	0.5982			

Table 9. Indicator weights of the artificial elements.

Indicator layer C	Eigenvector	Weighting value	Maximum characteristic root	CI	CI
Odor of rockery stackings C_{12}	0.5365	0.0796			
Odor of construction materials C_{13}	0.274	0.0407	5.0097	0.0024	0.002
Odor of sports trails C_{14}	1.5518	0.2303			
Odor of air pollution C_{15}	2.8252	0.4192			
Odor of garbage C_{16}	1.5518	0.2303			

5.2 *Comprehensive evaluation*

The evaluation level is divided into five levels, namely V{V1(2), V2(1), V3(0), V4(-1), V5(-2)}, where -2 means very poor, -1 means poor, 0 means average, 1 means good, and 2 means very good.

Combining online and offline questionnaires to collect the evaluation of visitors and park staff on the olfactory landscape of Hangzhou Grand Canal Asian Games Park, a total of 260 questionnaires were distributed, 246 were returned, of which 227 were valid, and the questionnaire recovery rate was 94.6%, and the effective rate of the questionnaire was 87.3%.

According to the weighting values of the criterion and indicator layers, the weighting set A1 = (0.5379 0.0542 0.1055 0.3025) for the criterion layer, A2 = (0.5023 0.1558 0.2900 0.0519) for the plant elements, A3 = (0.6154 0.3077 0.0769) for the animal elements, and A4 = for the natural abiotic elements (0.1174 0.2242 0.0602 0.5982), and artificial element weight set A5 = (0.0796 0.0407 0.2303 0.4192 0.2303).

Table 10. Importance of quantification value and meaning.

Criterion layer	Indicator layer	Number of people evaluated					Number of people evaluated as a percentage of the total number of people				
		−2	−1	0	1	2	−2	−1	0	1	2
B_1	C_1	12	23	58	87	47	0.0529	0.1013	0.2555	0.3832	0.2070
	C_2	5	17	64	79	62	0.0220	0.0749	0.2819	0.3480	0.2731
	C_3	11	11	88	61	56	0.0485	0.0484	0.3876	0.3876	0.2466
	C_4	8	58	20	83	58	0.0352	0.2555	0.0881	0.0881	0.2555
B_2	C_5	11	13	41	80	82	0.0484	0.0573	0.0572	0.1806	0.3612
	C_6	25	14	34	76	78	0.1101	0.0616	0.1497	0.3348	0.3436
	C_7	11	20	75	45	76	0.0484	0.0881	0.3303	0.3303	0.3348
B_3	C_8	29	6	35	91	66	0.1277	0.0264	0.1541	0.4008	0.2907
	C_9	30	12	76	33	76	0.1321	0.0528	0.3348	0.1453	0.3348
	C_{10}	13	26	73	73	36	0.0572	0.1145	0.3215	0.3215	0.1585
	C_{11}	27	15	74	37	74	0.0044	0.0660	0.3259	0.1629	0.3259
B_4	C_{12}	9	19	37	78	84	0.0396	0.0837	0.1629	0.3436	0.3700
	C_{13}	12	29	72	39	75	0.0528	0.1277	0.3171	0.1718	0.3303
	C_{14}	5	25	38	77	82	0.0220	0.1674	0.1674	0.3392	0.3612
	C_{15}	22	14	54	57	80	0.0969	0.0616	0.2378	0.2511	0.3524
	C_{16}	15	28	74	43	67	0.0660	01233	0.3259	0.1894	0.2951

$$R1 = \begin{pmatrix} 0.0529 & 0.1013 & 0.2555 & 0.3832 & 0.2070 \\ 0.0220 & 0.0749 & 0.2819 & 0.3480 & 0.2731 \\ 0.0485 & 0.0484 & 0.3876 & 0.3876 & 0.2466 \\ 0.0352 & 0.2555 & 0.0881 & 0.0881 & 0.2555 \end{pmatrix}$$

$$R2=\begin{pmatrix} 0.0484 & 0.0573 & 0.0572 & 0.1806 & 0.3612 \\ 0.1101 & 0.0616 & 0.1497 & 0.3348 & 0.3436 \\ 0.0484 & 0.0881 & 0.3303 & 03303 & 0.3348 \end{pmatrix}$$

$$R3=\begin{pmatrix} 0.1277 & 0.0264 & 0.1541 & 0.4008 & 0.2907 \\ 0.1321 & 0.0528 & 0.3348 & 0.1453 & 0.3348 \\ 0.0572 & 0.1145 & 0.3215 & 0.3215 & 0.1585 \\ 0.0044 & 0.0660 & 0.3259 & 0.1629 & 0.3259 \end{pmatrix}$$

$$R4=\begin{pmatrix} 0.0396 & 0.0837 & 0.1629 & 0.3436 & 0.3700 \\ 0.0528 & 0.1277 & 0.3171 & 0.1718 & 0.3303 \\ 0.0220 & 0.1674 & 0.1674 & 0.3392 & 0.3612 \\ 0.0969 & 0.0616 & 0.2378 & 0.2511 & 0.3524 \\ 0.0660 & 0.1233 & 0.3259 & 0.1894 & 0.2951 \end{pmatrix}$$

The first level integrated rubric was constructed and the results were as follows: plant elements: B1=A1×R1= (0.0897 0.1723 0.3602 0.3378), animal elements: B2=A2×R2=(0.3020 0.3448 0.0741), natural abiotic elements: B3=A3×R3=(0.0491 0.0794 0.5524 0.1684 0.1507), and artificial elements: B4=A4×R4=(0.5591 0.6974 0.3524 0.1884 0.2007).

Then, the second level of integrated evaluation is: B=A×R=A×[B1, B2, B3, B4, B5]T=(0.0465 0.0966 0.2267 0.3524 0.2766).

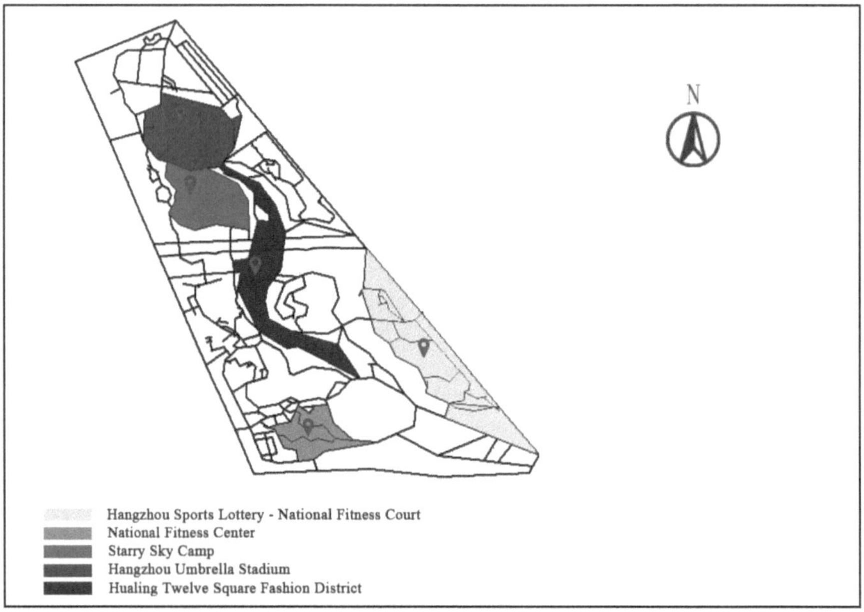

Figure 2. Degree of olfactory landscape preference in Hangzhou Grand Canal Asian Games Park.

5.3 *Analysis*

(1) As shown in Table 2, where the highest fitted values are for 500 meters, and the global integration degree, which represents the accessibility of the park to one-fifth of the whole, and the accessibility of the rest of the space is not so strong. The 500m and 3m selectivity, which are the most relevant to the study, have the second lowest fit values, and pedestrians may ignore the sense of smell while walking.

(2) From Tables 5–9, it can be seen that the weight of plant elements is 0.5379, which has an important position in the olfactory landscape of the Grand Canal Asian Games Park with more than half of the weight. The rest of the elements have the same weight value, among which the animal elements have the lowest weight value of 0.0542, indicating that it is the least important. Among the plant elements, the odor of trees has the highest weight value of 0.5023, indicating that the odor of trees occupies the most important position among the plant elements. In contrast, the odor of aquatic wetland plants has the lowest weight value and is more different from other elements. From the overall park condition, the aquatic wetland plants occupy a smaller area, which is consistent with the situation. Among the animal elements, the odor weight value of terrestrial animals was the highest at 0.6154 and the lowest at 0.0769 for amphibians, which may be since the number of terrestrial animals in the Grand Canal Asian Games Park is relatively high among the animal elements. In contrast, the number of amphibians is relatively low. Among the natural abiotic elements, the odor of soil has the highest weight value of 0.5982, the odor of rivers and artificial lakes has the average weight value, and the lowest weight value of surface flow wetlands is 0.0602, which is probably due to the small area of surface flow wetlands in the whole park. Among the artificial elements, air pollution odor, garbage odor, and sports trail odor, which may bring a negative olfactory sensation to visitors, have relatively high weighting values. In contrast, the construction materials' odor and rockery stacking odor have low weighting values, less influencing the whole park's olfactory landscape.
(3) According to the principle of maximum affiliation (Table 10), the overall olfactory landscape of Hangzhou Grand Canal Asian Games Park was evaluated as V2(1), i.e., "good."

6 CONCLUSIONS AND SUGGESTIONS

6.1 *Conclusion*

In this paper, based on the existing perceptual research related to olfactory landscape, field research was conducted in the Grand Canal Asian Games Park, and the crowd's main gathering and dispersal points were calculated using the spatial syntax. Then, based on the existing conditions of the Grand Canal Asian Games Park, a comprehensive olfactory landscape evaluation system was constructed with natural and artificial elements as the main indicators, and a questionnaire survey and AHP were used to evaluate the olfactory perception of the park. The comparison of the olfactory landscape preference of Hangzhou Grand Canal Asian Games Park and the crowd gathering and dispersion points show that the olfactory landscape nodes and the easy-to-reach points of the crowd in radius R3 and R500 basically coincide, and the paths walked by people basically overlap. Therefore, the overall evaluation of the olfactory landscape of the Grand Canal Asian Games Park is "good." However, the olfactory landscape points and spatial points are mainly concentrated in the northwest of the park, there is no concentrated olfactory landscape around the National Gymnasium, and there is a shift in the center of gravity and only a single center.

6.2 *Suggestions*

(1) Fully consider the distribution of olfactory landscape. The olfactory landscape should be designed according to the spatial division scale of the site and according to the limit distance of human olfaction (R3 m) and walking distance (R500 m), so that visitors can experience the enjoyment of the olfactory landscape in the park in all aspects and play a uniform and reasonable distribution effect.
(2) Build a systematic olfactory landscape in the park. In the design of an olfactory landscape, olfactory landscape points should be reasonably allocated according to the tour route; different landscape environments should be set up with different olfactory landscapes, using olfaction to reflect the functions of different environments and play the effect of moving to different

scenes. In the whole park, build a reasonable and rich olfactory landscape system with a sense of experience and hierarchy.

(3) Focus on the negative olfactory sensation in the landscape. From the evaluation results, it is clear that visitors pay much attention to the smell which affects the olfactory experience. In constructing an olfactory landscape, we should not only pay attention to the creation of a landscape that is beneficial to the olfactory experience, but also pay attention to avoiding bringing unpleasant smells to tourists, especially the nonsubjective artificial smell brought by the environment.

REFERENCES

Ba, M H. (2021). Research on audio-olfactory interaction in urban public open spaces. D. *Harbin Institute of Technology*. 000357.

Badach, J., Kolasinska, P., Paciorek, M. & et al. (2018). A case study of odour nuisance evaluation in the context of integrated urban planning. *J. Journal of Environmental Management.* (213): 417–424.

Bouchard, N. (2013). Le théâtre de la mémoireolfactive: Le pouvoir des odeurs à modeler notre perception spatiotemporelle de l'environnement. D. *University of Montreal.*

Chen, Y Q. (2020). Study of perception evaluation and planning strategy of Urban built environment based on smellmap – A case study of urban district in Chongqing. D. *Chongqing University.* 004044.

Chen, Y W. & Yuan, X M. (2017). Research progress on smellscape. *J. Chinese Landscape Architecture.* 33(02):107–112.

Diaconu, M., Heuberger, E., Mateus-Berr, R. et al. (2011). Senses and the City: An Interdisciplinary Approach to Urban Sensescapes. M. *LIT Verlag.*

Gao, X. & Yao, L. (2011). Preliminary study on the combinations of specific aromatic plants for hypotensive healthcare. *J. Chinese Landscape Architecture.* 27(04):37–38.

Henshaw, V. (2014). Urban smellscapes: Understanding and designing city smell environments. M. London: Routledge. 198.

Li, C Z., Pan, Y., Zhou, W., Pan H. & Lan, S R. (2022). Research on tourists' involvement, narration satisfaction, environmental education evaluation of urban park—A case of Fuzhou Jin'an Park. *J. Chinese Landscape Architecture.* 38(02):104–109.

Li, J G., Yang J., Zhou, J Y., Liu, J. & Yang, L. (2019). Mapping and localization of multiple odor sources with evidence theory in outdoor environments. *J. Robot.* 41(06):771–778+787.

Li, T., Wu, F W. & Li, L H. (2021). Research on quality optimization of public space in the historic district from an olfactory perspective. C. Spatial Governance for High-quality Development—*Proceedings of 2020 China Annual Conference on Urban Planning (2009 Urban Cultural and Genetic Conservation)*: 942–953.

Liu, D Z. (1956). An overview of Chinese housing. *J. Architectural Journal.* (04), 1–53.

Ma, X M., Zhang, Z B., Dong, J H., Gao, F. & Li, R H. (2021). Spatial distribution and supply & demand matching of parks—A case study of lanzhou city. *J. Journal of northwest Forestry university.* 36(06):289–296.

Rodaway, P. (2002). *Sensuous Geographies: Body, Sense, and Place.* M. New York: Routledge. 80.14.

Tuan, Y F. (1977). *Space and place: The Perspective of Experience.* M. Minneapolis: University of Minnesota Press.

Xiao, J L., Feng, H C. & Xie, H. (2021). From "smell" to "smellscape": a systematic review of smellscape research and design methods. *J. Journal of Human Settlements in West China.* 36(5): 7–14.

Yin, M., Yang, Z Y., Li, G Z. & Xue, J G. (2016). A tentative study on smell design in urban public space. *J. Urban Design.* 40(03):58–62.

Zhang, J J. (2019). The effects of ambient odors on pro-environmental behavior. D. *Beijing Forestry University.* 000374.

Zhang, M Y. (2019). The types and methods of applying olfaction to emotional design. *J. Design.* 32(15):72–75.

Advances in Civil Engineering and Environmental Engineering – bin Yahya & Bin Daud (Eds)
© 2023 The Author(s), ISBN: 978-1-032-39397-1

Research on municipal engineering construction based on aerial survey data

Yuhui Liu*, Xianzheng Fan, Jingxin Xu, Mingming Shuai & Yi Qin
China Railway Shanghai Design Institute Group Corporation Limited, Shanghai, China

ABSTRACT: In order to objectively and accurately evaluate the present situation and level of municipal engineering construction in the Guangdong-Hong Kong-Macao Greater Bay Area, and explore the direction of future municipal engineering construction according to the evaluation results. In this paper, the evaluation index system of municipal engineering construction in the Guangdong-Hong Kong-Macao Greater Bay Area is constructed by combining aerial survey data and urban attribute information data. The index system dynamically evaluates and analyzes the municipal engineering construction of 11 cities in the Guangdong-Hong Kong-Macao Greater Bay Area. The results show that:1) the overall level of municipal engineering construction in the Greater Bay Area is above the medium level, among which Hong Kong, Macao, Guangzhou, and Shenzhen rank high. Other cities might consider borrowing its ideas for municipal engineering. 2) The construction of a municipal civilization project is a first-level core index affecting municipal engineering construction. Per capita water resources, urbanization rate, and centralized treatment rate of urban domestic sewage are the second-level core indicators affecting port competitiveness.

1 INTRODUCTION

In recent years, with the continuous promotion of current construction in China, the number of municipal engineering projects has been increasing, and the role of municipal engineering in the city construction process has gradually become prominent. The construction of municipal works affects the quality of people's lives and the appearance of the city. These projects are not only significant for urban reform, development and construction (Feng 2022). With the continuous advancement of municipal engineering construction at home and abroad, scholars from all walks of life have done a lot of research on the evaluation of regional municipal engineering construction. The core objective of municipal engineering project management is to ensure the smooth implementation of the construction project, reduce the project construction cost, shorten the project construction period, ensure the quality and safety of the project, and meet the requirements of the relevant stakeholders (Zhao 2019). Energy saving, green and environmental protection technologies should be adopted in the construction of municipal projects to improve the quality of project construction, energy conservation, and environmental protection (Li 2020). At the same time, vigorously promoting the Internet of things is conducive to the development of municipal traffic engineering construction (strong 2019). Due to the characteristics of municipal engineering projects, the comprehensive evaluation system of municipal engineering has some technical difficulties in constructing an evaluation model, selecting an evaluation index, and determining the weight of the evaluation index. Among them, scientific and reasonable determination of index weight is important to determining whether the evaluation result is accurate (Zhang 2015). This paper summarizes the research status of municipal engineering construction and its evaluation index system at home and abroad, analyzes

*Corresponding Author: liuyuhui@sty.sh.cn

DOI 10.1201/9781003349563-47

the bottleneck problems and future development direction of municipal engineering construction, and builds a set of evaluation index systems suitable for municipal engineering construction based on aerial survey data and using analytic hierarchy process and entropy weight method. Taking 11 cities in the Guangdong-Hong Kong-Macao Greater Bay Area as examples, the evaluation index system is studied in order to provide reasonable references and suggestions for municipal engineering construction.

2 DATA AND METHODS

2.1 *Data*

The sample data of this study are obtained from UAV aerial survey data, 2019–2020 statistical yearbook, National Economic and Social Development Statistical Bulletin, 2019–2020 Water resources Bulletin of each city and special administrative region, as well as relevant information provided by the official websites of each city and special administrative region statistics bureau and water bureau.

2.2 *Methods*

2.2.1 *Evaluation index system construction:*

Based on a large amount of literature on the evaluation of municipal engineering construction, following the basic principles of accurate connotation description, comprehensive quantitative presentation, and strong universality of indicators, an evaluation index system for the level of municipal engineering construction in the Guangdong-Hong Kong-Macao Greater Bay Area was constructed. There are four first-level indicators (A1-A4) and 12 second-level indicators (B1-B12), including municipal economic engineering construction, municipal social engineering construction, municipal political engineering construction, and municipal environmental engineering construction, as shown in Table 1.

Table 1. Evaluation index system of municipal engineering construction level.

First-level indicators	Second-level indicators	Unit
Municipal economic engineering construction	B1 Per capita GDP	%
	B2 Urbanization rate	%
	B3 Engel coefficient	%
Municipal social engineering construction	B4 Urban medical insurance participation rate	%
	B5 Number of buses per 10,000 people	vehicle
Municipal political engineering construction	B6 Proportion of ecological townships above the provincial level	%
	B7 Number of companies audited for cleaner production	individual
	B8 Number of users connected to household environmental protection equipment	household
Municipal environmental engineering construction	B9 Water resources per capita	ton/p
	B10 Excellent rate of ambient air quality	%
	B11 Centralized treatment rate of urban domestic sewage	%
	B12 Coverage rate of noise standard area	%

2.2.2 *Evaluation index weight calculation:*

This paper uses the entropy method to determine the weight of the municipal engineering construction evaluation index. The entropy weight method is an objective method to determine the weight of indicators (Zhang 2010). The specific operation process is as follows:

(1) Data normalization processing: Since different evaluation indicators often have different dimensions and dimensional units, in order to eliminate the dimensional influence between the

indicators and unify the standards, it is necessary to normalize the data so that all indicators are in the same order of magnitude, which is suitable for comprehensive comparative evaluation. Therefore, for a research field, the normalization method of the initial matrix of p research indicators is as follows

$$X=(x_{ij})_{o\times p} \quad i=1,2,\ldots,o;\ j=1,2,\ldots,p \tag{1}$$

$$Y=(y_{ij})_{o\times p} \quad i=1,2,\ldots,o; j=1,2,\ldots,p \tag{2}$$

$$y_{ij}=\frac{\max(x_j)-x_{ij}}{\max(x_j)-\min(x_j)} \tag{3}$$

In the formula: X represents the initial data matrix, and Y represents the normalized benefit data matrix.

(2) Find the information entropy of each evaluation index: Generally speaking, the smaller the information entropy E_j of an index, the greater the degree of variation of the index value, the more information it provides, and the greater its role in a comprehensive evaluation. The greater the weight, the greater the weight. According to the definition of information entropy, the information entropy of a set of data

$$E_j=-\frac{1}{\ln m}\sum_{i=1}^{m}p_{ij}\ln p_{ij} \tag{4}$$

In this formula

$$p_{ij}=\frac{y_{ij}}{\sum_{i=1}^{m}y_{ij}} \tag{5}$$

In formulas (2) and (3), p_{ij} represents the proportion of the i-th city under the j-th index, and E_j represents the information entropy of the j-th index.

(3) Determine the weight of each index: according to the calculation formula of information entropy, calculate the information entropy of each index as E_1, $E_2 \ldots E_j$, and calculate the weight of each index through information entropy

$$W_j=\frac{1-E_j}{\sum_{j=1}^{n}(1-E_j)} \tag{6}$$

Among them, E_j represents the information entropy of the j-th index, and W_j represents the weight of the j-th index.

2.2.3 *Calculation of the level of municipal engineering construction*

(1) Construct a judgment matrix A reflecting the relationship between the two indicators. Among them, $A=(a_{ij})_{n*n}$. A represents the relative importance of all indicators of this level relative to the indicators of the previous level (the relative importance is determined according to the weight obtained by the above entropy method).

(2) Calculate the weight ranking of each city under a single index using the single-level ranking. While calculating the weight, the maximum characteristic root value (CI) can be obtained. The formula is

$$M=(m_i)=\frac{x(j,d)}{\sum_{j=1}^{n}(1-E_j)} \quad i,j=1,2,\ldots,n \tag{7}$$

$$CI=\frac{\lambda_{\max(p)}-n}{n-1} \tag{8}$$

Where M represents the weight of each port under a single index, X represents the eigenvector matrix of judgment matrix A, D represents the maximum eigenvalue column, $\lambda_{\max(p)}$ represents the maximum eigenvalue of judgment matrix A, and n represents the number of paired comparison indexes.

(4) The indicators are divided into five grades at equal intervals. The grade set W = {w1, w2, w3, w4, w5} = {Excellent, Good, Medium, Bad, Terrible} = {5, 4, 3, 2, 1}, using the weighted average M (., +) fuzzy operator to determine the level of municipal engineering construction.

3 TEST RESULTS AND DISCUSSIONS

First, we calculate the scores of each indicator of the 11 ports in the status evaluation system based on the calculation of statistical data and related formulas, and then calculate the objective weight value of each indicator through the entropy method and its formulas (1)-(6), as shown in Table 2.

Table 2. Evaluation index weight value.

First-level indicators	Weight	Second-level indicators	Weight	Comprehensive weight
Municipal economic engineering construction	0.234	Per capita GDP	0.329	0.077
		Urbanization rate	0.529	0.124
		Engel coefficient	0.142	0.087
Municipal social engineering construction	0.147	Urban medical insurance participation rate	0.502	0.074
		Number of buses per 10,000 people	0.498	0.073
		Proportion of ecological townships above the provincial level	0.461	0.033
Municipal political engineering construction	0.189	Number of companies audited for cleaner production	0.189	0.036
		Number of users connected to household environmental protection equipment	0.350	0.066
		Water resources per capita	0.396	0.169
Municipal environmental engineering construction	0.43	Excellent rate of ambient air quality	0.261	0.112
		Centralized treatment rate of urban domestic sewage	0.264	0.115
		Coverage rate of noise standard area	0.079	0.034

According to the entropy weight in Table 3, the evaluation levels of secondary indicators and primary indicators are calculated, and finally, the development potential level of each port is calculated.

The evaluation results of the municipal engineering construction level are shown in Table 9. The rankings are as follows: Hong Kong, Shenzhen, Guangzhou, Macau, Dongguan, Zhuhai, Foshan, Jiangmen, Zhongshan, Huizhou, and Zhaoqing.

In terms of municipal economic engineering construction, the ranking order is Hong Kong, Guangzhou, Shenzhen, Macao, Dongguan, Zhuhai, Zhongshan, Jiangmen, Huizhou, Foshan, and Zhaoqing. The weight of each index of municipal economic engineering construction is as follows: urbanization rate (0.529) > per capita GDP(0.329) > Engel coefficient (0.142). It can be seen that in order to improve the ecological, economic, and civilization construction of the port, on the basis of ensuring the material wealth of citizens, it is necessary to give priority to increasing the urbanization rate, increasing the per capita GDP, and stimulating other consumption modes to reduce the Engel coefficient.

From the perspective of municipal social engineering construction, the ranking order is: Hong Kong, Guangzhou, Shenzhen, Zhuhai, Jiangmen, Zhaoqing, Dongguan, Foshan, Zhongshan,

Table 3. Overall evaluation result of the level of municipal engineering construction.

City	Evaluation score	Evaluation level
Hongkong	4.43	Excellent
Macao	3.43	Good
Guangzhou City	4.31	Excellent
Shenzhen City	4.41	Excellent
Zhuhai City	3.31	Good
Foshan City	3.28	Good
Huizhou City	2.59	Medium
Dongguan City	3.35	Good
Zhongshan City	2.71	Medium
Jiangmen City	2.77	Medium
Zhaoqing City	2.48	Medium

Huizhou, and Macao. The weighting order of the municipal social engineering construction index is: urban medical insurance coverage rate (0.502)> the number of buses per 10,000 people (0.498). From the weight ranking, the difference between the two is not big. Therefore, if a city wants to improve the construction of municipal social engineering, it should not only strengthen the coverage of urban medical insurance coverage, but also plan the city's bus routes according to its own geographical conditions, so that the bus can serve more citizens.

From the perspective of municipal political engineering construction, the ranking is Hong Kong, Guangzhou, Shenzhen, Macao, Foshan, Dongguan, Zhuhai, Huizhou, Jiangmen, Zhaoqing, and Zhongshan. The order of weight of each indicator of municipal political construction was as follows: proportion of eco-villages and towns above province, city, and special administrative region (0.461)> the number of enterprises audited to implement cleaner production (0.350)> the number of users connected to household environmental protection equipment (0.189). Therefore, in improving the municipal political project, the city should give priority to developing the construction of villages and towns and promote the increase of the proportion of villages and towns. Cities also need to encourage the construction and development of cleaner production enterprises, and give certain policy and economic support when necessary.

From the perspective of municipal environmental protection project construction, the ranking order is: Macao, Hong Kong, Guangzhou, Shenzhen, Foshan, Dongguan, Dongguan, Zhuhai, Zhongshan, Huizhou, Zhaoqing, and Jiangmen. The order of weight of each indicator of ecological and environmental civilization was water resources per capita (0.396), centralized treatment rate of urban domestic sewage (0.264), excellent and good ambient air quality rate (0.261), and coverage rate of noise standard area (0.079). The ranking shows that municipal environmental protection projects should give priority to ensuring the stability of per capita water resources, improving the efficiency of domestic sewage treatment through policies and government investment, improving air quality by planting trees and reducing exhaust emissions, controlling the generation of noise, and eliminating or weakening the impact of noise as much as possible.

4 CONCLUSIONS

Based on the background of municipal engineering construction in Guangdong-Hong Kong-Macao Greater Bay Area, this paper uses AHP and entropy method to construct a comprehensive evaluation system from four aspects: municipal economic engineering construction, municipal social engineering construction, municipal political engineering construction, and municipal environmental engineering construction.

According to the ranking of the comprehensive evaluation system, the overall level of municipal engineering construction in the Guangdong-Hong Kong-Macao Greater Bay Area is at the upper middle level. The specific ranking is Hong Kong (excellent), Shenzhen (excellent), Guangzhou (excellent), Macao (good), Dongguan (good), Zhuhai (good), Foshan (good), Jiangmen (middle), Zhongshan (middle), Huizhou (middle), Zhaoqing (middle). From the perspective of single indicators, the influence of the four first-level indicators on the level of municipal engineering construction is ranked as follows: municipal environmental engineering construction, municipal economic engineering construction, municipal political engineering construction, and municipal social construction. Thus, municipal environmental engineering construction is the most important municipal engineering construction. Paying attention to environmental engineering construction is the most cost-effective choice in municipal engineering construction. Among them, the rate of urbanization is the main factor in municipal economic construction level, the urban medical insurance rate is the main factor in municipal social engineering construction level, and the proportion more than provinces, municipalities, and special administrative region ecological villages and towns is the main factor to municipal political construction level, and per capita water resources are the main factor to environmental engineering construction. Among the 14 second-level indicators affected by municipal engineering construction, the top three are per capita water resources, urbanization rate, and centralized treatment rate of urban domestic sewage. If the city performs well on these three indicators, then the level of municipal engineering construction will also be high.

ACKNOWLEDGEMENT

This paper is based on the fund "Research on the application of UAV tilt photography in the municipal survey of railway involved." Contract number: "Cooperation No.20-21".

REFERENCES

Feng Zhaoying. Discussion on strengthening construction management of municipal engineering and improving quality of municipal engineering [J]. *Construction of China*, 2022 (06): 22–24.

Li Yejian. Energy saving and green environmental protection technology in municipal engineering construction [J]. *Low-carbon World*, 2020,10 (05): 18+20.

Qiang Baixiang, Hu Tianli, Zhang Jiong. Construction status and development of municipal engineering Internet of things [J]. *Journal of Internet of Things*, 2019, 3(01): 106–110.

Zhang Sui, Zhang Mei, Chi Guotai. The science and technology evaluation model based on entropy weight and empirical research [J]. *Chinese Journal of Management*, 2010, 7(01): 34–42.

Zhang Xinliang. Research on comprehensive evaluation system of municipal engineering construction management [J]. *Jiangxi Building Materials*, 2015(17): 298.

Zhao Qi. Construction project management and analysis of municipal engineering [J]. *Shandong Industrial Technology*, 2019(18): 105.

Advances in Civil Engineering and Environmental Engineering – bin Yahya & Bin Daud (Eds)
© 2023 The Author(s), ISBN: 978-1-032-39397-1

The hazard of earthquake and solution directions

Linfei Li*
Xi'an Gaoxin No.1 High School, Xi'an, China

ABSTRACT: Seismicity is a complex natural phenomenon that has not been able to accurately predict the time, place, and intensity of earthquake, let alone prevent it from happening. The prevention and control of earthquake disaster risk is a very comprehensive work, involving basic investigation and detection, scientific understanding and modeling in various evaluation links, seamless connection between various investigation, evaluation and evaluation products, and relevant governance measures. It involves a wide range of disciplines such as geoscience, engineering science, management science, social behavior, which requires scientific research institutions, professional departments, and governments at all levels. Only when social organizations form a joint force, they can produce effective results in reducing losses caused by earthquake disasters. In this academic paper, some kinds of earthquakes, as well as many hazards of earthquakes, are discussed. It also focuses on some effective measures and solutions for earthquake prevention. These measures can effectively alleviate the damage caused by earthquakes in the future.

1 INTRODUCTION

1.1 *Various kinds of seismic belts*

Several countries and regions worldwide are located on three of the biggest seismic belts, which are plagued by earthquakes frequently. They are the circum-Pacific seismic belt, Eurasian seismic belt, and ocean ridge seismic belt, as shown in Table 1. They are distributed on the junctions of different continental plates. The epicenters are dense within the seismic zone and scattered outside. For example, Japan is located in the Pacific Rim seismic belt. China's Sichuan belongs to the Tibetan Plateau seismic belt, a branch of the Eurasian continental seismic belt. The seismic ridge belt is distributed in the global ocean ridge axis and is relatively weak. Earthquakes in the first two earthquake zones often bring great disasters and losses to human beings. In addition to the direct impact of human casualties and economic losses, earthquakes can also cause water pollution, soil pollution, air pollution, and other environmental problems, as well as soil erosion, farmland degradation, habitat loss of animals, biodiversity degradation, and other ecological disasters. There are about 5 million earthquakes worldwide every year, of which 15% occur on land, and 85% occur in the ocean. Most earthquake disasters are caused by land earthquakes (Li 2009). The main cause of ground shaking (that is, earthquakes) is that the earth's plates squeeze and collide, resulting in plate edges, internal dislocation, and rupture (Aleotti & Chowdhury 2009).

1.2 *Damages caused by earthquake*

An earthquake is a sudden natural disaster that can cause all kinds of damage. A disaster chain can be induced by various secondary disasters caused by environmental pollution. In 2011, Japan experienced a devastating earthquake, and the tsunami caused more than 187,000 houses to be

*Corresponding Author: 15000640150@xs.hnit.edu.cn

DOI 10.1201/9781003349563-48

Table 1. Three major earthquake zones in the world.

Seismic belts	Involved countries	Percentage of earthquakes (%)
Circum-Pacific seismic belt	United States, Canada, Mexico, Columbia, Chile, New Zealand, Fiji, Indonesia, Philippines, China, Japan,	80
Eurasian seismic belt	Indonesia, China, India, Pakistan, Nepal, Afghanistan, Iran, Turkey	15
Ocean ridge seismic belt	Russia, Iceland	5

destroyed. The debris from these houses and discarded cars formed about 23 million tons of solid waste. Huge amounts of waste will take several years to clean up. Earthquakes also bring ecological disasters. Take the Wenchuan earthquake in 2008 as an example. Wenchuan earthquake epicenter is located in Chengdu and Qionglai mountains which have rich ecological resources, and the earthquake caused a serious influence on the ecosystem environment (Lan 2011). It destroyed forests, grassland, and cropland, and damaged living environment of wild animals. Wenchuan earthquake-induced landslides during the collapse of the quake lake and other secondary disasters in Sichuan, Shanxi, and Gansu caused at least 1221 ecological changes.

Strong destructiveness and instantaneous suddenness are the prominent characteristics of earthquake disasters. The social and environmental problems caused by the Wenchuan earthquake not only include a large number of casualties and property losses but also include great psychological damage. Among them, the casualties and property losses that people can directly observe with the naked eye are easy to get social attention, while the psychological damage that is not intuitive after the earthquake can easily be ignored by people (Song et al. 2012). Earthquakes can have some adverse effects on psychological behavior. These effects are from bad stress responses and stress disorders. These effects can replicate in emotional, cognitive, physical, and behavioral disorders. If these symptoms can not be effectively treated or alleviated, they will produce post-disaster syndrome. It will have a long-term and lasting impact on the physical and mental health of people in these disasters.

1.3 *Aim of this paper*

My country is one of the countries seriously threatened by geological disasters. Hutchinson concluded that China loses 1–2% of its GDP to geological disasters every year (Xu 2009). In this paper, several hazards caused by earthquakes will be given. Some measurements and specific theories will be mentioned to get out of these troubles. Measures will be divided into three main parts according to different stages of the earthquake: planning to select the size, design construction, operating maintenance, and emergency preparedness. The method of a combination of systematic natural recovery and manual intervention is adopted, with natural recovery as the main method and manual intervention as the auxiliary method (Guo et al. 1999). Clary believes the disaster management process has four stages: predisaster prevention, preparation, disaster response, and postdisaster recovery and reconstruction (Zhang 2022).

2 SITE SELECTION OF CONSTRUCTION

2.1 *Selecting site scientifically*

It is essential to make site selection of construction scientifically. The risk of an earthquake is greatly reduced if people live in a safer area before an earthquake. So, builders need to make sense of the site before construction to prevent future problems. Specific actions can be taken, such as reducing carrier exposure. It is possible to collect data on carrier exposure and then use these data to solve the problem.

2.2 *Reducing exposure*

Based on the house building census results, people can determine the distribution of exposure degree of earthquake-bearing bodies, and provide the information service platform of national ground motion parameter zoning to provide data query service for housing and construction departments.

The national seismic parameter zoning information service platform provides data query services for housing and construction departments. For instance, 12 million historical seismic fortification queries were completed during the nationwide pilot period, which supported the nationwide housing and building census (Zhou et al. 2009). Real landforms will be combined with datasets. The basin is inhospitable, and there are plate crossings. On the contrary, people should build more residential buildings on the plain.

3 DESIGN CONSTRUCTION

3.1 *Reasonable seismic fortification*

In the process of designing and constructing buildings, we need to use some prevention and control measures, such as reasonable earthquake resistance, to reduce the vulnerability of buildings. The exposure data of the disaster-bearing body have just been solved in the previous section, but there are still some crucial factors, such as vulnerability. Vulnerability has always been the responsibility of the seismological department, which has traditionally been a strength. This census work to collect before work, and then will write the national seismic active fault survey database, national field drilling database, update the evaluation model of a major part of seismic risk evaluation.

3.2 *Reduce vulnerability*

Establish the national macro site classification figure, given the national macro site classification of the surface seismic hazard map. The seismic risk index is the peak acceleration of ground motion and response spectrum value. Considering the support of earthquake disasters and comprehensive risk assessment, the seismic risk map of four probability levels is given.

4 OPERATING MAINTENANCE

4.1 *Retrofitting and reinforcing houses*

The operation and maintenance also need to reinforce and improve seismic facilities. For example, before building, the foundation is alternately stacked with wood and earth to reduce the magnitude of the earthquake. In the aftermath of an earthquake, new technologies can be applied to enhance emergency equipment, facilitate earthquake-resistant rail and emergency transportation, and upgrade and protect sewage systems with new materials. These technologies are applied to prevention and control measures. These are effective measures to prevent and control earthquake disaster risks in some cities.

4.2 *Reducing vulnerability*

We must focus on resilience, which means that we improve the resilience of our systems against external shocks. Only by strengthening urban construction in different aspects, such as economic development, cultural development, transportation development, and so on, the city can be better equipped to cope with these natural disasters. The fundamental purpose of risk assessment is to identify the earthquake disaster risk base, identify the key hidden dangers and high-risk areas of the earthquake disaster, and provide basic support for risk prevention and control. To this end, this census will also carry out earthquake disaster risk prevention zoning. The preliminary planning has two categories of products: First is the national scale building reinforcement priority zoning map. Considering building seismic fortification status and hazard identification and risk assessment results, and the condition of scale seismic risk a decade or more, the county-level administrative region is used to reinforce the urgency and priority, guiding the work plan for the subsequent seismic strengthening of national housing. The second is to carry out seismic active fault avoidance zoning.

5 EMERGENCY PREPAREDNESS

5.1 *Emergency preparedness*

Postdisaster reconstruction is a long-term and complex systematic project, which cannot be completed overnight. Therefore, comprehensive investigation and planning are needed. Postdisaster relief is particularly important in earthquake-stricken areas. It is an indispensable part of the social assistance system and plays a vital role in social stability after a disaster. Earthquake safety is not only an important aspect of people's safety needs but also an important part of people's better life needs. In the past decade, earthquakes have occurred frequently in China. The Wenchuan earthquake, Ya'an earthquake, and other earthquakes have seriously affected people's daily life and the national economy's healthy development. Geological disasters seriously restrict the development of the national economy, threaten the safety of people's life and property, and become an important factor restricting the sustainable development of China's social economy. Data show that China is one of the countries with the most active seismic activity and the most serious earthquake disaster in the world. Historically, almost all provinces, autonomous regions, and municipalities directly under the central government have been hit by earthquakes of magnitude or above. The earthquake caused casualties and serious damage to buildings and other secondary disasters, causing huge losses to the country and the people. Earthquake prevention and disaster reduction work has become all levels of government attach great importance to work (Ma & Wang 2003). So, measures must be taken in the disaster areas after the earthquake. It is the government's duty and the people's expectation to reduce disaster losses and safeguard the safety of people's lives and property. The earthquake is a major disaster and a special test. To solve this problem, we can take the following measures, such as resettling the victims and rebuilding their homes, comforting the psychological trauma of the victims, and revitalizing the development of urban functions in the disaster areas. Reconstruction of disaster-affected areas includes restoration of employment, environmental improvement, water quality improvement, construction of subways and roads, improvement of medical conditions, etc. At the same time, we must help resettle the life of victims and help them make plans so that they live safely.

5.2 *Aid for people and develop the economy*

The first aspect is about improving people's lives in disaster areas. To build disaster area housing, give disaster area people stable and stable housing. Since 1978, about 800 million yuan of disaster relief funds have been allocated, and about 100,000 households have been rebuilt. Therefore, we must work hard on the resettlement of victims. Reconstruction work should be planned, arranged, and started early so the affected people can move into new homes as soon as possible (Chen et al. 2021).

The second aspect concerns the psychological comfort of people in disaster areas, which is a tricky problem. From the current situation, the psychological problems of the people in the earthquake area are the most important because he is afraid. There are a lot of psychological obstacles caused by the problem that should be more serious than real radiation. This social problem should be solved jointly by the government, society, and family rather than by a single force. These people need some real help. First, the government should subsidize the people in the quake-stricken area and send them a large number of materials for initial relief. And the expense is not only to ensure their livelihood, but also to make the poor people feel safe and help them in relieving their pressure, which is a good policy. Second, some psychological consultants can be arranged for postdisaster psychological counseling of victims. Then according to their actual situation, psychological counseling, psychological intervention, and other work for them. Coupled with warmth and tolerance in the family, the psychological problems of victims can be further solved.

The last point is to revive the economies of the affected areas. This one will be easier to do after the first two. The first is to arrange employment for people in disaster areas. We should not only provide them with employment opportunities but also ensure that their income is guaranteed. This can effectively prevent the spread of poverty and provide a means of self-reliance rather than material assistance to these people. All in all, we can solve our economic problems by ourselves. To do a better job of that, to organize our disasters better. Second, we should develop the functions of the cities in disaster-stricken areas, such as economic and cultural development. We can vigorously develop an economic model suitable for local conditions. For example, we should vigorously develop tourism and the agricultural industry. That way, we can put the money to good use and solve the underlying problem, which we hope to do better. Alternatively, scattered victims of the disaster can be organized to carry out production, sales, and various household sideline businesses to carry out production and disaster relief work. It is of great significance to promote the sustainable utilization of natural resources and the rapid development of the social economy (Li 2009).

6 FUTURE DIRECTION

The future development trend is technology and management, or the organic combination of different measures. Combine these improvement measures, and strengthen the popularization of environmental disaster knowledge, improving people's awareness of disaster prevention and reduction. At present, human beings can't prevent and control geological disasters completely. Still, it is necessary to understand and master the basic knowledge of the occurrence and development of geological disasters, as well as the environmental disasters that may be brought about, to take effective prevention and control measures, or take timely actions of refuge and disaster reduction to reduce losses, which can be done.

Earthquake disasters are the result of natural laws, and the occurrence of disasters is bound to have a huge impact on our lives and the environment. We should thoroughly explore the impact of earthquakes on the local environment and resources, do a good job of disaster prevention and post-disaster reconstruction, minimize losses and effectively maintain the ecological environment.

Earthquake disasters are inevitable because they are partly influenced by nature. But many earthquakes are determined by human behavior. So natural disasters and human actions are the two elements of earthquakes, but the more critical element is human behavior. Prevention and control of earthquake disaster risk is a very comprehensive work that involves more than just basic investigations and probes, and assessments. Ecological monitoring can not only provide basic information for the scientific understanding of earthquake disasters but also provide an effective decision-making basis for the formulation and modification of long-term ecological protection, restoration, and reconstruction programs (Ma 2003).

The scientific understanding of the link also involves the seamless connection of all kinds of investigation and evaluation products and relevant management measures. Of course, there are many disciplines involved, such as geophysics, engineering, management, and even social behavior. And this requires the joint efforts of scientific research institutions, professional departments, governments at all levels, and social organizations to produce effective results.

We should be confident about the future. Now that we know the harm of earthquakes is great, we have come up with many measures to solve and deal with such disasters. It is necessary to implement these measures and methods strictly. The more effective the measures, the smaller the costs, otherwise our future will be very worrying. If there is danger, we must take measures to avoid it. So that when the next crisis comes, we can be confident enough to use these measures to help the country through this crisis.

7 CONCLUSION

This paper discusses the types and causes of earthquakes. It focuses on the methods of reducing earthquake disasters, including how to select the location of buildings, how to design, how to operate better and maintain, how to deal with disasters, and how to rescue and psychological counseling the people affected by the earthquake. Finally, the paper discusses the development direction of earthquake disaster response. This paper puts forward some useful suggestions, which have positive reference significance for mitigating earthquake disasters.

REFERENCES

Aleotti, P., Chowdhury, R. Landslide hazard assessment: Summary review and new perspectives [J]. *Bull Eng Geol Env.* 1999, 58:21–44.

Chen, P.Y., Zou, X.Y., Zhong, Y.T., Wang, L., Zheng, X.W. Evaluation of post-earthquake development of the most severely affected area in wenchuan earthquake. *Journal of Chengdu Normal University*, 2021, 37(7):116–124.

Guo, X.D., Chen, L.D., Fu, B.J. Effects of land use/land cover changes on regional ecological environment. *Advances In Environmental Science*, 1999, (06):66–75.

Lan, D.X. A Comparative study of characteristics of post-earthquake reconstructions abroad and the experience of "5·12" post-earthquake reconstructions in China. *Journal of Chengdu University.* 2011, 4:6–8.

Li, Y.X. Classification and prevention of geological disasters. *Exploration Engineering of Western China.* 2009, (4):42–46.

Ma, T., Wang, Y.J. Ecological environment monitoring and its development in our country. *Sichuan Environment*, 2003, 22(2):42–43.

Song, H.W., Xia, F., Liu, J.C. Preliminary study on causes and secondary disasters of Wenchuan earthquake in Sichuan Province. *Groundwater*, 2012, 34(2): 174–176.

Xu, F. Study on first aid measures for psychological injury in earthquake disaster. *Journal of a Clinical Emergency*, 2009, 10(2): 127–128.

Zhang, F. Study on the evolution path and diffusion mechanism of earthquake secondary disasters – a case study of Lushan earthquake in Sichuan Province. *Huabei Natural Resource*, 2022, 1:53–55.

Zhou, W.S., Mao, F., Liu, Z.E., Li, Q., Fu, Q. Research and application of planning support system based on 3S technique for post-disaster reconstruction after Wenchuan Earthquake in China. *IEEE International Symposium on Geoscience and Remote Sensing* (IGARSS), 2009.

Advances in Civil Engineering and Environmental Engineering – bin Yahya & Bin Daud (Eds)
© 2023 The Author(s), ISBN: 978-1-032-39397-1

Compression damage in high-toughness cementitious composites after dry-wet cycles

Renyi Fang & Bo Fan
Sichuan Jiuma Expressway Co., Ltd., Chengdu, China

Lei Yu*
Beijing Xinqiao Technology Development Co., Ltd., Beijing, China

ABSTRACT: Compression and ultrasonic pulse tests were carried out on high-toughness cementitious composites (HTCC) after dry-wet cycles. Special focus was centered on the stress-strain relation and the corresponding variation of wave velocity under different water immersion cycles. The results show that the initial P wave velocity and S wave velocity of HTCC was kept almost unchanged at the early stage of immersion. As the immersion period extended, P wave velocity during loading gradually exhibited a "compacting" characteristic. On the other hand, HTCC, at the early immersion stage, experienced a steep drop in S wave velocity during loading. However, the steep drop degree tended to be moderate with the immersion cycle, and the associated failure mode was gradually translated from a nonsteady state into a steady state. It turns out that the ultrasonic pulse method can effectively identify the compression damage and characterize the failure mode of HTCC.

1 INTRODUCTION

High-toughness cementitious composites (HTCC) have attracted extensive attention due to their excellent tensile properties and crack control ability (Li 1993; Xu & Cai 2009; Zhang et al. 2009). The ductility and toughness of HTCC are achieved by forming multiple cracks after the first cracking, and the width of a single crack can be controlled within 100 μm. Many efforts have been devoted to the performance improvement and application of HTCC (Qian et al. 2009; Wang et al. 2013, 2018, 2021; Zhang et al. 2017). And the application field has extended to hydraulic engineering and marine engineering (Liu et al. 2016).

Nondestructive testing of concrete properties usually deploys the ultrasonic pulse method. Liu et al. (2016) used the ultrasonic method to evaluate the wave velocity change of sulfate-eroded concrete in the process of compression and summed up the development characteristics of wave velocity with loading. Through the ultrasonic test during loading, Hisham et al. (Qasrawi & Marie 2003) obtained the variation of the wave velocity in the concrete and explored the evolution of the internal damage in the concrete. Wang et al. (2015) carried out the ultrasonic test on concrete with different strengths and obtained the relationship between wave velocity and compressive strength. Suryanto et al. (2015) used the ultrasonic velocity test to study the self-healing of HTCC under natural conditions and confirmed the feasibility of ultrasonic velocity to characterize the healing process of HTCC.

*Corresponding Author: yuleimabel@163.com

DOI 10.1201/9781003349563-49

However, the existing research usually only selects a single P wave velocity; thus, it is difficult to grasp the damage evolution of HTCC materials accurately. If the ultrasonic pulse technology can be introduced into the HTCC study, the P wave and S wave velocity could be comprehensively utilized, which would be beneficial to reveal the failure mechanism of HTCC under dry-wet cycles.

This study simulates the dry-wet environment of HTCC through laboratory tests and conducts HTCC compression tests and ultrasonic pulse tests after different immersion periods to obtain compression stress-strain curves and the variation of the P wave and S wave velocity during loading. First, the development of the initial P wave and S wave velocities of HTCC with the immersion period is investigated. Based on this, the evolution of P wave and S wave velocities with the development of stress-strain relationships is analyzed, and the compressive failure mechanism of HTCC under dry-wet cycles is explored.

2 TEST PROCEDURE

2.1 *Materials and mix proportion*

Using P.O 42.5 cement as the major cementitious material of HTCC and mixing it with Class I fly ash and expansion agent; the water reducer is a high-efficiency polycarboxylic acid water reducer; the mixing water is tap water. The reinforcing fibers are polyvinyl alcohol (PVA) fibers commonly used in high-tough cementitious materials. The diameter of the circular section of the fiber is 39 μm, the length is 12 mm, and the volume fraction is 2.0 %. The mix proportion (by mass ratio) of matrix is *m*(cement): *m*(fly ash): *m*(expansion agent): *m*(water): *m*(water reducer)=1.00: 0.29: 0.14: 0.36: 0.70: 0.01.

2.2 *Dry-wet cycle test*

Considering the immersion saturation of the specimen and the absence of coarse aggregate in HTCC, this paper selected *a* ϕ 50 mm×100 mm small-sized cylindrical specimen (CES-82 2008). Three specimens were prepared for each group, and the dry-wet cycle test was started after curing under standard conditions (temperature 20±2°C, relative humidity > 95%) for 28 days. The end surface of the specimen is sealed with paraffin before immersion.

The dry-wet cycle regime is: the samples are placed vertically in the container and completely immersed in water. After seven days of immersion, the samples are taken out and placed at ambient temperatures for natural drying for eight days, with 15 days as a dry-wet cycle (Guo et al. 2018). Thus, the exposure environment of the material is closer to the actual environment. After selecting dry-wet cycles of 0 (28 days of standard curing), 2, 6, 8, and 14 times (corresponding to 0, 30, 90, 120, and 210 days of immersion, respectively), ultrasonic pulse tests of HTCC specimens under axial compression are carried out, respectively.

2.3 *Compression and ultrasonic pulse test*

The uniaxial compression test is carried out on an MTS material testing system. The loading rate is set as 0.10 mm/min. The ultrasonic pulser and receiver are embedded in the upper and lower cell, respectively, and the gap between the cell and the specimen is coupled with Vaseline. The velocity of the primary wave (P wave) and the second wave (S wave) was simultaneously measured at 10 s intervals during loading. The axial strain is measured by LVDT sensors mounted on both sides of the specimen, and the radial strain is measured by a radial sensor surrounding the specimen, as shown in Figure 1.

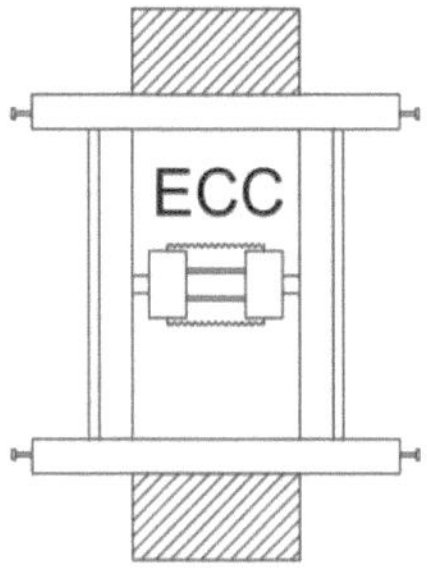

Figure 1. Set-up of uniaxial compression.

3 RESULTS AND DISCUSSION

3.1 *Variation of mechanical parameters and wave velocity*

Within 90 days of immersion, the changes of the P and S wave velocities were marginal, which were stable at about 4.2 km/s and 2.4 km/s, respectively. After 120 days of immersion, the wave velocities of the P wave and the S wave were reduced simultaneously, and both reduction rates were around 25%. The variation of compressive strength and elastic modulus with immersion age are shown in Figure 2. With the increase of immersion age, the compressive strength of HTCC experienced a change of first increase and subsequent decrease. The compressive strength of HTCC maximized at about 90 days of immersion, which was about 50% higher than before immersion, and then began to decrease in a small range. The compressive strength dropped to 66.5 MPa at 210 days but was still higher than before immersion. The development of elastic modulus is different from the strength change. The elastic modulus at the early immersion stage did not change much and was stable at around 20 GPa. The elastic modulus reached its highest after 120 days of immersion but dropped to 12.6 GPa after 210 days of immersion. At the later stage of immersion, cement hydration tends to be completed. The dry-wet cycle is dominant, which further expands the internal pores of the concrete, resulting in a decrease in the elastic modulus and strength of the material.

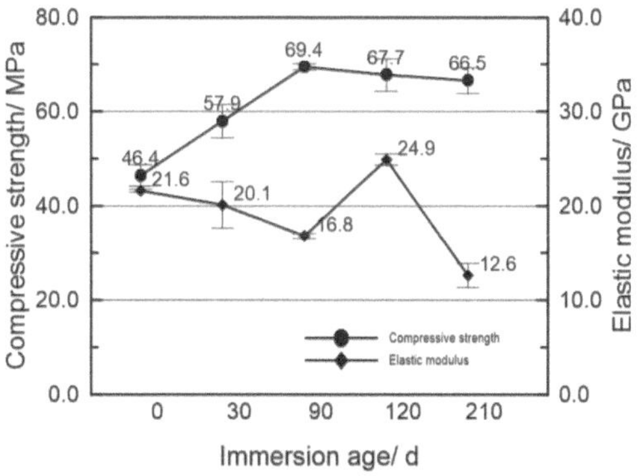

Figure 2. Variation of compressive strength and elastic modulus of HTCC under different immersion ages.

3.2 *Evolution of S wave velocity during loading*

The development of S wave velocity with the axial strain of HTCC at immersion ages of 0 and 210 days is shown in Figure 3. There existed obvious differences in the variation law of S wave velocity under different immersion ages. At the early stage of immersion, the S wave velocity of virgin HTCC specimens experienced a two-stage change with stable development and then a rapid decline. The internal damage of the HTCC at the initial stage of loading had not been formed yet, and the S wave velocity developed steadily with a small increase. When the strain increased to a certain level, the S wave velocity suddenly decreased. This indicates that the damage to the HTCC material at the early stage of immersion is relatively sudden. Even though the stress-strain curve (shown in Figure 4) shows no obvious post-peak stress drop, the changes in the S wave track the evolution of the internal structure of the HTCC. With the further increase of the immersion age, the development of the S wave velocity changed significantly. After 210 days, the sharp drop of the wave velocity completely disappeared, and the S wave velocity directly accessed the slow drop segment after the stable development.

According to the development of radial strain, it roughly encountered three stages: slow increase, speed increase, and rapid increase. At the initial stage of loading, the radial strain increased linearly with the increase of axial strain; after that, the radial strain deviated from the linearity, the growth rate increased rapidly, and gradually surpassed the axial strain in magnitude. Note that the slow growth segment of radial strain was synchronized with the stable development segment of S wave and P wave velocity, and the end of this stage arrived ahead of the stress peak point, after which the wave velocity started to decrease remarkably. It can be seen that the development of wave velocity is closely related to the various characteristics of the radial strain. It is effective to use the wave velocity method, especially the S wave velocity, to identify the development of stress damage in HTCC materials and to characterize the failure mechanism of HTCC at different immersion ages.

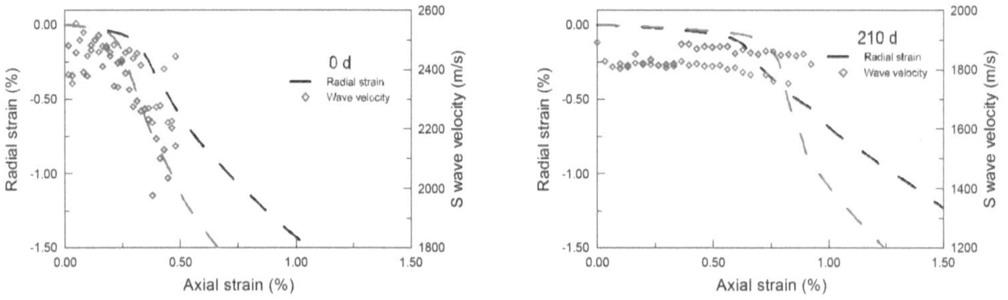

(a) 0 d (b) 210 d

Figure 3. Development of radial strain and S wave velocity with the axial strain of HTCC at different immersion ages.

3.3 *Variation of P wave velocity during loading*

The axial compressive stress-strain relationships of HTCC at 0 and 210 days of immersion age are shown in Figure 4, and the corresponding development of P wave velocity with axial strain is also depicted in Figure 4. Prior to the immersion, the axial compressive stress-strain curve of HTCC had a relatively gradual descending section. After 30 days of immersion, the descending section of the curve began to show a certain degree of stress drop, and the stress drop became slower after 210 days. Note that the ascending segment of the curve after 210 days has slight compaction characteristics, and the stiffness decreases significantly. During the dry-wet cycle, the specimen is repeatedly taken out from the container and wiped. The soluble alkali on the surface of the specimen is gradually dissolved and dissipated. In each drying process, the specimen is repeatedly subjected to the carbonization process. Calcium carbonate crystals are formed in the interior of the near-surface pores, which further expands the pores, leading to variations in stress-strain behavior.

Differences existed in the P wave velocity development characteristics under each immersion age. Prior to immersion, the P wave velocity went through three stages: steady development, rapid

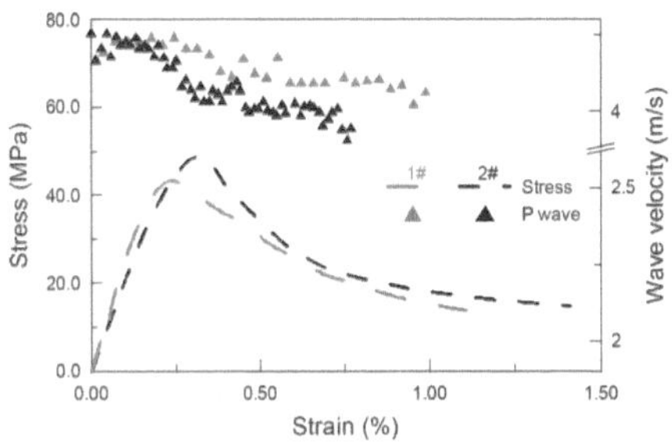

(a) 0 d

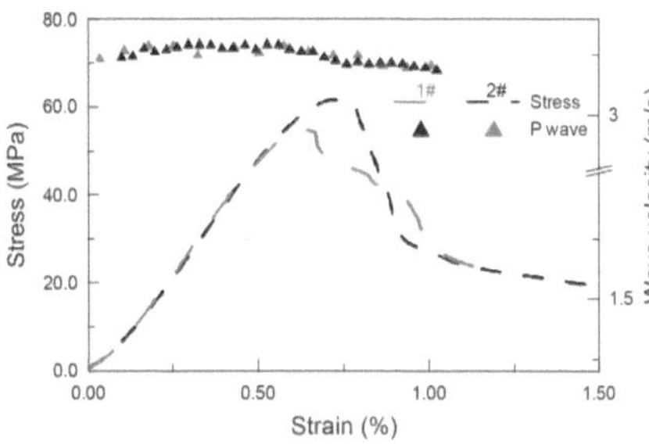

(b) 210 d

Figure 4. Compressive stress-strain relationship and P wave velocity development of HTCC at different immersion ages.

decline, and slow decline. In the stable development segment, the P wave velocity was always stable at about 4.2 km/s, which indicates that no significant stress damage was generated at this stage, and then the wave velocity began to decrease with a relatively large magnitude. Note that the onset of the wave velocity drop was ahead of the stress peak point, which indicated that the wave velocity measurement might play an effective role in identifying stress damage and early warning of ultimate stress. After the strain reached about 0.50 %, the wave velocity dropped below 4.1 km/s and began to decrease slowly with a small amplitude.

With the extension of the immersion age, the changes in the P wave velocity are mainly reflected in two aspects. First, the wave velocity compaction characteristics in the stable development section gradually appeared, especially after 210 days of immersion. The P wave velocity at the initial stage of loading significantly increased with the increase of the axial strain, and the wave velocity increasing section was almost synchronized with the stress compaction section. Second, the drop of the wave velocity in the fast descending section was significantly reduced. After 210 days of immersion, the fast descending section even completely disappeared, and the P wave velocity directly entered the slow descending section after experiencing a stable development section. With the increase of the dry-wet cycle, although the elastic modulus and strength of HTCC decrease, the failure mode after the peak stress has gradually translated from a nonstable state into a stable state.

4 CONCLUSIONS

The uniaxial compression test of HTCC under the action of the dry-wetting cycle was carried out, and the ultrasonic test during the loading process was simultaneously deployed. The conclusions are as follows:

The initial P wave and S wave velocities of HTCC showed similar development characteristics with immersion age. The wave velocity kept almost unchanged within 90 days, followed by a decrease of about 25% after 120 days. The development of ultrasonic velocity was consistent with the change of elastic modulus, but its correlation with compressive strength was weak.

The S wave velocity at the early stage developed stably at the beginning but exhibited a steep drop when the strain increased. The steep drop of wave velocity tended to be moderate with the extension of immersion age. The development of wave velocity was closely related to the variation of radial strain. The wave velocity measurement has been verified in the identification of stress damage in HTCC under different immersion ages.

The P wave velocity was stable at about 4.2 km/s at the onset of loading. After the strain reached 0.50 %, the wave velocity decreased to 4.1 km/s. With the increase of the dry-wet cycle, the wave velocity compaction characteristics gradually appeared in the stable development segment of the P wave velocity. The drop in wave velocity significantly decreased in the fast-falling segment.

ACKNOWLEDGEMENT

The authors gratefully acknowledge the support of the Transportation Technology Project of Sichuan Provincial (2019-ZL-18).

REFERENCES

CES-82. (2008). Recommendations for design and construction of high-performance fiber reinforced cement composites with multiple fine cracks. S. *Japan Society of Civil Engineers.*

Guo Jinjun et al. (2018). Experimental study on fracture properties of modified concrete under sulfate corrosion under dry-wet cycles. *J. Hydraulic Engineering*. 49(4), 419–427.

Li, V.C. (1993). From micromechanics to structure engineering-the design of cementitious composites for civil engineering application. *J. Structure Mechanics and Earthquake Engineering*. 10(2), 37–48.

Liu Hezhi et al. (2016). Influence of micro-cracking on the permeability of engineered cementitious composites. *J. Cement and Concrete Composites*. 2016(72), 104–113.

Liu Juanhong et al. (2016). Variation characteristics and mechanism of the acoustic wave and acoustic emission of concrete sulfate corrosion damage. *J. Engineering Science*. 38(8), 1075–1081.

Qasrawi, H.Y. & Marie, I.A. (2003). The use of USPV to anticipate failure in concrete under compression. *J. Cement and Concrete Research*. 33(12), 2017–2012.

Qian S. et al. (2009). Introduction of transition zone design for bridge deck link slabs using ductile concrete. *J. ACI Structural Journal*. 106(1), 96–105.

Suryanto, B. et al. (2015). Smart cement composites for durable and intelligent infrastructure. *J. Procedia Engineering*. 125, 796–803.

Wang Xueping et al. (2015). Experimental study on the relationship between ultrasonic velocity and compressive strength for different grades of concrete. *J. Concrete*. 2015(12), 34–37.

Wang Zhenbo et al. (2013). Application of ductile fiber reinforced cementitious composites in jointless concrete pavements. *J. Composites Part B: Engineering*. 2013(50), 224–231.

Wang Zhenbo et al. (2018). Mechanical properties and crack width control of hybrid fiber reinforced ductile cementitious composites. *J. Building Materials*. 21(2), 216–221+227.

Wang Zhenbo et al. (2021). Tensile properties and microstructure of ECC under sulfate attack. *J. Huazhong University of Science and Technology* (Natural Science Edition). 49(7), 31–36.

Xu Shilang & Cai Xiangrong. (2009). Basic mechanical properties of ultra-high toughness fiber-reinforced cementitious composites. *J. Hydraulic Engineering*. 40(9), 1055–1063.

Zhang Jun et al. (2009). Engineered cementitious composite with the characteristic of low drying shrinkage. *J.Cement and Concrete Research*. 39(4), 303–312.

Zhang Jun et al. (2017). Design and construction of jointless cement concrete pavement. *J. Harbin Institute of Technology*. 49(3), 68–73.

Advances in Civil Engineering and Environmental Engineering – bin Yahya & Bin Daud (Eds)
© 2023 The Author(s), ISBN: 978-1-032-39397-1

Study on spatial differentiation of comprehensive development potential among counties and districts in Lhasa

Yizhou Han*
Geographic Information Department, Tibet Dezhong Geographic Information Co., Ltd, Tibet, China

Kai Zhang
Institute of International Education, Henan Agricultural University, Zhengzhou, China

ABSTRACT: Urban development potential is an important index to measure and evaluate whether a city can achieve revitalization and rapid growth in the future. While the relationship between the related affecting factors and the urban development potential can't be analyzed and figured out by the current evaluation methods and systems. Therefore, the research on evaluation methods of urban development potentials is of great importance. Based on quantitative analysis and hierarchical clustering, this paper classifies and evaluates the development potential of Lhasa districts and counties and visualizes the spatial distribution of urban development potential by selecting five factors affecting urban development: economic scale, resources and environment, population potential, public service capability, and spatial connection. The comprehensive development potential of Lhasa is extremely uneven among districts and counties. Chengguan District is much higher than other districts and counties. Duilong Deqing District has a medium potential, while other districts and counties have a poor potential. The overall spatial distribution is higher in the southeast traffic corridor, with the Chengguan area as the pole and spreading outward. Overall, comprehensive development potential is highly positively correlated with population potential and spatial connection potential.

1 INTRODUCTION

The urban development potential can evaluate the development status and development expectations of the city. What's more, the determination of development can be regarded as the foundation of constructing the optimal modern cities' development strategies and regional development policies. Therefore, establishing a comprehensive evaluation of the city's development potential and optimizing the spatial structure of the city based on the territorial distribution of the urban development potential has significant importance in promoting urban growth (Wang & Xu 2012). Recently, many scholars have made some analyses and discussions about urban development potential. Wang Feng and Xu Yilun selected a total of 16 indicators from 8 aspects representing the economic development, social environment development level, industrial structure, characteristic resources, traffic location conditions, infrastructure security, etc., to measure the development potential of 30 townships in Linquan County, Anhui Province, through principal component analysis (Wang & Xu 2012). Ma Renfeng, Wang Xiaochun et al. constructed a regional development potential evaluation index system composed of a target layer, indicator layer, and indicator item, and analyzed the spatial differentiation characteristics of county development potential in Yunnan Province by using the systematic clustering method (Ma et al., 2009a, 2009b, 2011). Zhang Ge et al. selected the economic growth, village land planning, and other four factors affecting the city development as

*Corresponding Author: 201911081315@std.uestc.edu.cn

DOI 10.1201/9781003349563-50

the research object and utilize 24 kinds of evaluation indicators, such as each villager's income, distance from the downtown and traffic arteries, natural resources and effect of the natural disasters, to make the evaluation of the city growth potential and city planning in Chuanfangyu Town, Jizhou District, Tianjin (Zhang & Zho & Zhu 2022). Dong Jieyun et al. selected 15 indicators from six aspects of the village's natural environment, resource endowment, settlement status, location conditions, economic development level, and facility conditions to construct the evaluation index system of village development comprehensive potential and then determined the weight of each indicator of rural development comprehensive potential by AHP. Then, the entropy weight method and linear weighted sum method are used to calculate the comprehensive development potential index of each village to evaluate the development potential of the village in the county (Dong, Zhang & Shao 2022). Li Fuxiang, Wang Cheng et al. explored the potential of urban development from sustainable development (Li et al. 2022; Wang et al. 2021).

But the majority of the research and analysis always take mainland China as the study area. At the same time, this paper chooses Lhasa, located in the western highlands of China, as the study area. The 92% area of Tibet sits in an extremely fragile ecological environment with an altitude of over 4000m, and many natural disasters frequently occur in Tibet (Li & Zheng 2017). Moreover, due to the remote location of Tibet, those Tibetans have been in a state of economic backwardness for a long time. Until the total liberation of Tibet in 1952, the central government started to financially support Tibet to develop advantageous industries and boost urbanization in relatively remote areas. Therefore, over the past 60 years, Tibet experienced an unprecedented acceleration in growth (Kong & Zhu 2021). However, compared with those developed cities, Tibet's economy is still far from the mainland cities in China. This is because there are large numbers of Tibetans living scattered in the farming and pastoral areas, leading to the low urban growth level in Tibet. And cities and towns in Tibet are not only low in numbers and far apart from each other's, but their radiation-driving ability is weak (Fang & Li 2015). So, while exploring Tibet's development potential, we combine the actual stage of urban growth in Tibet based on the rules of choosing the urban development index and construct an evaluation system and model that can relatively accurately describe and analyze the development potential of Lhasa.

2 STUDY AREA AND MATERIALS

2.1 *Study area*

Lhasa, located in the middle area of the Tibetan Plateau and the north side of the Himalayas, is the capital city of the Tibet Autonomous Region in China, with a geographical location between 91°06′ E, 29°36′ N. The altitude of Lhasa is at the level of 3650m, and five counties and three districts (Chengguan District, Linzhou County, Dangxiong County, Nimu County, Qushui County, Duilong Deqing District, Daz District, Mozhu Gongka County,) are under its jurisdiction. Lhasa occupies 30000 square kilometers and has a population of 56w. At the same time, it's the political, cultural, economic, and technological education center of Tibet.

2.2 *Sources of the data*

Taking the eight counties in the modern area of Lhasa as the research area, the sources of economic data used in this paper all come from the Lhasa Statistical Yearbook 2018; the statistics about the population in Lhasa are collected from the seventh National Census Main Data Bulletin in Lhasa, the Lhasa Statistical Yearbook 2018, the Guiding Opinions on Accounting for Overall Planning of Land and Space in Cities and Counties of the Tibet Autonomous Region, the data of the cultivated land and waters both come from the Main Data Results of the Third National Land Survey and the traffic route data is from OSM (Open Street Map, https://www.openstreetmap.org) and POI (Point of Interest) with the vector illustration of administrative divisions coming from the Amap(https://lbs.amap.com/). Ultimately, the remote sensing images used in this paper are all from USGS (United States Geological Survey). All image information is shown in Table 1.

Figure 1. Study area.

Table 1. Information for the data sources of the remote sensing imagery.

Date of image acquisition	Product ID	Related extracted indicators and variables
2021-11-12	LC08_L1TP_137039_20211112_20211124_02_T1	NDVI, WET, NDBSI, LST, RSEI
2021-11-12	LC08_L1TP_137040_20211112_20211124_02_T1	NDVI, WET, NDBSI, LST, RSEI
2021-11-07	LC09_L1TP_138038_20211107_20220119_02_T1	NDVI, WET, NDBSI, LST, RSEI
2021-11-07	LC09_L1TP_138039_20211107_20220119_02_T1	NDVI, WET, NDBSI, LST, RSEI
2021-11-07	LC09_L1TP_138040_20211107_20220119_02_T1	NDVI, WET, NDBSI, LST, RSEI

3 METHODOLOGY

Through asking for advice from the experts and theoretical theories, we choose five mainly affecting factors, including economic scale (gross domestic product), natural resources and environment, population potentials, public facility capacity, and traffic capacity, as the influence factors to calculate the values of the urban development potentials. In this paper, we use the hierarchical clustering method to evaluate the regional urban development potentials in Lhasa and classify the potential types of each district. The hierarchical clustering method can simultaneously make the classification depending on their quantitative results and characteristics and simply quantify the evaluation of the urban development potentials of the whole study area. Moreover, the software for processing the statistical information (Statistical Product Statistical Solutions) is also used to calculate the urban development potential. The distance and relationship between the five variables will be generated using the Euclidean distance measure method. And the sum of the five indicators is calculated by the squared deviation method. In the end, after the integration of the five affecting factors, the spatial discrimination of the urban growth potentials will be obtained. To further analyze and study more clearly, we divided some factors into five grades. Different categories correspond to different values of influencing factors, as shown in Table 2.

Table 2. Level division description.

Level	Arable land per capita (m^2/person)	Water area per capita (m^2/person)	Tourist attraction	Population density	The intensity of population movement (%)	City gravity (per thousand people)
1	0 ~ 600	0 ~ 600	10 ~ 20	0 ~ 10	<= 0	0 ~ 20
2	600 ~ 1200	600 ~ 1200	20 ~ 30	10 ~ 20	0 ~ 10	20 ~ 40
3	1200 ~ 1800	1200 ~ 1800	30 ~ 40	20 ~ 30	10 ~ 20	40 ~ 60
4	1800 ~2400	1800 ~2400	40 ~ 50	30 ~ 40	20 ~ 0	60 ~ 80
5	2400 >	2400>	50 >	40 ~	30 >	

3.1 *Economic Scale Index*

The economic scale index represents the development scale and strength of the region, which can be calculated as the formula shows as expressed.

$$E = perGDP \times k \tag{1}$$

Where the per GDP indicates the gross domestic product for everyone in Lhasa, and k represents the growth rate of GDP over the past ten years.

By integrating the GDP per capita and GDP growth rate, we generate the line chart of the economic scale index between each district in Lhasa (Figure 2).

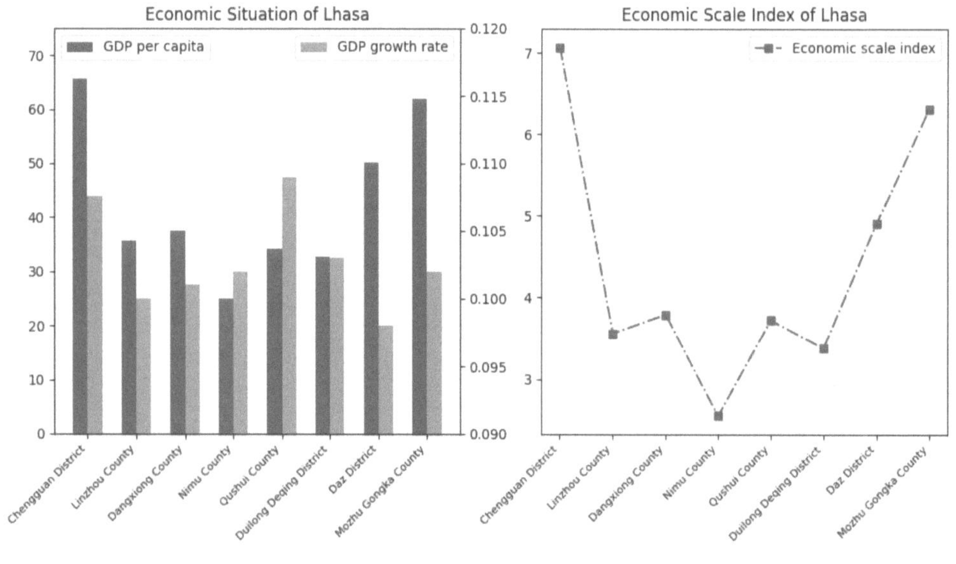

Figure 2. Economic situation of lhasa and economic scale index of Lhasa.

3.2 *Resource and Environment Index*

Land, water, and mineral resources, which can be developed in the future and play a role in urban growth, could be represented by the Resource and Environmental Indicator. The Resource and Environmental Indicator are calculated by the values of the cultivated land resource, water resource, tourism resource, and ecological index. The formulas are shown as follows.

$$REI = (CLRL + WRL + TRL) \times EI \tag{2}$$

Where REI represents the resource and environment indicators, CLRL represents the per capita cultivated land resource level, WRL represents the per capita water resource level, TRL represents the tourism resource level, and EI represents the ecological index. The hierarchical classification methods of CLRL, WRL, and TRL are shown in Table 2 Level division description.

RSEI (Remote sensing ecological index) was also used to reflect EI(ecological index). The formula of the IRSEI is expressed as follows.

$$RSEI = f\ (Wet, NDVI, NDBSI, LST) \tag{3}$$

Formula 3 is used to process Landsat images to obtain RSEI maps, and the results are shown in Figure 1. The images used are shown in Table 1. The specific processing methods are described in detail in Xu Hanqiu's article: < A remote sensing index for assessment of regional ecological changes > (Xu 2013).

Due to the error brought by the river and snow during the ecological index calculation, we extracted the water body of Lhasa before the calculation of EI for a more accurate analysis. After obtaining the values of EI in Lhasa, we put the snow area and river back. The following Figure 3 pictures of the spatial distribution of the weights of EI in Lhasa.

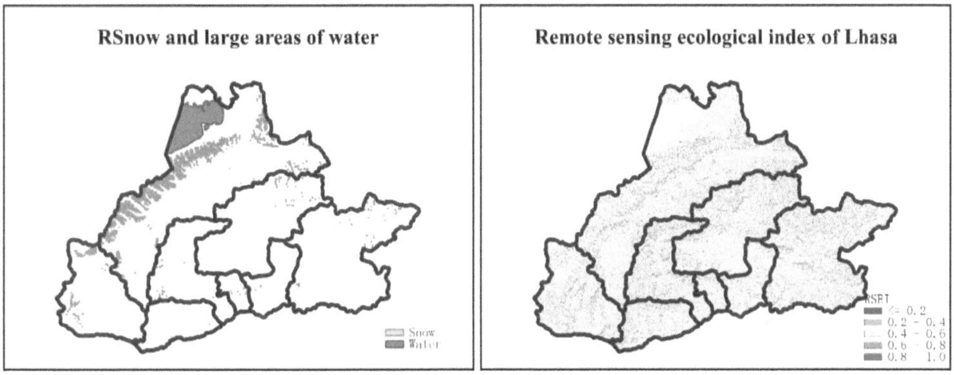

Figure 3. Remote sensing ecological index of Lhasa.

The RSEI's mean value of each district and county represents EI. And after acquiring the values of these three involved indicators (arable land per person, water area per person, and Number of POI in scenic spots), rank them. The resource and environmental index could be obtained using grading results (CLRL, WRL, TRL) and RSEI averages. The result is shown in Figure 4.

3.3 *Population potential index*

The population potential index indicates the potential size of the regions that can continue to accommodate how many populations with the current population situation. The values of the population potential index can be expressed by the region's total population and the density of the area. We take the complexity of the natural environment into consideration, so we choose the population density to describe the population potential. The formulas are shown as follows.

$$P = \frac{(PDL + DL)}{2} \tag{4}$$

Where the P means the population potential index, the PDL in the formula indicates the grading of the population intensity in each district, and the DL means the grading of the population flow

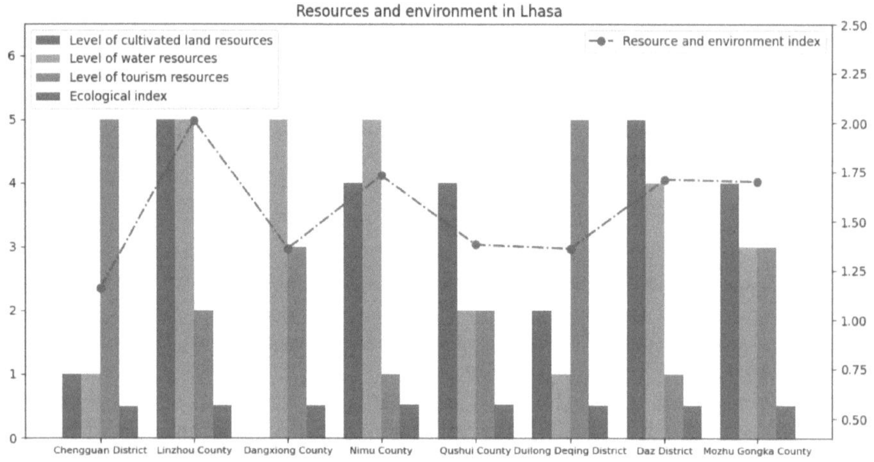

Figure 4. Resources and environment in Lhasa.

intensity. The classification method of PDL and DL is shown in Table 2 Level division description.

$$D = \frac{PP}{A} \tag{5}$$

Where D means the population flow intensity, PP indicates the permanent population in the region, and A means the square kilometers of the region.

Figure 5 shows the values of the population potential index of each district in Lhasa.

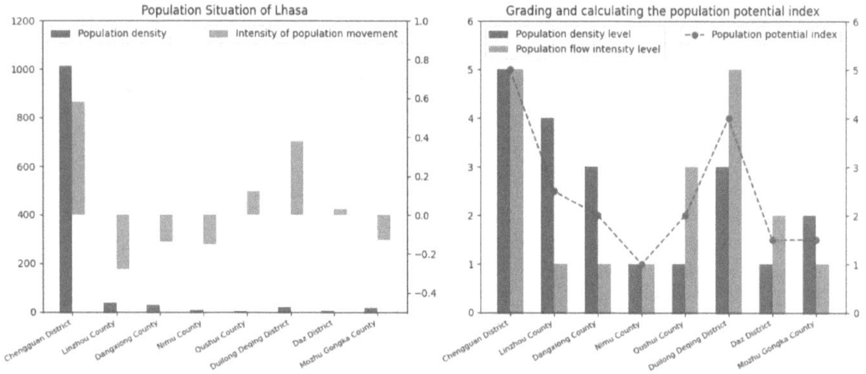

Figure 5. Population potential.

3.4 *Public service capability index*

Infrastructure and public services are tightly associated with human life, which can be reflected by the Point of Interest (POI) with high efficiency. Point of Interest can represent the real geographic entities, including common geographic places such as schools, hospitals, gas stations, etc. We collect the data of POI as a total of 668693 POI points in Lhasa in 2021. And we divide the POI data into five categories corresponding to the various facilities for different purposes.

And through integrating these different kinds of POI data in the Qgis, we generate the map of the distribution of the different categories. Through Qgis processing, the number of infrastructures

Table 3. POI data classification and corresponding influence weight coefficient.

Poi Indicates the data class number	Category name	Instruction	Weight
1	Commercial housing Accommodation services Residential district	Housing supply	0.30
2	Health care services	Medical facilities	0.30
3	Science, education and cultural services	Education facilities	0.30
4	Life service facilities Public facilities Social institutions and social organizations Sports and Leisure Services	Life service	0.05
5	Shopping service Food and beverage service	Shopping service	0.05

owned by each district and county can be obtained. The public service capacity index can be obtained with the weight of various facilities in Table 3. The result is shown in Figure 6.

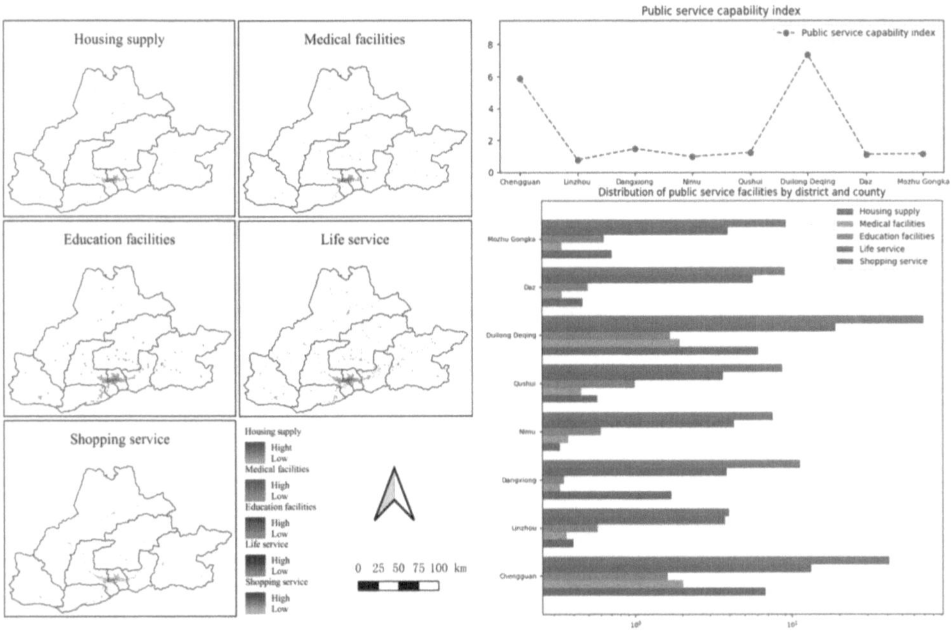

Figure 6. Distribution of public service facilities and public service capability index.

3.5 *Spatial connection index*

This paper uses the gravity model to analyze the spatial connection between each district as the formula shows as follows:

$$R_{i,j} = \frac{\sqrt{P_i V \times P_j V_j}}{D_{ij}^b}, \ (i \neq j) \tag{6}$$

R_{ij} means the gravity between the $city_i$ and $city_j$. P_i and P_j indicate the population in $city_i$ and $city_j$, V_i and V_j mean the GDP in each place. D_{ij}^b represents the Euclidean distance between the $city_i$ and $city_j$, and b means the friction coefficient. In most circumstances, when the values of the b are 1 or 2, the relationship between each region can get a detailed explanation. According to the real situation of Lhasa, b in this study is assigned 2 (Liu et al. 2014).

Based on the grading of the city gravity shown in Table 2, we can calculate the values of the spatial connectivity index depending on the following formula:

$$SPI = DN \times CGL \tag{7}$$

The SPI means the traffic capacity index, DN means the density of the network, and CGL means the city gravity level. The result is shown in Figure 7.

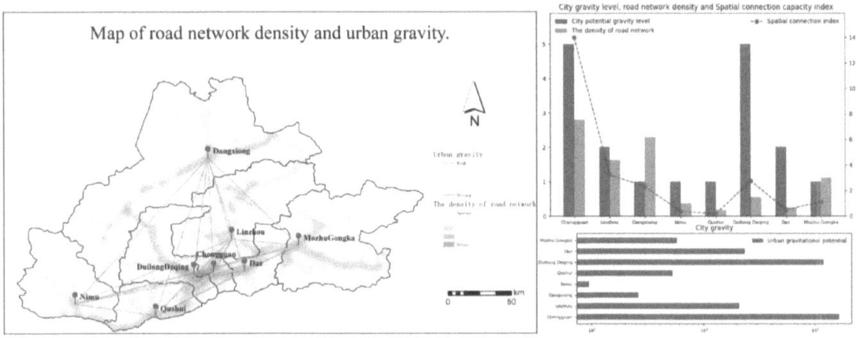

Figure 7. City gravity level, road network density, and spatial connection index.

3.6 *Comprehensive measurement of the comprehensive development potential of districts and counties*

Hierarchical clustering is one of the methods to evaluate the regional development potential and classify the types. It can simultaneously complete the quantitative evaluation of the development potential of all districts and counties in Lhasa and classify them according to the quantitative results. In the process of hierarchical clustering in this study, the statistical analysis software package SPSS25 for Windows was used for hierarchical clustering calculation. The Euclidean distance measure was used to measure the distance between individual variables, and the distance between classes was calculated by the sum of squared deviations method.

However, before hierarchical clustering, it is necessary to normalize the five influencing indices, and the formula is as follows. The normalized results are shown in Figure 8.

$$y_i = \frac{x_i - x_{min}}{x_{max} - x_{min}} \tag{8}$$

Then, based on the five normalized district and county development potential impact indexes, process the eight districts and counties in Lhasa by hierarchical clustering, and the development potential differentiation results of Lhasa districts and counties were obtained.

4 RESULT

4.1 *Classification of the comprehensive development potential of districts and counties*

Using SPSS to carry out hierarchical cluster analysis on the five main influencing factors adopted in this paper, the results can be obtained, as shown in Figure 9.

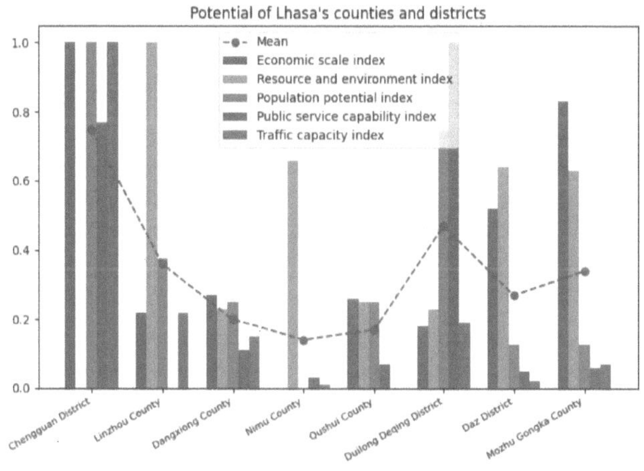

Figure 8. Potential of Lhasa's counties and districts.

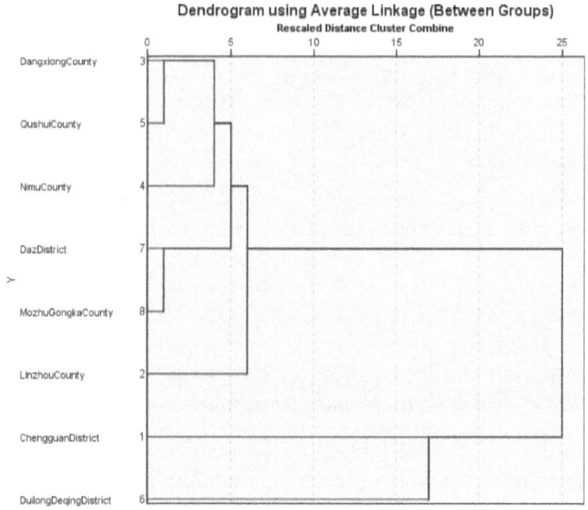

Figure 9. District and county clustering results of Lhasa City.

According to the dendrogram obtained from the clustering results, it can be seen that the difference between Dangxiong County and Qushui County is very small. The difference between Daz District and Mozhu Gongka County is very small, too, and their respective difference distance is only about one. The distance between the six districts and counties in Lhasa, except Chengguan District, is about 5, which is also not a big number. But for Chengguan District and Duilong Deqing District, the difference between them is about 17, and the difference between them and the other six districts and counties is 25. Therefore, according to the clustering results, eight districts and counties in Lhasa were divided into three categories. Chengguan District and Duilong Deqing District each belong to one category, and the other six districts and counties belong to one category.

4.2 *District and county comprehensive development potential level assessment*

After normalizing the quantitative values of the five main development potential influencing factors selected in this paper, this paper also calculates the mean value of the normalized values of the five main influencing factors. These mean values comprehensively reflect the level of urban development potential to a certain extent. The result is shown in Figure 8.

The mean value of the Chengguan district is 0.75. This is the highest value and about one-third higher than the second value, Duilong Deqing District, which is 0.47. The mean value of other districts and counties is between 0.14 and 0.36.

According to these values and the classification results above, this paper evaluates the comprehensive development potential of eight districts and counties in Lhasa. The evaluation results are shown in Figure 10. The comprehensive development potential level of Chengguan District is high, that of Duilong Deqing district is medium, and that of other districts and counties is low.

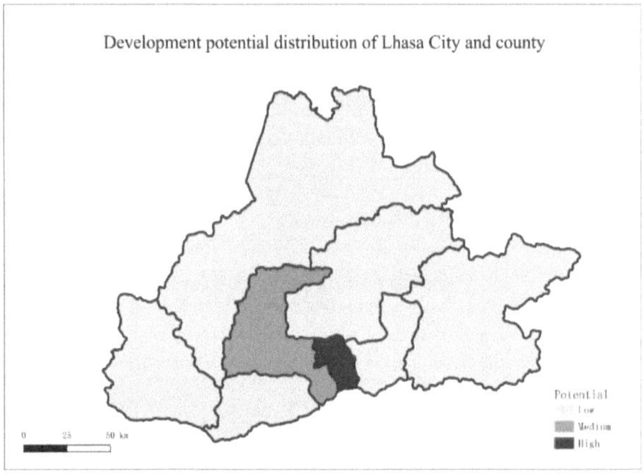

Figure 10. Development potential distribution of Lhasa city and county.

5 DISCUSSION

5.1 *Spatial distribution of Lhasa districts and counties' development potential*

Chengguan District has the best overall development potential, Duilong Deqing District is second, and other districts and counties are worse. However, a comprehensive evaluation of the five main influencing factors showed that Dazi District and Mozhugongka County were relatively good in other districts and counties. These districts and counties are distributed along the transportation lines in the southeast of Lhasa as a whole, forming a spatial distribution with the overall development potential of Lhasa districts and counties being high in the southeast traffic corridor and low in the northwest and spreading outward with Lhasa as the pole.

Besides, the development of urban and county areas in Lhasa is extremely uneven. Compared with other districts and counties, except for Duilong Deqing District, Chengguan District is far ahead, but Chengguan District and Duilong Deqing District only account for one-sixth of Lhasa City. This shows that most areas of Lhasa are still in the stage of very inadequate development.

5.2 *Characteristics of main influencing factors in districts and counties*

In the process of quantifying the resource and environment index, it is found by analyzing RSEI that the ecological and environmental quality of Lhasa is generally good and evenly distributed. Even in Chengguan District, an area with a high degree of urbanization, there is no sudden decline in environmental quality, as shown in Figure 3. It means that Lhasa has taken good environmental protection in urban development and has not significantly damaged the environment.

Through the other four influencing factors, it can be seen that all kinds of service facilities, population resources, and economic development levels are clustered in and around Chengguan

District, with Chengguan district as the pole, and spread along Duilong Deqing District, Chengguan District, Daz District, and Mozhu Gongka in a linear manner. As the development pole of the whole of Lhasa, Chengguan district is also the district with the highest spatial connection, much higher than other districts and counties, as shown in Figure 7. However, the whole town cluster is only connected with the surrounding counties, such as Duilong Deqing District, Dazi District, and Linzhou County, at a medium to the upper level. This shows that the Chengguan area, as a growth area, has a small radiation range and a weak ability to drive other towns. Although there are many reasons for this situation, such as the scattered distribution of towns and the division of mountains, it also reflects the weak coordination ability of Lhasa's urban system.

According to hierarchical clustering and normalization results of main influencing factors, the urban area has the smallest area and the largest population, so its resources and environmental endowments are relatively the lowest. However, the other four indexes except public service ability are the highest, and its public service ability is also at a high level. Duilong Deqing District, which has the strongest public service ability, also has advantages in population resources but does not have advantages in other aspects. The other six counties are in the development of a relatively backward stage. Daz District and Mozhu Gongka County have good overall development, and their characteristics are very similar. Their economic scale and endowment of environmental resources are at a high level. Still, their population resources, public service ability, and spatial connection between cities are low, hindering their development.

5.3 *Correlation analysis of main influencing factors*

Pearson correlation coefficients of main influencing factors and means were calculated using the values in Figure 8, and the results are shown in Table 4.

Table 4. Correlation analysis.

	Mean	Economic scale index	Resource and environment index	Population potential index	Public service capability index	Traffic capacity index
Mean	1.00	0.68	–0.42	0.91	0.76	0.91
Economic scale index	0.68	1.00	–0.31	0.41	0.23	0.64
Resource and environment index	–0.42	–0.31	1.00	–0.61	–0.65	–0.52
Population potential index	0.91	0.41	–0.61	1.00	0.88	0.85
Public service capability index	0.76	0.23	–0.65	0.88	1.00	0.60
Traffic capacity index	0.91	0.64	–0.52	0.85	0.60	1.00

According to the results, the economic scale, population potential, public service capability, and traffic capacity are positively correlated. They are negatively correlated with the Resource and Environment index. This is because resources and environment are limiting factors for urban development, and their carrying capacity has a maximum value. With the development of the city, the population gradually gathers, the development and utilization of the environment gradually increase, and the available area and per capita resources gradually decrease. However, other factors are positive factors, which will further promote urban development by improving its capacity. To some extent, the comprehensive development potential of Lhasa is more closely related to the population potential and spatial connection potential.

6 CONCLUSION

This paper identifies the complex and multi-factor comprehensive development potential into five main influencing indicators, 15 subdividing influencing factors and constructs the evaluation index system of district and county comprehensive development potential, starting from the main influencing factors of urban development potential. And taking Lhasa as an example, the index quantification and hierarchical clustering method were used to evaluate eight districts and counties in Lhasa. The results show that Chengguan District, as the development pole of Lhasa City, is far ahead of other districts and counties, but its radiation driving ability is still insufficient. The development potential of districts and counties except for Chengguan District and Duilong Deqing District is relatively low. Daz District and Mozhu Gongka County, as relatively good regions, have the relatively good economic capacity, but their public service ability is their weakness, which needs to be made up. The cities and towns in the northwest region of Lhasa have poor foundations and conditions in all aspects, and their ability to connect with the outside world is not strong, so they may need to find a new development path suitable for them. As a common problem in all districts and counties, population resources also need to find other development paths to overcome one of the shortcomings.

REFERENCES

Dong, J Y., Zhang, X R. & Shao, W. 2022. Evaluation of the comprehensive potential of county village development based on GIS and AHP hierarchical analysis. *Journal of Nanyang Institute of Technology*. (02): 89–96. doi: 10.16827/j.cnki.41-1404/z. 02: 015.

Fang, C L. & Li, G D. 2015. Particularities, gradual patterns, and countermeasures of new-type urbanization in Tibet, China. *Bulletin of Chinese Academy of Sciences*. (03): 294–305. doi: 10.16418/j.issn.1000-3045. (03): 002

Kong, S H. & Zhu, H Y. 2021. A Research on the economic connection and spatial structural evolution of seven prefecture level cities in Tibet. *Journal of Tibet University*. (Social science edition) (04): 139–148. doi: 10.16249/j.cnki.1005-5738. (04): 018.

Li, F X., Liu, D F., Kong, X S. & Liu, Y L. 2022. Multi-scenario evaluation of county-scale development potential based on shared socioeconomic pathways and dynamic simulation approach. *Journal of Geo-information Science*. (04): 684–697.

Li, Z B. & Zheng G T. 2017. The predicament and mode of new-type urbanization development in ethnic minority areas. *Ethno-National Studies*. (05): 27-41+124.

Ma, R F., Zhang, H Y, Sheng Y F & Wang X C. 2009. Evaluation method of regional development potential at provincial scale. *Research on Development* (03): 18–23. doi: 10.13483/j.cnki.kfyj. 007.

Liu, J Y., Yang, H L., Song, Q. & Fan, X X. 2014. Spatio-temporal pattern evolution of the interaction among the cities in central plains economic zone. *Scientia Geographica Sinica*. (09): 1060–1068. doi: 10.13249/j.cnki.sgs. (09): 005.

Ma, R F., Shen, Y F., Wang X C. & Liu, X T. 2009a. Study on index of evaluation on the county comprehensive development potential at province level—A case from the division of main-functional zones of Yunnan Province. *World Regional Studies* (03): 53–60.

Ma, R F., Wang, X C., Li, W J. & Ma, J J. 2011. Spatial disparity of county comprehensive development potential in Yunnan Province. *Scientia Geographica Sinica* (03), 344–350. doi: 10.13249/j.cnki.sgs. (03): 012.

Ma, R F., Zhang, H Y, Sheng Y F & Wang X C. 2009b. Evaluation method of regional development potential at provincial scale. *Research on Development* (03): 18–23. doi: 10.13483/j.cnki.kfyj. 007.

Wang, C., Ren, M J. & Fan, R R. 2021. Study on sustainable development capacity of villages and towns and its types based on the framework of potential-support-resilience. *Journal of Natural Resources*. (12): 3069–3083.

Wang, F. & Xu, Y L. 2012. Evaluation of urban developmental potential based on principal component analysis: Taking example for Linquan County. *Agriculture Jiangxi (03)*, 170–175. doi: 10.19386/j.cnki.jxnyxb. (02): 052.

Xu, H Q. 2013. A remote sensing index for assessment of regional ecological changes. *China Environmental Science*. (05): 889–897.

Zhang, G., Zhou, S M. & Zhu F J. 2022. Study on Development Potential Evaluation and Type Identification of Town Villages: A Case Study of Chuanfangyu Town, Jizhou District, Tianjin. *Modern Agricultural Science and Technology*. (12): 189–192.

Advances in Civil Engineering and Environmental Engineering – bin Yahya & Bin Daud (Eds)

Study on the application of BIM+GIS technology in the resilience transformation of old community

Jing Liang, Tengfei Wu*, Xiaoli Xu & Tong Liu
Northeast Petroleum University, Daqing, Heilongjiang, China

ABSTRACT: From the perspective of resilience, we analyze the problems of old communities in urbanization and use BIM+GIS technology to help old communities complete resilience transformation. The BIM model and GIS model are integrated to help the community assess and maintain the safety of buildings and lifeline systems by relying on BIM 3D visualization and simulability. The accessibility of emergency shelters and planning of the best evacuation path were analyzed by using GIS simulation. The community selection of the best location for emergency security facilities was assisted using optimal site analysis. To ensure no blind spots in the community, the location of emergency lighting, security surveillance, and safety signs were selected using the model's 3D visual field analysis. The ultimate goal is to improve the resilience of the old community.

1 INTRODUCTION

At present, China is still in the process of continuous urbanization. In the context of rapid urbanization, cities are constantly exposed to many problems in their development. In the "Proposal of the Central Committee of the Communist Party of China on Formulating the 14th Five-Year Plan for National Economic and Social Development and Visionary Goals for 2035," released in 2020, it is proposed to promote a new type of urbanization with people at its core, strengthen the renovation of old communities and community building, and build resilient cities (CPC Central Committee, 2020-2022). As a sub-system of the city, the community is an important component of society. Moreover, it is a unit that directly faces various problems from the environment, politics, economy, and society, and the issue of sustainable development of the community will certainly become one of the important factors for the city to reach sustainable development. According to research, old communities account for a large proportion of China's communities, and existing old communities generally suffer from poor emergency and disaster prevention facilities and poor road access, making the renovation of old urban communities an undoubted priority in building resilient cities.

With the development of information technology, BIM and GIS are becoming more and more closely integrated and play an increasingly important role in the development of smart cities (Liang & Huang 2013; Wang et al. 2016; Zhang, 2014). Combining the two enables GIS to move from macro to micro and BIM to move from indoors to outdoors. The accuracy of analysis and the quality of modeling can be significantly improved compared to individual applications. In the context of "Internet+," there are more and more points of integration between resilient cities and smart cities. The application of BIM+ GIS technology to the construction of resilient communities undoubtedly provides a new way of thinking about how to improve the resilience of communities.

*Corresponding Author: wtf163163@163.com

DOI 10.1201/9781003349563-51

2 OVERVIEW OF BIM AND GIS

Building Information Modelling (BIM) is a product information model that integrates various information about a construction project based on three-dimensional digital technology and is a digital representation of the physical and functional characteristics of the project facilities (Zhang 2009; National Institute of Building Sciences 2013). Due to its unique visualization, coordination, simulation, optimization, and precise control, BIM can provide the basis for all decisions throughout the life cycle of a project, from design to demolition and coordination operations.

Since its introduction, Geographic Information System or Geo-Information system (GIS) has used its powerful ability to integrate, store, edit, analyze, share and present geographic information to provide a powerful analytical and supplementary decision-making tool for urban planning and design work, thereby improving the scientific and implementable nature of the program. With the continuous research and development of GIS, 3D GIS has replaced two-dimensional GIS as the mainstream of research in the field of GIS and has become an important supporting technology for the construction of smart cities.

Along with the development of information technology, Chinese scholars have increasingly researched the integration of BIM and GIS technology (Figure 1), and the integrated application of BIM and GIS is an important research direction for the combination of the two. Through the analysis of the development history of GIS and BIM, we realize that GIS without BIM stops outside the building; BIM without GIS becomes a "building in the air." (Xu 2000). The combination of BIM and GIS will play a greater role in the construction of resilient communities. Using the advantageous characteristics of BIM and GIS and powerful analysis capabilities can provide auxiliary decision-making, rational planning, and efficient management for the construction of resilient communities, providing a new way of thinking about the construction of resilient communities.

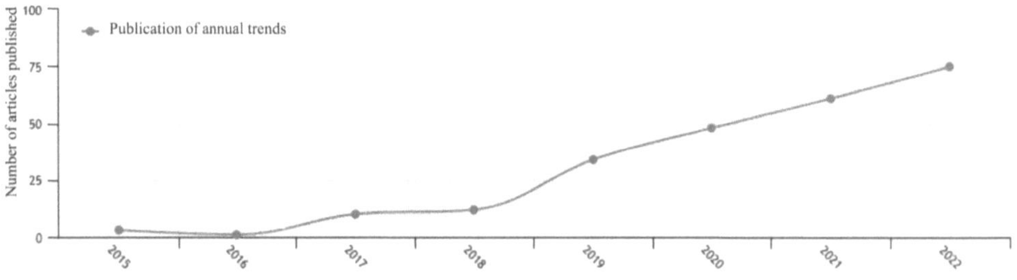

Figure 1. Statistics on the number of articles published on the topic of BIM+GIS technology (data from CNKI).

3 THE CURRENT SITUATION IN OLD COMMUNITIES-AN EXAMPLE FROM A DISTRICT IN DAQING

Over the past few decades, China's economy has grown at a rapid pace, while the level of urbanization has also increased. In the process of rapid urbanization, a large number of family buildings, resettlement houses, and commercial houses have been created. Today, these housing estates have generally been built for more than 20 years, and they are close to reaching the end of their useful life. According to Chinese standards, the durability of the main structure of residential buildings is 50 years, but statistics show that the average life expectancy of buildings in China is less than 30 years (Cao 2011). According to preliminary statistics from China's Ministry of Housing and Urban-Rural Development, as of May 2019, nearly 170,000 old communities need renovation, involving more than 42 million households and a building area of about 4 billion square meters (Tianjin Real Estate News 2019). According to on-site field research, old communities have problems in terms of disaster prevention. They are mainly manifested in the following areas.

3.1 *Degradation of the performance of the main body of the building and lifeline systems*

Most of the older communities were built in the 1980s and 1990s, 20 to 30 years ago, so older communities commonly suffer from structural wear and tear of buildings, aging equipment, and rusting pipes (Figure 2). When a disaster strikes, a sufficiently stable building structure can provide sufficient time for residents to evacuate, but these aging pipes and building structures in older communities make it difficult to keep buildings and lifeline systems stable, which can seriously affect the robustness of a community's response to a disaster.

Figure 2. Current building conditions in older communities.

3.2 *Spatial confusion in road traffic*

As older communities were built over a longer period of time, the prevalence of cars was relatively low when the communities were built. As a result, the roads in the community are generally narrow, dominated by nonmotorized lanes and footpaths, and there is a lack of consideration for parking facilities. Since the 21st century, the prevalence of cars has increased rapidly, and vehicles are often parked along the streets (Figure 3), which increases congestion within the community. Eventually, this can lead to failure to meet fire protection requirements. On the other hand, the roads in older communities are also generally in a dilapidated or even impassable condition due to age and disrepair. This can significantly reduce the accessibility of places of refuge in the community, thus affecting evacuation in times of disaster.

Figure 3. Current status of roads in older communities.

3.3 *Inadequate redundancy of emergency security facilities*

Older communities are built to a lower standard in disaster prevention, so most of them have insufficient redundancy or even an outright lack of emergency protection facilities to meet the requirements of disaster prevention. In addition, the lack of dedicated management in most older communities has led to the encroachment of public spaces and green spaces, indirectly leading to a confusing situation in the community's disaster preparedness space. Without adequate emergency protection facilities, the community will not be properly housed in the event of a disaster, which will make the community very vulnerable and unstable.

3.4 *Inadequate safety management*

Most older communities lack professional property management. As a result, there are many problems with the health and security management of the community, such as a lack of signage facilities or unclear signage, old security monitoring equipment, and a lack of understanding of disaster preparedness and emergency plans by residents. A simple and clear signage system can provide residents with the correct emergency information in the event of a disaster and is important for safe evacuation. Intelligent security monitoring with no blind spots can make a community much safer and contribute to the well-being of residents. Residents' knowledge of disaster preparedness and emergency plans determines the efficiency of evacuation in the event of a disaster.

4 APPLICATION OF BIM+GIS TECHNOLOGY IN THE PROCESS OF RESILIENCE TRANSFORMATION OF OLD COMMUNITIES

The research on the integration of GIS and BIM mainly includes three aspects: data integration, functional integration, and integrated application, among which the integrated application is based on data and functional integration. This paper focuses on the role that the integrated application of BIM and GIS in the process of disaster prevention and resilience enhancement of old communities. There are several main application directions for BIM+GIS technology in the process of resilience transformation of old communities.

4.1 *Robustness assessment and maintenance of building structures and lifeline systems*

When a disaster strikes, the robustness of the building structure is the most important indicator of disaster resilience. Only when the building structure is stable enough can residents have enough time to evacuate safely. After a disaster, a robust and reliable lifeline system is an important indicator of disaster resilience that can help communities rebuild quickly. Improving the robustness of building structures and lifeline systems is one of the most important aspects of improving disaster resilience in older communities. Older communities are generally built over 20 years ago and are approaching or have already reached the end of their useful life. The building structure and lifeline system may suffer from localized or structural deterioration, or even from the emergence of dangerous buildings. BIM can be used to carry out 3D modeling of the building structure and integrated pipelines, so BIM's visualization and information technology advantages can be used to help the relevant departments carry out assessment and maintenance management. The use of BIM's 3D visualization to simulate the construction environment during the maintenance phase can effectively avoid "errors and omissions" that are not readily apparent in the traditional 2D model, such as collisions between internal components and potential intersections between heating and gas pipes and electrical pipework (Figure 4).

4.2 *Create safe and orderly life paths*

The roads in older communities are generally narrow, and private cars facing parking pressure choose to park along the streets, which can cause congestion in the safe evacuation routes for residents when a disaster strikes, greatly reducing the accessibility of evacuation facilities. BIM+GIS technology can be used to help managers plan the best evacuation routes and develop emergency evacuation plans with its powerful and best path analysis and navigation capabilities. On the other hand, the 3D visibility of BIM is used in the renovation phase to reshape the alleyway space within the old neighborhood and assist in the renovation of the road traffic system. Illegal buildings in

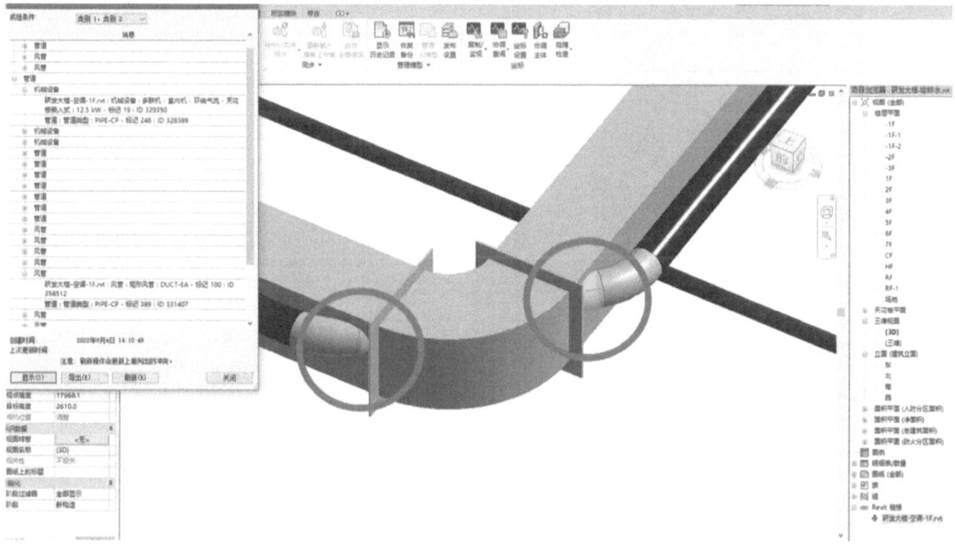

Figure 4. Pipe crash test.

the community are demolished, cut-off roads are opened, and parking areas are rationalized, thus improving the accessibility of evacuation sites.

4.3 *Modification of emergency support facilities to support decision-making*

Most of the old communities are unable to meet the requirements of the evacuation facilities or even are missing, which makes the redundancy of the emergency safeguard facilities in the community far from adequate. The visualization of BIM + GIS can be used to assist in the rational planning of public spaces. The best site analysis function can be used to analyze the location of the emergency security facilities, cost estimates can be made for the reconstruction of basic security facilities, and different options can be compared to select the most economical and reasonable planning solution. On the other hand, the modeling process will attach the components of the infrastructure to the 3D model for easy maintenance and management at a later stage.

4.4 *Scientific analysis to aid community safety management*

To address the lack of signage or unclear signage in older communities and blind spots for security surveillance equipment, SuperMap's 3D GPU analysis can be used to analyze the viewable area of the community and help security managers choose the best location for emergency lighting, security surveillance, and signage within the community (Figure 5). The 3D visibility of BIM+GIS can help the authorities to demonstrate emergency plans to residents and educate them about disaster preparedness. This will improve the adaptability of the community to the disaster situation.

5 RESULT

The combination of BIM and GIS technology can be applied to the resilience transformation of old communities, which can fully exploit the advantages of GIS in spatial analysis and geographic simulation and the advantages of BIM in 3D model visualization and whole life cycle management. This paper analyses the resilience problems of old communities and further proposes that BIM+GIS technology can be used in the process of resilience transformation of old communities

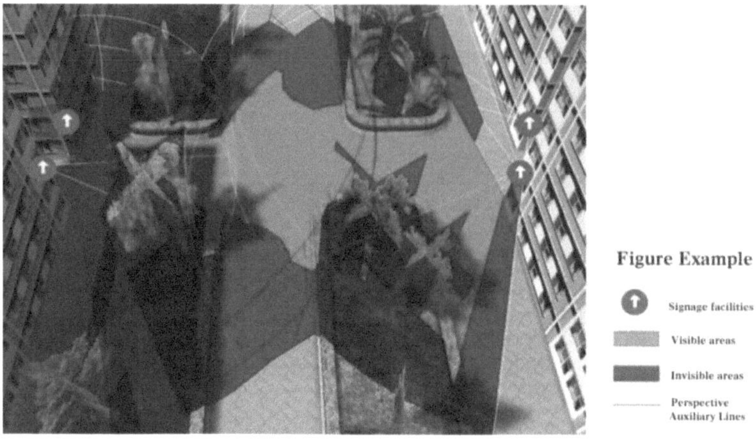

Figure 5. Visibility analysis of signage facilities.

to design and manage the maintenance of building structures and lifeline systems, assist in the scientific location and redundancy assessment of emergency facilities, simulate and optimize the accessibility of the community, help management departments to promote disaster prevention and formulate emergency plans, and improve the intelligence of emergency preparedness and enhance community adaptability. The results of the study can provide guidance and theoretical reference for the construction of resilient communities.

ACKNOWLEDGMENTS

This work is financially supported by the Heilongjiang Province basic research project foundation in university (2020QNW-01).

REFERENCES

Cao Suya. Renovation and Regeneration – Research on renewal Mode of old residential areas in Gulou District, Nanjing [J]. *Sichuan Architecture*. 2011 (4): 47–49.

CPC Central Committee. The central committee of the communist party of China to develop the national economy and social development of the 14th five-year plan and 2035 vision. [EB/OL]. HTTP://http://www.gov.cn/zhengce/2020-11/03/content_5556991.htm.2020-11-03/2022-05-18.

Liang Jun, Huang Qian. *Geographical information world*, 2013, 20(01):81–86+102.

National Institute of Building Sciences. United states national building information modeling standard version 1 – Part 1: Overview, principles, and methodologies [M/OL]. [2013-10-17]. http: //www.wbdg.org/pdfs/NBIMSv1_p1.pdf.

Tianjin Real Estate News. Housing: 170000 old village renovation Involving hundreds of millions of inhabitants. [EB/OL]. https://tj.focus.cn/zixun/49c57706f630c3a0.html, 2019-07-02/2019-07-02

Wang Lingli, Dai Chenguang, Ma Rui. Research on application of GIS and BIM integration in urban architecture planning [J]. *Geospatial Information*, 2016, 14(06):75–78+8.

Xu Qing. Three-dimensional terrain visualization [J]. *Beijing: Surveying and Mapping*. Publishing House, 2000.

Zhang Lin. Research on the application of BIM and GIS technology in site analysis [J]. *Sichuan Building Research*, 2014, 40(05):327–329.

Zhang Yang. Research on Building Engineering Information Integration and Management based on BIM [D]. *Beijing: Tsinghua University*, 2009.

Advances in Civil Engineering and Environmental Engineering – bin Yahya & Bin Daud (Eds)
© 2023 The Author(s), ISBN: 978-1-032-39397-1

Road detection and grouting effect analysis based on ground penetrating radar

Wenxiu Wu* & Xiaoyong Zou
Jinhua Highway and Transportation Administration Center, Jinhua, China

Wenyou Xu
Transportation Bureau of Wuyi County, Jinhua, China

Longting Chen
Highway Business Development Center of Pujiang County, Jinhua, China

Jing Liu
Dongyang Highway and Transportation Administration Center, Dongyang, China

Dapeng Li
Zhejiang Institute of Communications Co., LTD, Hangzhou, China

ABSTRACT: The detection of road internal distress is very important for the development of the road maintenance scheme. Therefore, the analysis of road internal distress detection evaluation and maintenance scheme based on ground penetrating radar (GPR) was carried out. Three-dimensional (3D) GPR was used for road detection, and then GPR data processing, image detection, internal distress diagnosis, and evaluation were conducted. The sections that still show dense cracks or churning after milling resurfacing or repairing were selected as typical sections, and preliminary analysis was carried out from the aspects of radar images of pavement internal structure and internal distress detection results. According to the grouting effect of the distress detection and GPR image, the status quo of road maintenance was analyzed, and the future maintenance scheme of the road section was finally put forward. This paper provides relevant criteria for the non-destructive detection of pavement internal distress, so as to provide guidance for the evaluation of the internal condition of pavement structure and pavement maintenance decisions.

1 INTRODUCTION

How to strengthen the pavement maintenance according to different roads' characteristics and improve the level of maintenance decision, so as to maintain the long-term performance of pavement surface (Li et al. 2021; Liu et al. 2019, 2021; Ren et al. 2021), is one of the major problems faced by highway construction and maintenance departments. The scientific maintenance plan and decision must be based on accurate highway condition detection (Liu et al. 2021, 2022; Wang et al. 2022, 2020). More attention should be paid to the hidden distresses inside the pavement which have been in use for a long time (Chen et al. 2022; Guo et al. 2016; Liu et al. 2021, 2022, 2022, 2022, Xu et al. 2021).

In recent years, GPR, as a non-destructive testing technology, has gained great popularity in the field of highway non-destructive testing because of its characteristics such as fast speed, high precision, high portability, and low traffic impact (Liu et al. 2022, 2022; Wang et al. 2022). However, in the practical application at home and abroad, the following problems have not been

*Corresponding Author: 277182929@qq.com

DOI 10.1201/9781003349563-52

effectively solved, specifically, mainly in the following aspects (Dinh et al. 2018; Xiong et al. 2021): (1) At present, the accuracy of GPR pavement detection technology is different, the efficiency of distress recognition is low, and different distresses are difficult to distinguish (Li et al. 2021). (2) There is no subdivision of distress in radar data analysis (Liu et al. 2022, 2021, 2019; Zhang et al. 2022), which leads to the same detection and analysis methods for different uses and lack of pertinence; meanwhile, due to the lack of reasonable and effective evaluation index, it is impossible to quantitatively evaluate the pavement structure state.

Therefore, in view of the above problems, it is urgent to carry out research on the evaluation index and method of highway internal distress detection based on GPR. A fast method for the diagnosis of internal distress in GPR data processing image recognition is to be presented and an evaluation system is to be established. It provides a relevant judgment basis for GPR non-destructive detection of hidden internal distress under the highway, then evaluates the internal state of highway pavement structure, and provides a basis for pavement maintenance decisions.

2 GPR PAVEMENT DETECTION

2.1 *Detection principle*

To detect the asphalt pavement on the road section, three-dimensional (3D) GPR was used to explore the internal structure of the pavement, so as to master the internal distress of the pavement structure and provide basic data for the treatment and repair of pavement distress. When the receiving antenna moves, continuous GPR reflected waves can be received, and the received signals can be analyzed and processed to form the radar analysis image along the direction of the antenna moving, that is, the longitudinal profile map. The abscissa indicates the distance along the pavement surface, and the ordinate indicates the depth. In addition, the scanned data can be processed into horizontal sections. In the detection process of a 3D big data detection system, electromagnetic waves have continuously emitted underground. When electromagnetic waves encounter different distress in the propagation process, characteristic echoes of specific distress will be generated according to the types of distress. The distress detection by GPR is mainly based on the characteristic signal of electromagnetic wave echo.

2.2 *Checkout equipment*

The 3D GPR system was used to detect the internal distress of pavement structure. The 3D GPR system can be used for 3D imaging, which greatly reduces the misjudgment of the internal structure of the pavement in the past 2D imaging condition. As shown in Figure 1, the 3D GPR system used

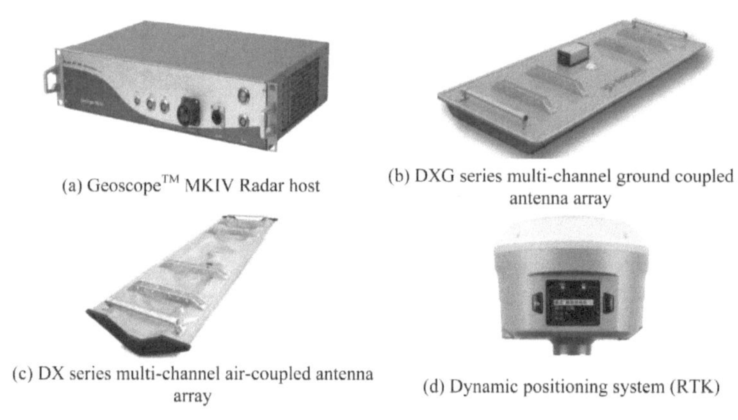

(a) Geoscope™ MKIV Radar host

(b) DXG series multi-channel ground coupled antenna array

(c) DX series multi-channel air-coupled antenna array

(d) Dynamic positioning system (RTK)

Figure 1. Schematic diagram of GPR detection equipment.

in this detection mainly includes Geoscope^{TM} MKIV radar host, DXG series multi-channel ground coupled antenna array, DX series multi-channel air-coupled antenna array, and dynamic positioning System (RTK). The main equipment parameters of 3D GPR are shown in Table 1.

Table 1. Parameters of 3D GPR equipment.

Parameters	Index
Detection depth	8m
Detection speed	60km/h
Transverse sampling interval	0.071m
Longitudinal sampling interval	0.025m~0.1m
Vertical sampling interval	0.004m
Advantages	Wide coverage, no breakage

2.3 *Detection scheme*

The detection section was the K7+000~K9+000 section of the North Line of Jinlan on the S210 provincial highway in Jinhua City, with a total length of about two kilometers. In view of the grouting condition, GPR detection collected detection data from a single lane, in turn, covering all lanes. The detection vehicle was followed by the vehicle closely to ensure the safety of the detection personnel and equipment. The detection process should be closed according to the highway maintenance safety Operation code (JTG H30-2015).

3 ANALYSIS OF DETECTION RESULTS

3.1 *Detection image analysis*

The sections with dense cracks or muddied sections still appearing after milling resurfacing or repairing were selected as typical sections, and the preliminary analysis was carried out from the aspects of radar images of pavement internal structure and internal disease detection results.

The first was to detect the internal structure of the left road section, as shown in Figure 2(a). At position K8+250-K8+300 on the left of the road, the radar image inside the pavement structure shows that grout is mainly distributed at the base level (20cm-50cm depth) in the grouting area. Due to the large dielectric difference between the grouting material and the road material, the radar wave coverage overlaps, indicating that the overall grouting effect is good.

The second is to detect the internal structure of the right road section, as shown in Figure 2(b). At the position of K8+900-K9+000 on the right side of the road, the radar image inside the pavement structure shows that grout is mainly distributed in the base, bottom base, and soil base (20cm-120cm depth) in the grouting area. The dielectric difference between the grouting material and the road material is large, but the coverage overlap degree of radar reflection wave is medium, indicating that the overall grouting effect is medium.

Finally, the 3D radar detection results are summarized in Table 2 according to the grouting status of the detected road sections.

3.2 *Analysis of maintenance scheme*

According to the detection results, the main distresses of the road section are the looseness and crack of the base, and the overall grouting effect is moderate. Considering the road traffic and maintenance cost, it is suggested that some areas with poor grouting effect should be reinforced by grouting, and the areas with good efficiency should not be treated temporarily. The specific pavement maintenance scheme is shown in Table 3.

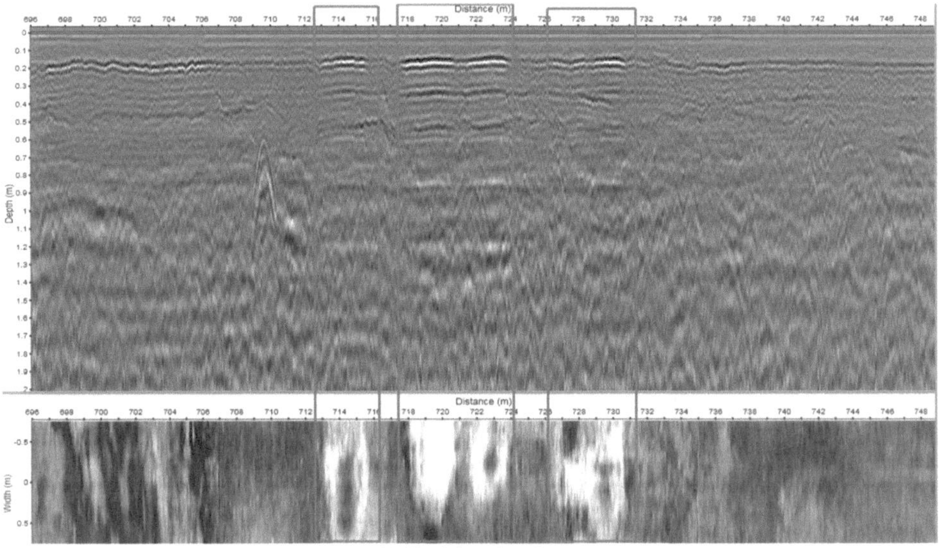

(a) The left road surface

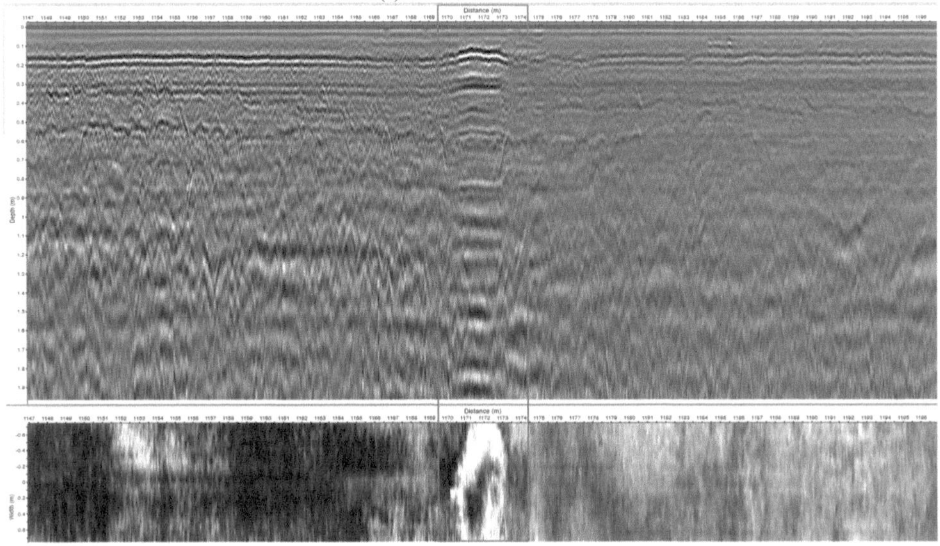

(b) The right road surface

Figure 2. GPR image of internal structure condition.

Table 2. Results from K7+000~K9+000 of Jinlan North Line.

Stake number	Direction	Total area detected/m^2	Area of grouting area/m^2	Average height of the top surface of the grouting area/m	Grouting rate/%
K7+500~K9+000	The right road section	2250	7.65	0.18	0.34
K7+700~K9+000	The left road section	2250	194.55	0.21	8.65

Table 3. Maintenance scheme of K7+000-K9+000 left section of Jinlan North Line.

Sequence number	Stake number	Distance from the center/m	Initial depth/m	Area/m^2	Grouting effect	Maintenance scheme
1	K7+673.2	1.75	0.16	4.8	Good	/
2	K7+709.1	1.75	0.15	5.1	Good	/
3	K7+723.2	1.75	0.17	5.4	Ordinary	Resonance crushing petrified laminate grouting
4	K7+729.2	1.75	0.17	4.6	Good	/
5	K7+739.6	1.75	0.17	2.8	Good	/
6	K7+767.9	1.75	0.17	8.7	Ordinary	Resonance crushing petrified laminate grouting
7	K7+791.6	1.75	0.39	16.5	Ordinary	Base grouting
8	K7+804.6	1.75	0.38	10.3	Ordinary	Base grouting
9	K7+817.7	1.75	0.38	14.1	Ordinary	Base grouting
10	K7+840.3	1.75	0.38	6.0	Good	/
11	K7+845.6	1.75	0.17	3.6	Good	/
12	K7+857.8	1.75	0.40	9.6	Ordinary	Base grouting
13	K7+869.2	1.75	0.40	10.5	Ordinary	Base grouting
14	K7+920.4	1.75	0.12	6.6	Good	/
15	K7+925.5	1.75	0.14	5.5	Good	/
16	K7+930.3	1.75	0.15	5.8	Ordinary	Resonance crushing petrified laminate grouting
17	K7+945.1	1.75	0.16	5.0	Good	/
18	K8+171.7	1.75	0.17	6.8	Good	/
19	K8+180.8	1.75	0.17	10.3	Good	/
20	K8+187.9	1.75	0.17	5.3	Good	/
21	K8+629.7	1.75	0.17	18.8	Good	/
22	K8+659.5	1.75	0.19	5.7	Good	/
23	K8+665.7	1.75	0.22	4.3	Good	/
24	K8+671.9	1.75	0.22	8.7	Good	/
25	K8+689.7	1.75	0.19	5.7	Good	/
26	K8+704.6	1.75	0.18	3.9	Good	/

4 CONCLUSIONS

The detection, evaluation, and maintenance of road internal distress based on GPR were analyzed. The main conclusions are as follows.

1) The 3D GPR image of road internal distress condition was analyzed. Combined with the detection image, there are highlighted areas in the radar reflected waves, indicating the existence of different degrees of cavitation and other distress.
2) The 3D GPR image after road grouting was analyzed. Due to the large dielectric difference between the grouting material and the road material, the overall grouting effect can be judged according to the coverage overlap of the radar reflection wave.
3) Combined with the maintenance project of the Jinhua S210 provincial road section, the GPR testing road detection scheme was designed. The testing road was tested and evaluated, and the corresponding maintenance suggestions were given according to the testing results.

ACKNOWLEDGMENTS

This study was sponsored by the Key Science and Technology Research Project of Jinhua, China under Grant 2021-3-176.

REFERENCES

Chen, Y. H., Gu, X. Y., Liu, Z., and Liang, J. (2022). "A fast inference vision transformer for automatic pavement image classification and its visual interpretation method." *Remote Sens-Basel*, 14(8).

Dinh, K., Gucunski, N., and Duong, T. H. (2018). "An algorithm for automatic localization and detection of rebars from GPR data of concrete bridge decks." *Automat Constr*, 89, 292–298.

Guo, S. L., Yan, F., Zhu, P. M., and Xiu-Zhong, L. I. (2016). "Numerical study on the response of ground penetrating radar wave field to crack width." *Progress in Geophysics*.

Li, S. W., Gu, X. Y., Xu, X. R., Xu, D. W., Zhang, T. J., Liu, Z., and Dong, Q. (2021). "Detection of concealed cracks from ground penetrating radar images based on a deep learning algorithm." *Constr Build Mater*, 273.

Li, X. F., Liu, H., Zhou, F., Chen, Z. C., Giannakis, I., and Slob, E. (2021). "Deep learning-based nondestructive evaluation of reinforcement bars using ground-penetrating radar and electromagnetic induction data." *Comput-Aided Civ Inf.*

Liu, Z., Chen, Y. H., Gu, X. Y., Yeoh, J. K. W., and Zhang, Q. P. (2022). "Visibility classification and influencing-factors analysis of airport: A deep learning approach." *Atmospheric Environment*, 278.

Liu, Z., Gu, X. Y., and Dong, Q. (2019). "Multi-Scale 3D Display of the Internal Quality of the Pavement Based on BIM." *Proc., 19th COTA International Conference of Transportation Professionals (CICTP) – Transportation in China 2025*, 4265–4273.

Liu, Z., Gu, X. Y., Dong, Q., Tu, S. S., and Li, S. W. (2021). "3D visualization of airport pavement quality based on BIM and WebGL integration." *J Transp Eng B-Pave*, 147(3).

Liu, Z., Gu, X. Y., Ren, H., Wang, X., and Dong, Q. (2022). "Three-dimensional finite element analysis for structural parameters of asphalt pavement: A combined laboratory and field accelerated testing approach." *Case Stud Constr Mat*, 17.

Liu, Z., Gu, X. Y., Ren, H., Zhou, Z., Wang, X., and Tang, S. (2022). "Analysis of the dynamic responses of asphalt pavement based on full-scale accelerated testing and finite element simulation." *Constr Build Mater*, 325.

Liu, Z., Gu, X. Y., Wu, C. Y., Ren, H., Zhou, Z., and Tang, S. (2022). "Studies on the validity of strain sensors for pavement monitoring: A case study for a fiber Bragg grating sensor and resistive sensor." *Constr Build Mater*, 321.

Liu, Z., Gu, X. Y., Wu, W. X., Zou, X. Y., Dong, Q., and Wang, L. T. (2022). "GPR-based detection of internal cracks in asphalt pavement: A combination method of DeepAugment data and object detection." *Measurement*, 197.

Liu, Z., Gu, X. Y., Yang, H. L., Wang, L. T., Chen, Y. H., and Wang, D. Y. (2022). "Novel YOLOv3 model with structure and hyperparameter optimization for detection of pavement concealed cracks in GPR images." *Ieee T Intell Transp.*

Liu, Z., Gu, X., and Wang, L. (2021). "Research on information management of Airport pavement quality based on BIM and GIS integration." *Green and Intelligent Technologies for Sustainable and Smart Asphalt Pavements*, CRC Press, 432–436.

Liu, Z., Gu, X., Chen, Y., and Chen, Y. "System architecture and key technologies for the whole life cycle of smart road." Proc., *Journal of Physics: Conference Series, IOP Publishing*, 012105.

Liu, Z., Wang, S. Y., Gu, X. Y., Li, Z. G., Dong, Q., and Cui, B. Y. (2022). "Application of a novel EWMA-phi chart on quality control in asphalt mixtures production." *Constr Build Mater*, 323.

Liu, Z., Wu, W. X., Gu, X. Y., Li, S. W., Wang, L. T., and Zhang, T. J. (2021). "Application of combining YOLO models and 3D GPR images in road detection and maintenance." *Remote Sens-Basel*, 13(6).

Ren, H., Gu, X. Y., and Liu, Z. (2021). "Analysis of mechanical responses for semi-rigid base asphalt pavement based on mls66 accelerated loading test." proc., 20th and 21st joint COTA international conference of transportation professionals – advanced transportation, *Enhanced Connection*, 732–742.

Wang, D., Liu, Z., Gu, X., Wu, W., Chen, Y., and Wang, L. (2022). "Automatic detection of pothole distress in asphalt pavement using improved convolutional neural networks." *Remote Sens-Basel*, 14(16).

Wang, L. T., Gu, X. Y., Liu, Z., Wu, W. X., and Wang, D. Y. (2022). "Automatic detection of asphalt pavement thickness: A method combining GPR images and improved Canny algorithm." *Measurement*, 196.

Wang, S., Al-Qadi, I. L., and Cao, Q. (2020). "Factors impacting monitoring asphalt pavement density by ground penetrating radar." *Ndt&E Int*, 115.

Xiong, C. L., Yu, J. M., and Zhang, X. N. (2021). "Use of NDT systems to investigate pavement reconstruction needs and improve maintenance treatment decision-making." *Int J Pavement Eng.*

Xu, Z., Zhou, Q., Chen, G., Wang, Y., Yang, X., and Liu, Z. "An effectiveness study of spc control charts on the quality control for construction in asphalt pavement." *Proc., IOP Conference Series: Earth and Environmental Science*, IOP Publishing, 012115.

Zhang, J., Lu, Y., Yang, Z., Zhu, X., Zheng, T., Liu, X., Tian, Y., and Li, W. (2022). "Recognition of void defects in airport runways using ground-penetrating radar and shallow CNN." *Automat Constr*, 138, 104260.

Advances in Civil Engineering and Environmental Engineering – bin Yahya & Bin Daud (Eds)
© 2023 The Author(s), ISBN: 978-1-032-39397-1

Ground temperature measurement and geological genesis analysis of Nige high geotemperature tunnel

Weiyu Tang, Chenchen Jiang, Shao qiang Zhang & bokuan Li
Power China Road Bridge Group Co., Ltd, Beijing, China

Ping Zhang* & Kaixin Li
College of Resources and Safety Engineering, Chongqing University, Chongqing, China

ABSTRACT: In order to reveal the temperature distribution pattern and geological causes of high geothermal tunnels, this paper adopts the method of field test and geological analysis to study the temperature distribution characteristics and geological causes of high geothermal tunnels in Nige. The results show that: the temperature of the tunnel wall and air temperature increase with the increase of tunnel distance, while the relative humidity of air shows a trend of increasing and then decreasing due to the increase of temperature; the temperature of overrun horizontal borehole and lateral horizontal borehole increases with the increase of borehole distance. The heat source of the Nige tunnel mainly comes from Nige hot spring, Laohutan hot spring, and Yashadi hot spring. The heat conduction channels are faults, secondary faults, contact zone fractures, structural fractures, dissolution fractures, and karst pipelines.

1 INTRODUCTION

With the continuous development of the Chinese economy, Chinese highway construction mileage is rapidly increasing, the advantages of tunnels through the mountains are becoming more and more obvious, and a large number of high-ground temperature tunnels have emerged in recent years. High ground temperature produces high heat damage deteriorates the construction, affects the construction progress, and also weakens the strength of the surrounding rock. Therefore, the study of high ground temperature tunnels has a more important practical significance.

In recent years, many scholars have conducted relevant studies on issues related to high-ground temperature tunnels. Li et al. (2016) explored the influence of temperature on tunnel envelope structure and proposed related measures. Zheng et al. (2018) compared and analyzed the settlement and stress change law of tunnel envelope support structure under different temperature conditions and found that the lining stress was positively correlated with the envelope temperature. Wang et al. (2016) studied the change law of high-temperature tunnel envelope and the simulation results showed that the safety coefficient of tunnel envelope was negatively correlated with temperature. Wu et al. (2019) studied the effect of temperature on tunnel lining structure by using FLAC3D software. Wu et al. (2020) studied the deformation law of tunnel surrounding rock under high-temperature conditions with the help of ANSYS software simulation and derived the change law of tunnel surrounding rock deformation with time. Wang et al. (2019) studied the initial support of high ground temperature tunnel by field test and numerical simulation based on the Sangzuling high ground temperature tunnel. From the above studies, it can be seen that there are more studies on thermal insulation and lining of high geothermal tunnels, but there are no reports on the field temperature testing and geological causes of high geothermal tunnels. Therefore, it is of great

*Corresponding Author: zhangpgut@163.com

 DOI 10.1201/9781003349563-53

theoretical significance and engineering value to investigate the field temperature distribution characteristics and geological formation causes of high geothermal tunnels in Nige for the long-term stability of such tunnels.

In this paper, relying on the Nige high geothermal tunnel in Honghe Prefecture, Yunnan Province, using field tests and geological analysis, we study the characteristics of the temperature distribution and geological causes of the Nige high geothermal tunnel site, with a view to providing reference and learning for the design and construction of similar tunnel projects.

2 PROJECT OVERVIEW

Nige Tunnel is located in Kele Village, Honghe Prefecture, with a total tunnel length of 3366m and a maximum burial depth of 643m. The two-way double tunnel construction model is adopted. The tunnel crosses the Triassic Middle Tectonics Jielao Group tuff and Yanshan Stage intrusion granite, the maximum temperature of the limestone section is 65.1°C, and the maximum temperature of the granite section is 88.8°C. The Yashadi, Nige, and Laohutan hot springs are distributed around the tunnel site and the hot spring water temperature is 51.8~86.6°C, as shown in Figure 1.

Figure 1. Diagram of Nige tunnel entrance.

3 RESULT ANALYSIS

3.1 *Test and analysis of ground temperature in Niger tunnel*

3.1.1 *Temperature Analysis of air and rock wall in different excavation sections*

From the entrance to the palm face, the wall and air temperature and humidity tests are carried out every 100m. The measurement result is shown in Figure 2.

As can be seen in Figure 2, both the wall temperature and the air temperature of the tunnel increase with the distance of the tunnel, and temperatures around the palm face may rise sharply. The relative humidity of the air increases first and then decreases due to the increase in temperature, and the humidity is closely related to water inflow.

As shown in Figure 3, there is some high temperature in the ponding of the tunnel. It can be concluded that there is a large amount of ponding area in the right line of the entrance section of nig tunnel, and the temperature of the ponding is higher than that of air and wall, both above 30°C and up to 40°C. As the larger the ponding area is, the higher the construction ambient temperature is, the ponding in the tunnel should be discharged in time during the construction process.

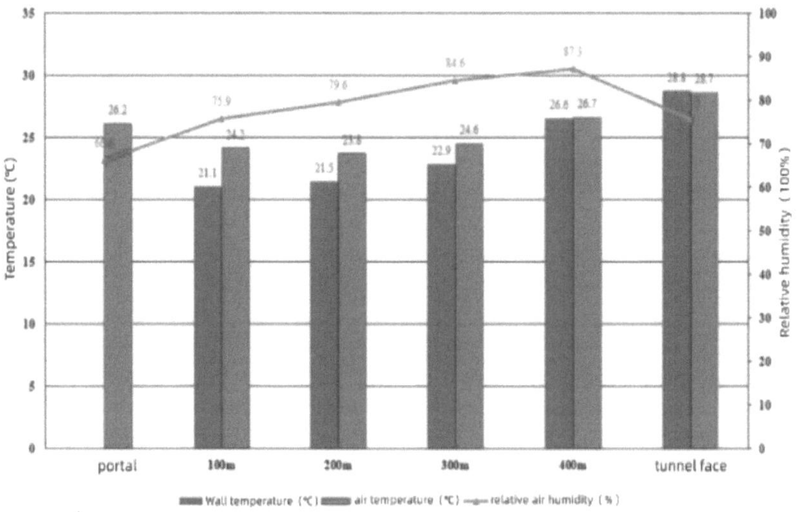

Figure 2. Temperature and humidity distribution characteristics of different distances of tunnel.

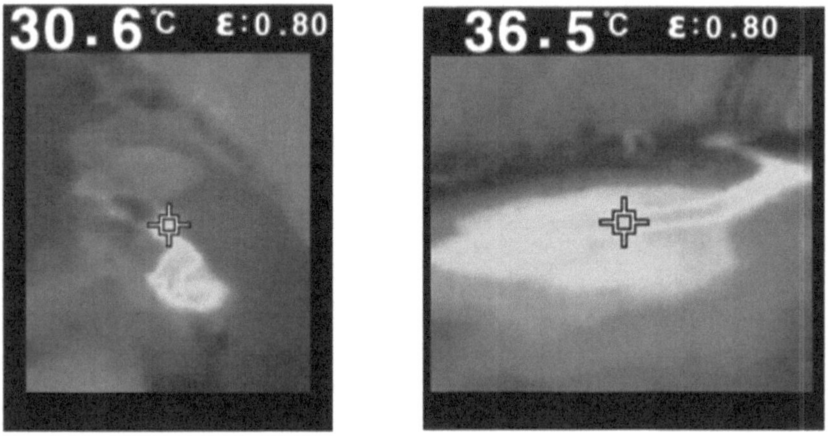

Figure 3. Temperature distribution of ponding at the entrance section of the Nige tunnel.

3.1.2 *Temperature analysis of advanced horizontal drilling*

In order to better obtain the temperature in front of the tunnel face, it is necessary to carry out advanced temperature measurement drilling for the Nige tunnel. In the temperature measurement advance drilling, a drill hole is arranged in the center of the tunnel excavation cross-section, the drill hole depth is 20m, and the drill hole diameter is Φ70mm. The layout of the temperature test scheme is shown in Figure 4.

It can be seen from Figure 5 that the temperature in the range of 0~4m is about 54.4~56.6°C, and the temperature distribution in the two orifices is basically the same. The measuring hole is in the range of 2~20m in length, and the temperature rises slowly. However, within a distance of about 2m from the orifice, the temperature changes rapidly, changing to about 13°C. In addition, the temperature in the left hole is higher, the highest temperature in the 20m deep hole reaches 59°C, while the temperature in the right hole is 56°C in 20m deep hole. Within 5m of the orifice, the temperature difference between the two orifices is only about 1°C.

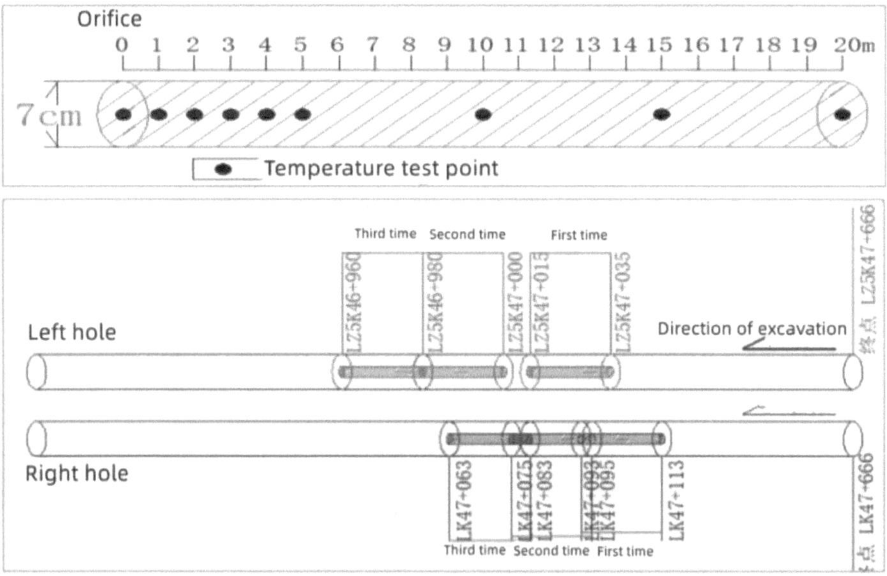

Figure 4. Test scheme of advanced horizontal hole temperature.

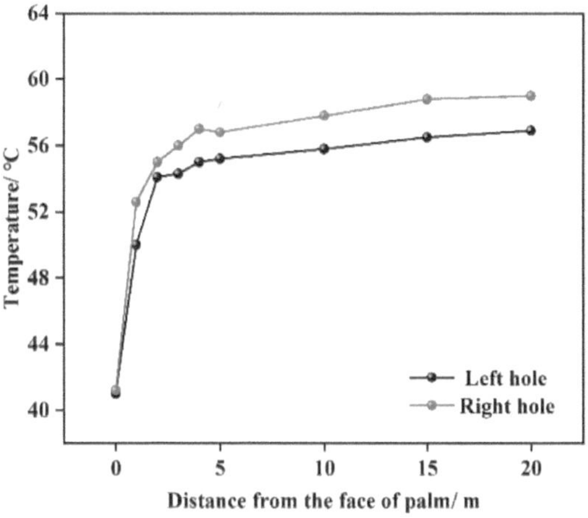

Figure 5. Temperature variation curve of advanced horizontal hole.

3.1.3 *Horizontal borehole geothermal test*

In order to study the temperature change of the surrounding rock in the attachment of the secondary lining, a lateral drilling test was carried out at the position of the secondary lining, and the height of the lateral horizontal drilling was about 1.3m. As shown in Figure 6, the drilling depth is about 3m and the hole diameter is about 5cm. The measuring points are arranged at positions of 3m, 2m, and 40cm, and the influence of concrete pouring on heat transfer can be judged according to the measurement results.

As shown in Figure 7, the maximum temperature of the rock at 3m is 50.2°C, the average temperature is 50.1°C, the maximum temperature of the rock at 2m is 49.1°C, and the average temperature is 49.0°C. The maximum temperature at 0cm is 46.8°C, and the average is 46.2°C. According to the measurement results, in the pouring layer, the temperature can drop rapidly in

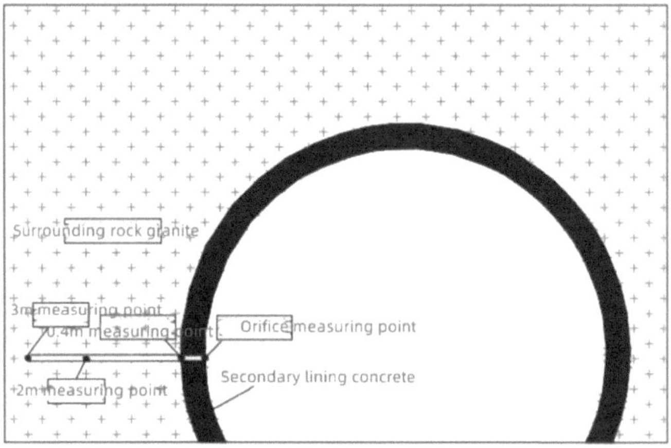

Figure 6. Schematic diagram of lateral horizontal drilling.

a short distance, while the temperature of the surrounding rock does not change much. It can be seen that pouring concrete can partially isolate the heat from the mountain and conduct it into the tunnel, which is why the temperature of the tunnel near the entrance is relatively low, while the temperature of the rock wall has risen sharply.

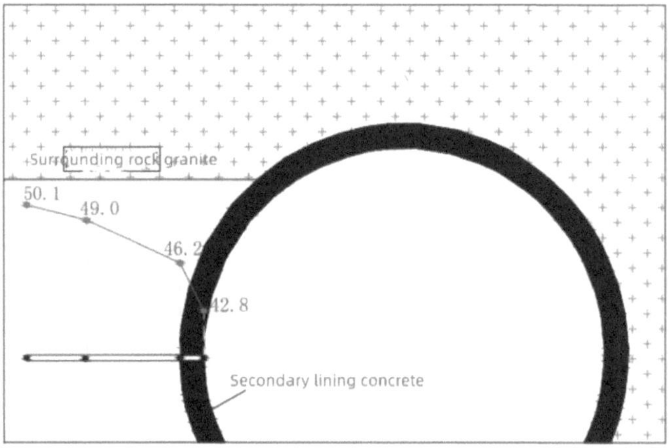

Figure 7. Temperature distribution of lateral drilling.

3.2 *Geological genesis analysis*

3.2.1 *Heat source investigation and cause analysis*

In order to find out the heat source that caused the high temperature of the tunnel, the geologists investigated a series of hot springs around the Jian (Ge) Yuan expressway line. Focus on the investigation and analysis of three hot springs near Nige Tunnel, the three hot springs are Nige hot Spring, Laohutan hot Spring, and Yashadi hot Spring. The three hot springs are distributed around the entrance and exit of the Nige Tunnel. The high ground temperature of the tunnel is consistent with the heat source of the hot spring. The hot spring is the discharge outlet of the geothermal heat source and is connected to the underground heat source. The geographic location is shown in Figure 8.

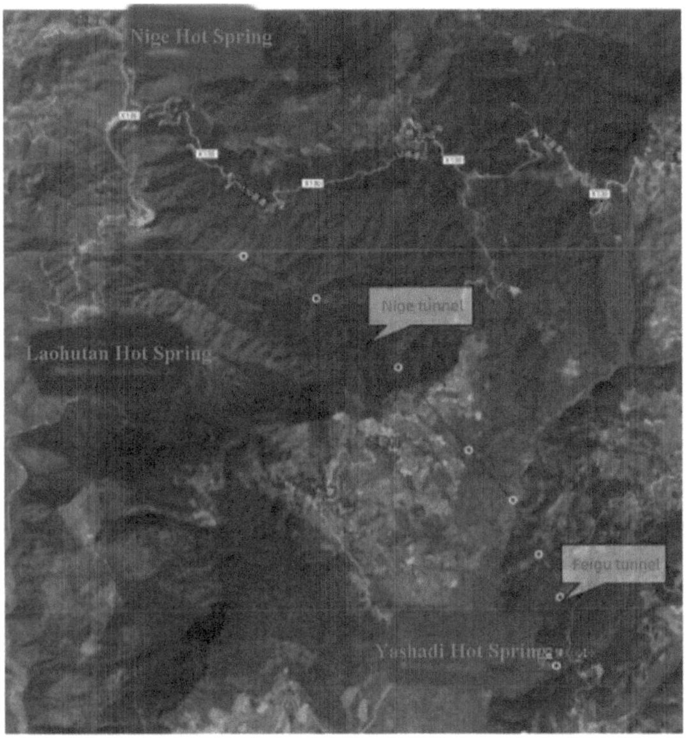

Figure 8. Geographical distribution of hot springs.

(1) Nige Springs

The outlet of Nige Hot Spring is located on the left bank of the Longcha river at the exit of the Nige tunnel (Figure 9). The measured water temperature of the hot spring outlet is 67.9°C, the water volume is about 1.06L/s, the outlet elevation is 908m, and the water vapor smells of a small number of rotten eggs.

Figure 9. Water temperature test of Nige hot spring outlet.

(2) Tiger Beach Hot Spring

The outlet of the Laohutan hot spring is located on the right bank of the Longcha river at the exit of the Nige tunnel (Figure 10). The water temperature of the hot spring outlet is 53.8°C, the water volume is about 1.24L/s, the outlet elevation is 762m, and the water vapor has a small smell of rotten eggs.

Figure 10. Water temperature test of Laohutan hot spring.

(3) Yashadi Hot Spring

The outlet of the Yashadi hot Spring is located on the left bank of the Jasha river at the entrance of the Nige tunnel (Figure 11). The measured water temperature of the hot spring is 87.9°C, the water volume is about 2.20L/s, and the height of the water outlet is 940m. It is a rising spring. There is a small amount of yellow material on the wall of the water outlet, and the water vapor has a small amount of rotten egg smell.

Figure 11. Water temperature test at Yashadi hot spring outlet.

3.2.2 *Analysis of hydraulic and thermal connection of Nige tunnel*

The hydraulic and thermal connection of the Niger Tunnel is shown in Figure 12. It can be concluded that the thermal anomaly existing in the deep crust of the Nige tunnel conducts upward through the fault Fn1 and the fault Fn2. Part of the heat source is exposed to the surface to form hot springs along the Longcha river and along the Jiasha river, while another part of the heat source is transmitted to the superficial section of the crust, passing through secondary faults fn1, contact zone cracks, structural cracks, dissolution cracks, and karst pipelines or rock masses, etc. The secondary channel conducts discretely. In the contact zone between limestone and granite, the heat source transmitted to the superficial surface of the crust is mainly conducted through secondary small faults, dissolution fissures, karst pipelines, or contact zones.

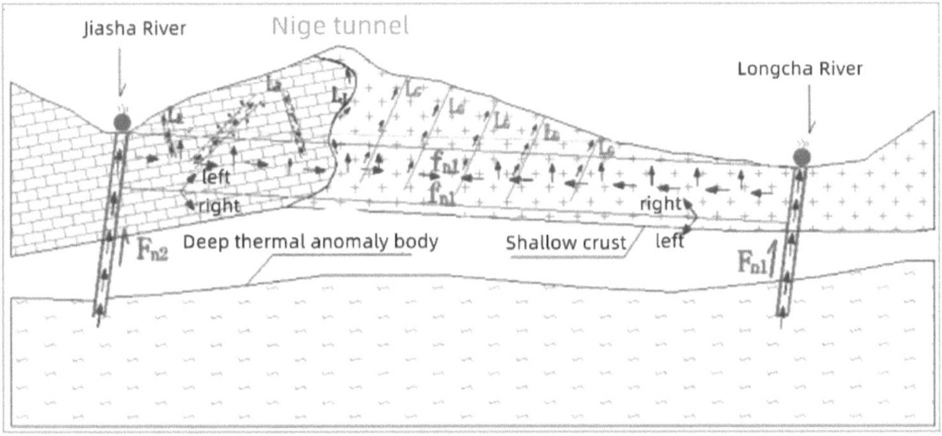

Figure 12. Schematic diagram of the analysis of the causes of Nige tunnel.

4 CONCLUSION

In this paper, the method of the field test is used to analyze the temperature distribution characteristics of air, rock wall, advanced horizontal borehole, and lateral horizontal borehole in different excavated tunnel sections, explore the heat source of the Niger high geothermal tunnel, and analyze the hydraulic and hydraulic characteristics of the Niger tunnel. The conclusions are as below:

(1) Both the wall temperature and air temperature of the tunnel increase with the increase of the tunnel distance, while the relative humidity of the air first increases and then decreases due to the increase of the temperature.
(2) The temperature of both the advanced horizontal drilling and the lateral horizontal drilling increased with the increase of the drilling distance.
(3) The heat sources of the Nige tunnel are mainly Nige hot spring, Laohutan hot spring, and yashadi hot spring. The heat sources are dispersed through secondary channels such as faults, secondary faults, contact zone fractures, structural fractures, dissolution fractures, and karst pipelines or rock masses.
(4) This study found geothermal sources and heat conduction channels, which can lay the foundation for geothermal exploitation and establish geothermal heat storage bases here.

ACKNOWLEDGEMENTS

This work was financially supported by the National Natural Science Foundation of China (No. 51774057).

REFERENCES

Li Guoliang, Cheng Lei, Wang Fei. Study on key technology for the construction of high ground temperature tunnel [J]. *Railway Standard Design*, 2016, 60(06): 55–59.

Wang Mingnian, Tang Xinghua, Wu Qiujun, et al. Temperature field variation rules of rock and support structure in high rock temperature tunnel[J]. *Journal of the China Railway Society*, 2016, 38(11): 126–131.

Wang Mingnian, Wang Qiling, Hu Yunpeng, et al. Mechanical characteristics analysis of initial support of high geothermal tunnel [J]. *Journal of the China Railway Society*, 2019, 41(11): 116–122.

Wu Biao, Peng Xuejun, Yuan Chao, et al. Discussion on lining structure design and construction technology of high ground temperature tunnel [J]. *Journal of Hunan University of Science and Technology* (Natural Science Edition), 2019, 34(02): 18–24.

Wu Zhaofeng, Hu Huirong. Study on the deformation and mechanical behaviors of the tunnel lining structure induced by high temperature in the fire scenario [J]. *Modern Tunnelling Technology*, 2020, 57(06): 101–106.

Zheng Wen, Liu Naifei, Liu Xiaoping. Mechanical characteristics of supporting structures for high geo-temperature tunnel [J]. *Coal Geology & Exploration*, 2018, 46(06): 138–143.

Advances in Civil Engineering and Environmental Engineering – bin Yahya & Bin Daud (Eds)

Analysis of the effect of dewatering and excavation on adjacent bridge piles in the foundation pit

Xin Ning*
School of Civil Engineering, Lanzhou Jiaotong University, Lanzhou, China

ABSTRACT: Based on Biot consolidation theory and finite element software Midas GTX, a numerical model of seepage stress coupling simulating foundation pit excavation and dewatering was established to study the influence of deep foundation pit excavation on the neutral point of lateral friction resistance, negative friction resistance and pile axial force of adjacent bridge piles considering multiple dewatering projects. The results show that precipitation has a great influence on the negative friction resistance of the bridge pile side. When the excavation depth of the foundation pit is within 15.8m, the settlement of the bridge pile foundation is less affected by the construction of the foundation pit. However, when the excavation depth exceeds this range, the settlement of the bridge pile foundation caused by each construction gradually increases.

1 RESEARCH BACKGROUND

With the rapid development of economic construction and urbanization, large-scale projects continue to appear, and the area and depth of the foundation pit requirements are becoming higher and higher. As is known to all, excavation of the foundation pit will cause obvious deformation of subsoil around it and have a corresponding influence on the performance characteristics of adjacent structures (Guo 2017; Jin 2018). When the excavation of the foundation pit has a significant influence on the performance characteristics of the adjacent foundation, it may even lead to the destruction of the adjacent structure and the instability of the foundation.

At present, most of the research focuses on the response law of surrounding soil or adjacent existing buildings after excavation (Jin 2005, 2004; Li 2004). There are few studies on the influence of friction resistance on bridge pile foundations in the construction process. Therefore, taking a deep foundation pit as an example, this paper established a three-dimensional finite element model of stress-seepage coupling, studied the influence of foundation pit excavation and stepwise dewatering on bridge pile foundation, and simulated the influence of different drainage materials on foundation pit engineering.

2 NUMERICAL ANALYSIS MODEL

MIDAS GTS/NX finite element software is used to conduct three-dimensional finite element simulation analysis of foundation pit excavation and bridge pile foundation (Li 2007, 2014). In order to consider the impact of foundation pit excavation on the surrounding environment, a THREE-DIMENSIONAL solid calculation model was established with a length of 210m, a width of 205m, and a soil layer thickness of 75m. The structural load is the uniform load that transforms the bridge deck and vehicle load into the pier (Wang 2010, 2018).

In the calculation model, the modified Mohr-Coulomb model was used to simulate the basic physical properties and mechanical properties of soil. The model obeyed the Mohr-Coulomb shear

*Corresponding Author: ningx16@lzu.edu.cn

DOI 10.1201/9781003349563-54

failure criterion, and the soil adopted a four-node/eight-node solid hybrid grid. Soil layer parameters are determined according to the geotechnical investigation report of the site, as shown in Table 1.

Supporting pile, prestressed anchor cable, crown beam, cap-cap, bridge pier, and bridge pile foundation are linear elastic constitutive models, supporting pile is equivalent to slab element underground continuous wall simulation, anchor cable is simulated by truss element, cap-cap and bridge pier are simulated by the solid element, and bridge pile foundation and crown beam are simulated by beam element. The pile interface element and pile tip element are applied between the bridge pile foundation and soil to simulate pile-soil interaction. Due to the large difference in stiffness between equivalent ground wall, cap, and soil, a contact element is established in the contact part of the ground wall and soil to simulate the actual contact effect.

The stress-seepage coupling analysis function of MIDAS GTS/NX is applied in the analysis of the construction stage. The excavation, support, and multiple precipitation simulation of the foundation pit are divided into 13 steps.

Table 1. Physical and mechanical properties of soil layer.

The soil	The thickness/m	Modulus of elasticity/ MPa	The permeability coefficient/ $(m \cdot s^{-1})$	Poisson's ratio	Bulk density/ $(kN \cdot m^{-3})$	Cohesive force/kPa	Angle of internal friction/(°)
Miscellaneous fill	7.0	18	1.4×10^{-11}	0.38	18.0	10.0	23.0
The pebble	5.0	135	5.2×10^{-4}	0.30	20.5	3.0	40.0
Strong weathered rock	3.0	120	5.8×10^{-6}	0.35	20.5	25.0	32.0
The weathered rock	50.0	165	5.8×10^{-7}	0.36	21.0	30.0	35.0

3 CALCULATION RESULTS AND ANALYSIS

Excavation of the foundation pit will change the distribution of the original soil stress field, and the unloading of the soil in the pit will inevitably lead to the movement of the soil outside the pit (Wu 2019; Yang 2019). When the settlement deformation of the soil near the bridge pile caused by excavation is greater than that of the pile itself, the negative friction resistance between the soil and the bridge pile will occur, which will have an adverse effect on the bridge.

In order to study the influence of foundation pit excavation on girder bridge pile foundation, the displacement, friction resistance, and internal force of bridge pile foundation under various construction and precipitation conditions were extracted and analyzed.

(1) Displacement of bridge pile foundation

As shown in Figure 1, after the completion of foundation pit excavation, the horizontal displacement of pile top of no. 1 and No. 2 piles near the foundation pit is roughly equal in numerical value. The maximum horizontal displacement of the pile top of the No.1 pile foundation is 24.30mm, that of the No. 2 pile foundation is 24.10mm, that of the No.3 pile foundation is 23.16mm and that of the No. 4 pile foundation is 22.80mm. With the progress of dewatering in the foundation pit, the water head difference increases, and the horizontal displacement of the bridge pile foundation increases under the influence of dewatering conditions.

As can be seen from Figures 2–3, after construction step 9, that is, after the fifth precipitation, the settlement value of the center of the bridge pile and pier increases significantly. This shows that when the foundation pit excavation dewatering project is within a certain depth range of 5.8m, as shown in Figure 1, the impact of the construction in the foundation pit on the settlement of the bridge pile foundation is not great. However, when the excavation depth exceeds this range, the impact of each construction on the settlement of the bridge pile foundation gradually increases. This is because the settlement value of soil outside the pit increases with the increase of excavation depth, resulting in negative friction resistance on the side of the pile foundation and accelerating the settlement of the pile foundation. At the same time, due to the influence of precipitation, the water head difference between inside and outside the pit causes the soil outside the pit to produce downward seepage stress, which further accelerates the soil settlement.

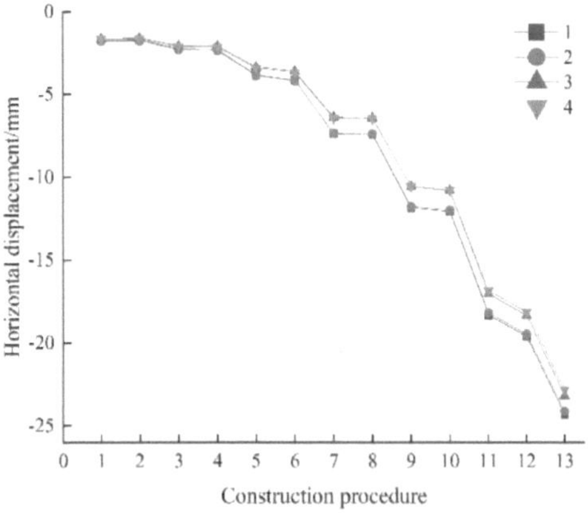

Figure 1. Horizontal displacement of pile top at each construction stage.

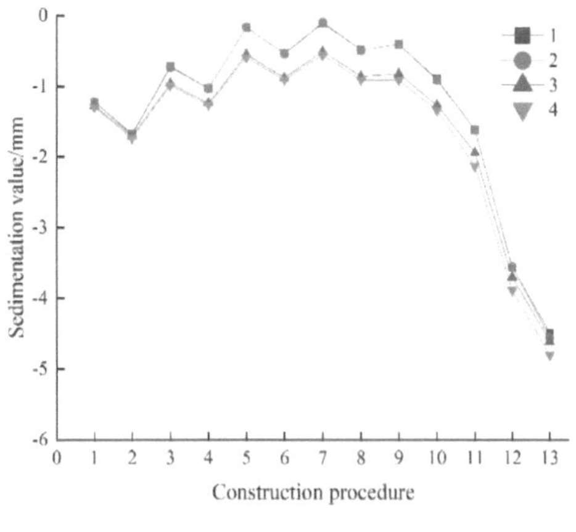

Figure 2. Pile bottom settlement at each construction stage.

(2) Pile lateral friction and axial force

The no.1 pile foundation near the foundation pit is taken to study the distribution changes of pile lateral friction and axial force under construction and dewatering conditions. As shown in Figures 4–6, with foundation pit excavation, the neutral point of pile lateral friction moves downward along the pile body and the negative friction increases. After the completion of foundation pit excavation, the neutral point of lateral friction resistance of the bridge pile foundation is about 17.3m along the pile length, and the negative friction resistance value of the pile side is about 2772.66kN. It can be seen from the figure that foundation pit precipitation has a certain influence on pile lateral friction, but it is not obvious. With the increase of foundation pit excavation depth, especially after construction step 6, the influence of foundation pit excavation and precipitation on the lateral friction of the bridge pile foundation becomes more sensitive. The changing trend of the axial force of the bridge pile foundation is consistent with that of the neutral point of pile lateral friction resistance and has a linear relationship. The maximum axial force of the bridge pile foundation after the completion of foundation pit excavation is 7443.18kN.

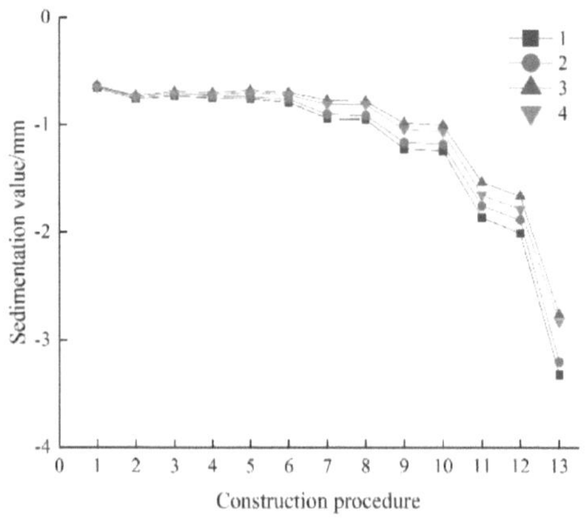

Figure 3. Settlement of pier center at each construction stage.

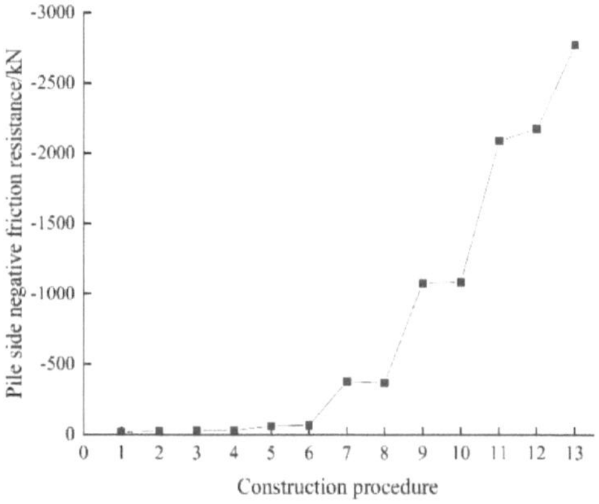

Figure 4. Pile side negative friction resistance at each construction stage.

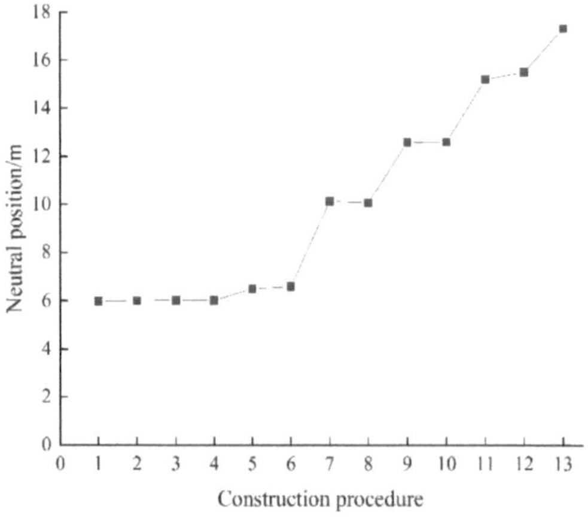

Figure 5. Location of neutral points in each construction stage.

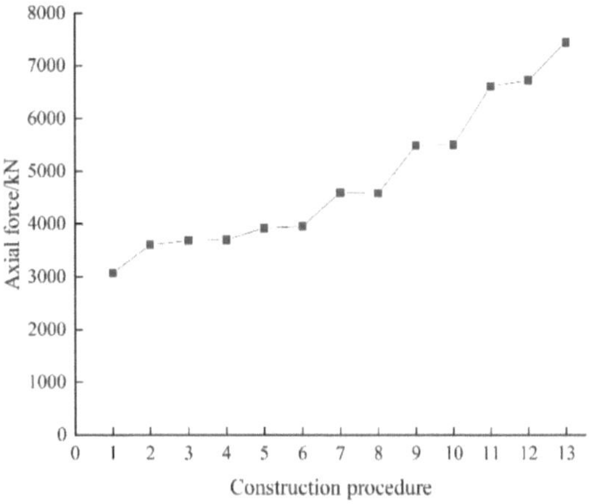

Figure 6. The axial force of the pile in each construction stage.

4 CONCLUSION

In this paper, taking a deep foundation pit excavation project as an example, the finite element method is used to establish a stress-seepage coupling model of foundation pit, pile, soil, and dewatering. The construction process of the foundation pit is simulated and analyzed, and the influence of foundation pit excavation and dewatering on adjacent bridge piles is studied. The following conclusions are drawn:

(1) Within a certain depth of foundation pit excavation, the impact of pit construction on the bridge pile foundation is small, and beyond this range, each construction will have a great impact on the bridge pile foundation. In the construction process, after the foundation pit excavation to a certain stage, we should pay attention to the influence on the bridge pile foundation.
(2) the foundation pit precipitation surface settlement has a certain influence on the pit outside, and each time should not be too great fractional precipitation. Precipitation change will cause the head pressure inside and outside the pit, and the pit seepage field soil produces high pressure, causing the pit surface settlement to be too large, increasing the bridge pile foundation pile's negative friction resistance, and exerting an adverse impact on the bridge pier.
(3) The position of the neutral point of negative friction resistance on the pile side has a linear relationship with the additional axial force of the pile foundation, and the maximum axial force is 7443.18kN.

REFERENCES

Guo Quanyuan. Analysis of the influence of foundation pit excavation on viaduct piers of adjacent intercity railway [J]. *Railway Construction Technology*, 2017(10):1–6.

Jin Xiao-Fei, Huang Yun-Tian, Yin Han-Qing, Liang Shu-Ting, Zhu Xiao-Jun. Seepage stress coupled finite element analysis method for deep foundation pit excavation [J]. *Building Construction*, 2018, 48(S1): 804–807.

Jin Xiaorong, Yu Jianlin, Zhu Shaochen, et al. Analysis of settlement behavior of surrounding soil caused by dewatering of foundation pit [J]. *Rock and Soil Mechanics*, 2005, 26 (10):7.

Li Daozhen. Research and calculation of negative friction of pile foundation [D]. *Ocean University of China*, 2004.

Li Ke. Research on the Influence of deep foundation pit excavation on Adjacent Bridge piles [D]. *Tianjin University*, 2007.

Li Yonggang. Numerical analysis of deformation and stability of deep foundation pit with pile-anchor support [J]. *Subgrade Engineering*, 2014(4):5.

Wang Cui, YAN Shuwang, Zhang Qibin. Research on Influence Mechanism and Control Measures of Deep foundation pit Excavation on Adjacent Bridge Piles [C]// *Proceedings of the 11th National Conference on Rock Mechanics and Engineering*. 2010.

Wang Hongxia. Analysis of Negative friction Effect and Calculation Method of Pile Foundation [J]. *Fujian Building Materials*, 2018(9):3.

Wu Hao, Yao Shunyi, Xue Xun-qiang, et al. Analysis of the influence of foundation pit dewatering and excavation on adjacent bridge piles based on Seepage Stress coupling [J]. *Highway Transportation Technology*, 2019(10).

Yang Yicai, Xu Jing, Wang Liwei, et al. *Engineering and Construction*, 2019, 033(002):262–265,273. (in Chinese)

Advances in Civil Engineering and Environmental Engineering – bin Yahya & Bin Daud (Eds)
© 2023 The Author(s), ISBN: 978-1-032-39397-1

Research on the impact of old neighborhoods' renovation based on thermal comfort

Jie Tang
Tongji Architectural Design (Group) Co., Ltd, Shanghai, China

Jiawang Bao*
Shanghai Tongji Planning and Design Institute Co., Ltd, Shanghai, China

ABSTRACT: Under the background of frequent extreme high-temperature weather in China, the renovation of old neighborhoods, as an important step to improve the urban living environment, must pay more attention to the impact of the thermal environment. This study takes Huangshan-Ercun Neighborhood as an example, which is an old neighborhood in Jinyang-Xincun Sub-district, Pudong New District, Shanghai. Combining with the possibility of old neighborhoods' renovation, the software ENVI-met is used to simulate four different renovation plans. And the distribution changes of thermal environmental parameters and thermal comfort UTCI value in the current situation and four different renovation plans are compared, so as to judge the advantages and disadvantages of different plans and analyze the impact of old neighborhoods' renovation on thermal comfort. On this basis, the old neighborhoods' renovation strategies based on the improvement of thermal comfort are summarized.

1 INSTRUCTIONS

In the context of the increasing global greenhouse effect, the frequent occurrence of extremely high-temperature weather in summer has affected the livability of urban neighborhoods in China. The old neighborhoods' renovation, as an important step to improve the urban living environment, should pay more attention to the influence of the thermal environment.

Based on field investigation and empirical judgment, this study puts forward several renovation plans combined with the feasibility of old neighborhoods' renovation. However, the implementation effect of different plans cannot be observed actually, so it needs to be helped with software simulation to judge the advantages and disadvantages. Relevant common software includes EVNI-Met, FLUENT, Pheonics, STAR-CD, CFX, and so on. (Zeng et al. 2019). Among them, ENVI-Met has a high degree of imitation of micro-climate at the block and neighborhood scale, and a high degree of calculation optimization, which facilitates timeliness of experimental results output (Helge et al. 2018). Therefore, ENVI-Met is used for simulation in this study.

In addition, this study focuses on the outdoor space of old neighborhoods, so it does not involve the structural reconstruction and indoor renovation of buildings. And it mainly studies the hot and humid climate in summer, so the renovation of old neighborhoods under other geographical and climatic conditions is not to be discussed.

2 RESEARCH OBJECTS AND RESEARCH METHODS

2.1 *Research objects*

This study focuses on the comprehensive impact of different renovation plans on thermal comfort in old neighborhoods. Therefore, Huangshan-Ercun Neighborhood, as an old neighborhood in

*Corresponding Author: 18817596075@126.com

DOI 10.1201/9781003349563-55

Jinyang-Xincun Sub-district, Pudong New District, Shanghai, is selected as the research object (Figure 1). Its renovation is rich in content, rapid in progress, and representative of its own spatial characteristics (Shanghai Municipal People's Government 2018).

Figure 1. The location of the Huangshan-Ercun Neighborhood.

2.2 *Research methods*

2.2.1 *Thermal comfort index evaluation model*

At present, the internationally recognized thermal comfort index evaluation model is the UTCI model organized by the World Meteorological Organization (WMO), which takes into account all important heat exchange mechanisms. Although it shows minor differences when comparing the results of some high-level models, it is superior to other empirical or mechanistic evaluation models in terms of evaluation procedures at present (The International Society of Biometeorology 2012). Therefore, this study uses the UTCI model to evaluate thermal comfort, and the formula is as follows:

$$UTCI = UTCI = Ta + Offset(Ta, TMRT, Uwind, Pvapour) \tag{1}$$

Among them, *Ta* represents the air temperature, *TMRT* represents the average radiation temperature, *Uwind* represents the ground wind speed, and *Pvapour* represents the air humidity. These four thermal environment parameters jointly affect the thermal comfort UCI value. In hot summer, the higher the UTCI value, the lower the human thermal comfort.

2.2.2 *ENVI-met simulation*

(1) Introduction of ENVI-met

The developer of ENVI-met is Michael Bruse's team from the University of Mainz, Germany. The software is based on CFD (Computational Fluid Dynamics) and thermodynamic modeling, and includes five sub-modules of atmosphere, radiation, soil, plants, and buildings (Helge et al. 2018), which can realize the coupled calculation of wind, heat, humidity, and sunlight at the block and neighborhood scale. ENVI-met uses the grid method for calculation. Due to the limited computing power, in order to reflect the thermal environment parameters as accurately as possible, the grid of the core area near the ground in the horizontal direction should be smaller. In the vertical direction, the near-ground grid should also be more detailed, and the high-altitude area can be appropriately enlarged.

(2) Simulation methods

The longitude, latitude, season, date, and simulation duration of the simulated location (Huangshan-Ercun Neighborhood) are input into ENVI-Met. By setting the grid and inputting the material, height, and other model information of each component, the internal buildings, plants, and soil of the Huangshan-Ercun Neighborhood are modeled. After calculation, the simulation results are compared with the actual situation. If the conclusion is consistent, it indicates the accuracy of the modeling and external data (Paula et al. 2019). After the feasibility is proved, the changes in the thermal environment in the neighborhood after different renovation plans are further simulated. The differences in spatial distribution characteristics of thermal environment parameters and thermal comfort UTCI value in the current situation and different plans are compared, so as to provide theoretical basis and data support for improving the thermal comfort of old neighborhoods.

3 ANALYSIS OF ENVI-MET SIMULATION RESULTS

3.1 *Feasibility analysis of simulation*

According to the climate of Shanghai (Shanghai Local Chronicle Office 2008), August 19, 2021 is selected as the simulation date. Using the daily data from Pudong Meteorological Bureau, the highest air temperature was 32°C; the highest time period was 13:00–14:00; the surface air humidity was 52%–70% throughout the day; the wind direction was mainly southeast wind; the average ground wind speed (wind speed at 10 meters) was 3m/s, and the surface roughness is 0.01. The above are input into ENVI-met as external conditions for simulation. Modeling of Huangshan-Ercun Neighborhood, the average radiation temperature can be obtained by inputting external conditions and calculations, which is basically the same as the distribution of black bulb temperature in the field survey on that day, and the average radiation temperature is directly related to the black bulb temperature. On this basis, the current situation and different renovation plans can be simulated in the following.

3.2 *Simulation of different renovation plans*

According to the current common methods of old neighborhoods' renovation at home and abroad, with other external conditions remaining unchanged, the following 4 renovation plans of the neighborhood are simulated respectively. And the study compares the thermal environment parameters (air temperature, average radiation temperature, air humidity, ground wind speed) and thermal comfort UTCI value of each plan, so as to summarize the advantages and disadvantages of each plan.

- Plan 1: Replacing all residential gable walls with vertical greening, and using roof greening.
- Plan 2: Adding the vegetation in the neighborhood parking lot, for example, planting deciduous trees in the parking spaces near the south side of the residential buildings, and changing the underlying surface of some parking spaces to grass-planting bricks.
- Plan 3: Changing the asphalt pavement to cement pavement to simulate the pavement before the renovation, so as to judge intuitively whether it is necessary to stop the operation of some neighborhoods that have not yet carried out the "changing white to black" work.
- Plan 4: Simulating dark walls, so as to reduce the reflectivity of the residential buildings' walls on the first floor.

3.3 *Comparison of thermal environment parameters*

3.3.1 *Air temperature*

Figure 2 shows the air temperature distribution diagrams of the current situation and Plan 1, Plan 2, Plan 3, and Plan 4 from left to right. In the current situation, the air temperature on the north side of the neighborhood is basically above 29.05°C, while the low-temperature areas basically surround the river bank on the east side.

Compared with the current situation, the air temperature of Plan 1 increases in a small area, and the increase rate is about 1°C. It is considered that the transpiration effect of gable walls' vertical greening increases the air temperature. Compared with the current situation, the air temperature of Plan 2 and Plan 3 decreases in a large area. In Plan 2, a large number of trees have a good shading effect. In Plan 3, the specific heat capacity of the cement pavement is high, which reduces the heating of air near the ground. However, Plan 4 has no obvious change compared with the current situation. To sum up, it can be judged that the air temperature near the ground is mainly affected by surface temperature. Therefore, the ground air temperature can be effectively reduced by shading or improving the specific heat capacity of the pavement material. But overall, the cooling rate is not outstanding.

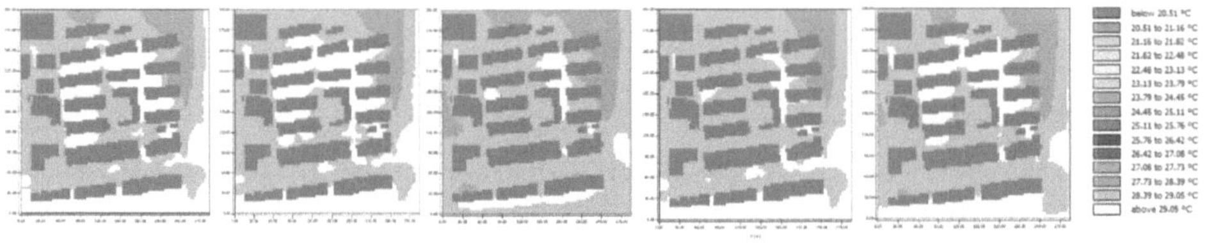

Figure 2. The air temperature distribution of the current situation and each plan.

3.3.2 *Average radiation temperature*

Figure 3 shows the average radiant temperature distribution diagrams of the current situation and Plan 1, Plan 2, Plan 3, and Plan 4 from left to right. In the current situation, the average radiant temperature under the shadow of residential buildings and trees is quite different from that under direct sunlight. The average radiant temperature under the shadow of the buildings and trees is basically below 42.24°C, while that under direct sunlight can reach above 58.24°C.

According to the maximum and minimum temperatures, except for Plan 3, the other plans all show a decreasing trend in the average radiation temperature compared with the current situation to a certain extent. In Plan 2, the shading effect of trees significantly reduces the average radiant temperature of the neighborhood. In comparison, Plan 1 shows no obvious difference because the vertical greening planted on the gable walls is in the shade during this period, but the gable walls close to the sunny side show a certain downward trend. Compared with the current situation, Plan 4 has a small decrease along the south wall side, which is about 0.8°C. In Plan 3, due to the increase in the reflectance of cement pavement, the average radiation temperature generally increases, which is about 0.6°C. To sum up, it can be judged that the most effective way to reduce the average radiation temperature is to provide shadow areas. In addition, due to the limitation of software accuracy, the cooling advantages of closing to vertical greening and dark walls cannot be effectively reflected.

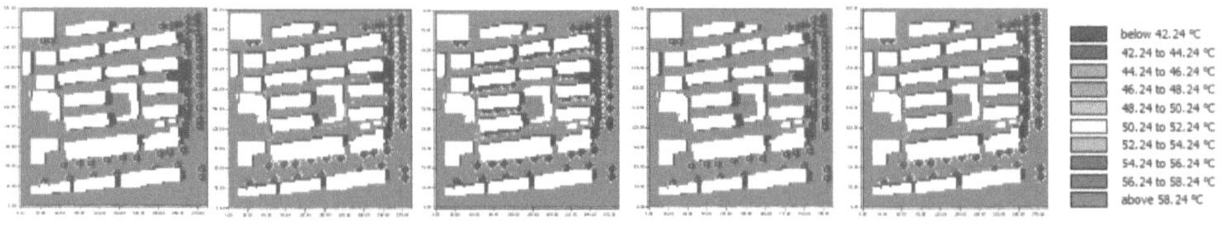

Figure 3. The average radiant temperature distribution of the current situation and each plan.

3.3.3 *Air humidity*

Figure 4 shows the air humidity distribution diagrams of the current situation and Plan 1, Plan 2, Plan 3, and Plan 4 from left to right. In the current situation, the river surface and the east side of the neighborhood near the river have relatively high air humidity, which is above 58.24%. While the air humidity inside the neighborhood is lower significantly, basically between 50.24% and 56.24%.

Due to the increase of a large quantity of vegetation in Plan 1 and Plan 2, the air humidity increases significantly during the high-temperature period, and the transpiration effect of trees in Plan 2 is more obvious than that in Plan 1. In Plan 3, the air humidity in some areas decreases, and Plan 4 shows no obvious air humidity changes.

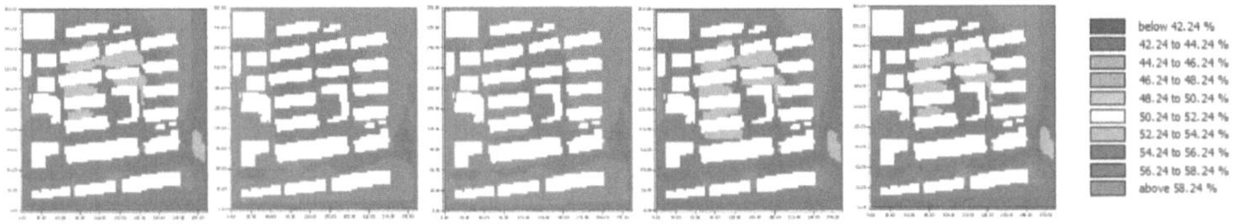

Figure 4. The air humidity distribution of the current situation and each plan.

3.3.4 *Ground wind speed*

Figure 5 shows the ground wind speed distribution diagrams of the current situation and Plan 1, Plan 2, Plan 3, and Plan 4 from left to right. The current average wind speed is 1.97m/s, the highest is 4.76m/s, and the lowest is 0.39m/s. The north-south residential roads are the main ventilation corridors, which have an obvious narrow tube effect. Considering that the southeasterly wind prevails in Shanghai in summer, the wind speed of the east-west road is significantly reduced due to the tall tree canopy blocking the wind inlet.

Compared with the current situation, the ground wind speed of each plan shows no significant changes. Among them, although Plan 2 adds many trees in the parking lots, these added trees cannot significantly reduce the wind speed of the parking space. Because the parking place is mainly distributed on the east-west road in the neighborhood, the current situation has shown a low wind speed. To sum up, it can be judged that the distribution of wind speed in neighborhoods is mainly affected by the wind direction, the neighborhood spatial shape, and tall trees. If the tall trees are densely distributed in the main wind inlet, the wind force is significantly reduced. If they are distributed in the weak wind, the wind resistance effect is not obvious.

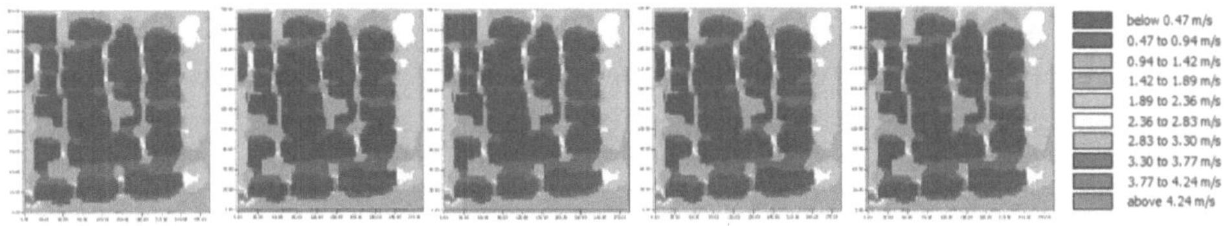

Figure 5. The ground wind speed distribution of the current situation and each plan.

3.3.5 *Comparison of thermal comfort UTCI value*

By integrating the thermal environment parameters of the current situation and Plan 1, Plan 2, Plan 3, and Plan 4, and calculating the UTCI value, the differences in thermal comfort caused by the current situation and different plans are analyzed. Figure 6 shows the UTCI value distribution diagrams of the current situation and Plan 1, Plan 2, Plan 3, and Plan 4 from left to right.

Under the current situation, the UTCI value is the highest at 51.3°C and the lowest at 22.5°C. The UTCI value under the shade of trees decreases significantly, and the UTCI value near the river bank keeps basically at 29.72 – 32.08°C. However, the UTCI value on the large exposed road between the residential buildings changes drastically and remains above 48.6°C basically.

In the simulation plans, planting tall trees in Plan 2 increases a large number of effective shading areas, which leads to a significant decrease in the UTCI value. However, Plan 1 and Plan 4 show no obvious cooling effect. Only the UTCI value near the gable wall decreases slightly, and the trend shows that the impact of wall renovation on thermal comfort is mainly reflected in the micro-scale. Plan 3 restores the road conditions before the renovation, and the thermal environment deteriorates significantly, which indicates that the implementation of "white to black" work is very necessary for general.

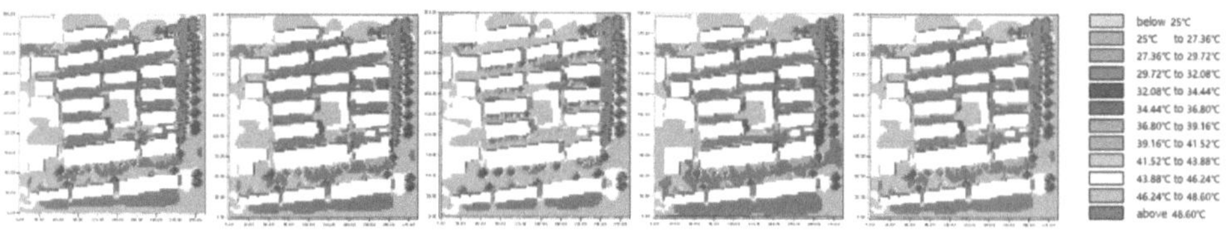

Figure 6. UTCI value distribution of the current situation and each plan.

4 SUMMARY AND DISCUSSION

In recent years, extremely high-temperature weather has become more frequent in China, and the public health problems caused by it have been gradually paid attention to. The hardware facilities in the old neighborhood are relatively poor. The average age of the residents is older and they spend a long time in the neighborhood, so they are more sensitive to the thermal comfort of the outdoor environment (Li & Li 2022). Based on the spatial characteristics of old neighborhoods and the population structure and activity characteristics of the residents, suitable space renovation can effectively improve thermal comfort, while reducing the accidental incidence of thermal damage, energy consumption, and anthropogenic heat generation.

Through the ENVI-met simulation, this study finds that the supplementary planting of tall trees in Plan 2 has the most obvious improvement in thermal comfort in old neighborhoods. While the impact of wall renovation in Plan 1 and Plan 4 is more microscopic, planting vertical greening and painting dark walls fail to affect the larger-scale thermal environment. In addition, Plan 3 shows that the current "white to black" work on the road surface can indeed significantly improve thermal comfort in the absence of shading conditions. Based on the above analysis conclusions, three suggestions for the improvement of old neighborhoods' renovation can be summarized: Firstly, it is still necessary to protect the existing tall trees as much as possible. while the neighborhood lacks green shade coverage, asphalt pavement is an effective method to improve thermal comfort. Secondly, vertical greening and wall painting have improved significantly at the micro-scale, and especially, vertical greening can make up for the lack of greening in old neighborhoods. Thirdly, for neighborhoods close to or with large water bodies inside, the cooling effect of the water bodies should be actively used. At the same time, it is necessary to prevent the interference of overcrowded trees from the cool wind blowing into the neighborhood.

ACKNOWLEDGMENTS

Supported by Major State Basic Research Development Program (2018YFC0704800)

REFERENCES

Helge, S., Jenny, L., David, H., Peter, B., Michael, B. & Jan, E. (2018). Modeling transpiration and leaf temperature of urban trees: A case study evaluating the microclimate model ENVI-met against measurement data. *J. Landscape and Urban Planning*. 174.

Li, Y.H. & Li, Q. (2022). Research on outdoor thermal comfort evaluation and regulation range in the transition season of old residential areas in Guangzhou. *J. Architectural Science*. 38(02), 71-81.

Paula, S., Helge, S., Denise, H. S. D., Michael, B. (2019). Calibration process and parametrization of tropical plants using ENVI-met V4 – Sao Paulo case study. *J. Architectural Science Review*. 62(2).

Shanghai Local Chronicle Office. (2008). *Volume II-Natural Environment*. A/OL.

Shanghai Municipal People's Government. (2018). *Notice on Printing and Distributing the "Three-Year Action Plan for the Construction of 'Beautiful Homes' in Shanghai Neighborhoods*. A/OL.

The International Society of Biometeorology. (2012). *Universal Thermal Climate Index Homepage*. A/OL.

Zeng, S.C., Tian, J. & Zeng, J. (2019). Research on ventilation efficiency and optimal layout of typical residential modules based on CFD simulation. *J. Journal of Architecture*. 2019(02), 24–30.

Advances in Civil Engineering and Environmental Engineering – bin Yahya & Bin Daud (Eds)

Study on space design of rural residence in the Hexi Corridor based on sustainable perspective

Yanjun Li*
School of Civil Engineering and Architecture, Xi'an University of Technology, Xi'an, China

ABSTRACT: At present, China is implementing the rural revitalization strategy, and rural settlements are being updated. How to reduce energy consumption and human damage to nature while improving the living conditions of villagers? On basis of the previous investigation, eight courtyards were constructed in the Hexi Corridor. Land use of rural residences was analyzed first, and building energy consumption and carbon emission of the eight constructed courtyards were analyzed by using the software DesignBuilder. The relationship between carbon emission and energy consumption was fitted in the study. The Land use, energy consumption, and carbon emission, caused by the space design of the eight constructed courtyards were quantitatively analyzed, and the sustainable development law of local rural residence was revealed. The study is conducive to local rural ecological construction.

1 GENERAL INSTRUCTIONS

Today's architectural design approaches do not adequately address the relationship between users' spatial, environmental and psychological experiences (Chowdhury 2021). Some studies show that passive techniques systems have a remarkable effect on the indoor air temperature, thermal comfort, and cooling/heating loads of the building (Mastouri 2017). Restoration of the traditional houses of rural areas, must in the future be defined in a way that enables the preservation of recognized general values and further improvement of environmental quality and climate resilience (Skataric 2021). There are many different requirements for thermal distribution according to local climate conditions. To satisfy these requirements there were derived several parameters (Wang 2015): compactness, building and nonbuilding density, courtyard proportions, heat transfer coefficient of structural elements, reflect coefficient of structural elements, and so on. Microclimate comfort control is available due to the quantitative evaluation of these parameters. Some scholars selected influencing factors through field survey and simulation analysis and constructed the prediction model of energy consumption in the heating season and natural lighting illuminance of rural residential buildings. The model can provide guidance for architects and rural residents, the model can improve indoor environmental quality and reduce the energy consumption of rural residential buildings (Zhen 2016). Several scholars have discussed different strategies to reduce energy consumption and improve the thermal comfort of buildings. However, it hasn't been found a study that is combined space design and sustainable development of rural human settlements. This paper explores various influencing factors caused by the change in space design. It makes a contribution to implementing sustainable rural construction in China.

2 OVERVIEW OF THE HEXI CORRIDOR

The Hexi Corridor is in the northwestern part of Gansu province, China, and it is a long and narrow strip running northwest-southeast. From viewpoint of administrative division, it is composed of Wuwei, Jinchang, Zhangye, Jiuquan, and Jiayuguan.

*Corresponding Author: 936069673@qq.com

DOI 10.1201/9781003349563-56

The Hexi Corridor has a long history. It was established as an administrative region in the Han dynasty. Then, it used to be the most important passage from ancient China to Xinjiang and Central Asia. Since Han and Tang dynasties, it became a part of the Silk Road. Thus, Hexi Corridor had played an important role. But it gradually declined after the 15th century. Nowadays, under China's "One Belt One Road" development strategy, the Hexi Corridor has been revived again.

The Hexi Corridor has a continental dry climate. It is dry throughout the year, and there are rare cloud cover and longer sunshine duration. There are more wind and sandstorms locally. The Hexi Corridor is a southward channel of cold flow in winter, so the winter here is very cold and long. The annual mean temperature is low. There is little rainfall but a big evaporation quantity locally. Table 1 shows the climate conditions of the Hexi Corridor.

Table 1. Climate conditions of the Hexi Corridor.

City	Average temperature (°C)	Rainfall (mm)	Evaporation (mm)	Sunshine hours (h)	Frost-free period (d)
Wuwei	7.8	60~610	1400~3040	2200~3030	85~165
Jinchang	4.8	139.8~351.7	2000.6~2094.2	2210.5~2981.6	134
Zhangye	7.6	122~370	2029	2869~3282	112~145
Jiuquan	3.9~9.3	39~300	2000~4000	3000~3300	127~158
Jiayuguan	7.3	85.3	2149	3030.7	130

(Source: Relevant circumstances of Hexi Corridor[EB/OL]. China Yearbook Full-text Database)

3 SPACE DESIGN OF RURAL RESIDENCE

3.1 *Characteristics of local courtyards*

Generally speaking, rural courtyards in the Hexi Corridor have characteristics of high enclosure walls, closed exteriors and open interiors, and green plantings in the courtyards. There is more wind and sandstorms locally, so high enclosure walls can reduce the impact of cold air in winter, hot air in summer, and sandstorm in spring. There is no window on the courtyard walls, and all windows and doors of the main rooms are opened inward courtyard. Due to the above two characteristics, local rural courtyards have a good defensive effect. In most cases, there is a flower bed in the courtyard. People usually plant some vegetables, flowers, fruit trees and so on. It is very dry locally, so the green in the courtyard can regulate air temperature and humidity. At the same time, it can improve air quality. Figure 1 shows the typical courtyard layout of the Hexi Corridor.

Figure 1. Typical courtyard layout.

3.2 *Courtyard design*

Various courtyard forms were constructed for comparative analysis. Through investigation, it was known that local vast majority of rural elderly lived with their children and enjoyed family care. Therefore, the most common type of family in the local rural area is a family of six members including grandparents, parents, and two children. Common courtyard forms in the Hexi Corridor area include siheyuan, sanheyuan, curved-ruler style courtyard, and newly built row house courtyard. When constructing a courtyard, it took into account not only the provision of sufficient living space for family members but also the change in courtyard form. These constructed courtyards had the same building area, height, and building construction practices, only courtyard space design was different. In these constructed courtyards, the rural residence had the same building area of 168.84 square meters and the same height of 3 meters.

Eight typical courtyards were constructed. The space design determined the shape, orientation, and layout of rural residences, and these factors directly affected building energy consumption and the indoor physical environment during operation. Table 2 shows some information on typical courtyards. Figures 2 ~ 9 show a schematic floor plan of eight typical courtyards respectively, and in these courtyard designs, the depth of the rural residence was determined to be 4200mm. Courtyard A was a local traditional courtyard siheyuan. The house was built along courtyard walls on four sides. There were no doors and windows on the outer walls, but doors and windows were opened to the inner courtyard. Courtyard B and Courtyard C were both sanheyuan, and Courtyard B was a vertical sanheyuan, but Courtyard C was a horizontal sanheyuan. The house was built along courtyard walls on three sides, and the gate was opened on the fourth side of the rectangular courtyard. Courtyard D, Courtyard E, Courtyard F, and Courtyard G were all curved-ruler style courtyards, the house was built along courtyard walls that are perpendicular to each other. The difference between these four courtyards was shown by the ratio of length to width of each courtyard. Courtyard H was a row house courtyard that was new in recent years. In this courtyard, two deep houses were concentrated and connected in parallel. In this courtyard form, garden space can be both available at the front and back of the house.

Table 2. Some information on typical courtyards.

Court yard	Courtyard name	Courtyard area(m^2)	Total land area (m^2)	Shape factor of house
A	siheyuan	56.16	225.00	0.86
B	vertical sanheyuan	83.16	252.00	0.86
C	horizontal sanheyuan	126.00	294.84	0.86
D	curved-ruler style courtyard, length/ width=3	200.78	369.62	0.86
E	curved-ruler style courtyard, length/ width=2	269.23	438.07	0.86
F	curved-ruler style courtyard, length/ width=1	323.99	492.83	0.86
G	curved-ruler style courtyard, length/ width=0.5	269.22	438.06	0.86
H	row house courtyard	235.16	404.00	0.67

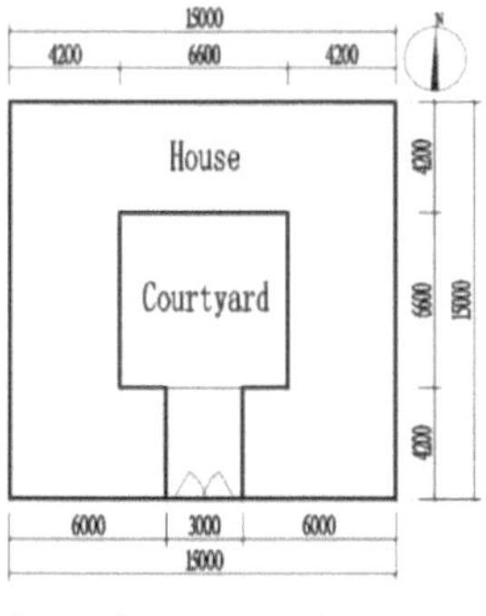

Figure 2. Layout of courtyard A.

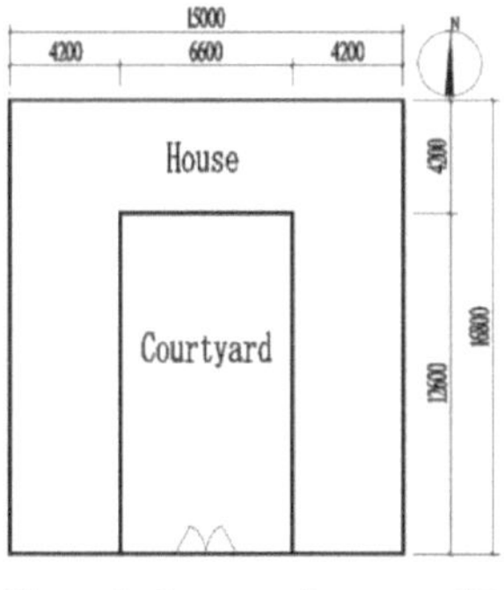

Figure 3. Layout of courtyard B.

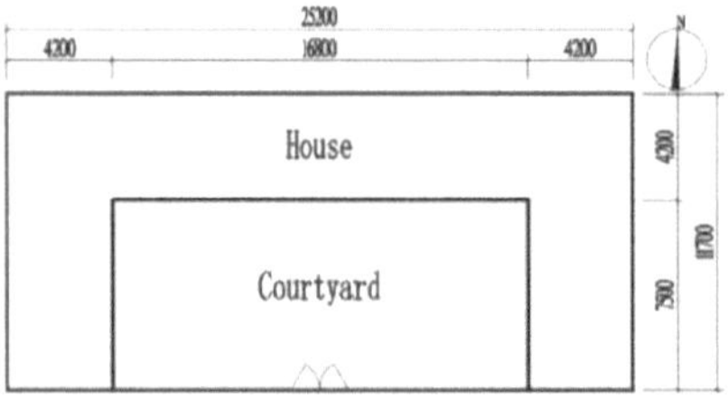

Figure 4. Layout of courtyard C.

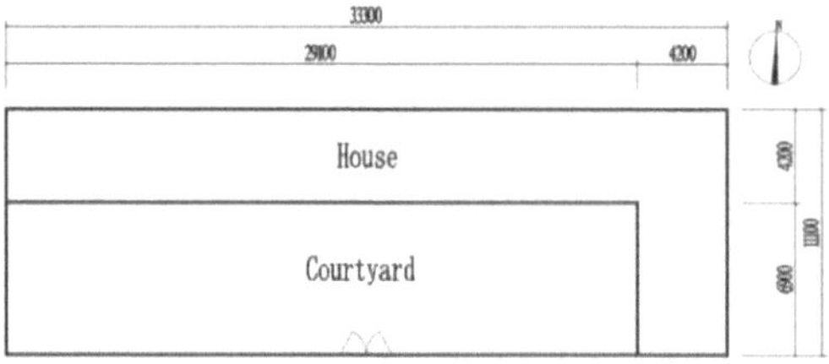

Figure 5. Layout of courtyard D.

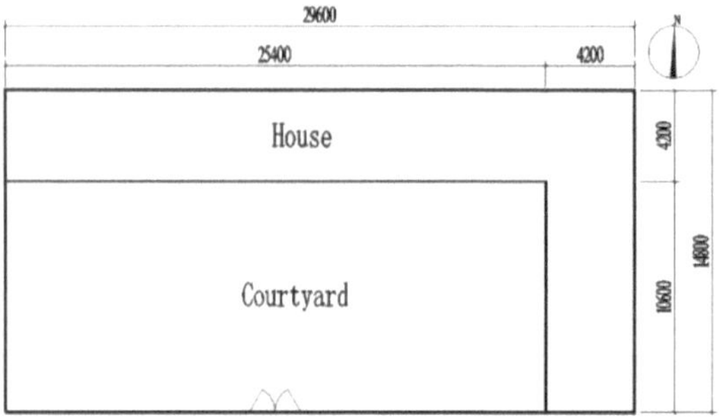

Figure 6. Layout of courtyard E.

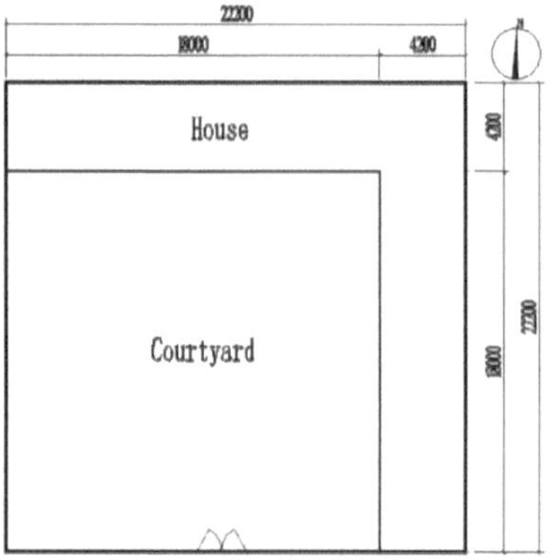

Figure 7. Layout of courtyard F.

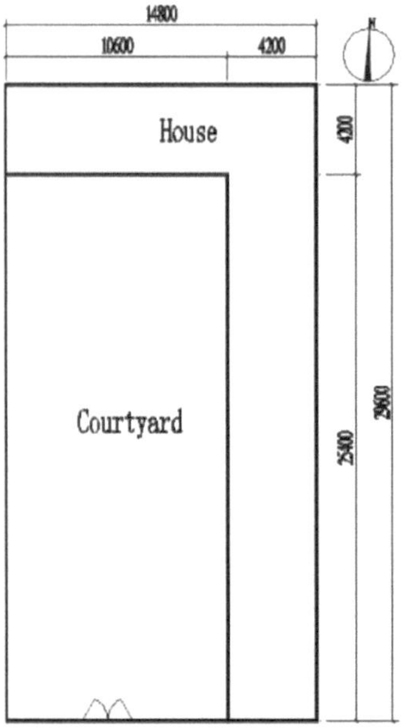

Figure 8. Layout of courtyard G.

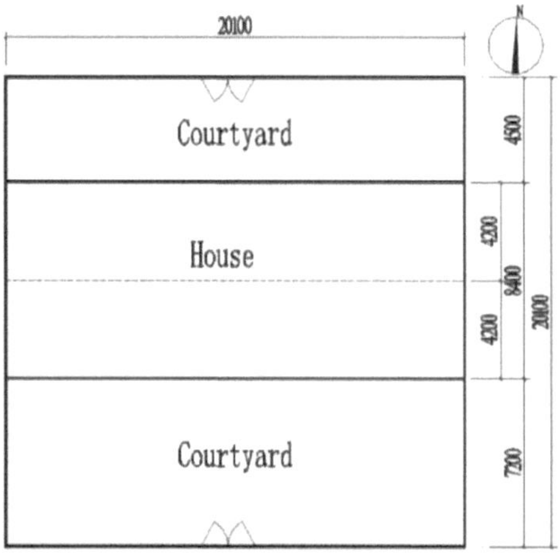

Figure 9. Layout of courtyard H.

Basic conditions of these courtyards, such as building envelope information, window-to-wall area ratio, building operation, etc., were all determined. Table 3 shows building envelope information of rural residences in typical courtyards. Each part building construction of the residence was designed, and the heat transfer coefficient of each part building construction was calculated. According to local traditional architectural style and characteristics, doors and windows of each room were opened to the inner courtyard, and windows were not opened on courtyard walls. For walls facing the inner courtyard, the window-to-wall area ratio was determined to be 30%. Although there is a living room, dining room, and kitchen in rural residences, the bedroom is the main part. In addition, the research focus of this paper was the impact of changes in courtyard space design. To simplify calculation and analysis, the overall operation of rural residences in the courtyard was determined to be consistent with the operation of the bedroom. All these basic conditions and information were set and simulated in the software DesignBuilder.

Table 3. Building envelope information of rural residences in typical courtyards.

Building component	Enclosure construction element (from interior to exterior)	Heat transfer coefficient($W/m^2 \cdot K$)
Flat roof	gypsum plastering 10mm+concrete, reinforced(with 2%steel) 100mm+MW glass wool 50mm+ bitumen felt 10mm	0.667
External wall	gypsum plastering 10mm+ brickwork 370mm+ mortar 10mm	1.434
Internal wall	gypsum plastering 10mm+ brickwork 240mm+ gypsum plastering 10mm	1.842
Ground floor	ceramic tile 20mm+foam slag 50mm+cast concrete 100mm+clay 100mm	1.723
External window	double wooden framed glass window,3+13A+3	1.960
Wooden door	painted oak 35mm	2.251

4 SUSTAINABILITY ANALYSIS OF SPACE DESIGN FOR RURAL RESIDENCE

4.1 *Land use analysis of rural residence*

The function and space design requirements of the courtyard directly affect the land use of the rural residence. As an outdoor space for rural families, the courtyard has various functions, including daily living, family reunion activities, agricultural production, and green planting. Therefore, a rural courtyard should have enough area and regular space. On one hand, these are requirements of courtyard use function and architectural culture, and on the other hand, these are requirements of building technology. In the design of rural courtyard space, technical requirements such as saving land resources, building sunshine, and natural ventilation need to be considered.

Land resources are very precious, so it is necessary to emphasize land use efficiency when discussing space design. Figure 10 shows land use analysis under different space designs when the building area is the same. The total land area consists of two parts, i.e., a courtyard area and a building area. As mentioned above, under the space design of various courtyard changes, the building covers the same 168.84 square meters. It can be seen from Figure 10 that the curves of total land area and courtyard area are completely parallel, when the building area is the same, the siheyuan and sanheyuan use less land, but the curved-ruler style courtyard uses more land. For the row house courtyard, the land area can be reduced by compressing the depth of the courtyard in front of and behind the house. In this case, there was no change in the space form of the courtyard. Figure 11 shows land use efficiency analysis. It can be found that under the condition of the same building area, the siheyuan and sanheyuan have higher land use efficiency than the curved-ruler style courtyard. In this study, the land use efficiency of these three courtyards is all higher than 0.5.

4.2 *Building energy consumption analysis of rural residences*

4.2.1 *Overview of basic conditions*

Software DesignBuilder was used to simulate energy consumption, and some basic conditions were set and determined before calculation. The eight typical courtyards were respectively constructed

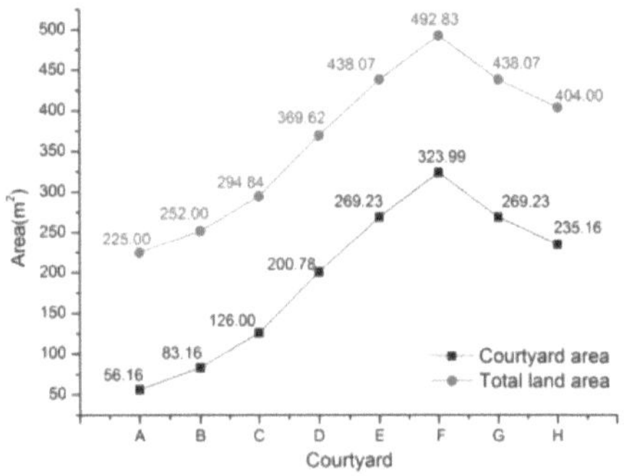

Figure 10. Land use analysis.

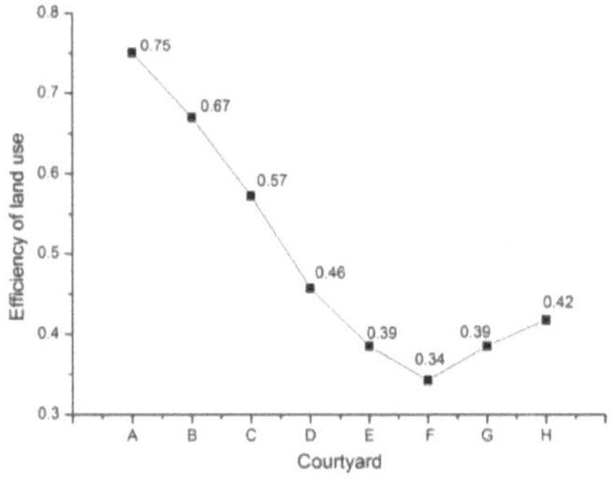

Figure 11. Land use efficiency analysis.

models, and these models were simulated separately under the climatic environment of Minqin, Jiuquan, Yumen, and Dunhuang. People density, building construction, window-to-wall area ratio, and building operation have been discussed above in 3.2. In addition to these, there was more information to be clear. The heating set point temperature was 18°C, and the cooling set point temperature was 26°C. In the simulation, the Fan-coil unit was chosen as an HVAC template. Natural gas was chosen as heating fuel, and the heating system CoP was set as 0.830. Electricity from the grid was chosen as cooling fuel, and cooling system CoP was set as 1.670. In natural ventilation, the outside air definition method was chosen by zone, and outside air was set as 3ac/h. In mechanical ventilation, the outside air definition method was chosen as min fresh air (sum per person and per area), and fan type was chosen as intake, meanwhile, total efficiency was set as 70%. Instantaneous DHW only was chosen as DHW type, and electricity from the grid was chosen as DHW fuel, simultaneously DHW CoP was set as 0.85. Target illuminance was set as 100lux, and lighting energy was set as $5W/m^2$. Indoor equipment gains were set as $5W/m^2$.

4.2.2 *Gas consumption analysis*

The Hexi Corridor is located in severe cold and cold zones of China Building Thermal Design Zoning Diagram, and it is very cold and it has a long duration in winter, so the heating energy consumption of buildings is very huge. Figure 12 shows gas consumption in Minqin, and Figure 13 shows gas consumption in Dunhuang. Both figures clearly expressed the monthly consumption of natural gas for heating systems by the eight constructed courtyards. From the analysis of curve changes in the figures, it can be found that, in general, the heating energy consumption of Courtyard A was the highest, followed by Courtyard B and Courtyard C, the middle of four curved-ruler

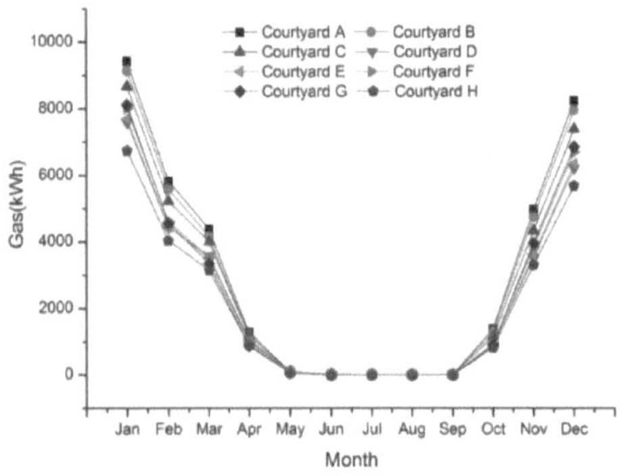

Figure 12. Gas consumption in Minqin.

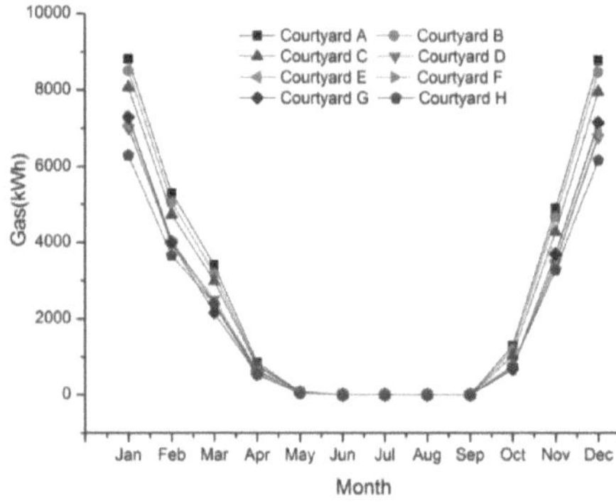

Figure 13. Gas consumption in Dunhuang.

style courtyards, and the least energy consumption of Courtyard H. Of course, there were slight differences between two figures due to different climatic conditions. In March, among these eight courtyards, Courtyard H still maintained the lowest gas consumption for the heating system in Minqin, but Courtyard G became the lowest gas consumption for the heating system in Dunhuang. Because the monthly consumption law has been summed up, limited to the length of this paper, the monthly consumption figures of Jiuquan and Yumen were skipped.

Although the monthly consumption of natural gas for heating systems was very detailed, the gas consumption during the heating season was closer to actual consumption for the whole year, and it should be analyzed and explored. In this study, the heating season was defined as the period from November of the previous year to March of the following year. Figure 14 shows natural gas consumption during the heating season in typical regions of the Hexi Corridor. It can be found that the natural gas consumption patterns of the eight constructed courtyards were very similar in various regions. For the eight constructed courtyards, natural gas consumption was divided into three intervals, and all courtyards consumed more energy than 21000kWh. In high energy consumption intervals, Courtyard A as siheyuan had the most energy consumption, Courtyard B as vertical sanheyuan had the second highest consumption, and Courtyard C as horizontal sanheyuan had the third highest consumption. In the moderate interval, Courtyard D, Courtyard E, Courtyard F, and Courtyard G were all curved-ruler style courtyards, the energy consumption of each courtyard varied with a ratio of length to width of the courtyard. In general, when the ratio was larger, the energy consumption was smaller. Among all courtyards, courtyard H as a row house courtyard consumed the least energy in all study regions. As shown in Table 2 above, the shape factor of the house in courtyard H was 0.67, but in the other seven courtyards, the shape factor of the house was 0.86. Therefore, it was consistent with the building energy efficiency design theory that the energy consumption of courtyard H in the heating season was the smallest among the eight constructed courtyards.

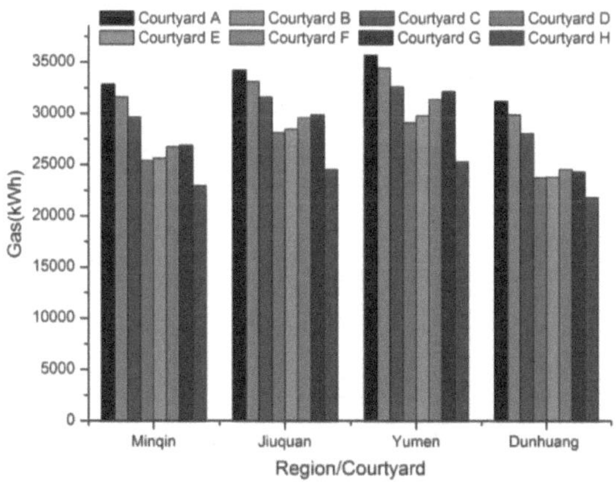

Figure 14. Gas consumption during the heating season.

4.2.3 *Electricity consumption analysis*

Although current rural residences are not necessarily equipped with summer air conditioning systems and mechanical ventilation equipment, it is necessary to compare the energy consumption of various courtyard space designs, and at the same time, the development status of the rural settlement must be considered. So some necessary building equipment was installed in the eight constructed courtyards. In this study, electricity consumption was mainly in summer cooling, lighting, DHW, indoor equipment, mechanical ventilation, etc. From analysis of life behavior, the electricity consumption of lighting, DHW, and indoor equipment fluctuated little from month to month, and the biggest change was the energy consumption of air-conditioning cooling. Figure 15 and Figure 16 show electricity consumption in Minqin and Dunhuang respectively. It can be clearly seen from the figures that from January to April, and from October to December, the electricity consumption of the eight constructed courtyards is stable However, from April to October; electricity consumption varied dramatically and peaked during the period. It's obvious that the air conditioning cooling system consumed lots of electricity. Analyzing these figures, it can be found that among the eight constructed courtyards, Courtyard G consumed the most electricity consumption in summer, followed by Courtyard F in second place.

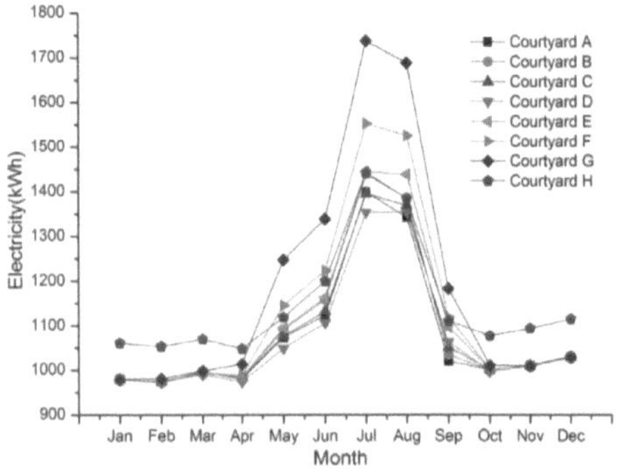

Figure 15. Electricity consumption in Minqin.

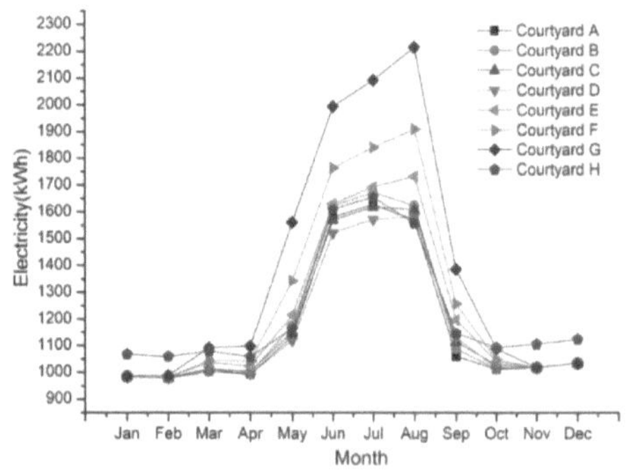

Figure 16. Electricity consumption in Dunhuang.

Figure 17 shows annual electricity consumption in typical regions of the Hexi Corridor. It can be found that the annual electricity consumption of courtyard F, courtyard G, and courtyard H was relatively high, and the annual consumption of the other five courtyards was relatively moderate. Among all constructed courtyards, Courtyard D as a curved-ruler style courtyard, in which the length-to-width ratio of the courtyard was 3, had the least annual electricity consumption. It can be inferred that from the perspective of saving electricity consumption, among the eight constructed courtyards, courtyard D was the most worthy of promotion.

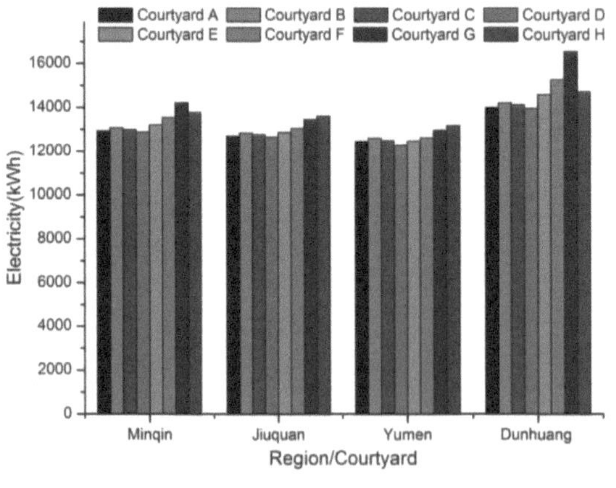

Figure 17. Annual electricity consumption.

4.2.4 *Solar gains analysis of exterior windows*

The Hexi Corridor is located in an area rich in solar energy resources in China, so the application of solar energy in buildings should be explored. Solar energy can be used to produce domestic hot water all year round, but a more important building application is to heat houses directly during the heating season. Figure 18 shows solar gains in exterior windows during the heating season in typical regions of the Hexi Corridor. It can be found that during the heating season, Courtyard A and Courtyard B had the lowest solar gains exterior windows, both below 5000kWh; but Courtyard D and Courtyard E had the highest solar gains exterior windows, both close to or above 9000kWh; while the other constructed courtyards had moderate solar gains exterior windows. Courtyard D and Courtyard E was both curved-ruler style courtyard. For constructed courtyards in this study, the length-to-width ratio of Courtyard D was 3, and length to width ratio of Courtyard E was 2. What they had in common was that more parts of the house were oriented north-south, and external windows of this part were only opened on the south wall facing the inner courtyard. In all study regions, solar gains in exterior windows of Courtyard D were more than double that of Courtyard A, and in Minqin and Dunhuang, the ratio was nearly three times.

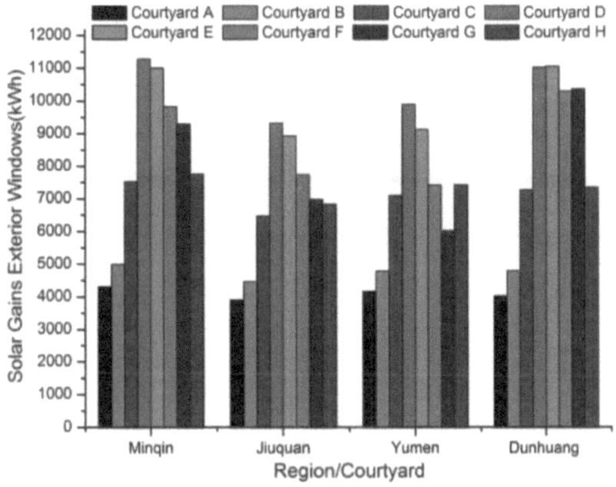

Figure 18. Solar gains exterior windows.

4.3 *Carbon emission analysis of rural residences*

4.3.1 *CO_2 emissions analysis*

At present, China is implementing a carbon peaking and carbon neutralization strategy, so CO_2 emissions should be studied during building operations. Figure 19 and Figure 20 show CO_2 emissions in Minqin and Dunhuang respectively. It can be seen that from January to April, and from October to December, the variation of CO_2 emissions was similar to the natural gas consumption during the heating season, and compared with other courtyards, courtyard A, courtyard B, and courtyard C had more CO_2 emissions. However, from April to October, courtyard G and courtyard F had more CO_2 emissions.

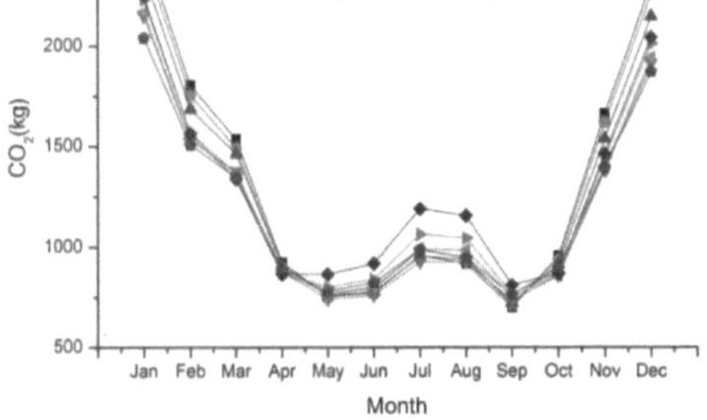

Figure 19. CO_2 emissions in Minqin.

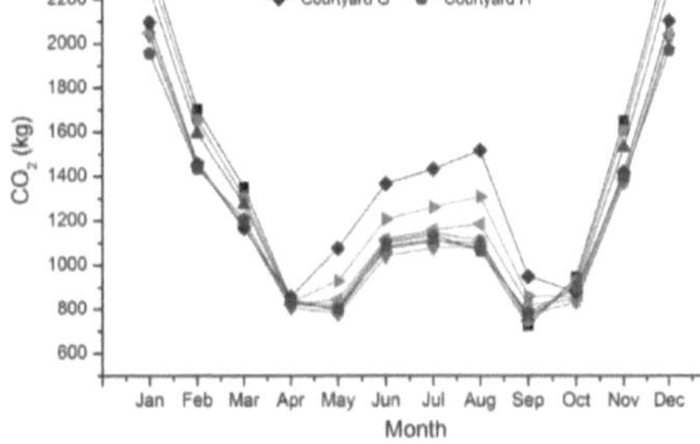

Figure 20. CO_2 emissions in Dunhuang.

Figure 21 shows annual CO_2 emissions in typical regions of the Hexi Corridor. It can be found that although there were some differences due to climatic conditions, the same rules of CO2 emissions existed across regions. Among the eight constructed courtyards, Courtyard D and Courtyard H emitted less CO_2 throughout the year, but Courtyard A and Courtyard B emitted more CO_2.

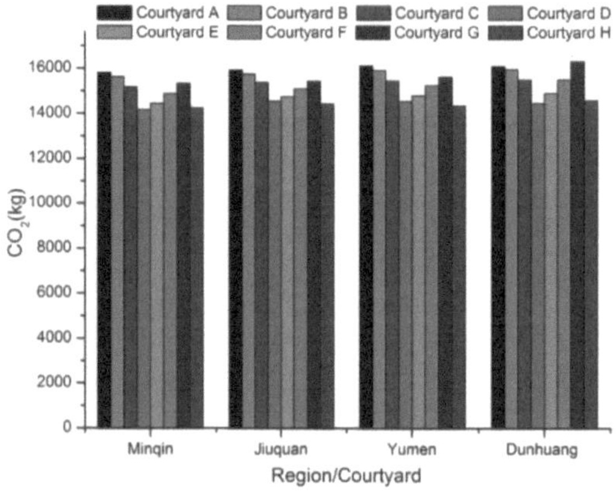

Figure 21. Annual CO_2 emissions.

4.3.2 *Fitting analysis of carbon dioxide, natural gas and electricity*

In this study, the main energy forms were natural gas and electricity, and it was clear that CO_2 emissions were related to the use of natural gas and electricity. Table 4 shows energy consumption and CO_2 emissions data in typical regions of the Hexi Corridor. In the table, CO_2 emissions (M_{co2}) and electricity consumption (M_{elec}) were annual data, and natural gas consumption (M_{gas}) was heating season data. These data were fitted and calculated in Origin software, and the relationship between CO_2 emissions, heating gas consumption, and electricity consumption can be obtained. Table 5 shows the functional relationship between energy consumption and CO_2 emissions. By analyzing the fitting function formula, it can be inferred that compared with natural gas, electricity consumption was the main factor affecting CO_2 emissions in this study.

Table 4. Energy consumption and CO_2 emissions data in typical regions of the Hexi Corridor.

Region	Courtyard	M_{co2}(kg)	M_{gas}(kWh)	M_{elec}(kWh)
Minqin	Courtyard A	15802.08	32835.04	12932.44
	Courtyard B	15624.10	31591.36	13079.37
	Courtyard C	15164.51	29643.99	12995.34
	Courtyard D	14169.03	25343.25	12875.92
	Courtyard E	14435.91	25612.40	13207.70
	Courtyard F	14876.20	26709.89	13541.66
	Courtyard G	15335.76	26862.49	14215.30
	Courtyard H	14252.15	22924.89	13769.75
Jiuquan	Courtyard A	15927.13	34206.17	12691.45
	Courtyard B	15757.39	33083.80	12824.66
	Courtyard C	15379.51	31569.01	12754.23
	Courtyard D	14542.30	28089.64	12646.88
	Courtyard E	14739.66	28425.10	12846.25
	Courtyard F	15105.64	29563.98	13031.57
	Courtyard G	15430.43	29857.48	13446.85
	Courtyard H	14427.73	24505.51	13601.47

(*continued*)

Table 4. Continued.

Region	Courtyard	M_{co2}(kg)	M_{gas}(kWh)	M_{elec}(kWh)
Yumen	Courtyard A	16113.13	35681.18	12447.04
	Courtyard B	15918.08	34410.13	12598.16
	Courtyard C	15443.65	32592.69	12485.16
	Courtyard D	14539.96	29087.42	12280.77
	Courtyard E	14794.25	29763.98	12456.94
	Courtyard F	15237.84	31358.01	12610.78
	Courtyard G	15618.60	32142.48	12953.44
	Courtyard H	14347.04	25265.29	13165.73
Dunhuang	Courtyard A	16098.31	31182.25	13993.17
	Courtyard B	15956.67	29879.53	14207.03
	Courtyard C	15496.15	28014.55	14117.62
	Courtyard D	14459.12	23705.62	13957.83
	Courtyard E	14893.22	23779.32	14581.81
	Courtyard F	15507.84	24529.85	15262.31
	Courtyard G	16322.01	24294.52	16549.71
	Courtyard H	14590.36	21790.05	14714.07

Table 5. The functional relationship between energy consumption and CO_2 emissions.

Region	Regression equation	R-Square	F
Minqin	$M_{co2} = 694.31731 + 0.21107M_{gas} + 0.63169M_{elec}$	0.99959	8534.26261
Jiuquan	$M_{co2} = 134.19600 + 0.21713M_{gas} + 0.65751M_{elec}$	0.99828	2034.74629
Yumen	$M_{co2} = -52.37508 + 0.21594M_{gas} + 0.67706M_{elec}$	0.99791	1670.42925
Dunhuang	$M_{co2} = 119.40539 + 0.21346M_{gas} + 0.66572M_{elec}$	0.99961	9062.80157

5 CONCLUSIONS

The conclusions drawn from the analyses of this study are as follows:

1) When the building area is the same, Courtyard A as a traditional courtyard form uses less land and has higher land use efficiency, but it has the highest heating energy consumption and more CO_2 emissions.
2) Courtyard B and Courtyard C have some similarities, including higher land use efficiency, more heating energy consumption, and more CO_2 emissions. In comparison between Courtyard B and courtyard C, Courtyard C has more advantages, it has less heating energy consumption and CO_2 emissions than Courtyard B, while Courtyard C has considerable solar gains in exterior windows.
3) Courtyard D and Courtyard E have some similarities, including lower land use efficiency, moderate heating energy consumption, and less CO_2 emissions. And they have the highest solar gains exterior windows, both close to or above 9000kWh. Comparison between them, Courtyard D has more advantages.
4) Courtyard H as a row house courtyard has the least heating energy consumption and less CO_2 emissions.
5) Electricity consumption is the main factor affecting CO2 emissions in this study.
6) Among the eight constructed courtyards, horizontal sanheyuan like Courtyard C, curved-ruler style courtyard like Courtyard D, and row house courtyard like Courtyard H, these three courtyard forms have more comprehensive advantages and should be promoted in local rural construction.

ACKNOWLEDGEMENTS

This work was financially supported by the Shaanxi Provincial Department of Housing and Urban-Rural Development (2015-K74), Shaanxi Provincial Department of Science and Technology (2022JQ-555), and Xi'an University of Technology (118-211417). The author wishes to express his most sincere appreciation for these three units.

REFERENCES

Chowdhury S, Noguch M, Doloi H (2021). Conceptual parametric relationship for occupants' domestic environmental experience. *J. Sustainability*. 13(5), 2982:1–28.

Mastouri H, Benhamou B, Hamdi H, Mouyal E (2017). Thermal performance assessment of passive techniques integrated into a residential building in a semi-arid climate. *J. Energy and Buildings*. (143): 1–16.

Skataric G, Spalevic V, Popovic S (2021). The vernacular and rural houses of agrarian areas in the zeta region, Montenegro. *J. Agriculture*. 11(8), 717:1–35.

Wang Zhu, Tyu Vet An (2015). Residential area parameters. *J. Architecture & Culture*. (07):160–162.

Zhen Meng, Sun Cheng, Jia Yongheng (2016). Study on coupled environment design of rural residential buildings in severe cold regions of northeast China. *J. China Illuminating Engineering Journal*. 27(4):113–122.

Advances in Civil Engineering and Environmental Engineering – bin Yahya & Bin Daud (Eds)
© 2023 The Author(s), ISBN: 978-1-032-39397-1

Research on the motion of flying gangue and design of protective device in steeply dipping coal seam

Ming Liu, Bohao Luan & Lin Sun*
School of Environment and Safety Engineering, Liaoning Petrochemical University, Fushun, China

ABSTRACT: The motion of flying gangue in steeply dipping coal seam has certain complexity and randomness. The accurate simulation of its motion has important significance for the safe and efficient mining in steeply dipping coal seam. Taking the No.5 coal seam of 2130 coal mine as the research background, using Rhino software establish the 3D model of ellipsoid flying gangue formed by roof caving and the working face respectively, the Abaqus software was applied to divide mesh for the models, and the energy tracking method was used to simulate the motion of flying gangue in three-dimensional space. Finally, an automatic intelligent device for preventing flying gangue was designed by using the grating to accurately identify the dangerous source. The results show that the motion of flying gangue is mainly affected by the shape of flying gangue, the environmental parameters of working face, and the initial collision attitude between flying gangue and the bottom plate of working face, in addition, the debris formed after collision and fracture have the possibility of generating secondary disaster of flying gangue. The application of the energy tracking method can provide a certain theoretical basis for the prediction and protection of flying gangue disaster.

1 INTRODUCTION

The steeply dipping coal seam is widely distributed in China, the steeply dipping coal seam generally refers to the coal seam with the dip angle between 35° and 55°, which is difficult to mine and has low stability of working face. With the development of deeply inclined coal seam mining and combined mining technology, the number of steep coal mines in China has increased year by year, and therefore the coal production, as well as the economic benefits, has increased. At the same time, the flying gangue disaster, a disaster occurred in deeply inclined coal seam mining, has become increasingly prominent. The flying gangue is the coal or rock caused by roof caving, coal wall caving and other reasons at the mining process of steeply dipping coal seam, causes sliding, collision and splashing in the working face, the collision of the flying gangue with people or equipment would occur instantly (Liu et al. 2016; Wu et al. 2020, 2013; Wang & Jiao 2016; Wu et al. 2019, 2014; Yun et al. 2017). Therefore, this paper will take ellipsoidal flying gangue of the roof leakage as researching object, using energy tracking method to study the motion law of flying gangue in steeply dipping coal seam. Because this method does not need to simulate test parameters for many times, it has a fast calculation speed, and can directly and accurately simulate the three-dimensional motion for the collision of flying gangue and bottom plate of working face as well as the motion after collision and fragmentation, which can provide a reliable theoretical basis for the prediction and protection of flying gangue disaster in the steeply inclined coal seam.

*Corresponding Author: sunlin0626321@163.com

DOI 10.1201/9781003349563-57

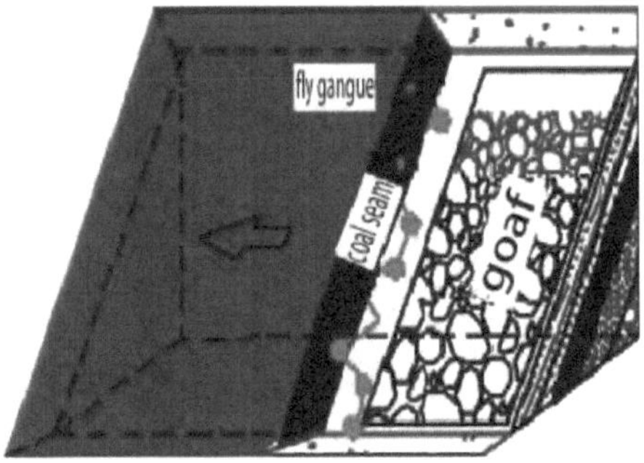

Figure 1. Motion mode of flying gangue.

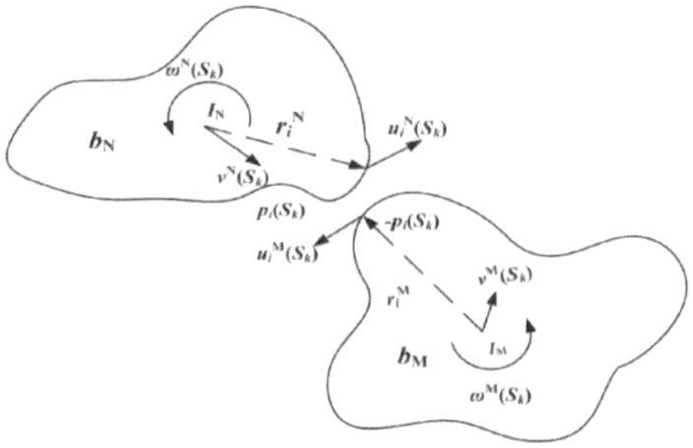

Figure 2. The models of two colliding bodies.

2 BASIC PRINCIPLE OF ENERGY TRACKING METHOD

The main motion modes of flying gangue are shown in Figure 1. The energy tracking method is based on the principle of energy iteration. According to the elastic potential energy absorbed by the flying gangue contact collision point at the moment of collision, a group of impulses applied to the point of collision are calculated, so as to release energy continuously until the motion stops.

Figure 2 shows the specific principle. The two blocks b_N and b_M appear contact collision. $v^N(s_k)$ and $v^M(s_k)$ are respectively linear velocities of the two block mass centers; $\omega^N(s_k)$ and $\omega^M(s_k)$ are respectively the angular velocities of the two blocks; r_i^N and r_i^M are respectively vectors from the mass center of two blocks to the contact point. $u_i^N(s_k)$ and $u_i^M(s_k)$ are the velocities at the contact point. The relative velocity of two colliding blocks $u_i(s_k)$ at the contact point are defined is

$$u_i(s_k) = u_i^N(s_k) - u_i^M(s_k) \tag{1}$$

The change of relative velocity is defined as

$$\Delta u_i(s_k) = u_i(s_{k+1}) - u_i(s_k) \tag{2}$$

Two impulses $p_i(s_k)$ and $p_i(s_k)$ are applied on two collision blocks to prevent the collision block from embedding, the process can be expressed as

$$\mathrm{p}_i(s_k) = K_i^{-1} \Delta u_i(s_k) \tag{3}$$

Where the collision matrix K_i, which is constant, non-singular, symmetric, and positive definite, is defined as

$$K_i = \left(\frac{1}{m^N} + \frac{1}{m^M}\right) I - \left(\tilde{r}_i^N \left(I^N\right)^{-1} \tilde{r}_i^N + \tilde{r}_i^M \left(I^M\right)^{-1} \tilde{r}_i^M\right) \tag{4}$$

Where I is the unit matrix, $\tilde{r}_i^N$ and $\tilde{r}_i^M$ are respectively cross product matrixes of r_i^N and r_i^M.

The energy tracking method divides multiple collisions into three stages for iterative calculation. The first stage is the energy absorption phase: according to the change of velocity at the collision point, a set of impulse data is obtained and applied to the collision contact point to gradually absorb energy. Kinetic energy decreases and the elastic potential energy increases at the collision contact point. The second stage is the energy loss stage: Stronge hypothesis was used to calculate the energy loss. And finally, the energy release phase: according to the calculated elastic potential energy, the maximum change in the relative normal velocity is obtained, and a set of impulses is calculated again and applied to the contact point of the collision to gradually release energy until the elastic potential energy is zero.

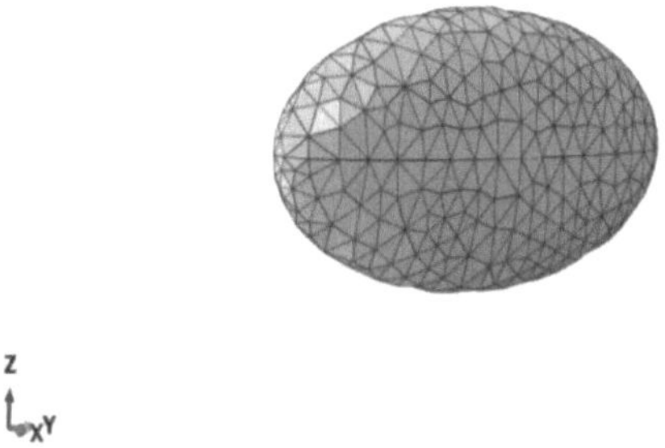

Figure 3. Discrete mesh model of flying gangue.

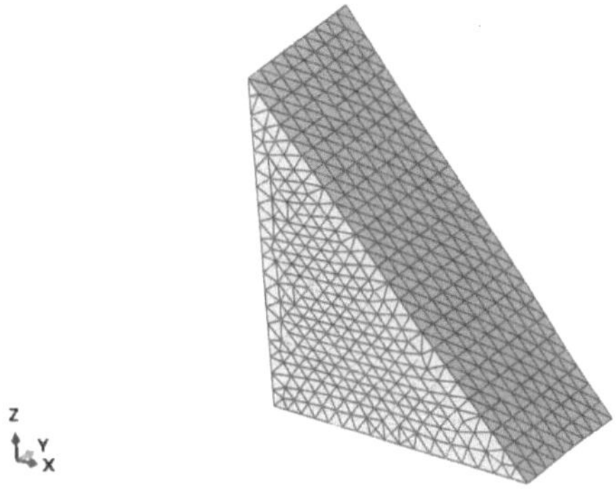

Figure 4. Discrete mesh model of working face.

3 3-D MOTION ANALYSIS OF FLYING GANGUE

In this paper, the No.5 coal seam of the 2130 coal mine of Xinjiang coking coal group at Aiweiergou is taken as the engineering simulation background. The model size of the working face is length × width × height = 120m×60m×140m, the average bearing ES of coal seam is 13°, the average inclination SN of coal seam is 44°, the collision recovery coefficient of bottom plate is 0.84, the friction coefficient is 0.4, the density of flying gangue is 2500kg / m^3, and the model size of flying gangue is an ellipsoid of 0.4m×0.2m×0.2m. After leaking from the roof, the ellipsoidal coal (rock) block forms flying gangue, which falls freely from the top of the working surface 2m high, and continuously bounces and rolls after colliding with the bottom plate of the working face.

The energy tracking method is used to simulate. Firstly, three-dimensional model of ellipsoidal flying gangue formed by roof leakage and working face of steeply dipping coal seam is established by Rhino software, and then the model is imported into Abaqus software for the grid generation, as shown in Figure 3 and Figure 4, the approximate global size of the global seed is 7, the shape of the element is tetrahedron, the geometric order of the element type is linear. Finally, the data file of three-dimensional discrete mesh model is input into the program of energy tracking method to simulate.

The results show that the flying gangue collides with the bottom plate of the working face ten times during the falling process. The first free-falling collision occurred after roof leakage, and the collision posture was shown in Figure 3. The first collision attitude determines the subsequent motion track. After flying for a long distance after the ninth collision, the gangue was not bounced after the last collision with the bottom plate at the lower end of the working face. The landing point was (121.737,10.9152,0). The maximum lateral displacement of flying gangue is 19.2848m, and the lateral displacement mainly depends on the shape of flying gangue and the strike inclination of the working face.

The energy tracking method can be used to calculate the velocity, angular velocity and energy changes of flying gangue at any time during the whole motion, as shown in Figures 5-7.

It can be seen from figure 5 that the velocity of the center of mass of flying gangue is mainly affected by collision. It is assumed that when the flying gangue colliding with the bottom plate of the working face, the flying gangue and the bottom plate of the working face will only produce normal deformation and tangential deformation will not occur. The collision process is divided into two stages, namely compression stage and rebound stage. The process from the contact between the flying gangue and the bottom plate of working face to the contact area that produces the maximum elastic-plastic deformation is a compression stage. The normal kinetic energy of flying gangue is converted into the deformation energy of the contact zone between flying gangue and the bottom plate of the working face, and the normal velocity of flying gangue will gradually decrease to zero. When the elastic-plastic deformation of the contact area reaches the maximum value, the

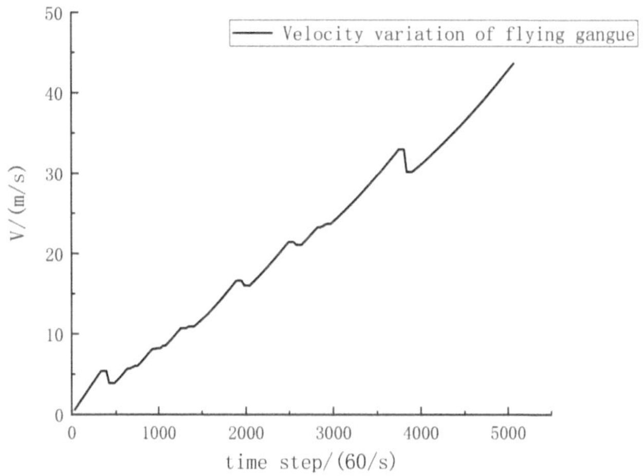

Figure 5. Velocity of flying gangue motion.

impact force of flying gangue will reach the maximum value. The process from the maximum elastic-plastic deformation position to the residual plastic deformation position in the contact area is a rebound stage. The elastic deformation of the contact area between the flying gangue and the bottom plate of working face began to recover gradually, and the normal velocity of the flying gangue would gradually increase from zero. Therefore, the velocity of flying gangue will fluctuate. During the motion of flying gangue, most of the gravitational potential energy of flying gangue is converted into kinetic energy of flying gangue, and only a small amount of energy is lost. Therefore, the overall trend of velocity of flying gangue is still increasing, and the maximum velocity can reach 43.75108m/s.

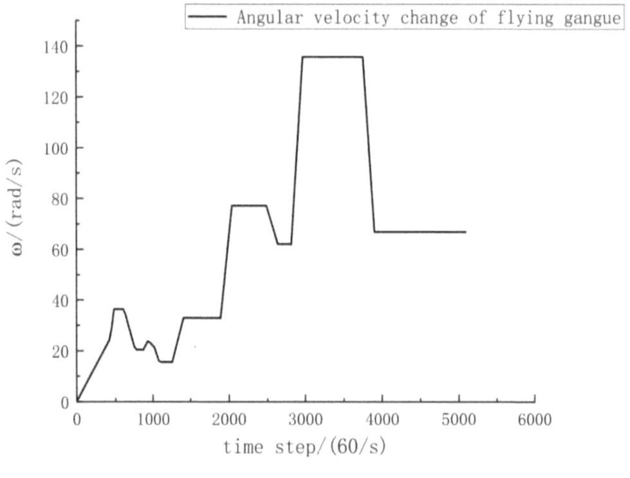

Figure 6. Angular velocity of flying gangue movement.

Figure 6 shows that the angular velocity of flying gangue is mainly affected by the collision contact friction and the distance from the collision contact point to the center of mass. During the collision, the work done by tangential friction was converted into rotational kinetic energy of flying gangue, and the angular velocity of flying gangue increased. Since the flying gangue has no contact with the bottom plate of the working face, the angular velocity of the flying gangue remains unchanged during the motion between the two collision times. During the collision between the flying gangue and the bottom plate of the working face, the angular velocity decreases when the distance between the collision contact point and the center of mass increases from small to large. Otherwise, the angular velocity will increase. The maximum angular velocity of the flying gangue is 135.6865rad/s in the whole motion process.

It can be seen from Figure 7 that the change trend of the total kinetic energy of flying gangue is a quadratic parabola, which is similar to the quadratic change trend of the velocity of the center of mass. It can be seen that the translational kinetic energy of flying gangue accounts for a large proportion of the total kinetic energy. As the time goes, the total kinetic energy of flying gangue increases significantly, and the maximum total kinetic energy is 6315.376J.

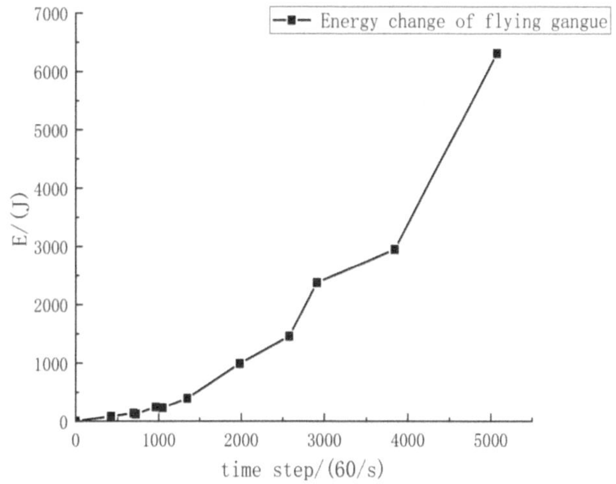

Figure 7. Change in energy of flying gangue.

4 DESIGN OF PROTECTIVE DEVICE FOR FLYING GANGUE

By simulating the motion trajectory of flying gangue, combining the number of bouncing and the location of falling points, it was determined that a key protective device was installed between the third and fourth collision positions, that is, between 126.457m and 125.824m. In order to meet the actual working condition of working face in steeply dipping coal seam, the barrier device should be arranged at the lower end of the working face. The GER1 of SNS flexible protecting mesh is adopted as the executive component of the flying gangue protection device. The steel wire grille is SO/2.2/50, the working radius of the protecting mesh is 1m, the maximum height is 2m, the rotation angle is 90°, and the grating response time is within 20ms.The basic structure of the protective device for flying gangue is shown in Figure 8.

This device is divided into five parts. The first part is the starting component. The first motor adopts DC power to provide power for the coal scraper, and the second motor adopts a three-phase asynchronous motor to provide power to the protective device. The second part is the early warning component, namely the safety grating. In order to meet the design requirements, the LSL18-4J-4J safety light curtain sensor was selected. The third part is the transmission component, which runs through the chain drive, and then a set of bevel gears change the direction of torque to send the torque drive to the executive device for interception. The coal scraper is reciprocated by the ball screw. The fourth part is the executive components, namely the eccentric disc and the SNS flexible protecting mesh. When the torque is transmitted by the transmission device, the eccentric disc drives the protective mesh to rotate, intercepting and capturing the flying gangue. The fifth part is

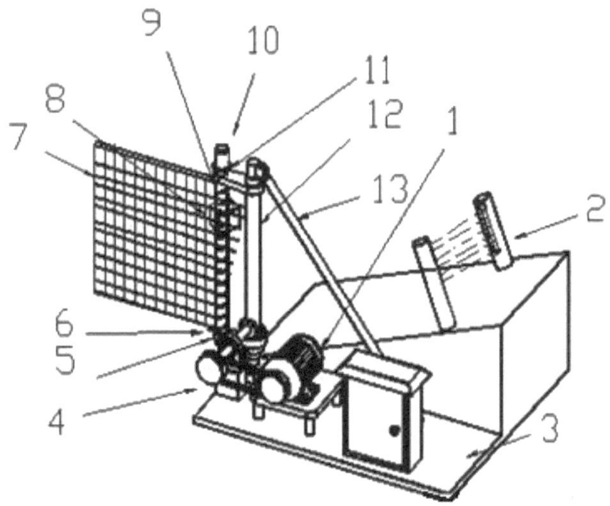

1 AC motor; 2 Grating; 3 Control box and working plate; 4 Transmission device; 5 Open limit of protective device; 6 Lower limit of ball screw d; 7 Protecting mesh; 8 Coal scraper; 9 Upper limit of ball screw; 10 DC motor; 11 Closing limit of protective device; 12 Spindle; 13 Supporting rod

Figure 8. Schematic diagram of protective device for flying gangue.

the limiting part, which is the limit switch. When the protective mesh rotates at a certain angle and touches the limiting device, it will immediately stop at the current position for interception.

When the interception action is finished, it will be controlled by time relay, and then it will move back to the original position. There is still a limit switch in the original position, and it will stop after touch. This set of limit switches also serves as a start switch for the first drive motor. When the protecting mesh is fully closed, the coal scraper moves from bottom to top, and when the protecting mesh is fully opened again, the coal scraper moves from top to bottom. The length of the moving segment of the coal scraper is limited by another set of limit switches.

5 CONCLUSION

The lateral migration of flying gangue is greatly influenced by the shape of flying gangue and the strike inclination of coal seam. If the lateral displacement is large, the flying gangue may collide with the support column, scraper and other equipment, or fly out of the equipment gap into the pedestrian passage, causing gangue injury. Therefore, the lateral protection of flying gangue between the machine road and the pedestrian passage should not be ignored. In view of the low degree of automation and poor protective effect of the equipment against flying gangue in steeply dipping coal seam, an automatic and intelligent device was designed to accurately identify the hazard source of flying gangue by grating. The equipment can effectively solve the problem of poor protection caused by uncertainty in the occurrence of flying gangue disaster.

ACKNOWLEDGEMENTS

This work was financially supported by the FuShun Revitalization Talents Program (FSYC202107004), and the National Natural Science Foundation of China (51604213).

REFERENCES

Liu M, Wu Y P, Xu G. 2016 Research on rule of flying gangue in steeply dipping seam mining. *Coal Technology*, vol. 35, pp. 17–18.

Wang J A, Jiao J L. 2016 Criteria of support stability in mining of steeply inclined thick coal seam. *International Journal of Rock Mechanics and Mining Sciences*, vol. 82, pp. 22–35.

Wu Y P, Hu B S, Xie P S. 2019A new experimental system for quantifying the multidimensional loads on an on-site hydraulic support in steeply dipping seam mining. *Experimental Techniques*, pp. 1–15.

Wu Y P, Liu K Z, Yun D F, Xie P S, Wang H W. 2014 Research progress on the safe and efficient mining technology of steeply dipping seam. *Journal of China Coal Society*, vol. 39, pp. 1611–1618.

Wu Y P, Yun D F, Xie P S, Wang H W, Lang D and Hu B S. 2020 Progress, practice and scientific issues in steeply dipping coal seams fully-mechanized mining. *Journal of China Coal Society*, vol. 45, pp. 24–34.

Wu Y, Xie P, Wang H. 2013 Theory and practices of fully mechanized longwall mining in steeply dipping coal seam. *Mining Engineering*, vol. 65, pp. 35–41.

Yun D F, Liu Z, Cheng W D, Fan Z D, Wang D F, Zhang Y H.2017 Monitoring strata behavior due to multi-slicing top coal caving longwall mining in steeply dipping extra thick coal seam. *International Journal of Mining Science and Technology*, vol. 27, pp. 179–184.

Advances in Civil Engineering and Environmental Engineering – bin Yahya & Bin Daud (Eds)

Field test design and parameter optimization of 10 MPa-Scale pressure consolidation grouting

Yong Huang*
Guangdong Hydropower Planning & Design Institute Co., Ltd, Guangzhou, China

Wei Lu*
China Institute of Water Resources and Hydropower Research, Beijing, China

Chunhua Song
Guangdong Hydropower Planning & Design Institute Co., Ltd, Guangzhou, China

Weiquan Zhao
China Institute of Water Resources and Hydropower Research, Beijing, China

ABSTRACT: Based on the project of Yangjiang pumped-storage station, a field test of 10MPa scale pressure of consolidation grouting was done after systematic grouting test design, in order to evaluate the feasibility and choose suitable parameters for consolidation grouting of high-pressure tunnel with about 800m water pressure. The results reveal that, because of the heterogeneity of rock mass, the unit grouting amount of cement in test sections with the same types of surrounding rock was significantly different and chemical grouting was necessary; higher pressure could get a better grouting effect, 10MPa pressure cement grouting for holes greater than 2m into surrounding rock and 8MPa pressure chemical grouting was feasible; row spacing of 2.5m and 2m was better for rock class II and III. For the fault zones, a suitable grouting process was from systematic holes to holes along and across the fault; using OPC as grouting material was better than wet ground cement.

1 INTRODUCTION

The concrete lining was most accepted in high-pressure tunnels of the pumped-storage station, as it considered the bearing capacity of surrounding rock which could save costs and reduce the construction period. High-pressure tunnels using concrete lining had strict requirements on the geological condition and consolidation grouting should be used to reinforce the surrounding rock for resisting inner water pressure and reducing water leakage. Usually, the main design parameters of consolidation grouting, such as grouting pressure, grouting material, grouting technology and planning of grouting holes, were determined by using engineering analogy methods or site tests (He et al. 2021; Hou & Hu 2001; Huang & Wei 2020; Liu et al. 2021; Luo et al. 2020; Ma & Li 2017; Sun & Ran 2004; Wang et al. 2017; Xu & Zhou 2015; Xiao et al. 2021).

Compared with other operating pumped-storage stations in the world, the pressure tunnel of Yangjiang pumped-storage station has the greatest hydrostatic pressure, which is about 799m water head. Based on the design code of specification for design of hydraulic tunnel (DL/T 5195-2004, specification for design of hydraulic tunnel, 2004), consolidation grouting pressure is required to use 1.2 times the hydrostatic pressure which was nearly 10MPa. In fact, grouting using such high pressure was beyond the engineering experience that has ever existed. For high-pressure tunnel of Yangjiang pumped-storage station, whose minimum horizontal principal stress of the surrounding

*Corresponding Authors: huang.y@gpdiwe.com and luwei@iwhr.com

DOI 10.1201/9781003349563-58

rock should range from 9.75 to 10.43MPa, a 10 MPa-scale grouting pressure would increase the risks of hydrofracturing and lead to the break of concrete ling and water leakage.

In this paper, a systematic field test in the testing tunnel was introduced, which had similar surrounding rock condition of the high-pressure tunnel of Yangjiang pumped-storage station. And the feasibility of grouting using 10MPa scale pressure was discussed based on analyzing the grouting laws in different types of the surrounding rock. Finally, suitable optimistic grouting parameters were raised by evaluating the influence of the grouting effect under different grouting pressure, grouting material, plan of grouting holes, and grouting procedure in fault zones. The suggested grouting parameters were used during the construction of the high-pressure tunnel of the Yangjiang pumped-storage station and its successful experiėnce could provide a reference to other projects.

2 PROJECT CONDITIONS

Yangjiang pumped-storage station was in Bajia mountainous area between Yangchun and Dianbai city, Guangdong province. The planned installed capacity of the project was a total of 2400 MW and 1200MW in the first stage. The general layout of the water conveyance and power generation system of the project was from the south to the north and the circular high-pressure headrace tunnel was from the vertical shaft to the underground powerhouse. As shown in Figure 1, in the vertical plan, the high-pressure tunnel was divided into five parts by 2 vertical shafts and 3 horizontal tunnels, and concrete lining was used in the shafts and tunnels. The maximum inner diameter, hydrostatic pressure, and PD value of the high-pressure tunnel were 7.5m, 799m water head, and 5993m^2, respectively.

The geography of the project area belonged to poor tectonic stability hilly area and the north terrain was lower than the south part. The surrounding rock of the high-pressure tunnel was medium-coarse-grained granite of the Yanshanian period and mainly slightly weathered to fresh except for some moderately to the strongly weathered area by the influence of the fault. In the fault area, the surrounding rock was mainly class III and locally class IV. The high-pressure tunnel was crossing about 10 small faults with a total length of about 700m. The width of the faults was about 0.1~1.1m with water outflow. And the angle between high-pressure tunnel and faults was about 60° (Yang 2017; Li et al. 2009).

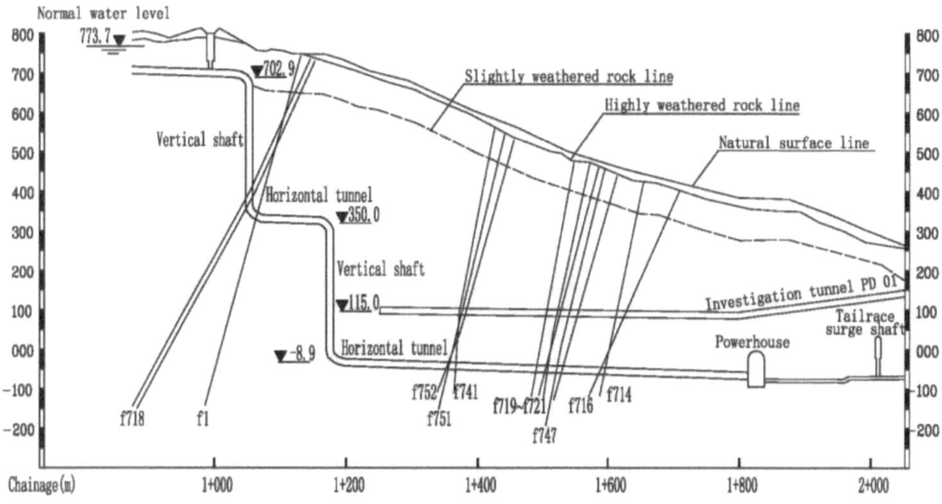

Figure 1. Vertical section water diversion and power system of Yangjiang pumped-storage station.

3 FIELD TEST DESIGN OF 10 MPA-SCALE PRESSURE GROUTING

3.1 *Objectives of field test*

For high-pressure tunnel under concrete lining and using bearing capacity of surrounding rock, the basic requirement was that hydrofracturing should be avoided. As 10 MPa-scale grouting pressure was greater than the horizontal principal stress of the surrounding rock and few experiences could be found for guidance, the purpose of the field test was to discuss the feasibility of 10MPa-scale grouting pressure, verify the basic design grouting parameters, estimate the grouting effect of different types of the surrounding rock, optimize the grouting parameters, and provide guidance and experience for consolidation grouting high-pressure tunnel of Yangjiang pumped-storage station.

3.2 *Basic Requirements of testing tunnel*

The 10MPa-scale pressure grouting test was an exploratory experiment. In fact, if the field test was carried out in the high-pressure tunnel, the results could provide more effective guidance for construction but it could also lead to irreversible breakings, such as the potential risks of ineffective grouting, moment of surrounding rock, hydrofracturing, and breaking of concrete lining. For the protection of the project, the suitable choice was selecting a similar tunnel, namely a testing tunnel.

Compared with the high-pressure tunnel, the basic requirements of the testing tunnel were as below: (1) the same lithology and strike direction of the rock stratum; (2) the types of the surrounding rock should include rock class II to class IV and also faults; (3) enough depth of the tunnel and convenience transportation were also needed. Following these requirements, from chainage of 0+890m to 1+026m of the engineering investigation tunnel PD01 was selected as the testing tunnel and the surrounding rock was coarse-grained granite, containing 3 faults, namely as f751, f752, and f718. The geological condition was 84.5m of rock class II, 41m of rock class III and 10.5m of rock class II/IV including faults area, respectively (Figure 2). The inner diameter of the testing tunnel and thickness of the concrete lining was the same as the high-pressure tunnel.

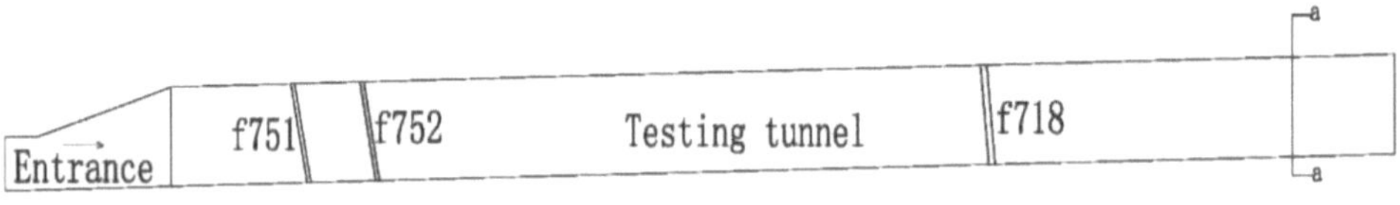

Class of Rock	II	II/IV		II						II/IV	III				
Section Number	14	13	12	11	10	9	8	7	6	5	4	3	2	1	
Chainage (m)	0+890.00	0+902.00	0+908.00	0+916.00	0+926.00	0+935.00	0+944.00	0+953.00	0+962.00	0+978.00	0+987.00	0+996.00	1+005.00	1+014.00	1+026.00

Figure 2. Geological condition of site test tunnel.

3.3 *Testing scheme*

The consolidation grouting test was not the same as other tests. First, due to the inhomogeneity of the surrounding rock, the characteristics of the test results were randomly distributed. Second, the grouting parameters including grouting pressure, grouting material, distance of grouting holes, and grouting process should be considered together with the mass conditions of the surrounding rock of the tunnel but for the limit of place and time, all the parameters could not be tested in each condition. Hence, a systematic grouting test design should be done before the field test.

Field test parameters of consolidation grouting for different surrounding rocks were shown in Table 1. The testing tunnel was divided into 14 sections based on the rock condition and each section was 6~16m in length.

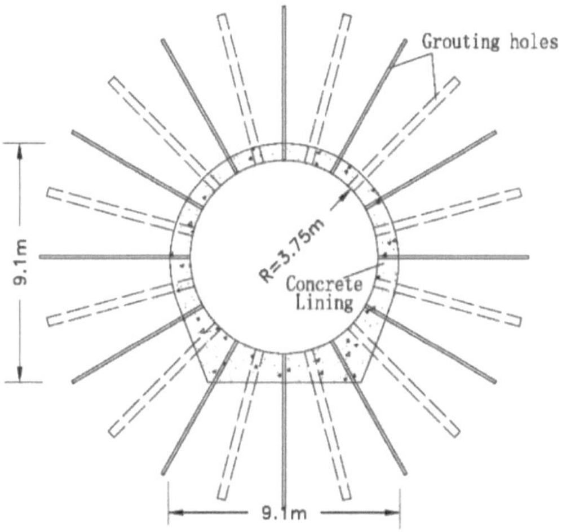

Figure 3. Cross-section a-a.

Table 1. Field test parameters in different surrounding rock types.

Rock types		Cement grouting: Row spacing (m)/ grouting process	Depth (m)	Pressure (MPa) 0~2m	2m~bottom	Material	Chemical grouting: Row spacing (m)	Depth (m)	Pressure (MPa)
Class II	Design	2.5	6	4.5	10	OPC	2.5	5	8
	Comparing	3	8,10	3,6	8	Wet ground cement	3	5	6
Class III	Design	2	6	4.5	10	OPC	2	5	8
	Comparing	3	6	4.5	10	Wet ground cement	3	5	6
Class II/IV	Design	2	6	4.5	10	OPC	2	5	8
	Comparing	3	6	4.5	10	OPC	3	5	8
Fault	Design	①→②→③	12	4.5	10	OPC	①→②→③	10	8
	Comparing	②→③→①	12	4.5	10	OPC	②→③→①	10	8

Note: ① is across fault holes, ② is systematic holes, ③ is along fault holes.

4 RESULTS AND DISCUSSION

4.1 *Unit grouting amount of each test section*

In fact, the grouting amount was discrete in testing sections with the same condition of surrounding rock because of the different distribution of cracks in the rock mass. But generally, bad rock condition needs more grouting materials. And after cement grouting, the chemical grouting amount was lower than 4kg/m except in fault zones (Figure 4). The results reveal that using high grouting pressure, cement grouting might be affected by both surrounding rock condition and grouting parameters and after large cracks in a rock mass were sealed by cement, chemical grouting could only be used in conditions of bad surrounding rock mass with lots of great fractures and small cracks that cement material could enter.

For the distribution of unit grouting amount of cement grouting holes, the better the condition of the surrounding rock was, the smaller grouting amount was needed though grouting parameters such as grouting pressure, grouting material, and grouting process were different in each testing section. For example, in rock class II, the frequency of grouting holes with grouting amounts lower than 5 kg/m was more than 85% but in fault zones, it was discrete and mostly from about 5 kg/m to 20 kg/m (Figure 5).

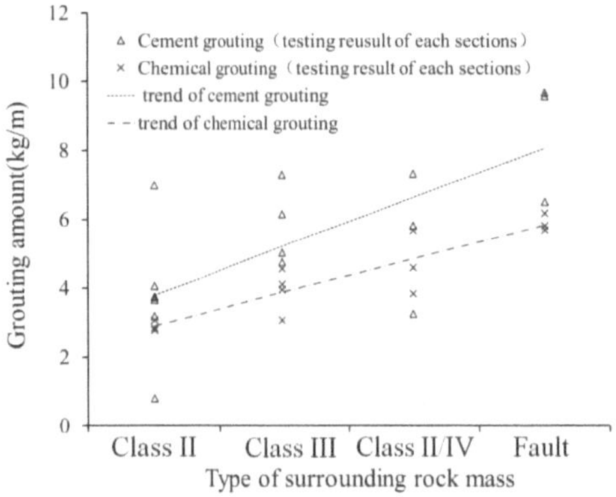

Figure 4. Average unit grouting amount of each test section under different types of surrounding rock.

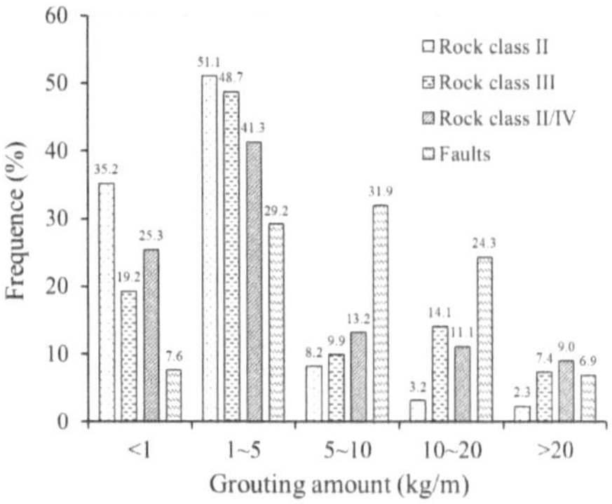

Figure 5. Distribution of unit grouting amount of each cement grouting hole.

4.2 *Feasibility of 10MPa-scale pressure grouting*

The test results of water pressure in rock class II and slurry leakage of grouting holes during the field test in each type of rock were shown in Tables 2 and 3 respectively. Based on the results, it was clear that for grouting holes of 2m depth in rock, 10MPa grouting pressure should not be used and the suitable water pressure and grouting pressure was less than 4MPa and 4.5MPa, respectively. Because the frequency of leaking holes was about 30%~50% and the largest elastic deformation of the concrete lining was about 126um when water pressure was greater than 4MPa and the frequency of slurry leakage holes was about 5%~10% when using 4.5MPa grouting pressure.

For grouting holes with 2~6m in rock, using 10MPa grouting pressure was allowable. Because even in the fault zone, the frequency of leakage holes was only about 5.2%. But for the water

Table 2. Test results of water pressure in rock class II.

Depth in rock (m)	Pressure (MPa)	Number of Testing holes	Number of leaking holes	Frequency (%)	Leaking pressure (MPa)	Lift deformation of concrete lining
0~2	<3	10	0	0	/	Without lift
	4	11	5	45.5	2.4~3.2	Maximum 40um
	4.5	24	7	29.2	4.1~4.5	Maximum 126um
	6	8	4	50	4.0~5.1	Maximum 12um
2~6	7	3	0	0	/	Without lift
	8	5	2	40	6.0~7.3	
	10	26	9	34.6	4.5~9.2	

Note: 1 Lift deformation change to zero after pressure released.

pressure test, the maximum water pressure should be lower than 8MPa. And after cement grouting, chemical grouting of all the holes using 8MPa pressure is feasible, which also indicates that cement grouting can strengthen the surrounding rock.

Table 3. Slurry leakage holes during the field tests.

Rock types	4.5MPa pressure cement grouting (from 0~2m in rock) Number of leaking holes	Frequency (%)	10MPa pressure cement grouting (from 2~6m in rock) Number of leaking holes	Frequency (%)	8MPa pressure chemical grouting (from 0~5m in rock)
Class II	31	9.9	8	2.4	Without leaking
Class III	30	14.5	0	0	
Class II/IV	17	8.6	4	1.9	
Fault	3	4.2	4	5.5	

4.3 *Influence of grouting pressure*

For rock class II, generally, the grouting amount of both cement and chemical grouting had increased and the permeability of the rock mass after grouting was reduced by using higher grouting pressure But the elastic moduli were nearly the same. Due to little disturbance of rock mass during the drilling and blasting process, the grouting amount increased under pressure from 4.5MPa to 3MPa for grouting holes of 2m depth in rock (Table 4).

Table 4. Grouting effect under different grouting pressure (rock class II).

Grouting material	Depth in rock (m)	Pressure (MPa)	Grouting amount (kg/m)	Elastic modulus after grouting (GPa)	Permeability of rock mass after grouting (Lu)
Cement grouting	2	3	10.34	40.19	0.68
		4.5	7.78	53.72	0.63
	2~6	8	3.67	47.9	0.58
		10	3.82	46.1	0.58
Chemical grouting	0~5	6	4.13	24.5	0.71
		8	4.57	26.6	0.54

4.4 *Influence of row spacing of grouting holes*

The field tests of both cement and chemical grouting in rock class II and III revealed that the smaller the row spacing was, the more the grouting amount was needed and the lower the permeability of rock mass after grouting would be (Table 5). For rock mass, considering the results of elastic modulus discrete and the permeability was a general reaction, the row spacing of 2.5m in rock class II and 2m in rock class III respectively was suggested.

4.5 *Influence of grouting material*

As shown in Table 6, comparing the grouting effect of grouting material using OPC and wet ground cement, the unit average grouting amount, frequency of leakage grouting holes, and permeability of rock mass after grouting were not changed much under the same condition of surrounding rock, except the elastic modulus of the rock mass. In fact, as cement slurry grouted in rock mass was

Table 5. Grouting effect under different row spacing of grouting holes.

Rock type	Row spacing (m)	Cement grouting amount (kg/m)	Chemical grouting amount (kg/m)	Elastic modulus after grouting (GPa)	Permeability of rock mass after grouting (Lu)
Class II	2.5	2.54	3.08	39.1	0.55
	3	2.27	2.85	48	0.63
Class III	2	7.0	4.57	24.8	0.79
	3	2.17	3.08	20.41	0.83

only about 4kg/m in rock class II and 6~7kg/m in rock class III, this could not lead to a great change of difference of elastic modulus of rock mass after grouting. Therefore, cement grouting using OPC as grouting material was suitable.

Table 6. Grouting effect under different grouting material.

Rock type	Grouting material	Grouting amount (kg/m)	Frequency of slurry leakage holes (%)	Elastic modulus after grouting (GPa)	Permeability of rock mass after grouting (Lu)
Class II	OPC	3.74	7.5	41.9	0.76
	Wet ground cement	3.66	10.4	48.61	0.63
Class III	OPC	7.3	15	17.8	0.47
	Wet ground cement	6.15	10.4	25.5	0.74

4.6 *Influence of grouting process in fault zone*

The grouting process in the fault zone was also tested. In the field test, for fault zone named f 718 and f 752, the grouting process was from systematic holes to along fault holes. And across fault holes, for fault named f 751, the grouting process was from across fault holes to systematic holes, and along fault holes. And the test results revealed that the grouting amount for both cement and chemical grouting in three faults using different process were about 6.51kg/m to 9.68kg/m and 5.78kg/m to 6.18kg/m, respectively and the grouting effect was better under the process of grouting from along fault holes to across fault holes, as the elastic modulus of the rock mass was higher and the permeability of rock mass was lower after grouting.

5 CONCLUSION

(1) It was necessary to select testing tunnel systematically and design field tests to study the feasibility of 10MPa-scale consolidation grouting pressure and optimize the grouting parameters as it was beyond the engineering experience that has ever existed.
(2) Due to the inhomogeneity of the surrounding rock, the average cement grouting amount was different even under the same condition of rock type and after cement grouting, chemical grouting was also needed to seal the micro cracks.
(3) It was better for grouting to use higher pressure. For cement grouting, the suitable grouting pressure was 4.5MPa and 10MPa for grouting holes of 2m in rock and other deep parts, respectively. And chemical grouting using 8MPa pressure was feasible.

(4) Suitable row spacing of grouting holes for rock class II and III was 2.5m and 2m. For the cement grout, compared with wet ground cement, using ordinary Portland cement as grouting material was more suitable.
(5) For faults, the suggested grouting process was from systematic grouting holes to along fault holes and finally across fault holes.

REFERENCES

DL/T 5195-2004, Specification for design of hydraulic tunnel. (2004) China, Beijing, *China Electric Power Press*.

HE Jiang, CHEN Nianhui, PAN Yibin. (2021) Design and practice of diversion tunnel surrounding rock consolidation grouting for Jinping II Hydropower Project. *GuangXi Water Resources & Hydropower Engineering*, 1:56–60.

Hou Jing, Hu Minyun. (2001) Discussion on some problems in design of high-pressure tunnel for hydro projects. *SHUILI XUEBAO*, 7:36–40.

Huang Wei, Wei Pengcheng. (2020) Key methods for non-cover-weight consolidation grouting of super high dam foundations. *Journal of Tsinghua University* (Science and Technology), 7:582–588.

Li Yongsong, Yin Jianmin, Ai Kai. (2009) Analysis on stability of high-pressure tunnel of Yangjiang Pumped-Storage Power Station. *Yangtze River*, 9:68–70.

Liu Pusheng, Chen Kaikun, Li Boyong, et al. (2021) Study on seepage and stability of high-pressure tunnel of Yangjiang pumped-storage power station. *Water Resources Planning and Design*, 1:105–110, 141.

Luo Chang, Feng Junchi, Dong Kangqian, et al. (2020) Experimental study on consolidation grouting of Sanhekou water control project. *Yellow River*, 10:112-115,119.

MA Junfeng, LI Faquan. (2017) Experimental studies on chemical grouting of anti-seepage and reinforcement for defective geological bodies in Shenzhen pumped storage power station. *China Building Waterproofing*, 10:30–32, 36.

Sun Jinku, Ran Maoge. (2004) High-pressure consolidation grouting for tunnel sections with unfavorable geology condition. *Chinese Journal of Rock Mechanics and Engineering*, 18:3196–3202.

Wang Qiang, Chen Zhao, Jin Lianghai. (2017) Experimental study on productive consolidation grouting of tailrace main tunnel of Qiongzhong pumped storage power station in Hainan Province. *Value Engineering*, 5:135–137.

Xiao Zhongli, Long Fang, Liu Quan, et al. (2021) Grouting quality control and analysis for the water diversion system of shenzhen pumped-storage power station. *Water Power*, 2:37–40.

Xu Qiyun, Zhou Guanghui. (2015) Experimental study on ultra-high-pressure grouting for small section tunnel in uncovered rock. Exploration engineering (*Tock & Soil Drilling and Tunneling*), 10:81–84.

Yang Lei. (2017) Analysis on engineering geological condition of high-pressure tunnel in Yangjiang pumped storage power station. *Guangdong Water Resources and Hydropower*, 4:40–44.

Advances in Civil Engineering and Environmental Engineering – bin Yahya & Bin Daud (Eds)
© 2023 The Author(s), ISBN: 978-1-032-39397-1

Research on mathematical model of tailing soil deformation under multi-field coupling

Xunzhen Zheng *
School of Civil Engineering, Changchun University of Architecture and Civil Engineering, Changchun, JiLin, P.R. China

Haichuan Jiang
JiLin Research and Design Institute of Building Science, Changchun, JiLin, P.R. China

ABSTRACT: The stability of the tailing dam will be affected by various factors if it was exposed to natural conditions for a long time. Therefore, the stability of the dam becomes a key concern. The deformation of tailing soil directly affects the stability of dam. In this paper, the coupling relationship of tailing soil deformation under the action of stress field, seepage field and chemical field is studied, the coupling mathematical model is established and the reliability of the model is verified.

1 INTRODUCTION

China is a big country with abundant and a complete range of mineral resources. With the mining of mineral resources, waste tailing are also discharged in the form of slurry and piled up in the constructed tailing ponds. As of March 2020, China is still the country with the largest number of tailing ponds in the world.

Plenty of stacked tailing not only occupy a large amount of land causing environmental pollution and waste of resources, but also bring the existence of hidden dangers and other problems attracting more and more attention. The tailing reservoir accumulates from the bottom and increases year by year. Under the action of dead weight, the pores between tailing particles gradually decrease and deformation occurs. The tailing soil can be regarded as in a stress field. Under the influence of consolidation, potential energy of water flow and osmotic volume force, tailing soil can be regarded as in a seepage field. Tailing particles have been affected by tailing water discharge and rainfall for a long time. Various chemical reactions occur between water chemical solution and tailing particles at all times, resulting in blockage of pore channels and corrosion of pore particles, affecting the permeability rate of the seepage field and consolidation of the application field. Tailing soil can be regarded as in a chemical field. The interaction of stress field, seepage field and chemical field on accumulated tailing soil easily leads to serious deformation of tailing as a whole, which is one of the causes of dam failure and also the focus of scholars (Feng 2003).

2 DEFORMATION MECHANISM OF TAILING SOIL UNDER MULTI-FIELD COUPLING

2.1 *Coupling mechanism of stress field and seepage field*

First of all, tailing soil in the stress field is the accumulation of granular porous media. The accumulation of particles and the dam body will be deformed under the action of force, affecting the shape and volume of porous media pore space, further influencing its permeability and porosity. Secondly, in the tailing soil in the seepage field, there is pore pressure in the pores, which will

*Corresponding Author: summer.6318057@163.com

DOI 10.1201/9781003349563-59

inevitably lead to the deformation of the solid skeleton, affecting the pore space and the medium stress, and ultimately it will affect the overall stability of the tailing dam.

2.2 *Main steps of chemical reaction between stress field and seepage field*

Chemical reactions in stress field and seepage field are inseparable from fluid reactants and solid reactants, and there are two processes of external mass transfer and internal diffusion in chemical reactions. In such a chemical reaction, fluid reactants by mass transfer first arrived outside the boundary layer and the outer surface of solid reactant, and then produce the solid product through the diffusion effect. In some chemical reactions such as precipitation, dissolution, adsorption, the separation of solid product will be more and more, then become the chemical reaction of solid reactant. Subsequent chemical reactions are carried out in accordance with the reverse process of the above reaction process (Shen 2011).

2.3 *Influence of water-chemical solution corrosion on mechanical properties of tailing soil*

Air, water, acid solution, alkali solution and salt solution will cause chemical corrosion to tailing soil, and then affect the mechanical properties of tailing soil, including shear strength and compress strength of tailing soil, so as to affect the stability of the dam body. The experimental results show that the concentration of H^+ and OH^- in solution will directly affect the effect of chemical corrosion, and the higher the concentration is, the more obvious the corrosion effect is. The erosion of the tailing soil environment also has a certain degree of influence on the macro and micro aspects of the tailing soil. The macro aspect affects the overall deformation, strength and stability of the tailing soil, and the micro aspect affects the pore space, skeleton volume, chemical reaction rate and consolidation effect of the tailing soil (Fesch 1998; Johnson 2003).

2.4 *Chemical silting in seepage process*

Our country has large quantities of tailing, in which the tailing pond of iron ore is the largest. The amount of acid solution, alkali and salt solution generated in the process of seepage water chemistry solution react with mineral of $Fe2^+$ and $Fe3^+$, and generating flocculant, clusters clogs, with chemical reaction continuing, precipitation increase gradually, for solution of distribution of pore space is narrow. Some pore channels are silted up gradually, which affects the permeability and permeability coefficient of tailing soil and hydrodynamic dispersion coefficient of chemical reaction.

3 MATHEMATICAL MODEL OF TAILING SOIL DEFORMATION UNDER MULTI-FIELD COUPLING

3.1 *Basic assumptions*

The seepage in tailing soil is unstable. The tailing soil is subjected to hydrostatic pressure, that is, the surface force of the outer surface of tailing soil is only normal stress component, excluding shear stress component. It is assumed that the solution through the tailing soil specimen is an incompressible Newtonian fluid, that is, the viscosity coefficient of the fluid is constant. There is an exponential relationship between permeability coefficient and porosity. The pore space of tailing soil refers to the pores between particles, without considering the pores inside particles and continuous bubbles and water droplets existing in pore space. The flow of solution in tailing soil satisfies linear Darcy's law. Considering the compressibility of pores in tailing soil, not considering the compressibility of solid particles in tailing soil.

3.2 *Basic equations of the coupling model*

3.2.1 *Equation of stress field*

According to the effective stress principle, the expression of tensor form, the constitutive equation of isotropic elastic body, the geometric equation under the premise of small deformation can be substituted into the equilibrium equation, and the equilibrium equation expressed by displacement can be obtained, which is the stress field equation.

$$(\lambda + G)\frac{\partial \varepsilon_V}{\partial x_i} - G\nabla^2 u_i + \alpha\frac{\partial p}{\partial x_i} = 0\ (i = 1, 2, 3) \tag{1}$$

3.2.2 *Seepage field equation*

Assuming that the seepage velocity in the three directions of x, y, z, and in the rectangular coordinate system is V_x, V_y, V_z, within the range of the selected element body, the mass conservation equation is listed as follows:

$$\frac{\partial}{\partial x}(\rho V_x) + \frac{\partial}{\partial y}\left(\rho V_y\right) + \frac{\partial}{\partial z}(\rho V_z) + \frac{\partial}{\partial t}[\rho\,(n_\circ + \Delta n)] = 0 \tag{2}$$

According to the hypothesis in the Boit equation, the fluid in the pores is incompressible, so the fluid density in the above equation is constant. After sorting out the above equation, it can be obtained:

$$\frac{\partial V_x}{\partial x} + \frac{\partial V_y}{\partial y} + \frac{\partial V_z}{\partial z} + \frac{\partial \Delta n}{\partial t} = 0 \tag{3}$$

Assuming that the tailing soil studied is isotropic and homogeneous porous medium, permeability is a function of the change of porosity of the medium, and the seepage continuity equation considering the change of permeability with porosity is:

$$\frac{\partial}{\partial x}\left[\frac{\kappa\,(\Delta n)}{\mu}\frac{\partial p}{\partial x}\right] + \frac{\partial}{\partial y}\left[\frac{\kappa\,(\Delta n)}{\mu}\frac{\partial p}{\partial y}\right] + \frac{\partial}{\partial z}\left[\frac{\kappa\,(\Delta n)}{\mu}\frac{\partial p}{\partial z}\right] = \frac{\partial \Delta n}{\partial t} \tag{4}$$

Where, κ is the permeability.

3.2.3 *Chemical field equation*

Through Reynolds transport theorem and Gaussian divergence theorem, ignoring some terms with minor effects, we assume that the density of solid skeleton does not change all the time, and only the saturated state is studied and the hydrodynamic dispersion coefficient is assumed to be constant D, and other influencing factors such as biodegradation and source/sink terms are not considered, it can be simplified as follows:

$$n\frac{\partial C}{\partial t} + C\frac{\partial n}{\partial t} + \nabla \cdot (VC) - D\nabla \cdot (n \cdot \nabla C) = 0 \tag{5}$$

Where, C is the dissolved concentration, V is the seepage velocity, and D is the hydrodynamic dispersion coefficient.

3.3 *Analysis of coupling relationship of three field equations*

3.3.1 *Analysis of coupling relationship between stress field and seepage field*

(1) Porosity

If the initial porosity of the porous medium is $n_\circ$, the change of porosity caused by deformation is Δn, and the porosity after deformation is:

$$n = n_\circ + \Delta n \tag{6}$$

Due to the total stress outside the porous media and the pore pressure inside the porous media, the pore structure of the media will change, which not only refers to the change of the pore volume, but also affects the direction of the pore channels. Assuming that the change of pore volume is only considered in the online elastic range, Δn also linearly depends on pore pressure p and total stress, so:

$$\Delta n = \frac{p}{R} - \frac{1}{H}\sigma_v \tag{7}$$

In the formula, the increase of porosity change is taken as positive. Since pore pressure p will cause pore expansion and increase pore volume, a positive sign is taken before p. R reflects the influence of pore pressure p on Δn, and R has the same unit as pore pressure p.

$$\Delta n = \frac{1}{Q}p - A\varepsilon_V \tag{8}$$

Where, $\frac{1}{Q} = \frac{1}{R} - \frac{\alpha}{H}, A = \frac{3\lambda+2G}{3H}$ are the mechanical parameter that reflects the deformation of porous media.

(2) Permeability coefficient

Zhao Yangsheng systematically studied the law of permeability coefficient of various coal samples through experiments, and obtained the empirical formula of exponential relationship between permeability coefficient and porosity change of coal samples. In this paper, it is assumed that the permeability coefficient and porosity of tailing soil also have an exponential relationship, and the following results can be obtained:

$$k = k_o e^{an} \tag{9}$$

Where, k_o is the initial permeability coefficient, a is a constant determined by experiment. Substituting Equations (6) and (8) into (9), it can be concluded that:

$$k = k_o' e^{a\left(\frac{1}{Q}p - A\varepsilon_V\right)} \tag{10}$$

Where, $k_o' = k_o e^{an_o}$.

According to the relationship between permeability and permeability coefficient, it can be obtained:

$$\kappa(\Delta n) = \frac{k_o \mu}{\gamma} e^{a\left(n_o + \frac{1}{Q}P - A\varepsilon_V\right)} \tag{11}$$

3.3.2 *Analysis of coupling relationship between chemical field and seepage field*

From the finally obtained seepage field equation (4) and chemical field equation (5), it can be seen that the two equations are coupled by the seepage velocity, that is, the seepage velocity is the coupling term of the seepage field equation and chemical field equation.

3.3.3 *Analysis of chemical field-stress field relationship*

Integrated a large number of relevant experimental literature, according to the modulus of elasticity over time and the change trend of acid and alkaline solution, and the relationship between the compression modulus and elastic modulus, while ensuring them consistent with the elastic modulus of soil orders of magnitude, iron ore tailing in Liaoning Fuxin is tentatively given soil elastic modulus changes with time, the solution of acid solution respectively, water and alkali solution, The fitting expressions of the relation between elastic modulus of tailing soil and time are as follows:

$$E_1 = 6.50214 \times 10^{-2}T^3 - 1.087T^2 + 0.9966T + 65.987 \tag{12}$$

$$E_2 = 3.53514 \times 10^{-2}T^3 - 0.734T^2 + 1.403T + 65.987 \tag{13}$$

$$E_3 = 2.77757 \times 10^{-2}T^3 - 0.627T^2 + 1.481T + 65.987 \tag{14}$$

Where, T is soaking time.

3.4 *Modeling and Solution*

For the convenience of calculation, the formula of 3D mathematical model is reduced to 1D model. The following equation is obtained:

$$\begin{cases} \frac{\partial}{\partial x}\left[\frac{\kappa(\Delta n)}{\mu}\frac{\partial p}{\partial x}\right] = \frac{\partial \Delta n}{\partial t} \\ (\lambda + G)\frac{\partial \varepsilon_V}{\partial x} - G\nabla^2 u + \alpha\frac{\partial p}{\partial x} = 0 \end{cases} \tag{15}$$

In order to uniquely determine the tailing soil deformation at different times and under different external conditions, giving the initial conditions and boundary conditions, the model is discretized by finite difference method, and solved by programming on Matlab platform. The axial deformation curves of tailings soil specimens with height of 60 cm in the three solutions along the height direction were obtained under the effect of external downward loads of 50 kPa, 100 kPa, 200 kPa and 800 kPa, as shown in Figures 1–4.

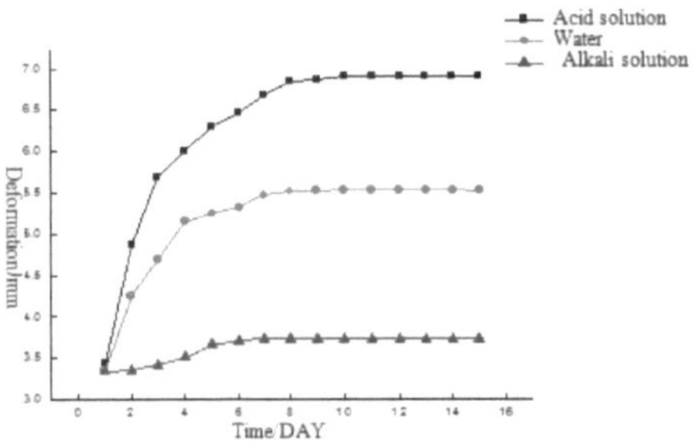

Figure 1. Deformation of soil column under 50kPa.

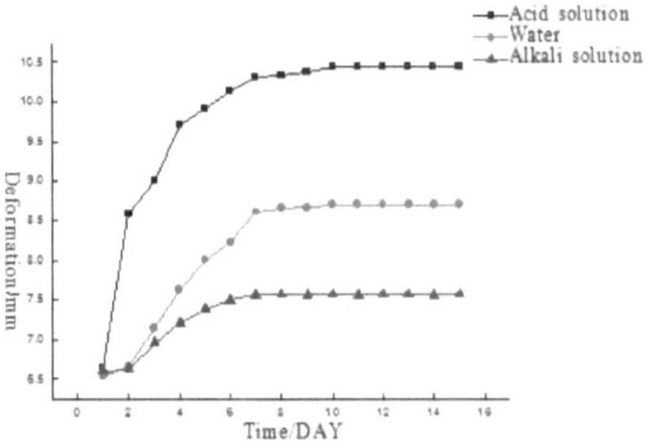

Figure 2. Deformation of soil column under 100kPa.

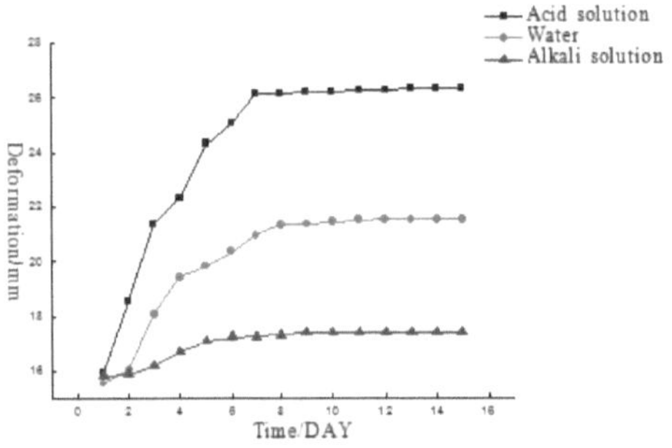

Figure 3. Deformation of soil column under 200kPa.

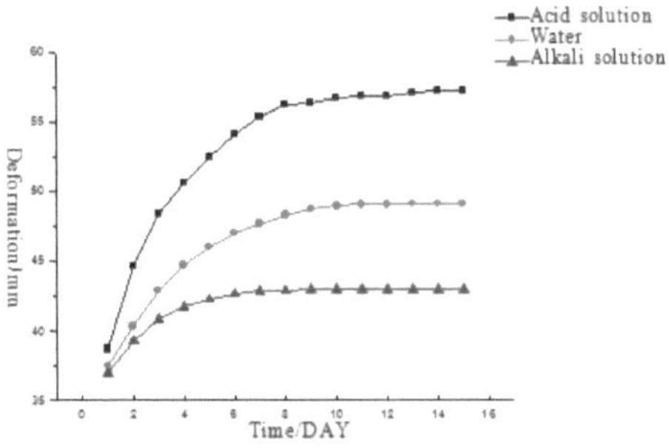

Figure 4. Deformation of soil column under 800kPa.

As can be seen from the above figures, the deformation amount of tailing soil column is constantly changing with time. With the increase of time, the amount of deformation gradually increases. When the time reaches a certain point, the curve does not change, indicating that the deformation of tailing soil specimen has reached a stable state. When the external load increases, the deformation of specimens in the three solutions increases to varying degrees. The heavier the external load is, the more the deformation increases. Under the four kinds of external loads, the deformation changes the most in acid solution, the second in distilled water, and the least in alkali solution. In acid solution, dissolution reaction mainly occurs, resulting in the formation of $FeCl_3$ and $FeCl_2$. The cementation between solid particles and the outer surface of particles are corroded by acid solution, resulting in the reduction of mechanical properties of solid particles and chemical damage. The higher the concentration of H+ in the solution, the more serious the corrosion of tailing soil and the greater the deformation, which is the same as the conclusion in literature (Bai 2010). In alkaline solution, the cementation reaction mainly occurs, in which Fe^{3+} and $(OH)^-$ combine to generate the new generation of material for cement $Fe(OH)_3$, filling intergranular pore, reducing the pore space and even closing pore channel. To prevent seepage effect, the compress stress under the action of compressible space is reduced, causing the overall relative deformation.

In order to verify the accuracy of the model, it is necessary to compare the calculated results with that of the tailing soil in the consolidation and permeability experiment (Liu 2013), where the settlement reaches a stable level. The application program recalculates the deformation of the specimen with a height of 20 mm. The comparison results are shown in Figures 5–6, which basically verifies the accuracy of the model.

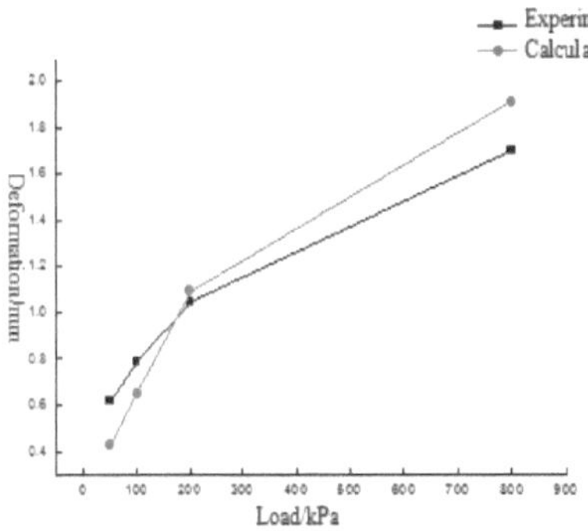

Figure 5. Deformation of soil column under acid solution.

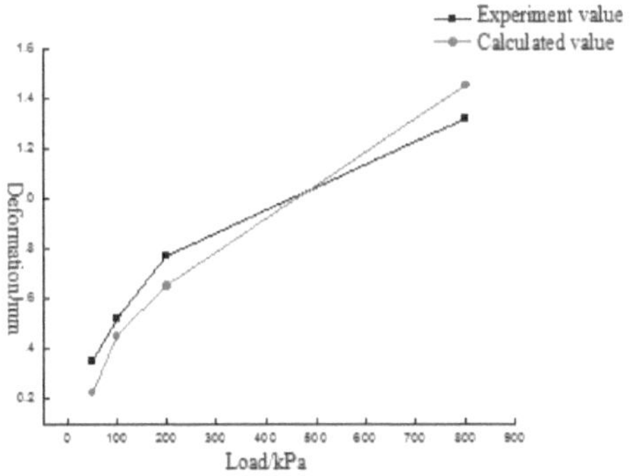

Figure 6. Deformation of soil column under alkali solution.

4 CONCLUSION

(1) Through the analysis of the main mechanism of tailing soil deformation chemical environment, it is concluded that affected by the external load and tailing water, pollution of the environment of the tailing soil, obvious change in mechanical properties,as well as change and mineral composition, strength of cement, etc., the settlement of dam body will be different, causing the instability of the whole structure of tailing dam;

(2) The relationship between the pore pressure - porosity -H^+ concentration and the nonlinear permeability coefficient - porosity -H^+ concentration of tailing soil deformation is established.

(3) It is concluded that under the same external load, acid solution has the greatest influence on the deformation of tailing soil, followed by water, and alkali solution has the least influence on the deformation of tailing soil. Under the condition of the same acidity and alkalinity of the solution, the larger the external load, the longer the time needed to stabilize the deformation.

REFERENCES

Bai, Y. P., (2010) Study on deformation mechanism of tailing dam under acid and alkali solution. *Fuxin:Liaoning Technical University*.

Feng, X. T., Liu J., Jing L., (2003) Research and application on coupled T-H-M-C processes of geological media in China-A review. *International Conference on Coupled T-H-M-C Processes in Geo-systems: Fundamentals Modelling*, (03):03:32–43.

Fesch C., Simon W., Stafan B. H., et al. (1998) Nonlinear sorption and non- equilibrium solute transport in aggregated porous media: experiments, process identification and modeling. *Journal of Contaminant Hydrology*, (31):373–407.

Johnson G. R., Gupta K., Putz D. K., et al. (2003) The effect of local-scale physical heterogeneity and nonlinear, rate-limited sorption/desorption on contaminant in porous media. *Journal of Contaminant Hydrology*, (64):35–58.

Liu, Y. W., (2013) Experimental study on mechanical properties and adsorption properties of tailing in hydrochemical environment. *Fuxin: Liaoning Technical University*.

Shen, L. F., Feng, X. T., Pan,P. Z., et al. (2011) Chemical kinetics dissolution mechanism of rock under stress. *Rock and Soil Mechanics*, 32(5):1320–1326.

Advances in Civil Engineering and Environmental Engineering – bin Yahya & Bin Daud (Eds)
© 2023 The Author(s), ISBN: 978-1-032-39397-1

On the multi-field coupling of thermo-hydro-mechanical-vapor in frozen soil

Zhenchao Teng*, Bo Li* & Yunchao Teng*
College of Civil Engineering and Architecture, Northeast Petroleum University, Daqing, China

ABSTRACT: This paper analyzes and studies the multi-field coupling theory and mechanism of frozen soil as well as the dynamic change process of temperature field, moisture field, and stress field in a freezing process according to the research results of scholars in various countries on frozen soil multi-field. The research on multi-physical field coupling of frozen soil is a complex, multi-physical field and multi-disciplinary field, this paper is mainly based on reviews from the aspects of Hydro-Thermo-Vapor-Mechanical (HTVM) field coupling. This summary is conducive to promoting the study of the soil freezing process and the frozen soil coupling mechanism in the cold area to promote the deep-level and multi-dimensional understanding of the multi-field coupling dynamic process of the soil freezing process.

1 INTRODUCTION

The freezing-thawing settlement is the most common frost damage caused by frozen soil in cold regions. It is primarily due to the changes in temperature, moisture, stress, and concentration fields of the frozen ground and interactions between multi-physical fields (Mu 1987). A variety of frost damage problems of the frozen soil originate in essence from multi-phase coupling (including solid, liquid, gas, and heat) in the porous media (Li 2001). The soil stability in the seasonally frozen ground is mainly controlled by the interactions and mutual influence between heat transfer, moisture migration, and phase change during the freeze-thaw cycles. For example, the temperature, moisture, and stress fields of the roadbed are dynamically changing. The coupling effects between these fields are the direct reason for many frost damage problems (Lai 1999). Therefore, it was necessary to systematically review the mechanisms behind the multi-physical field coupling of the frozen soil.

The multi-physical field coupling for the frozen soil presents a complex multi-disciplinary research sector that considers multiple physical fields simultaneously. The studies of multi-physical field coupling for the frozen soil are further divided into the following subsectors: thermo-hydro (TH) coupling, thermo-hydro-mechanical (THM) coupling, thermo-hydro-vapor-mechanical (THVM) coupling, and thermo-hydro-salt-mechanical (THSM) coupling. In addition, some scholars are committed to studying thermo-hydro-chemical-mechanical (THCM) coupling for rocks (Su 2010), which is a complex and dynamic process. Recently, the academic interests related to the frozen field are generally focused on the macroscopic strength properties and the thermo-hydro-mechanical coupled constitutive model. Strength criteria proposed considering the unique tension behavior and pressure melting of crystals and crystallization kinetics model of ice-water phase change. However, the constitutive experiments conducted so far rarely consider the mechanism of microscopic deformation due to the lack of special equipment. Therefore, there is an urgent need for a constitutive investigation on the minuscule scale (Zhang 2019).

*Corresponding Authors: tzcdqpi@163.com, 909069778@qq.com and 1124727135@qq.com

DOI 10.1201/9781003349563-60

A thorough study of the multi-physical field coupling for the frozen soil is highly significant for the safety and stability of lifeline engineering in cold regions, including building construction, railway operation, and oil pipelines. Researchers have conducted extensive 3D simulations and experiments on the multi-physical field coupling for the frozen soil but made only limited achievements given the broad range of engineering sectors involved in the multi-physical field coupling for the frozen soil. Therefore, it was crucial to probe deeper into the simulation and experiments of the multi-physical field coupled systems for the frozen soil.

2 CURRENT RESEARCH STATUS OF MULTI-PHYSICAL FIELD COUPLING FOR THE FROZEN SOIL

2.1 *Development of HT coupling of the frozen soil*

Coupling is divided according to physical model properties as shown in Figure 1 (Teng 2021).

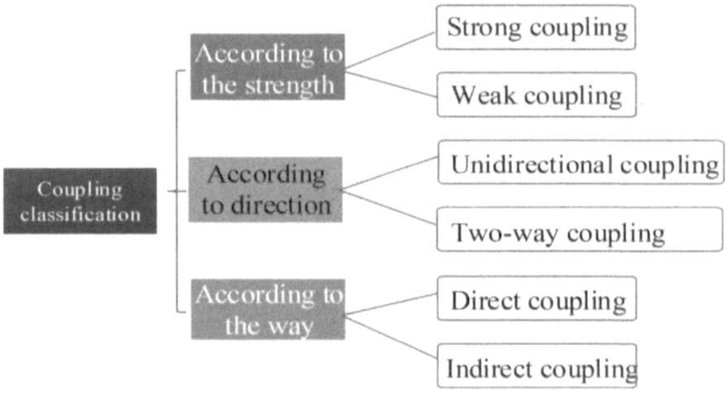

Figure 1. Coupling classification.

The TH coupling model was proposed for shallow loess by Zhang et al, and the model results agreed with the experimental results. However, the heat flow model did not consider the stress changes caused by the formation of ice lenses during the cooling process of soil (Zhang 2020).

It is assumed that the pore ice is immovable as the soil freezes and that the pore water migrates along with the negative water pressure and temperature gradient. Without considering the water vapor sorption isotherms, the parameter β is related to soil structure and porosity. In 2020, Lei L et al. proposed the seepage equation and the functional relation of the corresponding parameters based on the law of the conservation of energy (Lei 2020):

$$C_{21}\frac{\partial T}{\partial t} + C_{22}\frac{1}{1+e}\frac{\partial e}{\partial t} + \rho_l v\left(-\frac{k}{\gamma_l}\nabla P\right) + \rho_l V\left(-\frac{k\rho_l L}{\gamma_l T_0}\nabla T\right) + \rho_l \nabla(-k\nabla z) = 0 \quad (1)$$

where $C_{21} = n(\rho_i - \rho_l)\frac{\partial S_i}{\partial T}$, $C_{22} = S_i\rho_i + (1 - S_i)\rho_l$; k is the hydraulic conductivity of frozen and unfrozen soil, respectively (m.s^{-1}); ρ_l and ρ_i are the densities of water and ice, respectively (kg.m^{-3}); the meanings of ∇ and S_i have been explained for the temperature field above; n is poriness; γ_l is the unit weight of water; P_l and P_i are water pressure and ice pressure, respectively.

Similar to the case of the differential equations of temperature field, Xu Dong Zhang et al. (Zhang 2021) suggested, in 2020, that the hydrothermal transfer only occurred vertically. The discharge flow rate $q = f_{(z)}$ was derived by Taylor expansion. Based on the saturated soil model by Zhang et al., the proposed equation considered the ice-water phase change and the relative volume of different soil components.

$$\frac{\partial\{\rho_w\theta_u(1+e)dz_s dA + \rho_i[e - \theta_u(1+e)]z_s dA\}}{\partial t} = -\rho_w\frac{\partial}{\partial z}\left(k\frac{\partial\phi_z}{\partial z}\right)(1+e)dz_s dA \quad (2)$$

Where ρ_w and ρ_i are the densities of water and ice, respectively; e is the porosity ratio; θ_u is the volume content of unfrozen water; dz_s is the height of the soil particle; d_A is considered a constant.

So far, the more mature TH models for the frozen soil include the following: thermodynamic model (Lai 2009), the revised version of Harlan's model (Li 2001), segregation potential model (Su 2010), hydrothermal random coupling model (Yuanming 2010), and hydrothermal dynamic coupling models (Teng 2021). Numerical simulations have been conducted extensively for TH coupling for frozen soil. By contrast, the three-dimensional hydrothermal numerical simulation studies and the relevant experiments are few. It is necessary to conduct multi-field coupling studies involving the hydro-thermal-salt-vapor-chemical-mechanical aspects.

2.2 *Theory of THVM coupling*

Vapor migration is a common factor inducing the failure of road engineering (Li 2014). Water vapor migration is a crucial factor to be considered in the study of unsaturated soil freezing. Understanding the role of water vapor migration is of high importance for revealing the mechanism of THVM coupling. The primary THVM coupling process is shown in Figure 2 (Bai 2020).

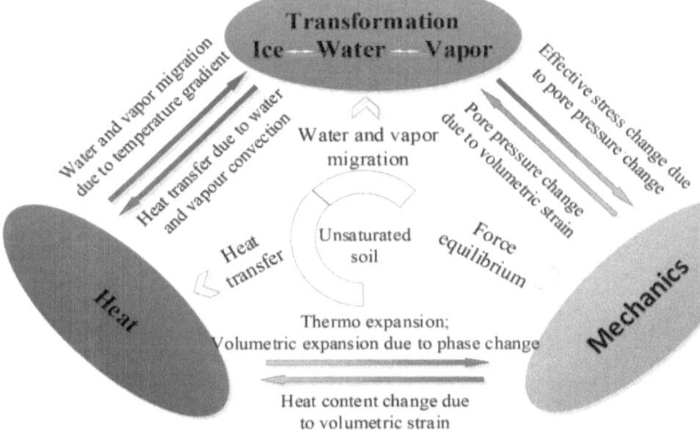

Figure 2. Coupled heat-water-vapor-mechanics mechanism of unsaturated soils.

Most of the existing models focus on the water vapor behaviors and heat transfer process but rarely consider the influence of deformation on the freezing of unsaturated soil. The studies on the latent heat of ice-water-vapor phase change, stress-strain relationship, water vapor migration mechanism, and heat transfer during the freezing process should be prioritized.

2.2.1 *Theory of THV coupling of unsaturated frozen soil*

The differential equation of the water field for unsaturated frozen water should consider the influence of ice-water phase change and the impact of water vapor on the migration and freezing of liquid water. Hansson (Hansson 2004), Sakai (Sakai 2009), and Zhang et al. (Zhang 2016) demonstrated that the earliest model proposed by Philip and De Vries was able to simulate the water-heat-water vapor behavior of unsaturated soil. Hence Bai R et al. Bai (2020) established the following equation controlling for the flow of unfrozen water and water vapor based on the modified Darcy's law and the equation of mass conservation:

$$\frac{\partial \theta_l}{\partial t} + \frac{\partial \theta_v}{\partial t} + \frac{\rho_i}{\rho_w}\frac{\partial \theta_i}{\partial t} = \nabla \cdot [K_{lh}\nabla(h+z) + K_{lT}\nabla T] + \nabla \cdot [(K_{vh}\nabla h + K_{vT}\nabla T] \tag{3}$$

Where θ_l, θ_v, and θ_i, are the volume contents of unfrozen water, equivalent vapor, and ice, respectively; ρ_w and ρ_i are the densities of water and ice, respectively; K_{lh} and K_{lT} are the isothermal and thermal-hydraulic conductivities for liquid phase fluxes due to gradients under h and T gradients, respectively; K_{vh} and K_{vT} are the isothermal and thermo-steam-hydro-mechanical conductivity, respectively; h is the pressure head; Z is the spatial coordinate.

2.2.2 *THVM coupling theory for unsaturated frozen soil*

Water adsorption of soil is a complex movement process involving multi-physical fields. At present, assuming a constant gas pressure is a common way to simplify the THM coupling model for unsaturated soil (Takagi 1970). The advantages and disadvantages of the THVM coupling model for unsaturated frozen soil are illustrated in Figure 3:

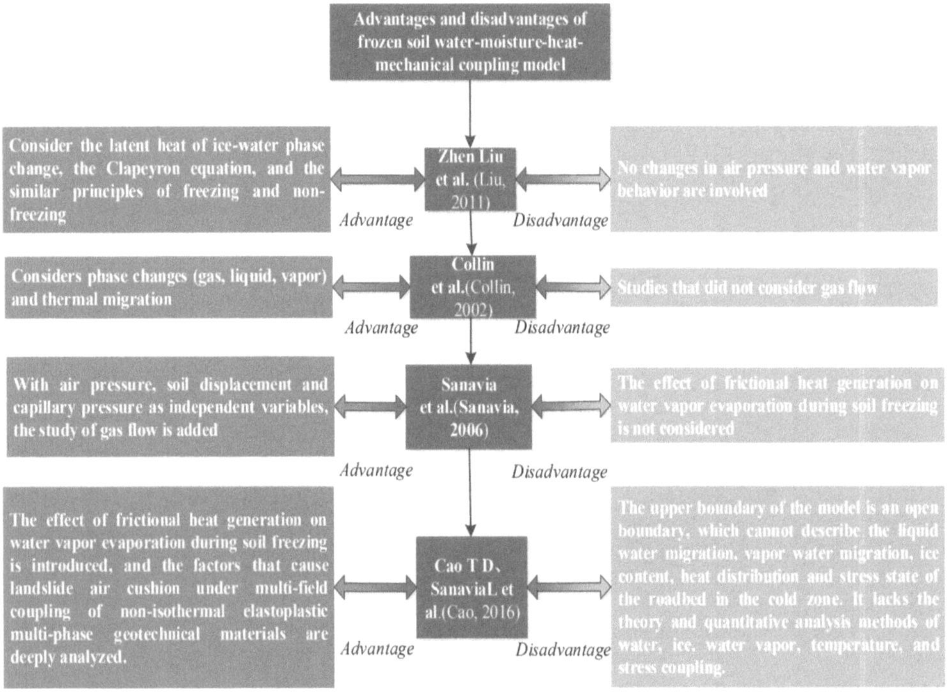

Figure 3. Advantages and disadvantages of the THVM coupling model for unsaturated frozen soil.

In the above studies, the upper boundary of the soil is usually assumed as an open boundary. However, the THVM coupling for unsaturated soil exposed to loading from above is not thoroughly studied. Besides, there seems to be a lack of studies on the THVM coupling model for unsaturated soil under the freeze-thaw cycles. The THVM coupling for the unsaturated soil in the seasonably frozen ground and cold region is poorly described.

3 CONCLUSIONS AND OUTLOOK

1. Based on the previous application of the coupling theory, mathematical model, and the premelting theory in the thermal coupling problem of frozen soil water. At present, the mathematical model of thermo-hydro-mechanical-vapor multi-field coupling interaction is not perfect, and the mathematical model of multi-field coupling interaction has not been completely established, which requires further theoretical derivation and experimental verification.

2. The formation of frozen soil involves a complex thermomechanical process. Therefore, a comprehensive description of the multi-field coupled differential equations, constitutive relations, and the multi-physical field coupling test simulation is necessary to promote further research in this field. Comprehensive models for multi-physical field coupling should consider the time-varying TH, THM, THVM, THSM, and THCM phenomena related to the constitutive relationship of the frozen soil.
3. The author believes that the study of using triaxial test devices in multiple physical field measurements should be strengthened at present, COMSOL is the most used software for multi-physical field coupling. Future research should continue to establish a multi-field coupled digital model in the multi-physical finite element software, using the water, steam, heat, stress, strain monitoring data, and the multi-field coupled environmental action simulation test data to verify the accuracy of the mathematical model.

REFERENCES

Bai, R., Lai, Y., Zhang, M., & Ren, J. (2020). Study on the coupled heat-water-vapor-mechanics process of unsaturated soils. *Journal of Hydrology*, 585, 124784.

Cao, T. D., Sanavia, L., & Schrefler, B. A. (2016). A thermo-hydro-mechanical model for multiphase geomaterials in dynamics with application to strain localization simulation. *International Journal for Numerical Methods in Engineering*, 107(4), 312–337.

Collin, F., Li, X. L., Radu, J. P., & Charlier, R. (2002). Thermo-hydro-mechanical coupling in clay barriers. *Engineering Geology*, 64(2–3), 179–193.

Hansson, K., Simunek, J., Mizoguchi, M., Lundin, L.C., van Genuchten, M.T., 2004. Water flow and heat transport in frozen soil: numerical solution and freeze-thaw applications. *Vadose Zone Journal*, 3(2), 693–704.

Lai, Y. M., Wu, Z., Zhu, Y., & Zhu, L. (1999). Nonlinear analysis for the coupled problem of temperature and seepage fields in cold regions tunnels. *Cold Regions Science and Technology*, 29(1), 89–96.

Lai, Y., Jin, L., & Chang, X. (2009). Yield criterion and elasto-plastic damage constitutive model for frozen sandy soil. *International Journal of Plasticity*, 25(6), 1177–1205.

Lei, D., Yang, Y., Cai, C., Chen, Y., & Wang, S. (2020). The Modelling of Freezing Process in Saturated Soil Based on the Thermal-Hydro-Mechanical Multi-Physics Field Coupling Theory. *Water*, 12(10), 2684.

Li Ning , Cheng Guodomg , Xu Xiaozu., 2001. The advance and review on frozen soil mechanics. *Advance in Mechanics*, 31(1) : 95–102.(in Chinese).

Li, Q., Yao, Y.P., Han, L.M., Hu, J., Peng, R., Wang, N.D., 2014. *Pot-cover effect of soil.Ind. Constr.* 44, 69–71 (In Chinese).

Liu, Z., & Yu, X. (2011). Coupled thermo-hydro-mechanical model for porous materials under frost action: theory and implementation. *Acta Geotechnica*, 6(2), 51–65.

Mu, S., & Ladanyi, B., 1987. Modelling of coupled heat, moisture and stress field in freezing soil. *Cold Regions Science and Technology*, 14(3), 237–246.

Sakai, M., Toride, N., Simunek, J., 2009. Water and vapor movement with condensation and evaporation in a sandy column. *Soil Sci. Soc. Am. J.* 73, 707–717.

Sanavia, L., Pesavento, F., & Schrefler, B. A. (2006). Finite element analysis of non-isothermal multiphase geomaterials with application to strain localization simulation. *Computational Mechanics*, 37(4), 331–348.

Su, B. Y., Zhang, W. J., Sheng, J. C., Xu, X. C., Zhan, M. L., & Liu, J. S. (2010). Study of permeability in single fracture under effects of coupled fluid flow and chemical dissolution. *Rock and Soil Mechanics*, 31(11), 3361–3366.

Takagi, S. (1970). An analysis of ice lens formation. *Water resources research*, 6(3), 736–749.

Teng, Z. C., Liu, X. Y., Liu, Y., Zhao, Y. X., Liu, K. Q., & Teng, Y. C. (2021). Stress–Strain Assessments for Buried Oil Pipelines Under Freeze-Thaw Cyclic Conditions. *Journal of Pressure Vessel Technology*, 143(4), 041803.

Yuanming, L., Yugui, Y., Xiaoxiao, C., & Shuangyang, L. (2010). Strength criterion and elastoplastic constitutive model of frozen silt in generalized plastic mechanics. *International Journal of Plasticity*, 26(10), 1461–1484.

Zhang, D., & Liu, E. (2019). Binary-medium-based constitutive model of frozen soils subjected to triaxial loading. *Results in Physics*, 12, 1999–2008.

Zhang, H., Zhi, B., Liu, E., & Wang, T. (2020). Study on Varying Characteristics of Temperature Field and Moisture Field of Shallow Loess in the Freeze-Thaw Period. *Advances in Materials Science and Engineering*, 2020.

Zhang, S., Teng, J., He, Z., & Sheng, D. (2016). Importance of vapor flow in unsaturated freezing soil: a numerical study. *Cold Regions Science and Technology*, 126, 1-9.

Zhang, X., Wu, Y., Zhai, E., & Ye, P. (2021). Coupling analysis of the heat-water dynamics and frozen depth in a seasonally frozen zone. *Journal of Hydrology*, 593, 125603.

Advances in Civil Engineering and Environmental Engineering – bin Yahya & Bin Daud (Eds)
© 2023 The Author(s), ISBN: 978-1-032-39397-1

Research of three-dimensional imaging sonar technology in the detection of underwater engineering project

Kewei Zhu* & Bing Xiao*
Fujian Chuanzheng Communications College Civil Engineering, Fuzhou, Fujian, China

ABSTRACT: A three-dimensional imaging sonar technique is employed to inspect the wharf structure underwater, create an underwater image of the concealed engineering, provide a visual display of the whole concealed project, and provide a basis for repairing any defects. This will provide hands-on experience for a similar underwater detection project.

1 INTRODUCTION

Damage and faults in the underwater part of water transport projects have an impact on the project structure's safety and reliability. Regular or irregular inspection, appraisal, and maintenance of the engineering structure are essential to ensure the integrity and safety reliability of the water transport engineering structure. Typical methods do not provide an effective investigation of underwater safety flaws due to their difficulty in locating, handling, and identifying (Li 2015). Underwater sounds, underwater robotic photography, multibeam bathymetry systems, and acoustic detection systems are some of the current methods for underwater examination (Qiu 2020). Researchers have conducted extensive research on sonar target detection (Han 2015), concluding that sonar is an effective way to achieve acoustic detection (Liu 2020). 3D imaging sonar technology is used for detection by using visual, underwater photography, or video images. With 3D images, it is possible to display underwater structures in a sand-filled environment with low visibility and complex underwater topography. The resulting images have the characteristics of large data volumes and multiple beams (Cho 2015), which are important in the quality control of underwater construction and the acceptance of port projects.

The paper takes as an example a wharf project, develops a standard size model, drops it into the wharf project waters, integrates the water environment, and inspection distance with model materials, and carries out experiments using 3D imaging sonar technology. To assist in the quality inspection and repairs of the wharf's underwater structures, the team utilizes 3D imaging sonar to achieve underwater full-coverage 3D detection (Yang 2020).

2 MAIN BODY

2.1 *Research design and methodology*

The three-dimensional imaging sonar technology involves transmitting a fan-shaped beam of ultra-high frequencies (Shi 2013), receiving the reflected signal from the target, and then analyzing that reflected signal. To determine the position of various parts of the target, techniques such as beam formation, pointing, amplitude, and phase detection are employed. Using this method, an underwater target's relative spatial position can be located finely and fine structure can be analyzed qualitatively and quantitatively.

*Corresponding Authors: 2271902550@qq.com and 284683895@qq.com

DOI 10.1201/9781003349563-61

As shown in Table 1, the BV5000-1350 3D sonar system includes the scanning sonar head, the head, the head sonar junction box, the cables, the software control platform, the power supply, and other accessories in this experiment (Zhang 2018).

Table 1. Key technical specifications of the BV5000 - 1350 (Yang 2015).

No.	Parameter name	Value
1	Sonar view	$1° \times 45°$
2	Scan range	$45° \sim 360°$
3	Range of tilt angles	$-65° \sim 65°$
4	Update frequency	40 Hz
5	Sound frequency	1.35 MHz
6	Max. scanning distance	30 m
7	Number of beams	256
8	Width of beams	$1° \times 1°$
9	Spacing of beams	0.178°
10	Time Resolution	0.031 m

To locate the target area, a sonar system BV5000- 1350 is installed in the water at the specified station location. The sonar head and head are mounted and fixed on a triangular bracket, and the bracket is sunk into the water (Church 2017). Because the BV5000-1350 can only measure 30 m and is unable to scan the entire target in one direction, multiple scanning stations must be set up and the instrument placed in the right position to obtain an overall scan of the target. At least three noncoincident targets must be placed at each station, with at least 10% of the image overlapping between two adjacent stations, based on the target's location, shape, and size. It is difficult to ensure stitching accuracy if there is too little overlap. Too much overlap will take more time and effort and may result in stitching errors.

2.2 *Concrete model experiment*

The experiments were conducted with a BV5000-1350 3D scanning sonar system, which was currently commonly used in the industry, and a homemade upright shore wall structure model was used to study underwater detection capabilities, as shown in Figure 1. For the model to simulate various widths of underwater components, each rectangle had a width of 1 cm, 2 cm, 5 cm, and 10

Figure 1. Front view of the home-made upright shore wall model.

cm; the thickness of the model was 10cm and its length was 40 cm. A model and a 3D sonar were placed in Meizhou Bay, Fujian Province, with a sound velocity of 1509 meters per second. They were attached to a bracket, as shown in Figure 2. With the sonar fixed on the stand, the distance between it and the model may be adjusted. For this experiment, the fixed distances were 5.8 meters, 1.22 meters, 1.25 meters, 1.0 m, 0.9 m, 0.8 m, and 0.6 meters. Based on comparing the different distances, the detection results were analyzed, and some of the acquired images were shown in Figure 3.

Figure 2. Measurement bracket for sonar and model fixation.

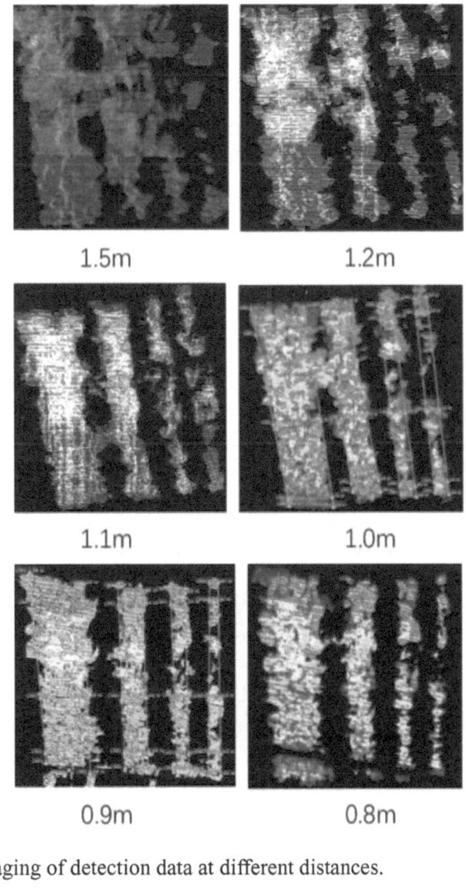

Figure 3. Point cloud imaging of detection data at different distances.

From Figure 3, it could be seen that the BV5000-1350 3D scanning sonar performs best at 1 m or more. Any distance less than 0.9m would result in mutual interference of the echo signals and a loss of reflected signals. For example, the signal at the bottom of the model was missing at a detection distance of 0.6 m. When the distance to the model boundary was greater than 1.1 m, the signal at the bottom of the model was absent.

Using the 3D sonar images, rectangular bodies of different widths were divided into 20 equal pieces, and their widths were measured for statistical purposes. As shown in Table 2, the measurements were performed at 0.9 m, 1 m, and 1.11 m. The inspection data showed that the actual resolution of the 3D sonar was 1mm, the 5 cm and 10 cm wide rectangles in the working condition models were accurate, boundaries were continuous, and the error between the average and actual value was less than 5%. While the boundaries of the rectangles of 1 cm and 2 cm were distinguishable, the error levels were large, and the true values were outside the confidence interval. The wider the rectangles, the better the detection.

Table 2. Measurements taken at distances of 0.9 m,1 m, and 1.11 m.

			95% confidence interval of the mean			
Detection distance (m)	Model width (cm)	Average (cm)	Down limit (cm)	Upper limit (cm)	Variance	The error between mean and true values (1%)
0.9	1	1.6	1.3	1.8	0.2	55.6
	2	2.6	2.3	2.9	0.3	29.4
	5	4.8	4.5	5.0	0.2	4.5
	10	9.8	9.4	10.2	0.5	2.1
1	1	1.5	1.2	1.7	0.2	47.5
	2	2.2	2.0	2.4	0.1	8.1
	5	5.2	5.0	5.4	0.1	4.5
	10	10.2	9.8	10.6	0.5	2.3
1.1	1	1.2	1.0	1.3	0.1	16.3
	2	1.8	1.6	2.0	0.2	10.2
	5	4.9	4.8	5.1	0.1	1.1
	10	9.7	9.1	10.4	1.5	2.6

2.3 *Quartz model experiment*

There were multiple quartz stone mosaics made of blocks that measured 2 cm × 40 cm, 5 cm × 40 cm, and 10 cm × 40 cm; the base plate measured 32 cm × 40 cm; the inner groove measured 5 cm; the depth measured 10cm; and the model was 10cm thick and 40 cm long, as seen in Figure 4. The sonar and model must first be attached to the measurement stand, as shown in Figure 5. Once the acoustic center of the BV5000 transducer was fixed to a fixed distance from the front plate of the model. In the acquisition software ProScan, the parameters of the BV5000 transducer were set to the upper and lower detection angles, scanning speed, the number of multi-point detections, sound velocity, etc. Finally, a model was placed in the inland river Minjiang River for detection. A few experimental images at varying distances were shown in Figure 6.

Based on Figure 6, the BV5000-1350 3D scanning sonar was most effective at around 1m. The findings were presented in Table 3. Data comparing the accuracy at different distances were presented in Table 4.

Figure 4. Front view of the home-made upright shore wall model.

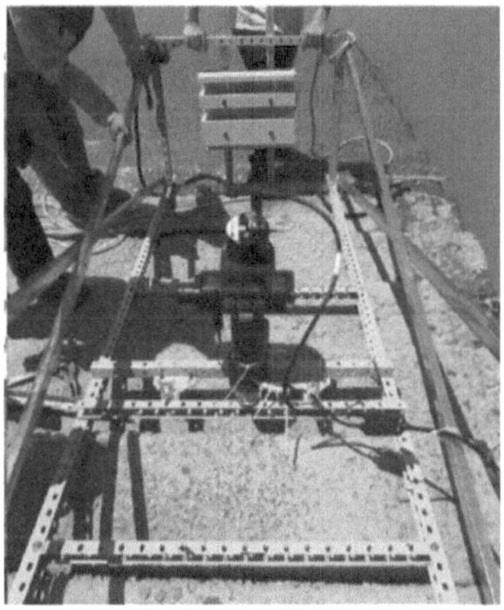

Figure 5. Sonar and model fixed to the measurement stand.

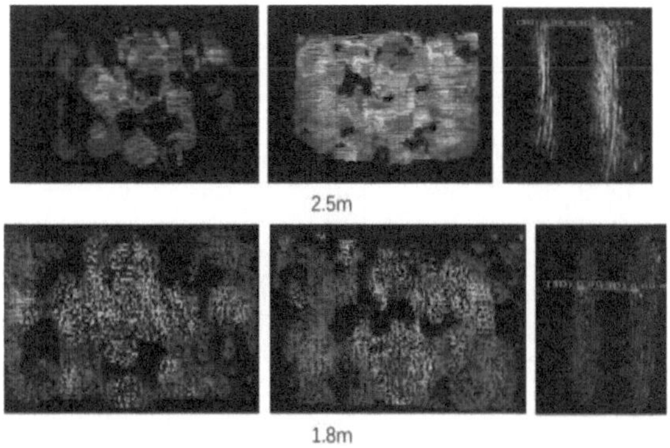

Figure 6. Images of experimental data at different distances.
(The left, middle, and right parts of the image represented the front, back, and front-back plate distances, respectively).

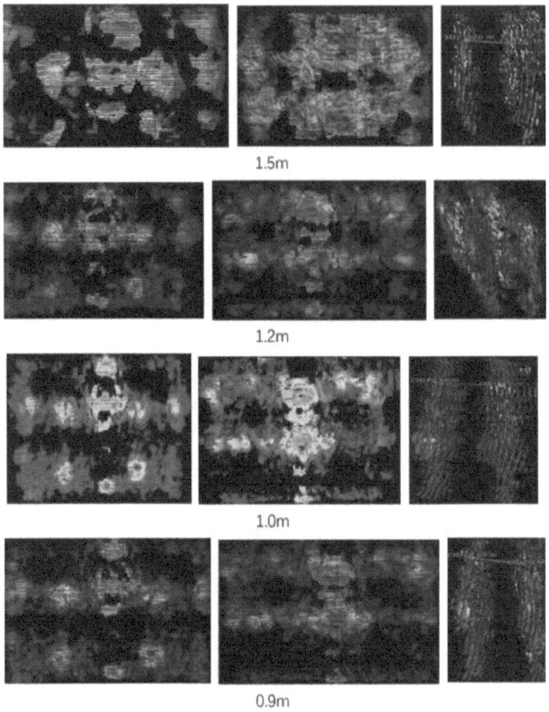

Figure 6. Continued.

Table 3. Detection conclusions for different center distances.

Distance (m)	Speed /Threshold	Imaging characteristics	Imaging resolution
2.5	0.5	Blocks with unclear boundaries	The distance between the front and rear panels was 9 cm.
1.8	5	Blocks with unclear boundaries	No data.
	0.5	Block boundaries were distinguishable but discontinuous	The front panel length and width were 10 cm; the front panel length was 43 cm, and the front panel width was 11 cm.
	1		The front panel length and width were 10 cm; the front panel length was 44 cm, and the front panel width was 11 cm.
1.5	0.5	Block boundaries were distinguishable but discontinuous	The distance between the front and back plates was 10cm; the length of the front plates was 3cm, 4cm, and 7cm respectively; the length of the back plates was 10cm and 7 cm.
1.2	1	Continuously distinguishable blocks (other than 2 cm)	The distance between the front and back plates was 10cm; the length of the front plates was 3 cm,7 cm, and 10 cm, respectively; the length of the back plates was 7 cm,7 cm, and 4 cm respectively.
	0.5		The distance between the front and back plates was 10cm; the length of the front plates was 2 cm to 3 cm,7 cm, and 8 cm, respectively; the length of the back plates was 8 cm,7 cm, and 4 cm, respectively.

(continued)

Table 3. Continued.

Distance (m)	Speed /Threshold	Imaging characteristics	Imaging resolution
1.0	0.5/120	Clear and continuous block boundaries	The distance between the front and back plates was 10cm; the length of the front plates was 1 cm~2 cm, 5 cm~6 cm, and 8 cm~9 cm, respectively; the length of the back plates was 6 cm, 6 cm, and 2 cm, respectively.
	0.5/200		The distance between the front and back plates was 10 cm; the length of the front plates was 2 cm, 5 cm, and 8 cm, respectively; the length of the back plates was 6 cm,7 cm, and 2 cm, respectively.

Table 4. Accuracy data sheet for different measuring distances.

		Front plate						Backboard					
No.	Distance (m)	2 cm	Difference	5 cm	Difference	10 cm	Difference	5 cm	Difference	5 cm	Difference	5 cm	Difference
1	1.5	5	3	7	2	3	7	10	5	7	2	/	/
2	1.2	2	0	5	0	9	1	6	1	6	1	2	3
3	1	2	0	5	0	8	2	6	1	7	2	2	3

2.4 *Results*

(1) A study is conducted using concrete and quartzite models in the Minjiang River and Meizhou Bay sea areas to assess the accuracy of 3D sonar technology. According to data collected, 3D sonar technology is most effective in detecting underwater concealed works at an optimal detection distance of 0.9 m to 1.1 m, achieving a resolution of 1 mm. Models with less than 1 m have many echoes and missing signals at the bottom, while models with more than 1 m have unclear boundary points.

(2) 3D sonar images have a resolution of 1mm under ideal conditions. 5 cm and 10 cm wide rectangular models under both conditions have accurate data, continuous boundaries, and an error of no more than 5% between the mean and the actual value. Similarly, for small and 2cm wide rectangular models, although they can be resolved, the boundaries are not continuous and the error is large, causing the true value to be outside the confidence interval. Therefore, when the resolution is constant and the object is larger, the conclusion is different.

(3) According to the results of the experiment, the scanning speed of 0.5° and 1° has little effect on block resolution accuracy; the threshold value of 500 to 2,000 does not affect block resolution. As a result of the manual setting and the model moving down the water, there is an error between the adjusted distance and the actual distance when setting the distance between the sonde and the model.

(4) In experiments, the amount of acoustic reflection is affected by the material used. For example, the use of quartzite models causes problems in obtaining reflection information at certain angles and specular reflections. For the field experiments, it is recommended that a model close to the material and surface roughness of the underwater structure be used, i.e., a general concrete casting model.

3 CONCLUSION

In addition to its high accuracy and clear imaging, 3D imaging sonar is also useful for providing a visual review of underwater structures and identifying defects. When the working mode of the system is added to the walk-around type, it will have a wider range of applications as technology develops and improves.

REFERENCES

Cho H, Gu J, Joe H. et al. (2015). Acoustic beam profile-based rapid underwater object detection for an imaging sonar. *Journal of Marine Science and Technology*. 20(1), 180–197.

Church I, Quas L, Willamson M. (2017). Multibeam water column data processing techniques to facilitate scientific bio-acoustic interpretation. *US Hydrographic Conference*. 1, 5–12.

Han Y. (2015). A Detection Method for Underwater Pipeline Tracker Based on Sonar Image. 2271902550@qq.com. 35, 66–69.

Li Bin, Jin Lijun, Hong Jia, et al. (2015). Application of three – dimensional imaging sonar technology in detection of underwater structure. *Journal of Water Resources & Water Engineering*. 26(3), 184–192.

Liu Hongxia. (2020). The application of multibeam sonar water column image in middle and bottom water target detection. Jinan, *Shandong university of science and technology*. 1–3 [in Chinses]

Qiu Anqi. (2020). Application of multi-beam bathymetric system in underwater obstacle detection. *Jiangxi Building Materials*. 12, 103–105. [in Chinses]

Shi Zhenwei, Liu Xiang,Zhang Jianfeng, et al. (2013). Application of 3D imaging sonar BV5000 in underwater measurement[J]. *Meteorological, Hydrological and Marine Instruments*. 3, 48–52. [in Chinses]

Yang F L, Xu F Z, Zhang K, et al. (2020). Characterisation of terrain variations of an underwater Ancient Town in Qiandao Lake. *Remote Sensing*. 12(268), 1–16.

Yang Zhi, Wang Jian-zhong, Fan Hong-xia, et al. (2015). Application of 3D imaging sonar system in detection of underwater detailed structure. *Water Resources and Power*. 33(6), 59–62, 47. [in Chinses]

Zhang Shun-yang, Zhang Zhuo. (2018). Application and comparison of multi- beam sounding and three-dimensional sonar system underwater part of The Wharf. *Jilin Water Resources*. 7, 40–43. [in Chinses]

Advances in Civil Engineering and Environmental Engineering – bin Yahya & Bin Daud (Eds)
© 2023 The Author(s), ISBN: 978-1-032-39397-1

Automatic slope disaster monitoring and early warning platform based on Kalman filtering model

Ao Yang*
Wuchang Institute of Technology, Wuhan, Hubei Province, China

Yabing Jin*
Shenzhen Municipal Bureau of Geology, Shenzhen, Guangdong Province, China

Boyuan Zhai*
Zhongyan Dadi Technology Co., Ltd., Beijing Municipality, China

Shaopeng Xiang*
Hubei Zhongdi Mingcheng Environmental Engineering Co., Ltd., Wuhan, Hubei province, China

ABSTRACT: Due to the problems and characteristics of traditional slope disaster monitoring, including static, intermittent, offline, easy interference, easy failure, full labor, etc., the three technologies of the Internet of Things, cloud computing, and the Kalman filtering model are now applied to slope disaster, and a set of automatic monitoring and early warning platform system suitable for the characteristics of slope disaster is developed. Through the platform, realize the visualization, continuous, real-time online, interference avoidance, no failure, and automatic monitoring methods, to meet the requirements of the accuracy, timeliness, and intelligence of the early warning and forecast of the monitoring objects, The collection, transmission, processing, early warning, and forecast of the monitoring data are all completed by the automatic monitoring system. We provide reasonable and reliable early warning and prediction threshold values and calculation methods, accurately determine the slope engineering early warning threshold value, and realize the prediction of the development trends of the monitoring data. Meanwhile, according to the size of the monitoring data and its changing trend, we also determine the early warning level and warn accordingly, and standardized the early warning and disposal process including the expert-assisted decision-making system, fast, timely, multichannel information release, and early warning disposal.

1 INTRODUCTION

Slope engineering with the global economic construction of rapid development and the continuous development and utilization of various resources will continue to increase. In recent years, disasters caused by slope engineering have emerged endlessly, and the landslide problem encountered in slope engineering has become one of the three major global natural disasters (Jin et al. 2019; Pecoraro Calvello Piciullo 2019). Therefore, to ensure the normal operation of various construction projects, the slope must be monitored. However, the current global slope monitoring generally adopts manual measurement (Qi et al. 2015), which is characterized by static, intermittent, offline, easy interference, easy failure, low precision, and full manual (Xu et al. 2008). Especially in a landslide, people cannot reach the danger monitoring area, so it is difficult to timely capture the

*Corresponding Authors: 408513947@qq.com; jinyabing25@sina.com; zhaiboyuan@126.com and 349925142@qq.com

DOI 10.1201/9781003349563-62

information before the slope disaster by using the traditional manual measurement method regularly, and it is impossible to achieve a systematic and comprehensive analysis of the monitoring data. And this information is exactly the most valuable.

With the development of electronic information technology and computer data operation algorithm, how to improve the traditional slope disaster monitoring method, in the slope disaster monitoring, cloud computing and regression analysis technology; realize the accuracy, timeliness, and intelligence; finally achieve the dynamic, continuous, online, less interference, difficult failure, high precision, visual monitoring purpose, are important contents of this paper. This paper mainly introduces the implementation and key technology of slope disaster automatic monitoring and early warning platform based on the Kalman filtering model.

2 SLOPE DISASTER

2.1 *Development status of automatic monitoring technology of side slope disaster*

Slope disaster monitoring before the 1990s was manual monitoring. China began to develop automation in the early 1990s. In 1993, with the help of a satellite communication system, the promotion and application of slope disaster automatic monitoring technology were started.

Fu Bingqing has implemented automatic monitoring (Fu 1993, 1997) of the bean sprout shed landslide control project in Wanquan County, Chongqing municipality.

Lei Mingtang and other officials have built the first karst collapse disaster monitoring station for automatic data collection in China, and the technical system and methods of karst collapse prediction are preliminarily established.

Li Yu et al. cited the water pressure sensor and the digital system of automatic collection and automatic transmission (Li et al. 2006).

Tang Yaming and others have classified and summarized the current situation of landslide monitoring technology according to different monitoring objects and methods (Tang et al. 2012).

The automatic monitoring of geological disasters in China in the past 30 years has made an inestimable contribution to the successful early warning and forecast of the global geological disaster field, but the proportion of automatic monitoring is still too low. For example, in 2004, in the statistical analysis of 715 geological disasters in China, automatic monitoring and forecast accounted for 9.8% of the total, and successful early warning and forecast of mass prevention and group accounted for 86.7% of the total (Liang Xu et al. 2008). It indicated that the accuracy of automatic monitoring, early warning, and forecast needs to be greatly improved. Therefore, it is of great significance to apply the Internet of Things, cloud computing, and Kalman filtering model-based technologies to slope disasters, and to develop a set of automatic monitoring and early warning platform system suitable for the characteristics of geological disasters.

2.2 *Problems existing in the popularization and application of slope disaster automation monitoring*

From the perspective of geological disasters, the popularization and application of automatic monitoring technology have major problems as follows:

1. Difficult to select disaster points. 2. Difficult numerical analysis and calculation. 3. Difficult to select alarm indicators. 4. Difficult to determine the threshold of early warning indicators. 5. Difficult to select monitoring equipment. 6. Difficult to protect site facilities. 7. More investment in instruments and equipment.

Therefore, to overcome the above problems, it is particularly necessary to develop a set of efficient and complete automatic slope disaster monitoring and early warning platform.

3 AUTOMATED MONITORING NETWORK SYSTEM

3.1 *Basic hierarchy structure*

The network construction of the slope disaster automatic monitoring and early warning platform adopts the hierarchical design scheme. The network structure is divided into four layers: the network is divided into the collection layer, transmission layer, storage layer, and application layer according to its functions.

The acquisition layer is the field collection part, including physical sensors and transmission media. The acquisition layer is responsible for sending the recorded data to the transport layer. The transmission layer is responsible for receiving the data from the acquisition layer and providing services to the storage layer and the application layers. The storage layer is data persistent, including database and file storage, and the receiving transfer layer data is imported to the database and stored. The application layer accepts the transmission layer data and provides services directly to the user.

3.2 *Application of Internet of Things Technology*

The Internet of things is radio frequency recognition ("RFID"), a communication technology. Specific targets can be identified through radio signals and read and write data can be used, and objects to the Internet are connected through the internet of a thing's domain name for information exchange and communication. To achieve the purpose of intelligent identification, positioning, tracking, monitoring, and management. As shown in Figure 1, the core and foundation of the internet of things is the internet, which is an extended and extended network based on the internet.

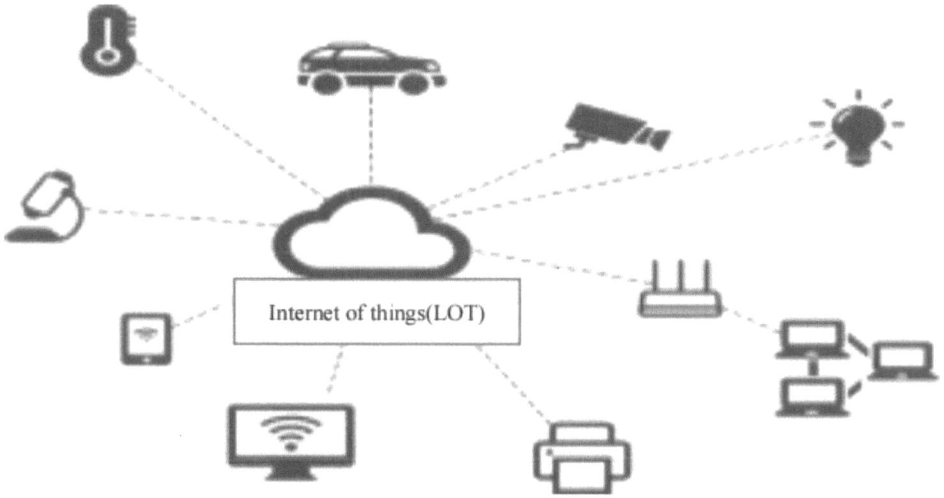

Figure 1. Schematic diagram of Internet of things.

The monitoring and early warning system platform mainly uses a wireless sensor network (WSN), namely wireless transmission (Huang 2012; Liu 2018; Liang 2011). That is, each monitoring equipment on the monitoring object site collects the data through the sensor, transmits the data information to the data center through the WSN, and the platform obtains the data from the data center for analysis. The schematic diagram of the monitoring platform WSN application is shown in Figure 2.

The cloud platform application services of this system are deployed based on the cloud server. The cloud server can achieve seamless expansion while having the high-security characteristics of cloud computing, DDoS protection, trojan detection, prevent violence and cracking, etc. Data storage can be deployed in distribution. According to the business development of the system,

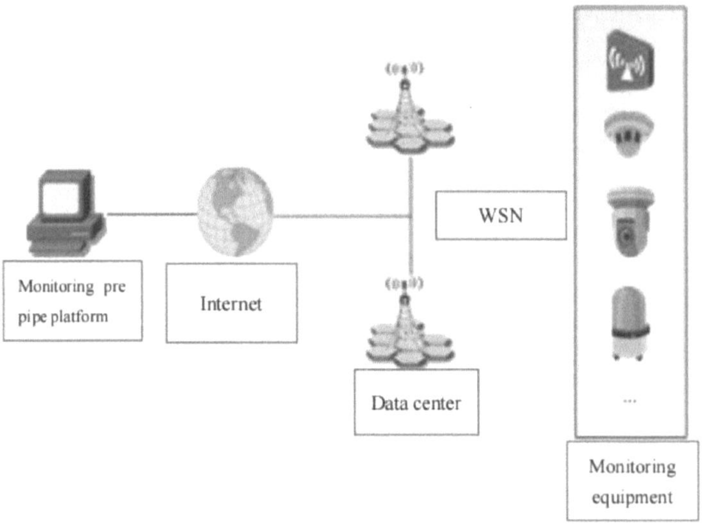

Figure 2. WSN application diagram of monitoring platform.

data access from the nearest distance can be realized to improve the overall response speed of the system.

4 AUTOMATIC MONITORING AND EARLY WARNING PLATFORM

4.1 *Basic requirements*

The monitoring and early warning platform, based on the internet of things, cloud computing, and the Kalman filtering model, is applied to slope disasters. It takes slope disasters in certain areas as the monitoring object, implements dynamic monitoring of the deformation damage information and disaster-induced factors information in the time and airspace, and visually presents real-time monitoring information. Through the relevant analysis and treatment of each monitoring factor, the stable state and change trend of the disaster variant are judged, and the point of possible disasters or engineering accidents in the future is predicted. Therefore, the system software required by the platform is easy to use, effective and reliable to make the early warning and forecast system meet the 7 × 24 hours of uninterrupted service requirements while ensuring the security of the system platform.

4.2 *Platform architecture*

The system platform adopts SOA architecture, mainly to solve the business integration and distribution of the platform in the internet environment. It is software architecture. The software makes the modules and services of the system more independent, more coupling, and more conducive to the development and maintenance of the platform.

The SOA design architecture is shown in Figure 3.

4.3 *Platform development and design*

4.3.1 *Development tools*

The system development uses Eclipse open-source code, which is a scalable development platform based on Java. It is not limited to the Java language; It can also support plug-ins for programming

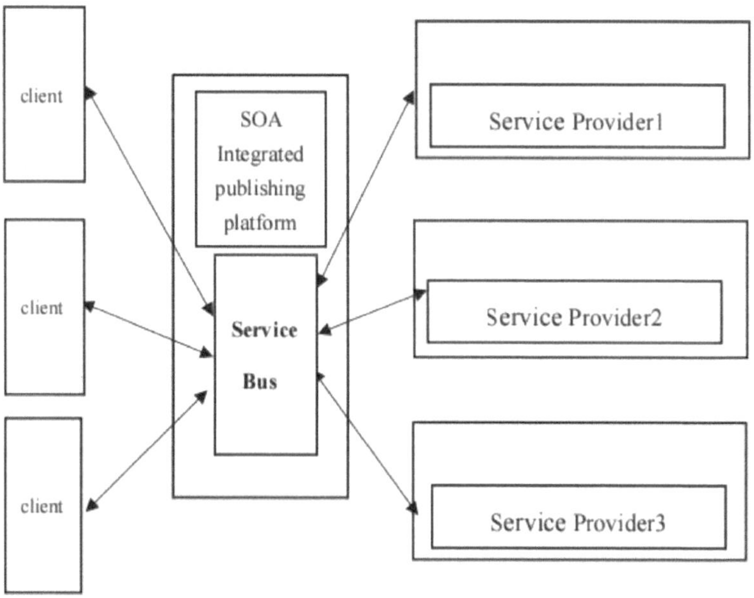

Figure 3. SOA design architecture.

languages such as C/C++, COBOL, PHP, and Android. The Eclipse framework can also be used as a basis for other application types unrelated to software development, such as content management systems.

4.3.2 *Functional module diagram*

The core of the function module is the early warning disposal module, including early warning processing, message push, map alert situation processing, and source. Each module is a fine molecular module. The platform can realize intelligent safety warnings of slope safety conditions under various working conditions.

After the alarm is successful, both the monitoring unit and the personnel on duty of the relevant owner in the monitoring area where the alarm situation is located can receive the message reminder (including the APP terminal message push and the WEB interface message popup) and the SMS reminder.

4.3.3 *Database design*

The platform adopts MyBatis as the persistent framework, and the MyBatis framework can greatly improve the SQL development efficiency in monitoring network applications. At the same time, in the application of the monitoring system, the more detailed SQL optimization of MyBatis can reduce the query fields and improve the software query efficiency. At the same time, it also achieved the data specification design, equipment information, and data sheet unification (Pecoraro Calvello Piciullo 2019).

4.4 *Commissioning and deployment*

The automatic monitoring and early warning platform for slope disasters is released in the form of incremental function in the version release. The test results can pass only if the data test is completed and the defect state of the functional module is all the final state. Test report points include test process, defect and self-repair records, test results analysis, test statistics, and analysis. Through the self-debugging of the system, the system operation status can be mastered in real-time to ensure the effective operation of the early warning platform.

4.5 *Safe operation and maintenance*

The safe operation and maintenance of the monitoring and early warning platform guarantee the orderly operation of the system, which is mainly divided into application disaster recovery and database disaster recovery maintenance. For the application disaster recovery part, A and B servers are set up for the application server and data platform damage repair, respectively. For the database synchronization of the database disaster recovery part, the database service regularly completes the data replication to backup for the automatic supplement of data loss. This function is realized through the main database outage SQL.

4.6 *Function display*

The functions of the automatic monitoring and early warning platform are displayed as (1) video system: the video system module provides the remote video monitoring function for the monitoring objects, and the monitoring personnel can watch the real-time monitoring video information. And supports multiple cameras on the screen. (2) Remote monitoring system: the remote monitoring system includes three sub-modules of real-time warning situation, data monitoring, and monitoring object distribution. The remote monitoring system is the main dependent module for the personnel on the duty in the system to monitor the monitoring objects. (3) Management system: personnel management, role management, equipment management, project management, contract management, monitoring and report management, SMS management, APP management, and emergency inspection management can be visualized and visually displayed on the interface. (4) Early warning system: this part is the core module of the whole system, which is mainly divided into the early warning information module, expert judgment module, manual alarm module, valve value early warning module, model early warning module, and AI early warning module.

5 KEY TECHNOLOGY—AUTOMATIC MONITORING DATA ANALYSIS AND EARLY WARNING

5.1 *Based on the Kalman filtering model*

"Filtering" is a tool for signal processing that is used to extract useful signals from discrete or continuous observed signals with noise interference. Kalman filtering method refers to a brand new linear dynamic recurrence method It is proposed under the principle of linear unbiased minimum variance estimation by using probability and mathematical statistics and using limited-time data as the calculation basis (Labiouse 2008). This method can be calculated according to the current moment observation value and the state valuation of the previous moment, with the help of the state transfer equation, according to the recurrence formula, and obtain the estimate of the current moment (Darvef 2011; Davis J C 2003)

5.1.1 *Implementation steps of the Kalman filter method*

(1) First determine the filter's initial value, including the initial value of the state vector $\hat{X}_0$ and its corresponding covariance array $\hat{P}_0$, observation noise covariance array R_k and covariance array of dynamic noise Q_k.
(2) Establish a Kalman filtering model to determine the state transfer $\emptyset_{k/k-1}$, dynamic noise Γ_{k-1} and observation matrix H_{k}.
(3) Complete the above preparation work is calculated $\hat{X}_{k/k-1}$, and get the forecast value $P_{k/k-1}$, forecast covariance matrix, gain matrix K_{k}.
(4) Input the observation data and conduct Kalman filtering to get the best-predicted value $\hat{X}_k$ and variance array P_k of the observation group.
(5) Return to (3) for recurrence calculation.

5.1.2 *Engineering examples*

The Kalman filter first calculates the forecast value from the initial value, calculates the Kalman gain, and then calculates the current point filter value according to the measured value of the point. The next point then calculates the forecast value from the filtered value of the previous point, and then the Kalman gain is calculated. According to the measured value of the point, the filter value of the current point is calculated, so that it is repeatedly pushed.

This paper takes the S1 monitoring data of the settlement of a slope in a city as an example.

The dynamic noise covariance is set to an infinitesimal value of 0.00000001, and the noise interaction in each monitoring period is negligible. The Kalman filtering model was run as shown in Figure 4. The filter value gradually tends to be the measured value with the increase in the number of data periods, and the forecast value also fluctuates during the period, which can achieve a very good forecast effect.

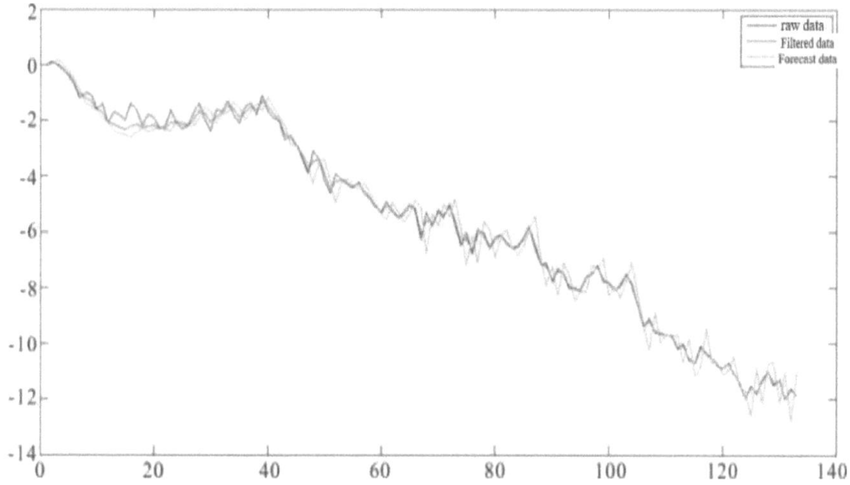

Figure 4. Filtering prediction effect of Kalman filtering model.

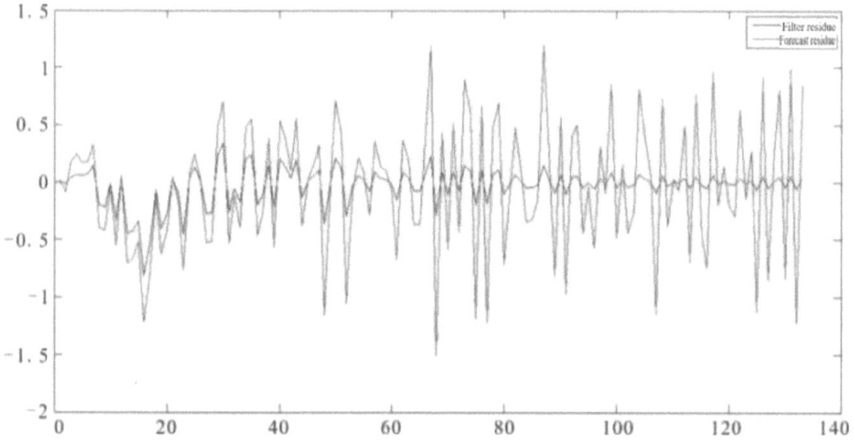

Figure 5. Filtering residual diagram of the Kalman filter model.

Figure 5 is the residual map of the Kalman filtering model. It can be seen from the residual map, the previous filter residual reaches 0.8 mm (the blue color line in the figure), and the back is infinitely close to 0, the convergence trend is stable, and the filtering effect is ideal. Although the forecast value is very sharp from the graph fluctuation, it fluctuates around 1mm, and the prediction effect is very good.

5.2 *Monitoring and early warning methods*

This slope disaster automatic monitoring and early warning are based on the monitored past data characteristics, based on the Kalman filter model for detailed data analysis and processing, to predict the trend of the monitored object safety, and to make an early warning of the possible unsafe trend and even disasters. From the early warning threshold value source (Xu & Cheng 2014), the slope disaster warning method is divided into two categories: one is directly according to the fixed threshold warning, and the other is based on the mathematical model of historical monitoring data analysis modeling and future monitoring data for a period, according to the future a point in time to predict the trend of early warning.

This system is a mathematical model warning based on the Kalman filtering method, which determines the warning level according to the predicted value size.

5.3 *Early warning of the prediction results based on the Kalman filtering model*

Kalman filtering is used to establish a mathematical prediction model for the existing historical monitoring value. The prediction model is fitted to deduce the numerical value of the monitoring value in the future time, reflecting the trend change and the predicted value in the future period. According to the predicted value, the slope disaster body may reach a certain threshold value in the future and will issue an alarm to the threatened area before it occurs. According to the monitoring and predicted values, the corresponding comparative relationship is established in Figure 6.

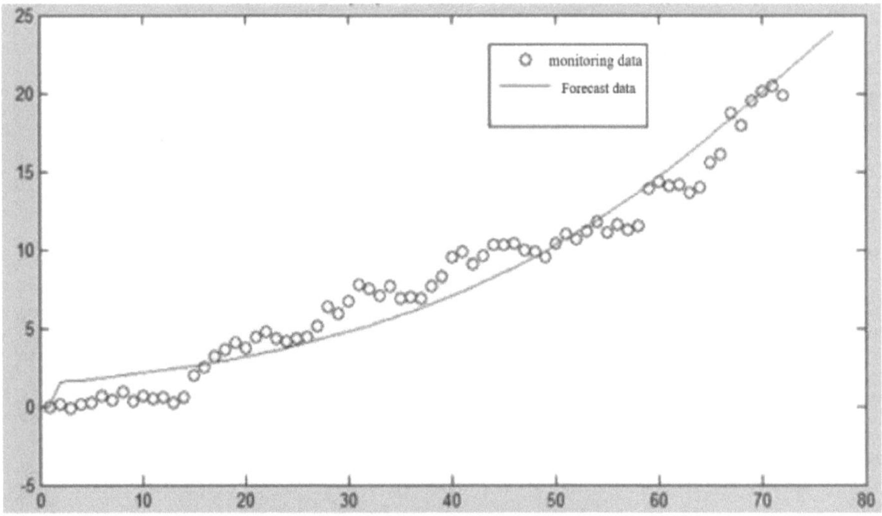

Figure 6. Comparison of monitored and predicted values.

Among them, the abscissa is the time, and the ordinate is the amount of monitored value change per unit period and per unit mm. According to the simulated curve, the predicted value of the monitoring value at a certain time point in the future is divided into the corresponding threshold according to the slope warning level to give early warning in advance. When the measured value has been obtained, the error of the predicted value is known. The new information is used for feedback calculation to revise and adjust the prediction model parameters. The prediction accuracy can be improved by constantly correcting the future prediction model parameters with the past prediction error.

5.4 *Division of early warning levels*

The slope disaster warning system is divided into attention, warning, warning, and warning at four levels. The corresponding warning type, stability coefficient, tangent Angle, deformation stage, slope status, and risk assessment are listed in Table 1.

Table 1. Classification of early warning level.

Grade	Pay attention to the level	Levels of warning	Levels of warning	Alert level
Alarm type	Blue	Yellow	Orange	Red
Stabilize sex coefficient	K>1.1	1.1≥K>1.05	1.05≥K>1	K≤1
Angle of contingence	$\alpha \approx 45$	$45° < \alpha < 80°$	$80° \leq \alpha < 85°$	$\alpha \geq 85°$
Be out of shape stage	Equal rapid change Form stage	The initial acceleration phase	Middle acceleration stage	Acceleration (temporary slip) stage
Side slope state	Stabilize	Basic stability	Understandable	Instability
Danger sexual evaluation	No danger	Low risk	In danger	High risk

According to the corresponding monitoring methods of different slopes, the alarm indicators corresponding to various monitoring factors can be selected simultaneously to determine the threshold value of different early warning levels based on the Kalman filter model analysis method.

6 CONCLUSIONS

The automatic monitoring and early warning platform of slope disaster developed by the three technologies of Internet of Things, cloud computing, and regression analysis method based on Kalman filtering model can obtain the following conclusions:

(1) Combined with computer software technology, Internet of things communication technology, and the development of slope disaster automatic early warning technology, can be real-time, dynamic, continuous, online, reliable, and visual automatic monitoring. The successful development of the automatic monitoring and early warning platform realizes that automatic monitoring to replace manual monitoring.
(2) The Kalman filtering model is adopted to improve the numerical analysis and calculation ability of slope engineering, provide reasonable and reliable early warning and prediction threshold and calculation method, accurately determine the slope engineering early warning threshold of slope engineering, and avoid the two extreme phenomena of "wolf coming" and "fish" with the small threshold.
(3) The application of the three major technologies based on the Internet of Things, cloud computing, and the Kalman filtering model in slope disasters effectively improves the rapid response ability of automatic monitoring, early warning and forecast, and provides a guarantee for the timely and accurate early warning of slope disasters.
(4) The system realizes the detailed analysis, processing, and calculation of the slope disaster monitoring data samples, judges the safety status and development trend of the slope disaster, and improves the accuracy and reliability of the early warning.
(5) Similar to equipment failure, or power failure, the monitoring information obtained is often the result of the superposition of the real situation and various interference factors, which not

only reduces the accuracy and accuracy of the information but also directly affects the early warning and judgment of geological disasters. The analysis of the monitoring data based on the Kalman filtering model avoids the influence of various uncertain factors on the monitoring results over time and realizes the prediction of the development trend of the monitoring data. At the same time, according to the size of the monitoring data and its change trend, the warning level can be determined and warned accordingly.

(6) A multilevel monitoring and early warning management system with customized service functions were created. The multilevel authority management center is set up to respond to the needs of the early warning levels and thresholds of different monitoring objects; and also to standardize the early warning and disposal process including the expert-assisted decision system, and realize the rapid, timely, and multi-channel information release and early warning disposal.

ACKNOWLEDGMENT

This work was supported by the Plan Project of Geological Engineering Academician Workstation of Shenzhen Geology Bureau of Guangdong Province (Grant No. 2013B090400025).

REFERENCES

Fu Bingqing. 1993. Experimental study on satellite transmission of Xintan landslide monitoring data [J]. *Journal of the Yangtze River academy of Sciences*, 10 (1): 20–27.

Fu Bingqing. 1997. Automatic monitoring and research of Douyapeng landslide control in Wanxian County [J]. *Chinese Journal of Geological Hazards and Prevention*, 8:69–75.

Gaetano Pecoraro, Michele Calvello, Luca Piciullo. 2019. Monitoring strategies for local landslide early warning systems [J]. *Journal of Landslides*, 16(2): 213–231.

Huang Jian, Jian Xinlei, Liu Renjie, et al. 2018. Research on monitoring and early warning model of geological disasters in Shenzhen [R]. *State Key Laboratory of science and technology of Chengdu.*

Huang Jian. 2012. Research on monitoring and early warning of geological disasters based on 3D WebGIS technology [D]. *Chengdu: Chengdu University of technology.*

Jin Yabing, Gong Shuyun, etc. Automatic monitoring technology and practice of slope and foundation pit. [J]. *Geological Publishing House*, 2019:1–2.

Li Yu, Lei Mingtang, Jiang Xiaozhen. 2006. *Monitoring of karst collapse in Litang*, Guangxi [J]. Karst, China, 25 (4): 341–346.

Liang Guilan, Xu Weiya, he Yuzhi, et al. 2008. Development and application of visual analysis system for slope engineering monitoring [J]. *Geotechnical Mechanics*, 24 (3): 849–853.

Liang Shuang. 2011. Research and implementation of SOA based cloud computing framework model [J]. *Computer Engineering and Applications*, 47 (35): 92–94.

Manzella, V. Labiouse. 2008. Qualitative analysis of rock avalanches propagation by means of physical modelling of non-constrained gravel flows [J]. *Rock Mechanics and Rock Engineering*, 41(1):133–151.

Nicotf, Darvef. 2011. Diffuse and localized failure modes: Two competing mechanisms [J] *International Journal for Numerical and Analytical Methods in Geomechanics*, 35(5) : 586–601.

Ohlmacher G C, Davis J C. 2003. Using multiple logistic regression and GIS technology to predict landslide hazard in Northeast Kansas, USA [J]. *Engineering Geology*, 69(3-4):331–343.

Qi Xing, Xu Qiang, Zheng Guang, et al. 2015. Dynamic early warning mechanical model of bedding rock and soil landslide induced by rainfall. *Disaster Science* 30 (3): 38–42.

Tang Yaming, Zhang Mao, et al.2012 Present situation and review of landslide monitoring and early warning at home and abroad [J]. *Geological Review*, 58:533–541.

Xu Qiang, Tang Minggao. et al. 2008. Study on spatiotemporal evolution and early warning and prediction of landslides [J]. *Journal of Rock Mechanics and Engineering*, 20 (6): 1104–1112.

Xu Yangqing, Cheng Lin. 2014. Analysis and processing of foundation pit monitoring data and prediction and early warning system [J]. *Chinese Journal of Geotechnical Engineering* 36 (S1): 219–224.

Advances in Civil Engineering and Environmental Engineering – bin Yahya & Bin Daud (Eds)

Determination of calculation parameters of bearing capacity of super-soft soil composite foundation reinforced by bamboo grid and sand cushion

Xiaoyun Ma & Ze Chen
CCCC Road & Bridge Special Engineering Co., Ltd. Wuhan, Hubei, China

Zenghui Yu*
China Construction Second Bureau Eighth Engineering Co., LTD, Jinan, China

Jihui Ding
CCCC Road & Bridge Special Engineering Co., Ltd. Wuhan, Hubei, China

Shengkang Yan
China State Construction Curtain Wall Co., Ltd., Wuhan, China

ABSTRACT: When the hydraulic fill soft soil foundation is deep soft soil foundation and the surface layer is super-soft soil, the reinforcement of sludge or super-soft soil foundation surface with bamboo grid can meet the requirements of personnel and construction equipment. The existing calculation parameters of bearing capacity of super-soft soil foundations reinforced by bamboo grid and sand cushion mainly rely on experience. In this paper, the load settlement and load strain laws are obtained through indoor model tests, and then the angle between the tension of bamboo grid and the horizontal plane and the stress diffusion angle of bamboo grid and sand cushion can be determined. The test results show: (1) when s/b is equal to 0.10 and 0.15, the angles between the bamboo grid tension and the horizontal plane are 6.74° and 7.93° respectively, which are far less than 20° by Yamanouchi; (2) The stress diffusion angle of bamboo grid and sand cushion is related to the settlement ratio and thickness width ratio. When the settlement ratio is certain, the stress diffusion angle increases with the increase of thickness to width ratio, and the spacing of bamboo grids also has an effect on the stress diffusion angle; when the thickness-to-width ratio is 0.39, 0.58 and 0.78, the mean value of the stress diffusion angle is respectively 14.2°, 2 6.5°, and 31°, and the maximum value of this test reaches 38.74°.

1 INSTRUCTIONS

Bamboo has the advantages of short growth period, high tensile strength and bending stiffness; strengthening silt or soft soil with bamboo grid as reinforcement has the advantages of stable structure, low price, convenient construction, energy conservation and environmental protection. Bamboo grid has high tensile strength and bending stiffness, and has a certain shear capacity (Mao 2021; Deng 2018; Zhu 2020). The bearing capacity of soft soil composite foundation with bamboo grid and sand cushion has been significantly improved after bamboo net reinforcement is used to reinforce the surface of silt or super soft soil foundation (Aazokhi 2015; Hegde 2015; Yuan Man 2014). How to determine the deformation of bamboo grid and the calculation parameters of bearing capacity of composite foundation after the surface treatment of soft soil foundation strengthened by bamboo grid mainly depends on experience. The existing calculation methods of bamboo grid

*Corresponding Author: 1976316086@qq.com

DOI 10.1201/9781003349563-63

reinforced soft soil bearing capacity are all based on the calculation methods of geotextile reinforced soft soil bearing capacity. Through the results of the load test and centrifugal model test, when it is sandy soil, clayey soil or clay, and compacted to a certain density, the diffusion angle can be taken as 30° (Lin 1992). It is suggested that the angle between the tensile force of bamboo grids and the horizontal plane should be taken as 20°, the tangent of the stress diffusion angle should be 0.5, and the diffusion angle should be less than 30°, resulting in a larger bearing capacity calculation result (Yamanouchi 1985). When China determines the geotextile reinforced foundation, it is suggested that the angle between the tensile force of geotextile and the horizontal plane may be 10°, and the stress diffusion angle is the same as that of Yamanouchi (Gong 2008). Only when the buried depth ratio is greater than 0.5 and the compression modulus ratio of the upper and lower layers is greater than 10, the stress diffusion angle is taken as 30°, and in other cases are less than 30° (GB50007 2011). The calculation method of bearing capacity of reinforced soft soil foundation is put forward in a three-dimensional stress state of geotextile, and the existing value of stress diffusion angle is too small (Bai 1994).

The existing research on the angle between the tensile force and the horizontal plane and the stress diffusion angle of bamboo grid is based on geotextiles. Geotextiles mainly bear tensile force, while bamboo grids in bamboo-grid-reinforced composite foundation are different from geotextiles. The tensile and bending strength of bamboo grid is very high, and the angle between the tensile force and the horizontal plane and the stress diffusion angle is different from geotextiles. Therefore, through the indoor model test of bamboo grid reinforced soft foundation, bamboo grid deformation and the parameters of bearing capacity under vertical load are analyzed, so as to provide a basis for a similar soft soil foundation treatment design.

2 CALCULATION PARAMETERS OF BEARING CAPACITY OF BAMBOO GRID REINFORCED SUPER SOFT SOIL FOUNDATION

2.1 *Angle between bamboo grid tension and horizontal plane*

Under the action of vertical load, the bamboo grid and geotextile are regarded as one, and bending deformation occurs. The deformation of bamboo grid refers to the deformation of geotextile, and common ones are circular surface and paraboloid. According to the paraboloid hypothesis, the expressions of bamboo strain and the angle between bamboo grid tension and horizontal plane are derived as follows (Bai 1994):

$$\varepsilon = 0.5\sqrt{1+(2S/a)^2} + 0.15(a/S)\ln\left[2S/a + \sqrt{1+(2S/a)^2}\right] - 1 \tag{1}$$

$$\sin\theta_h = 1/\sqrt{1+(0.5a/S)^2} \tag{2}$$

where ε is the strain of bamboo grid, θ_h is the angle between the bamboo net tension and the horizontal plane, S is the maximum vertical displacement of the bamboo grid (m), and its value is the settlement of the soft soil surface under the center point of the load action surface, and a is the geometric dimension of stress diffusion to the surface of bamboo grid (m).

2.2 *Stress diffusion angle of bamboo grid sand cushion*

Bamboo grid and sand cushion have the effect of stress diffusion. Based on diffusion theory, the bearing capacity of the composite foundation of bamboo grid-reinforced soft soil is shown as follows (Bai 1994):

$$f = [C_u N_c + KT\sin\theta_h/a](B + 2H\tan\alpha)(L + 2H\tan\alpha)/(BL) \tag{3}$$

where C_u is bearing capacity of reinforced foundation (kPa), T is the equivalent tension of bamboo net(kN/m), T is the undrained strength of saturated soft clay (kPa), N_c is the bearing capacity coefficient, $N_c = \pi + 2$, B the width of the load action surface(m), L the width of the load action surface(m), H the thickness of the sand cushion(m), α stress diffusion angle and K is the coefficient related to the foundation shape. K is generally 1.0~2.0; if $L/B \geq 10$, K=1; if $L/B = 1$, then K=1.

Using formula (3), when the geometric size of the load action surface is certain, the bearing capacity of the composite foundation after reinforcement and the bearing capacity of the soft soil before reinforcement is known, and the stress diffusion angle can be calculated by trial calculation method.

3 INDOOR MODEL LOAD TEST OF BAMBOO GRID REINFORCED SOFT SOIL FOUNDATION

3.1 *Test materials*

The bamboo grid in the experiment adopts bamboo from Jiangsu Province, with a diameter of 1.0cm~1.2 cm. The moisture content is 11.0%, and the length of the bamboo joint is 17.5cm~23.3 cm, with an average value of 22.43 cm.

The soft soil from the economic and Technological Development Zone of a city is taken as the sample. The gravity of soft soil layer is 17.4kN/m^3 ~18.3 kN/m^3, the water content is between 47.31% and 61.43%, the liquid index is greater than 1.50, and the plastic index is greater than 18. Saturate the soft soil on site with water, water content is greater than 60%. The gravity of sand is 16.5kN/m^3, the moisture content is 0.25%, and the friction angle is approximately 30°C.

The geotextile used in the test was the product of Hebei Shuntai Geotechnical Materials Ltd. The geotextile weights 150 g/m^2, with a longitudinal tear strength of 5.1 kN/m, a transverse tear strength of 5.2 kN/m, a longitudinal elongation of 65%, a transverse elongation of 66%, a bursting strength of 1 kN, and a longitudinal and transverse tear strength of 0.15 kN.

3.2 *Indoor load test*

Figure 1 shows the photos of geotextile soft soil and bamboo grid during and after the test. The size of the model box is 100 cm*100 cm*100cm. The soil layers in the model box from bottom to top are: 15cm thick saturated sand layer, 65cm thick soft soil layer, bamboo grid, and geotextile. Bamboo grid spacing 15 cm * 15 cm and 10cm* 10cm, and strain gauges are arranged on the tensile side of the bamboo grid. The width of the equivalent square load plate is 25.7cm, and the vertical load is applied by the hydraulic jack. After the test, take out the bamboo grid, and the deformation of the bamboo grid basically recovers.

(a) Load test (b) bamboo grid from soft soil

Figure 1. Photos in test.

3.3 *Bearing capacity analysis*

Figure 2 shows the settlement curve corresponding to different thickness of sand cushion. It can be seen from Figure 2 that when the thickness width ratio H/B=0.58, there are obvious nonlinear segments, which are not obvious in other cases. Strictly speaking, the bearing capacity of the foundation soil is not a fixed value. In addition to the physical and mechanical properties of the soil, it is related to the size, form and buried depth of the foundation, which is inversely proportional to the width B of the load plate. In this indoor model test, the size of the load plate is small, and the bearing capacity determined in strict accordance with the requirements of the specification is inconsistent with the indoor test of the small load plate. Therefore, the bearing capacity corresponding to different settlement ratios is mainly analyzed.

It can be seen from Figure 2 and Tables 1 and 2 that, under other conditions unchanged, compared with the composite foundation with a 15 cm-spacing of the bamboo grid, under the same settlement ratio, the deformation of the composite foundation with 10 cm-spacing of the bamboo grid decreases, and the bearing capacity increases by 15.6% to 31.9%. When the settlement ratio is constant, the bearing capacity of composite foundation increases with the H/B increase. When s/B=0.06, the spacing between bamboo grids is 15cm and 10cm, and the corresponding bearing capacity increases by 1.09~5.57 and 1.61~7.66 respectively compared with that before reinforcement; When s/B=0.10, the spacing between bamboo grids is 15cm and 10cm, and the corresponding bearing capacity increases by 1.15~4.97 and 1.65~6.32, respectively, compared with that before reinforcement; When s/B=0.15, the spacing between bamboo grids is 15cm and 10cm, and the corresponding bearing capacity increases by 1.36~4.88 and 2.00~5.88, respectively, compared with that before reinforcement.

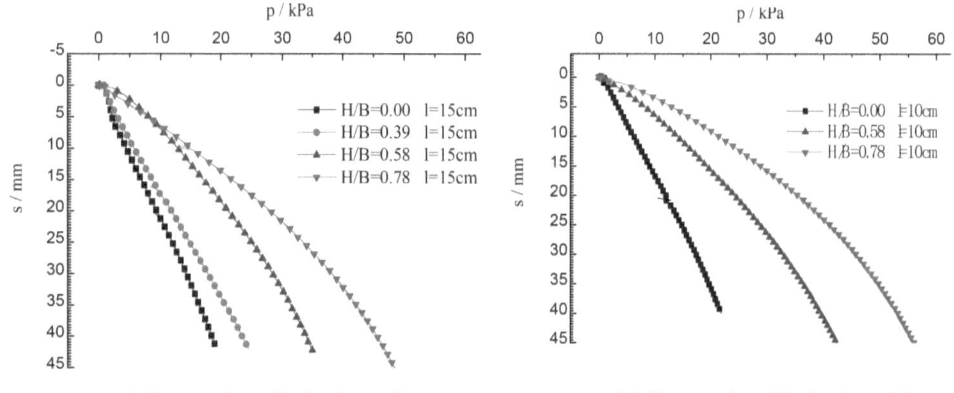

(a) 15cm-spacing of the bamboo grid (b) 10cm-spacing of the bamboo grid

Figure 2. p~s curve.

Table 1. Bearing capacity of composite foundation.

Working condition		Soft foundation	H/B 0.00	0.39	0.58	0.78
Space 15 cm	$p_{s/b=0.06}$	3.40	7.12	8.94	17.57	22.34
	$p_{s/b=0.10}$	5.72	12.27	15.30	25.52	34.23
	$p_{s/b=0.15}$	7.60	17.95	22.80	33.25	44.68
Space 10 cm	$p_{s/b=0.06}$	3.40	8.86	——	19.69	29.46
	$p_{s/b=0.10}$	5.72	15.15	——	29.51	41.85
	$p_{s/b=0.15}$	7.60	22.80	——	38.70	52.26

Table 2. Multiple of bearing capacity increase compared with soft soil foundation.

		H/B			
Working condition		0.00	0.39	0.58	0.78
Space 15 cm	$p_{s/b=0.06}$	1.09	1.63	4.17	5.57
	$p_{s/b=0.10}$	1.15	1.67	3.46	4.98
	$p_{s/b=0.15}$	1.36	2.00	3.38	4.88
Space 10 cm	$p_{s/b=0.06}$	1.61	——	4.79	7.66
	$p_{s/b=0.10}$	1.65	——	4.16	6.32
	$p_{s/b=0.15}$	2.00	——	4.09	5.88

3.4 *Deformation analysis of bamboo grid*

It can be seen from Figure 3 (a) that when the spacing of bamboo grids is 15cm and the load is 0.0kPa~6.0kPa, the strain value of bamboo grids is almost zero. The main reason is that under the action of load, the contact between bamboo grids and soft soil foundation is not close. When the load exceeds 6.0kPa, the strain of the bamboo grid basically increases linearly with the increase of the load, so it can be considered that the bamboo grid works within the online elastic range. Measuring points SN-1 and SN-3 are located directly below the load plate, respectively in the upper and lower bamboo grids. SN-4 and SN-5 are outside the load range. The strain value of the bamboo grid within the load plate range is significantly higher than that outside the load range, and the strain value of the upper bamboo grid is smaller than that of the lower bamboo grid. Figure 3 (b) shows the relationship between strain and load of different bamboo grid spacing when the thickness width ratio is 0.78. When the bamboo grid spacing is different, the shape of the strain and load relationship curve is basically the same, but when the load is the same, the strain with the bamboo grid spacing of 10cm is far less than the settlement with the spacing of 15cm. When the spacing between bamboo grids is small, the rigidity of bamboo grids increases and the deformation decreases. From Table 3, when s/B=0.10, the mean value of the strain under the load plate is 2327.5$\mu\varepsilon$, and the coefficient of variation is 0.14; When s/B=0.15, the average strain under the load plate is 3061.7$\mu\varepsilon$, and the coefficient of variation is 0.20.

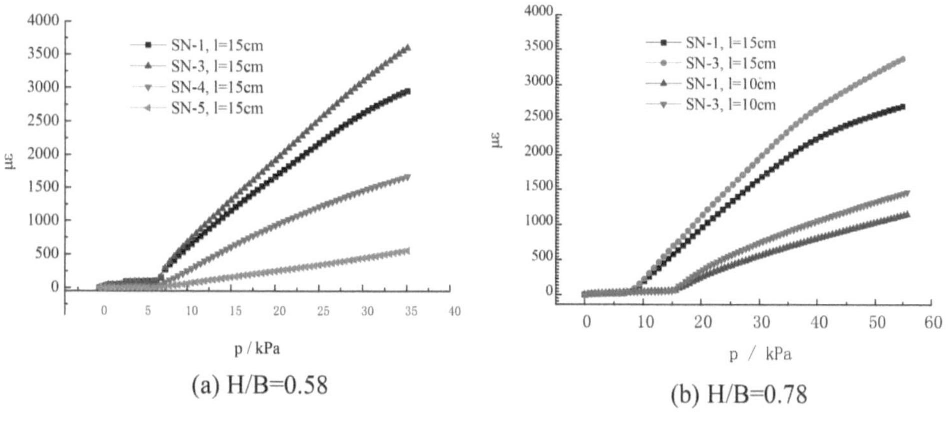

Figure 3. p~ε curve (H/B=0.58).

Substitute the strain value of bamboo net in Table 3 into formula (1) to calculate S/a, and then substitute S/a into formula (2) to calculate the angle (θ_h) between the tension of bamboo grid and the horizontal plane. It can be considered that θ_h is independent of H/B when s/B is constant.

Table 3. Strain value of 15 cm-spacing bamboo grid.

Working condition	No.	Strain/$\mu\varepsilon$ H/B= 0.00	H/B= 0.39	H/B= 0.58	H/B= 0.78	Mean value/ $\mu\varepsilon$	Variation Coef. Of Strain	mean value $\theta_h/°$
$p_{s/B=0.10}$	SN-1	2134.4	2071.4	2439.9	1937.1	2327.5	0.14	6.74
	SN-3	2708.8	2058.2	2894.1	2376.0			
$p_{s/B=0.15}$	SN-1	3024.4	3258.9	3017.7	2414.6	3061.7	0.20	7.93
	SN-3	4035.4	3573.6	3720.3	2916.9			

When the bearing capacity is equal to $p_{s/B=0.10}$, the average value of bamboo net is 6.74°; When the bearing capacity is equal to $p_{s/B=0.15}$, the average value of bamboo net is 7.93°.

3.5 *Analysis of stress diffusion angle*

According to the bearing capacity value given in Table 1, substitute it into formula (3) and calculate the stress diffusion angle by trial calculation, as shown in Table 4. It can be seen from Table 3 that the stress diffusion angle is related to the settlement ratio s/B and the thickness width ratio H/B. When the settlement ratio is constant, the stress diffusion angle increases with the increase of H/B.

Table 4. Diffusion angle of bamboo net and sand cushion (°).

Working condition	15 cm-spacing bamboo grid			10 cm-spacing bamboo grid		
	$\alpha_{s/B=0.06}$	$\alpha_{s/B=0.10}$	$\alpha_{s/B=0.15}$	$\alpha_{s/B=0.06}$	$\alpha_{s/B=0.10}$	$\alpha_{s/B=0.15}$
H/B=0.39	16.15	13.41	13.06	——	——	——
H/B=0.58	24.95	28.82	23.98	32.17	26.89	21.96
H/B=0.78	27.66	31.91	28.28	38.74	32.93	27.40

4 CONCLUSION

(1) Through the load test of indoor model, the influence of width thickness ratio and bamboo grid spacing on the bearing capacity of composite foundation are analyzed. With the decrease of bamboo grid spacing, the stiffness of composite foundation increases, and the bearing capacity increases accordingly. The bearing capacity of composite foundation increases greatly with the increase of width thickness ratio. When the bamboo grid spacing is 15 cm and width thickness ratio is 0.0~0.78, the bearing capacity of composite foundation corresponding to the settlement ratio is 0.10 increases by 1.15~4.98, and the bearing capacity of composite foundation corresponding to settlement ratio is 0.15 and increases by 1.36~4.88.

(2) The maximum strain occurs directly below the load plate, and the strain of the upper bamboo grid is less than that of the lower bamboo grid. The strain of 10 cm-spacing of bamboo grid is much smaller than that of 15cm spacing. The angle between the deformed bamboo grid and the horizontal plane is determined by the strain value. The magnitude of the included angle is related to the magnitude of the load and increases with the increase of the load. When s/B is equal to 0.10 and 0.15, the average value of the included angle is 6.74° and 7.93°, respectively.

(3) The stress diffusion angle of bamboo grid and sand cushion is related to the settlement ratio and thickness width ratio. When the settlement ratio is certain, the stress diffusion angle increases with the increase of thickness width ratio, and the spacing of bamboo grid also has an effect

on the stress diffusion angle. When the thickness width ratio is 0.39, 0.58, and 0.78, the mean value of stress diffusion angle is 14.2°, 26.5°, and 31°, respectively, but the maximum value of this test is 38.74°.

REFERENCES

Aazokhi W. (2014). Bamboo reinforcement in shallow foundation on the peat soil [J]. *Journal of Civil Engineering Research.* 4(3A): 96–102.

Aazokhi Waruwu, Husni Halim, et al. (2018). Bamboo grid reinforcement on peat soil under repeated loading [J]. *Journal of Engineering and Applied Sciences.* 13(8): 2190–2196.

Bai Bing (1994). Calculation of bearing capacity of reinforced soft foundation under three-dimensional stress state of geotextile [J]. *East China Highway.* 91(6): 54–58.

Deng You-Sheng, Cheng Zhi-He (2018). Compressive test study of bamboo structure [J]. *Science Technology and Engineering.* 18(27): 223–227.

GB50007 (2011), Code for design of building foundation [C]. *China Construction Industry Press.*

Gong Xiaonan (2008). Foundation treatment manual (3rd Edition) [M]. *China Construction Industry Press,* 2008.

Hegde A and Sitharam T G. (2015). Experimental and analytical studies on soft clay beds reinforced with bamboo cells and geocells [J]. *International Journal of Geosynthetics and Ground Engineering.* 1(2): 1–13.

Lin Kaiqiu, Bao Chenggang (1992). A new practical calculation method and test verifications of geotextile reinforced soft foundation [J]. *Journal of Yangtze River Scientific Research Institute,* (01): 47–56.

Mao Xiaofei, Hu Fuli, et al. (2021). Study on the stress distribution of Nanzhu bolt in the consolidation of ramming site [J]. *Chinese Journal of Applied Mechanics.* 38(1): 345–344.

Yamanouchi T. (1985). Recent development in the used synthetic geofabrics and geogrids [C]. *Symposium on Recent Developments in Ground Improvement Techniques,* Bangkok. 205–224.

Yuan Man, Ding Jihui, et al. (2014). The application of bamboo network reinforcement technology on hydraulic fill soft-soil foundation treatment [J]. *World Journal of Engineering & Technology,* 02(2): 68–72.

Zhu M Q, Zhang Z W, et al. (2020). Study on mechanical properties of bamboo and environmental effects in concrete. *Journal of Hunan University of Science and Technology* (Natural Science Edition), 35(3): 50–56.

Advances in Civil Engineering and Environmental Engineering – bin Yahya & Bin Daud (Eds)

Application of storm water management model in overload evaluation and reconstruction of urban stormwater pipes

Hui Zhao*, Jixiang Li*, Zongxin Li* & Jianxun Wang*
Harbin Space Star Data System Technology Co. Ltd, Harbin, China

ABSTRACT: As an important part of the urban infrastructure, the urban stormwater pipe network is responsible for rainwater drainage. To solve urban waterlogging after rainstorms and understand the effect of overload evaluation and reconstruction of urban stormwater pipes on urban stormwater control infrastructure, a new dynamic rainfall-runoff model is established in this paper by means of the storm water management model (SWMM). The overload of the stormwater pipe network in the study area for the recurrence periods of 0.5 years, one year, two years, and five years is simulated, respectively. According to the results of pipe overload evaluation, a method to determine the size of pipe renovation and expansion and improve the drainage capacity of the pipe network is proposed here. The simulation results show that the renovation and expansion of stormwater pipes based on the results of the overload evaluation of the stormwater pipe network are of great benefits. It cannot only reduce the full-flow length of pipes, but also improve the drainage performance of the stormwater pipe network.

1 INTRODUCTION

In recent years, the expanding footprint of urban areas, the significant changes in local hydro-meteorology, and the increasingly frequent outbreak of extreme weather come along with the acceleration of urbanization (Cai 2021). The urban drainage network has the insufficient coping capacity to convey and discharge the rainwater runoff water on the surface in a short period of time in extreme weather conditions such as persistent heavy rainfall or extremely heavy rainfall during the rainy season, which will make the network overloaded with water and lead to urban flooding, thus seriously affecting people's daily life and industrial production, and even threatening the life safety of urban residents and causing huge losses to the city.

In the context of frequent urban flooding, an urban drainage network model gradually becomes an effective tool essential for urban stormwater management and urban flooding analysis. The urban stormwater management model is established to analyze the overload of the stormwater pipe network, which can provide a basis for stormwater pipe renovation to help city managers and decision-makers improve the drainage capacity of the stormwater pipe network. Furthermore, a useful method to deal with urban flooding has been established.

2 INTRODUCTION TO SWMM MODEL

SWMM (Storm Water Management Model) is a dynamic rainfall-runoff simulation computer program developed by the US Environmental Protection Agency (US EPA) for single-event or

*Corresponding Authors: 381306372@qq.com, 404955939@qq.com, 624601761@qq.com and 546817208@qq.com

DOI 10.1201/9781003349563-64

long-term simulation of runoff quantity and water quality in urban areas, which can simulate the complete urban rainfall-runoff and pollutant movement processes, including surface runoff and water flow in drainage systems, stormwater and other storage treatment processes, and water quality impact assessment (Long 2016). The model can visually express the data input and output results, and is suitable for the analysis of urban area stormwater runoff simulation, as well as the planning and design of stormwater pipe networks and sewage pipe networks, which is an efficient and practical urban stormwater and flood management simulation model developed for rich and extensive research (Guo 2014).

SWMM has typical elements for an urban drainage system, including an atmospheric component (which generates precipitation as well as settled pollutants that reach the surface component), a surface component (which receives precipitation in the form of rain or snow from the atmospheric component), a subsurface component (which receives infiltration from the surface component), and a conveyance component (which contains a network of elements such as channels, pipes, pumps and regulators, and storage/treatment devices). It divides each sub-catchment into two types of impervious and permeable ground. After the rainfall exceeds the flow rate and volume of water that can be infiltrated from the surface, runoff production flow is formed at the surface. The model uses the nonlinear reservoir method for the calculation of surface runoff which will all enter the stormwater pipe network system. When the conveyed water exceeds the carrying capacity of the pipe network, the model will have overflow ponding at the corresponding nodes (rainwater wells, inlets), which is reflected as Internal flooding.

3 CONSTRUCTION OF SWMM

3.1 *Construction of SWMM geographic elements*

According to the requirements of SWMM construction, the GIS model database is constructed by combining the current situation and the information on the pipe network in the study area. The nodes, pipes, and discharge inlets of the stormwater pipe network model are extracted through GIS processing. In this study, the stormwater pipe network in the study area is generalized into 2,567 nodes, 2,558 pipes, and 24 outfalls. Based on the generalized map of the stormwater pipe network, the study area is divided into 2,562 sub-catchments in view of the distribution of pipe network inlets using a combination of manual division and Tyson polygon, see Figure 1.

Figure 1. Schematic diagram of the regional stormwater model research.

3.2 *SWMM parameter setting*

The SWMM calculation parameters mainly consist of deterministic and uncertain parameters. Deterministic parameters include nodal well bottom elevation, maximum depth, pipe diameter, pipe length, etc. Uncertain parameters include rainfall, sub-sink production, and sink flow-related parameters (characteristic width, depression storage volume, roughness coefficient, impermeability, infiltration coefficient, etc.).

The Chicago rainfall model was used as the storm rainfall model for this study, which is a heavy rainfall design model based on the local storm intensity formula for rainfall data and the generation of design return period rainfall data (Zhao 2021). The following storm intensity equations for the study area were used.

$$\mathrm{q} = \frac{166.67a\,(1 + Clg^{P})}{(t + b)^{n}} \tag{1}$$

where q = intensity of the rainstorm $\left(L/\,(S \cdot hm)^{2}\right)$; p = recurrence period (a); t = precipitation calendar time; a = rain force parameter (mm); C = rain force variation parameter; b = precipitation calendar time correction parameter (min); and n = rainfall attenuation index.

The generated design rainfall data for different rainfall return periods (p=0.5a, 1a, 2a, and 5a) are shown in Figure 2.

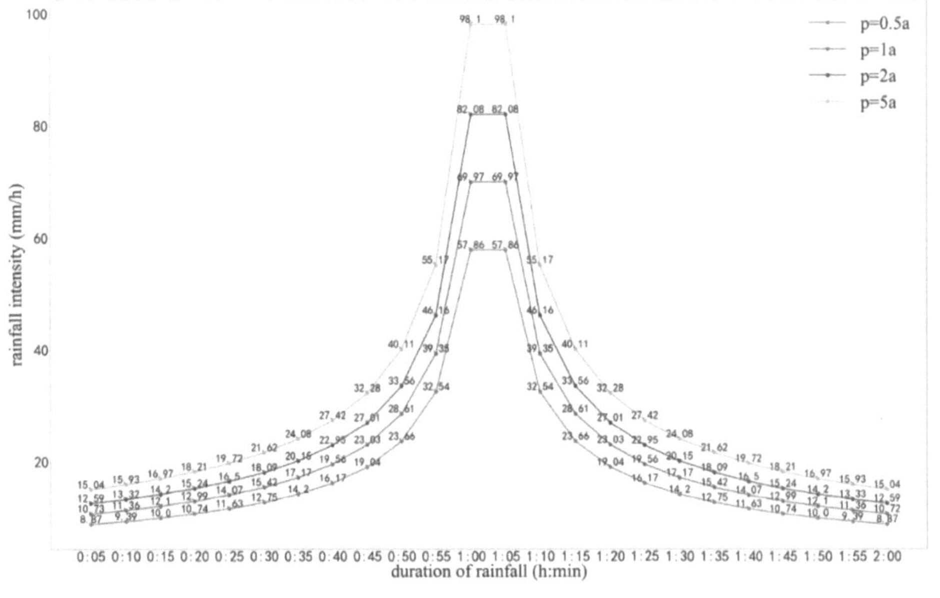

Figure 2. Rainfall process lines for different recurrence periods.

Sub-catchment area production and sink parameters, in this study, combined with the previous study results and the specific situation of this study area, rainfall storage in permeable and impermeable surface depressions with reference to Table 1 for selection; Horton infiltration formula for surface infiltration, the values of the parameters in the infiltration formula shown in Table 2.

3.3 *SWMM model building*

In this paper, a regional stormwater pipe network is examined as a whole, and a model of the stormwater pipe network in the study area is built, as shown in Figure 3, by constructing a correspondence between the GIS model database and the standard input file INP of SWMM.

Table 1. Parameters of rainfall storage in permeable/impermeable surface depressions.

Parameters	Permeable surface depression rainfall storage (mm)	Impervious surface depression rainfall storage (mm)
Value	7	2.5
Scope	2.56-7.62	1.27-2.56

Table 2. Surface infiltration parameters.

Parameters	Maximum infiltration rate (mm/h)	Minimum infiltration rate (mm/h)	Decay rate constant (hr^{-1})	Soil drainage days
Value	76.2	3.35	2	5
Scope	72.4-78.1	3.18-3.82	2-4	1-7

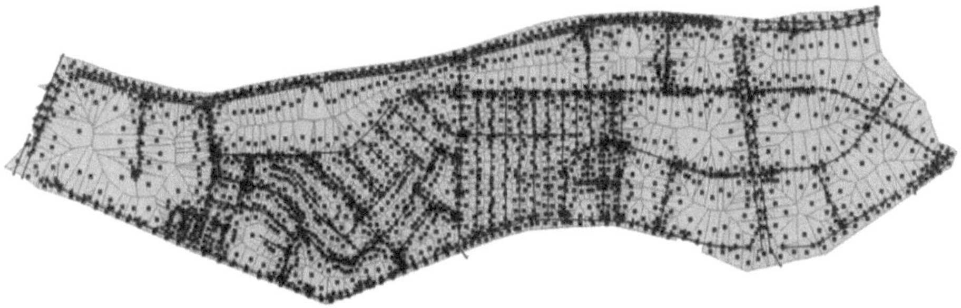

Figure 3. SWMM of stormwater pipe network in the study area.

The design rainfall data of different recurrence periods are selected for simulation calculation, and the percentage and the distribution of overloaded pipes in the pipe network are shown in Figure 4 according to the simulation results.

4 ANALYSIS OF PIPE OVERLOAD EVALUATION

4.1 *Modeling of pipe overload evaluation*

The overloaded pipes in the simulation results of SWMM can determine the distribution of overloaded pipes in the stormwater pipe network visually and specifically. However, it is not easy to compare only the single feature of the number of full flow hours upstream, the number of full flow hours downstream, or the number of full flow hours at both ends of the overloaded pipes in the simulation results. So, it is more practical to evaluate the overloading level of pipes by combining all the features, which can help determine the overloading level of overloaded pipes and provide a basis for the renovation of pipes.

It can be seen from the simulation results of different recurrence period design rainfall that with the increase of recurrence period, the number of overloaded pipe sections presented by the simulation results increases, both of which become positively correlated. According to the current

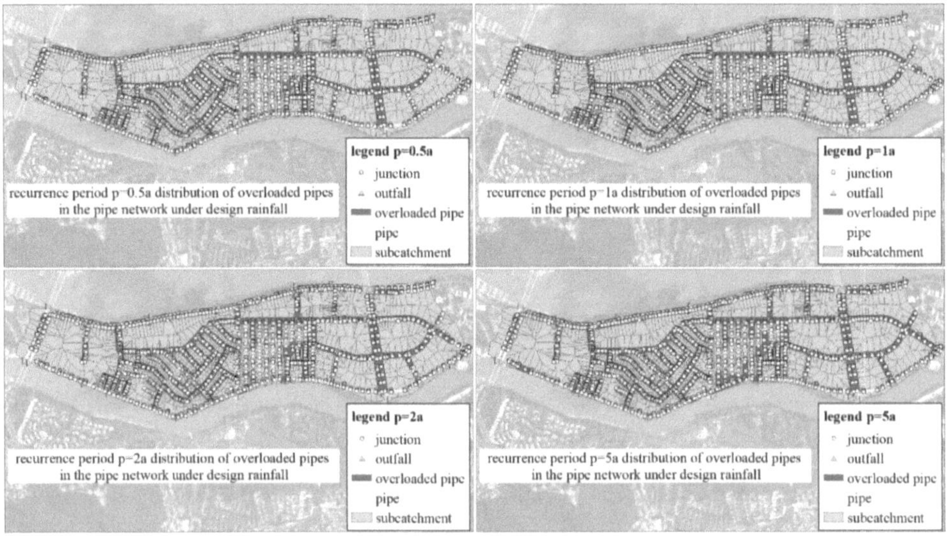

Figure 4. Distribution of overloaded pipes in the pipe network under different recurrence period design rainfall.

situation of the stormwater pipe network in the study area coping with the design of the recurrence period, the simulation results coping with the recurrence period p=1a are selected for the construction of the evaluation and analysis model of the overloading level of pipes.

The models and methods generally related to the evaluation rating are widely used for hierarchical analysis and entropy weighting. Hierarchical analysis can be used to determine the weight of the program level to the target level. This weighting method is more subjective than the entropy weighting method, and the latter yields more objective weights.

Evaluation of the overloading level of pipes (see Table 3 for the evaluation level) selected four indicators of the pipe network's upstream full flow hours, downstream full flow hours, full flow hours at both ends, and the inverse of the pipe diameter as evaluation indicators. The upstream and downstream full flow hours of the pipe network refer to the total length of time that the pipe network is in full flow to convey rainwater. The more the full flow hours the pipes take, the more serious they are overloaded to convey rainwater. The greater the inverse of the pipe diameter, indicating that the smaller the pipe diameter, the weaker the pipe's rainwater conveying capacity, which means it is more likely to have a full load of conveying rainwater. Therefore, the four indexes selected here are all positively related to the rainwater conveying capacity of the pipes. The entropy weighting method is used to determine the weights of each evaluation index, and then the comprehensive index method is used to establish the evaluation model of the overloading level of pipes.

Table 3. Scoring S of pipe overloading level evaluation (relative value no greater than 1).

Evaluation value of overloading level	Rating of pipe overload
0-0.2	Low
0.2-0.4	Relatively low
0.4-0.6	General
0.6-0.8	High
>0.8	High

4.2 *Analysis and evaluation of pipe overload*

4.2.1 *Dimensionless and standardization of the entropy method*

Standardized formula:

$$y_{ij} = \frac{x_{ij} - min\,(X_i)}{max\,(X_i) - min\,(X_i)} \tag{2}$$

The calculated standardized index results of the evaluation index of the overloading level of pipes are shown in Table 4.

Table 4. Standardized index data.

Pipes	Upstream full flow hours	Standardized values	Downstream full flow hours	Standardized values	Hours above full normal flow	Standardized values	Pipe Diameter	Standardized values
C1328	1.13	0.657	1.3	0.731	1.13	0.657	0.15	0.629
C1195	1.3	0.756	1.33	0.747	1.3	0.756	0.3	0.248
C1194	1.23	0.715	1.27	0.714	1.23	0.715	0.2	0.439
C1197	1.09	0.634	1.1	0.619	1.09	0.634	0.2	0.439
C1196	1.2	0.698	1.26	0.708	1.2	0.698	0.2	0.439
C1191	1.18	0.686	1.27	0.714	1.18	0.686	0.2	0.439
C1190	1.21	0.704	1.27	0.714	1.21	0.704	0.2	0.439
C712	1.24	0.721	1.5	0.842	1.24	0.721	0.4	0.153
C631	0.66	0.386	0.83	0.468	0.66	0.386	0.2	0.439
C441	0.86	0.501	0.96	0.541	0.86	0.501	0.4	0.153

4.2.2 *Indicator characteristic weight calculation by the entropy method*

Characteristic specific gravity calculation formula:

$$P_{\mathrm{ij}} = \frac{x_{ij}}{\sum_{i=1}^{n} x_{ij}} \tag{3}$$

where x_{ij} = the jth index value of the ith item.

The calculated weighting results of the evaluation index of the overloading level of pipes are shown in Table 5.

Table 5. Weighting of indicator characteristics.

Pipes	Upstream full flow hours	Feature weighting	Downstream full flow hours	Feature weighting	Hours above full normal flow	Feature weighting	Pipe Diameter	Feature weighting
C1328	1.13	0.002	1.3	0.002	1.13	0.002	0.15	0.004
C1195	1.3	0.003	1.33	0.002	1.3	0.003	0.3	0.001
C1194	1.23	0.003	1.27	0.002	1.23	0.003	0.2	0.002
C1197	1.09	0.002	1.1	0.002	1.09	0.002	0.2	0.002
C1196	1.2	0.003	1.26	0.002	1.2	0.003	0.2	0.002
C1191	1.18	0.003	1.27	0.002	1.18	0.003	0.2	0.002
C1190	1.21	0.003	1.27	0.002	1.21	0.003	0.2	0.002
C712	1.24	0.003	1.5	0.003	1.24	0.003	0.4	0.001
C631	0.66	0.001	0.83	0.001	0.66	0.001	0.2	0.002
C441	0.86	0.002	0.96	0.002	0.86	0.002	0.4	0.001

4.2.3 *Index information entropy calculation by the entropy method*
Information entropy formula:

$$E_j = -\frac{1}{ln\,(n)}\sum_{i=1}^{n} P_{ij} ln P_{ij} \tag{4}$$

The information entropy E of the three feature indicators is obtained by substituting the feature weight obtained in Step 1 into the information entropy calculation formula, and the results of the information entropy obtained are shown in Table 6.

Table 6. Information entropy of indicators.

Indicators	Upstream full flow hours	Downstream full flow hours	Hours above full normal flow	Pipe Diameter
Information entropy	0.948	0.981	0.948	0.968

4.2.4 *Index weights determination by the entropy method*
Weighting factor calculation formula:

$$W_i = \frac{1 - E_i}{k - \sum_{i=1}^{k} E_i} \tag{5}$$

The information entropy obtained from Step 2 was substituted into the weight calculation formula to find out the weight coefficients of the three characteristic indicators, as shown in Table 7.

Table 7. Indicator weighting factors.

Indicators	Upstream full flow hours	Downstream full flow hours	Number of hours of full flow at both ends	Pipe Diameter
Weighting factor	0.335	0.121	0.336	0.208

4.2.5 *Overloading level score calculation*
Composite index calculation formula:

$$S_i = \sum_{j=1}^{k} W_{ij} y_{ij}\ (i = 1, 2, \ldots, n) \tag{6}$$

The parameter weights of each index parameter and the standardized index are weighted and averaged to obtain the comprehensive evaluation index of the overloaded pipes. The comprehensive evaluation indexes are divided according to the divided evaluation levels, and the results are shown in Table 8.

5 PIPE RENOVATION

To deal with the urban flooding, measures can be taken, including selectively increasing green areas, laying permeable pavements to improve the permeability of the ground surface (Long 2016), and renovating the stormwater pipe network system, for example, adding storage facilities, adding stormwater inlets, changing the upstream and downstream elevations of pipes, increasing the depth of nodal stormwater wells, and increasing the pipe diameter. Based on the evaluation level of the overloaded pipes, the highly overloaded pipes are enlarged and renovated. The optimal pipe diameter size after renovation is determined based on SWMM calculation. The specific renovation steps are as follows.

Table 8. Evaluation score of the overloaded pipes.

Pipes	Score of overloading level	Rating
C1328	0.66	High
C1195	0.649	High
C1194	0.658	High
C1197	0.592	General
C1196	0.645	High
C1191	0.638	High
C1190	0.65	High
C712	0.618	High
C631	0.407	General
C441	0.434	General

5.1 *Step 1*

Prioritization of overloaded pipes based on the evaluated overloading levels.

5.2 *Step 2*

The overloaded pipes are in order of priority, in increments of 0.1 m for pipe diameter expansion and renovation (pipe diameter limited to the standard stormwater pipe diameter set).

5.3 *Step 3*

Performing another SWMM simulation and evaluation after the renovation to obtain a new set of overloaded pipes.

5.4 *Step 4*

Check whether the renovated pipe is still an overloaded pipe. If it is, proceed recursively to Step 2 and Step 3 until the pipe is no longer overloaded after renovation, and then proceed to the next overloaded pipe renovation.

The results of the overloaded pipe renovation before and after the overload state are shown in Table 9. Overloaded pipes total 692. The capacity of pipe diameter reconstruction and expansion in the state of rainwater discharge in non-full pipes ranges from 33%-633%. For the pipes with an excessive capacity for reconstruction and expansion, low-impact development should be considered to reduce the bearing pressure of pipes.

6 CONCLUSIONS

In this paper, SWMM was applied to simulate the stormwater pipe network system in an urban study area. Through the construction of the pipe overload evaluation model, the evaluation and analysis of overloaded pipes of the stormwater pipe network with 1-year design rainfall in the recurrence period, this study provides a reasonable basis for the renovation of overloaded pipes, and implements the simulation of overloaded pipe expansion and renovation based on SWMM. The example application demonstrates that SWMM is feasible in urban flooding prevention and drainage calculation, and drainage capacity judgment of the stormwater pipe network system (Zhu 2013). The method of pipe renovation based on the priority of pipe overloading level is feasible and has a certain reference value. Since only the evaluation and analysis of overloaded pipeline retrofitting study are carried out in the paper, the optimization plan can be offered in future studies in terms of

Table 9. Table of status before and after overloaded pipe renovation.

Overloaded pipes	Score of overloading level	Rating	Pipe diameter before modification (m)	Retrofitted pipe diameter (m)	Post-modification status
C1328	0.66	High	0.15	1.1	Normal
C1194	0.658	High	0.2	0.6	Normal
C1190	0.65	High	0.2	0.6	Normal
C1195	0.649	High	0.3	0.8	Normal
C1196	0.645	High	0.2	0.5	Normal
C1191	0.638	High	0.2	0.6	Normal
C712	0.618	High	0.4	1.35	Normal
C1197	0.592	General	0.2	0.5	Normal
C441	0.434	General	0.4	0.8	Normal
C631	0.407	General	0.2	0.5	Normal

comprehensive consideration of nodes, pipes, low impact development, and pipe maintenance and replacement costs.

REFERENCES

Jing Z., Jun L., Cui H. & Zhoushen Q. (2013). Application of SWMM model to urban flood control calculation in the pre-mountain plain of southwest China. *J. Hydropower Energy Science*, 12, 38–41.

Tian C. & Jia W. (2021). Application of drainage model and GIS fuzzy integrated evaluation in flood risk assessment. *J. China Water Supply and Drainage*, 11, 117–124.

Tianwei L., Yi L. & Wei X. (2016). Application of SWMM in rainfall and flood analysis of a city. *J. Water Resources Management*, 06, 1–4.

Xiaoqing Z. (2021). Simulation and validation of SWMM model for surface runoff in universities. *J. Science and Technology Innovation*, 20, 98–99.

Yunfei G. (2014). Research on urban storm waterlogging based on SWMM—A campus as an example, D. *Hunan University of Technology*.

Advances in Civil Engineering and Environmental Engineering – bin Yahya & Bin Daud (Eds)
© 2023 The Author(s), ISBN: 978-1-032-39397-1

Leakage prevention and treatment in water supply and drainage projects

Yuying Sun*
The First Ecological and Environmental Monitoring Station of the Xinjiang Production and Construction Corps Urumqi, Xinjiang, China

ABSTRACT: With the rapid development of social and economic development, people's living standards are improving day by day, and the demand for the living environment infrastructure is becoming increasingly high standard and strict requirements. It makes infrastructure construction a wide concern for all communities. The construction units must give full attention to the planning and construction of infrastructure so that the infrastructure can play its due role and value. Among the infrastructure projects, water supply and drainage engineering are relatively important infrastructure project, which has a crucial role and influence in ensuring the stable and harmonious development of society and protecting people's basic needs. Therefore, implementation and enforcement of the relevant measures of leakage prevention and treatment in water supply and drainage engineering ensure the stability and safety of water supply and drainage engineering. This paper explores the importance of leakage prevention and treatment in water supply and drainage projects, as well as also discusses the shortcomings of leakage prevention and prevention measures in water supply and drainage projects and the corresponding effective measures, with a view to providing reference and suggestions for better performance of leakage prevention and prevention measures in water supply and drainage projects.

1 INTRODUCTION

In the new era, the construction standards of all kinds of projects are becoming more standardized and stricter, and more attention is paid to the construction quality and environmental protection of infrastructure, which cannot have any adverse impact on the environment. Among them, the leakage prevention treatment and prevention work in water supply and drainage engineering is one of the important basic guarantees work in water supply and drainage engineering, which has a very important impact on the overall performance and effect of water supply and drainage engineering. Therefore, to ensure the construction quality of water supply and drainage engineering, the reliability of leakage prevention and prevention measures should be strengthened, ensuring the normal function of water supply and drainage engineering.

2 THE IMPORTANCE OF LEAKAGE PREVENTION TREATMENT AND PREVENTION IN WATER SUPPLY AND DRAINAGE ENGINEERING

Water supply and drainage project, as an important infrastructure project of guarantee, plays a very significant role in ensuring people's healthy and safe water use, improving flood control and drainage capacity, and avoiding sewage pollution and other aspects. Therefore, it is particularly critical to strengthening the stability, safety and reliability of water supply and drainage engineering. In many work links of water supply and drainage engineering, leakage prevention treatment and

*Corresponding Author: 444192811@qq.com

DOI 10.1201/9781003349563-65

prevention work play a vital role in the overall quality and effect of the project. First, water supply and drainage prevention and water supply can be planned and designed according to different water requirements, and pipeline materials and construction techniques are used to ensure that water supply, drainage, and sewage can improve the flood control and drainage capacity of water supply and drainage works.

3 CONSTRUCTION TECHNOLOGY OF LEAKAGE PREVENTION TREATMENT AND PREVENTIVE MEASURES FOR WATER SUPPLY AND DRAINAGE ENGINEERING

3.1 *Interface parts*

The interface part of the water supply and drainage engineering pipeline is prone to leakage problems, so leakage prevention treatment and prevention operation are needed. During the construction of the docking port, it shall check whether the actual performance of the rubber ring and pipe meet the construction standards of water supply and drainage engineering, including the materials, specifications, and performance parameters of the pipe, to ensure that the interface operation is effectively carried out in strict accordance with the established construction plan. Antileakage treatment of the interface parts includes both external interface and internal interface. Internal interface refers to the pipe hook, that is, when the pipe diameter is large, should be in the internal hook work, and ensure that this work must be synchronized with the pipe seat pouring, and complete the internal concave joint leakage treatment before the final solidification; the external interface according to the similar steps of the internal interface, and ensure that the pipe seat surface is in a basic smooth state. Only the external and internal interfaces are well sealed to ensure that the interface site leakage prevention treatment and prevention are reliable and effective.

3.2 *Inspection wells*

As an auxiliary facility of water supply and drainage works, once the well-room has leakage, the waterproof and seepage prevention effect of water supply and drainage works will be greatly affected. To improve the leakage prevention quality of the inspection well, the construction can be carried out according to the following steps: First, in the process of masonry inspection well, attention should be paid to the effective control of the ash seam mortar fullness, to prevent the ash seam between the bricks from leaking is not close enough. Second, the leakage problem of the inspection well mostly appears in the cracks of the plastering layer or the empty drum position, so it is necessary to improve the quality of the inspection well-plastering construction. Under normal circumstances, can according to the five-layer method of concrete construction. The first layer is cement slurry with a thickness of 2 mm and the second layer; the second layer is cement mortar of 5 mm-7 mm thickness with the third layer; third and four layers are cement slurry and cement mortar, the fifth layer of cement slurry before the fourth layer, and compacted together with the first four layers. In addition, check the contact position of the well cover plate and the brick surface is also prone to leakage problems, so when lifting the cover plate, it is necessary to ensure the fullness of the seat slurry, and the cover plate is fully wet before the slurry, to improve the tightness between the cover plate and the brick surface.

3.3 *Gulch excavation and trench slope treatment*

When there are no clear design requirements for trench excavation, it can be implemented according to the following data: $B=D0 + 2\ (b1 + b2 + b3)$. Formula B is the excavation depth at the bottom of the trench; D0 is the outer diameter of the pipe, and b1 is the working face width of the pipe side, which can be selected according to Table 1. b2 is the support thickness on the side of the pipe, ranging from 150mm to 200mm, and b3 is the thickness of the formwork of the concrete or reinforced concrete pipe and channel side poured on site.

Table 1. Working Face Width on One Side of Pipelines.

D0(mm)	Working Face Width on One Side of Pipelines b1 (mm)		
	Concrete Pipelines		Metal pipelines & Chemical Building Materials Pipelines
D0≤500	Rigid Joint	400	300
	Flexible Joint	300	
500 < D0≤1000	Rigid Joint	500	400
	Flexible Joint	400	

Displacement-controlled loading was adopted, loading the axial force first and keeping it constant after. The following standards can be used to implement the treatment of the trench slope: when the geological conditions are good, the soil texture is uniform, the groundwater level is below the elevation of the lowest surface of the trench, the excavation depth is within 5m, and the trench is not supported, the scarp slope of the trench slope can be treated according to the data in Table 2.

Table 2. Highest Grade of Side Slope of Trench (Depth within 5m).

Soil Classification	Grade of Side Slope (H: W)		
	No Load on Top of the Slope	Static Load on Top of the Slope	Moving Load on Top of the Slope
Intermediate-Density Sandy Soil	1:1.00	1:1.25	1:1.50
Intermediate-Density Gravel Soil (Sandy Soil as Filler)	1:0.75	1:1.00	1:1.25
Silt Treated with Stiff-Plastic Process	1:0.67	1:0.75	1:1.00
Intermediate-Density Gravel Soil (Cohesive Soil as Filler)	1:0.50	1:0.67	1:0.75
Silt & Clay Treated with Stiff-Plastic Process	1:0.33	1:0.50	1:0.67
Old Loess	1:0.10	1:0.25	1:0.33
Soft Soil (after Well-Points Dewatering)	1:1.25	–	–

3.4 *Trench backfilling*

The quality of trench backfilling will have a direct impact on the leakage prevention effect of water supply and drainage engineering. Once the construction operation is not carried out according to the engineering design requirements or technical specifications, there may be a series of leakage problems such as pipe top cracking, wipe belt collapse, pipe interface dislocation, and then cause serious engineering quality and safety accidents. Therefore, after the completion of the closed water test, the trench backfill operation must be carefully implemented. When carrying out the trench backfill construction operation, attention should be paid to the backfill of the top of the pipe concrete solidification and mortar has a certain strength structure, to avoid the direct impact of the pipe body as far as possible, and to ensure that there is no large hard gravel in the gravel. When the trench backfilling operation on both sides of the pipeline, the tamping operation should be matched, so that the backfill gravel can form a powerful whole, and the vault can play the protection role of diffusion and unloading, to effectively ensure the safety of water supply and drainage pipe (Chaoyang Shen 2019).

4 POSITIVE IMPACT ON ENVIRONMENTAL PROTECTION OF LEAKAGE PREVENTION AND TREATMENT FOR WATER SUPPLY AND DRAINAGE WORKS

Effective engineering work has an important and positive impact on the implementation and enforcement of environmental protection. First, it can effectively reduce the negative impact of water pollution on environmental protection. If the sewage flows into the soil or into rivers, lakes, and seas before being treated, it will cause a different degree of pollution to the environment and affect the effectiveness and quality of environmental protection. The pipeline contains a large amount of sewage, including domestic sewage, industrial sewage, other sewage, etc. If some of the sewage flows into the soil or rivers, lakes, and seas before being treated, it can cause different degrees of pollution to the environment and affect the effectiveness and quality of environmental protection. Second, the effective use of drainage works can reduce the negative impact of air pollution on environmental protection. A large amount of wastewater stored in the drainage pipe, after a long time of precipitation, fermentation, and other physical or chemical reactions, it will produce many toxic and harmful gases, which will cause air pollution. If the work is not truly implemented in place, it is very likely to lead to toxic and harmful gases out of the pipe, and cause various degrees of pollution to the surrounding air and influence the effectiveness of air protection in environmental protection. Implementation of prevention work, effective emission of waste gas and water, reflection of the function and value of water supply and drainage projects for environmental protection are to bring a positive impact on the role of local ecological environmental protection. Finally, the protection of plants is also crucial in environmental protection. The unreasonable discharge of wastewater and waste gas can have a certain impact on the growth of plants, and may even lead to the direct death of plants, causing large areas of plants to wither or die, which has a huge impact on ecological environmental protection. Therefore, we should implement and enforce leak prevention treatment and prevention for water supply and drainage, to ensure that wastewater and waste gas before the effective treatment, not indiscriminate discharge, play and reflect the role and value of water supply and drainage projects for environmental protection, to bring a positive impact on the local ecological environmental protection role.

5 DEFICIENCIES IN LEAKAGE PREVENTION TREATMENT AND PREVENTIVE MEASURES FOR WATER SUPPLY AND DRAINAGE PROJECTS

At this stage, there are still some deficiencies in the water supply and drainage project leakage prevention measures (Figure 1), which need to be optimized and improved in a targeted manner to ensure effective leakage prevention and treatment.

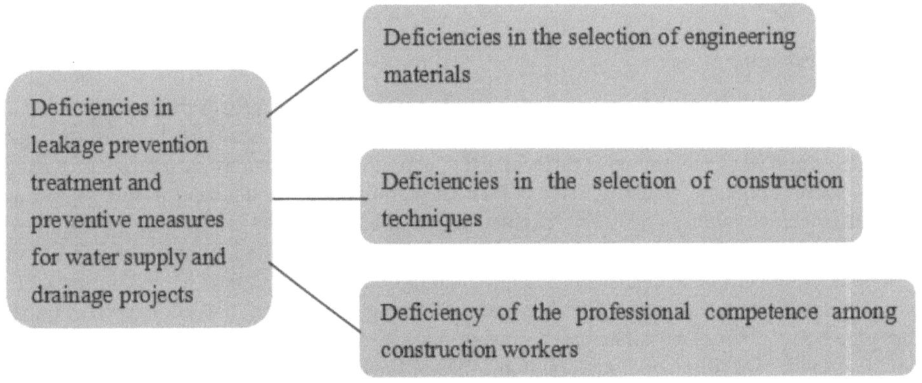

Figure 1. Major deficiencies in the treatment and prevention of leakage in water supply and drainage works.

5.1 *Deficiencies in the selection of engineering materials*

The selection of engineering materials will have a direct impact on the overall quality and effect of water supply and drainage projects, and will also influence the practicality and effectiveness of leakage prevention and treatment measures. If the selection of engineering materials is difficult to meet the actual construction requirements, it will leave huge hidden safety risks, and will also increase the difficulty of leakage prevention and treatment measures. Overall of the working, in the aspect of engineering materials selection, there are several factors that may affect the normal selection of engineering materials (Wang 2017). First of all, at the stage of procurement and selection of engineering materials, the deficiencies are reflected in the following aspects: they fail to implement and enforce the procurement and selection of engineering materials, and lack the necessary supervision and management to supervise the procurement and selection of engineering materials. For example, some valves, pipes, and connectors are not selected in accordance with the requirements of construction technology, which may lead to the leakage of engineering materials in the event of bad weather. Second, at the stage of using engineering materials, the deficiencies are: it fails to pay attention to the protection and management of engineering materials and lacks the necessary protection, which results in different degrees of damage to engineering materials, affecting the overall quality and effect of water supply and drainage projects, and laying a safety hazard for future water supply and drainage projects with leakage problems.

5.2 *Deficiencies in the selection of construction techniques*

The construction technology should be selected based on the construction technology regulations, to ensure that construction technology can play its due role. At present, in parts of the construction processes, there are still unreasonable and unscientific selections of construction technology, which not only affects the overall quality and effect of water supply and drainage projects but also delays the construction progress in some degree, and has a serious impact on the stability, safety, and reliability of water supply and drainage projects. In the selection of construction technology, if it fails to implement in accordance with the regulation of water supply and drainage projects, it may bring many safety risks for water supply and drainage projects, such as pipe leakage, pipe bursting, and so on; whereas, there are many reasons for pipe leakage, but the choice of construction technology is one of the main factors in the enforcement and prevention of leakage in water supply and drainage projects. For example, when the workers are welding pipe, there may several cracks between the pipe and flange, which will result in the tightness of the fillet, which does not correspond to the normal standard, or part of the galvanized pipeline may be fracture. And the selected construction technology is not practically used in accordance with the regulation, which may leave the potential safety risks in the future.

5.3 *Deficiency of the professional competence among construction workers*

The professional competence of the construction workers is a key factor in determining the quality and effectiveness of water supply and drainage projects. The higher the professional competence of the construction workers, the better the quality; and effectiveness of the project will be guaranteed and the project will be completed successfully and on time. Water supply and drainage engineering is a complex system project that involves a wide range of professional and comprehensive aspects. Different drainage requirements, different piping connection methods, as well as important and difficult points of the project, require the construction workers make scientific and reasonable judgments by using their work experience and combining it with the actual situation to ensure the smooth construction of projects. At the present stage, there are several problems in the construction of water supply and drainage projects, some construction workers lack professional ability, they have no experience and do not receive professional training and learning, and lack correct understanding of the projects. All that makes it difficult to effectively guarantee the quality of the projects. For example, some construction workers choose to conceal and not report the problems, lacking the awareness of safe production. Especially during pipe connection construction work, the pipeline articulation is not strong and tight enough which results in frequent leakage of the pipeline at the end of the drainage project, and may seriously affect the whole quality and effect of the drainage project.

6 EFFECTIVE MEASURES TO DEAL WITH AND PREVENT LEAKAGE IN WATER SUPPLY AND DRAINAGE WORKS

To improve the effectiveness of antileakage treatment and preventive measures, strengthening the supervision and management of all aspects of water supply and drainage project construction is quite necessary (as shown in Figure 2), thereby promoting the further improvement of the practicality and effectiveness of antileakage treatment and preventive measures.

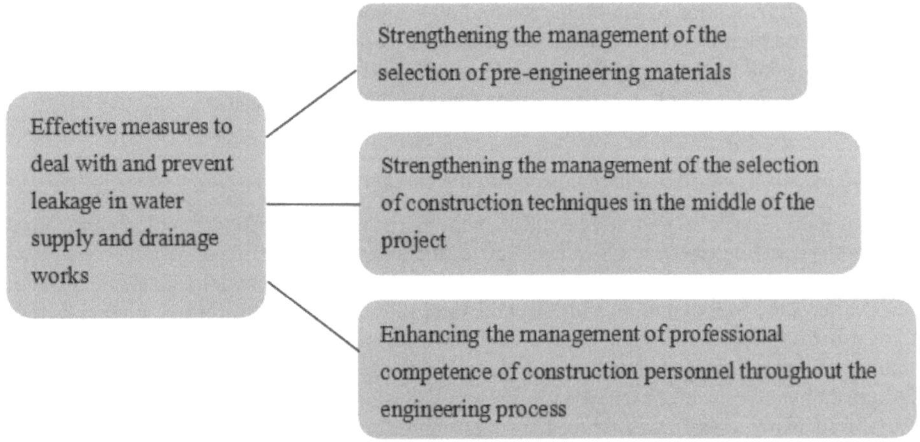

Figure 2. Main effective measures for leakage prevention treatment and prevention in water supply and drainage works.

6.1 *Strengthening the management of the selection of engineering materials*

The selection of engineering materials plays a very important role in influencing the effectiveness of preventive measures, therefore, the management of the selection of engineering materials should be strengthened to provide the necessary basic support and guarantee for seepage prevention treatment and preventive measures (Zhang 2021). First of all, in the engineering materials selection stage, the construction workers should be strictly in accordance with the construction of water supply and drainage engineering requirements for selection, to ensure the reliability of engineering materials from the source. While selecting the engineering materials, the management should arrange specific workers to supervise and manage the process, so as to avoid the selection workers having a weak awareness of safe production and selecting some engineering materials that are illegal or do not meet the standards, which affects the effectiveness of water supply and drainage leakage prevention treatment and preventive measures; Second, at the stage of using engineering materials, it is necessary to strengthen the supervision and control of engineering materials, improve the efficiency of the usage and avoid some defective engineering materials being used. In the actual construction process, the workers should report the first time when they found problems to reduce the potential safety risks in terms of engineering materials. For construction workers who use engineering materials that are restricted in law, the management should take necessary penalty measures toimprove awareness and further guarantee the effectiveness of the preventive measures against leakage.

6.2 *Strengthen the selection management of construction techniques*

The choice of construction technology promotes the smooth construction of water supply and drainage projects, as well as ensures the whole quality and effectiveness of the project. In the process of construction works and construction of water supply and drainage engineering, the choice of construction technology should be made in accordance with the actual construction needs to carry out specific, targeted, and purposeful selection, not only to ensure the practicality of construction technology but also to ensure the overall quality and effectiveness of construction and construction (Yuan 2018). For example, when connecting the interface of cast iron pipes to the mouth of the pipe, the workers must ensure that the pores around the interface remain vertical if the interface is not vertical, the workers should also correct it instantly to ensure compliance with the relevant technical requirements. In addition, they should strengthen the supervision and management of the construction process, and irregularities in the use of construction technology should be corrected in a timely manner to ensure that the drainage project leakage prevention treatment and preventive measures are practical and feasible.

6.3 *Improve professional competence of construction workers throughout the engineering process*

The professional ability of construction workers can guarantee feasibility and operability to a certain extent, and promote the whole quality and effect of water supply and drainage projects. Therefore, the overall management of the construction personnel's professional competence should be strengthened throughout the project (Zou 2019). First, the training and learning management of the professional skills of construction personnel should be strengthened, and regular training and learning should be conducted.

Construction workers can improve their professional skills and knowledge through training and learning professional skills,

It helps them to improve the overall quality and comprehensive ability and plays a positive role in promoting the feasibility and operability of measures to ensure antiseepage treatment and prevention work. Second, to improve the professional level of the water supply and drainage project, employing high-quality employees with professional capabilities could be an efficient step. The feasibility and operability of drainage leakage prevention and treatment require professional construction workers to take necessary construction operations so that quality is fully guaranteed. Therefore, the improvement of the professional ability of construction workers is of great practical significance to the effectiveness of drainage leakage prevention and treatment, and construction units should give full attention and training to the construction workers to facilitate the full development of the role and value of the professional competence of construction workers and its manifestation.

7 CONCLUSION

Summarizing the above data, to effectively meet the actual needs of social and economic development and people's lives, the corresponding preventive measures should be done at all kinds of infrastructure construction stages. By using these high-quality preventive and effective measures, on one hand, it is possible to improve the quality of infrastructure construction. On the other hand they are also effective measures for avoiding safety accidents and reducing the potential risk of mishandled infrastructures on people and surroundings. Pertaining to leakage prevention solutions and preventive measures for water supply and drainage projects, the management should strengthen the engineering materials, construction technology, and construction personnel management. It provides support and guarantees for the stability, safety, and reliability of the project, and also promotes the normal and smooth operation of the water supply and drainage project.

REFERENCES

Chaoyang Shen. A Rough narration about leakage prevention solution and preventive measures in water supply and drainage engineering [J]. *Urban Construction Theory Research (electronic version)*, 2019(09):77.

Dongping Wang. Anti-leakage solution and preventive measures in water supply and drainage engineering [J]. *Engineering Construction and Design*, 2017(09):113–114+118.

Dongshun Zhang. Research on leakage prevention solution and preventive measures in water supply and drainage engineering [J]. *China Plant Engineering*, 2021(18):219–220.

Feng Yuan. Preliminary investigation of anti-seepage solutions and preventive measures in water supply and drainage engineering [J]. *Building Materials and Decoration*, 2018(49):50–51.

Yaohai Zou. Leakage prevention solution and preventive measures in drainage engineering [J]. *Jushe*, 2019(07):24

Advances in Civil Engineering and Environmental Engineering – bin Yahya & Bin Daud (Eds)
© 2023 The Author(s), ISBN: 978-1-032-39397-1

Study on reinforcement measures and the effect of supporting for foundation pit with expansive soil and disintegration rock geological

Mingjie Chen* & Pingjie Li*
CCCC Fourth Harbor Engineering Institute Co., Ltd., Guangzhou, Guangdong, China
CCCC Key Lab of Environmental Protection & Safety in Foundation Engineering of Transportation, Guangzhou, Guangdong, China

ABSTRACT: In view of the expansive soil and disintegration rock geology in the Yangtze-Huaihe region, combined with the rainy weather conditions in local from April to August, the narrow and long deep foundation pit of ship lock is treated with multilevel slope caving, shotting concrete for temporary slope protection, laying steel mesh, pouring concrete secondary lining. And it used prefabricated ecological slope protection or turf slope protection at the top. The long-term monitoring data show that the cumulative value and change rate of vertical displacement and horizontal displacement are far less than the warning value. The conclusion is that the protection and reinforcement steps adopted for the slim deep foundation pit of ship lock can prevent the expansive soil and disintegration rock from being affected by rainwater for a long time which ensures the stability of slope in the foundation pit. The application effect is excellent.

1 INTRODUCTION

The deformation law of narrow and long deep foundation pit is complex, dangerous, and highly concerning. At present, some studies on narrow and long deep foundation pit involve the deformation characteristics analysis (Chen 2015) and deformation control (Tang 2016) of expansive soil in deep foundation pit engineering. The research on disintegration rock mainly focuses on the failure mechanism, disintegration characteristics (Kan 2017; Zi 2017) and related reinforcement measures (Liang 2016; Liu 2015) of disintegration rock. But there is no research on the reinforcement and protection measures and supporting effects in narrow and long deep foundation pit with the geological conditions of expansive soil and disintegration rock.

This paper is based on the foundation pit project of a ship lock in the major livelihood project in which water will be transported from Yangtze River to Huaihe River in future. This paper further verifies the effectiveness of waterproof and protective reinforcement measures for foundation pit slope under the geological conditions of expansive soil and disintegration rock and the climate of Jianghuai River Basin by monitoring the relevant indicators of temporary foundation pit for a long time. The deformation law of narrow and long deep foundation pit is unique, and the geological conditions of expansive soil and disintegrating rock in Jianghuai area further increase the risk of engineering construction. In addition, the climate of Huaihe River Basin is rainy and river system is developed. It is of great significance to study the reinforcement and protection measures and reinforcement effect of foundation pit slope under specific conditions.

2 PROJECT PROFILE

The range of mileage for the lock project in the Jianghuai communication section of the Yangtze-Huaihe River Diversion Project (Anhui Section) is J29 + 530 ~ J32 + 100. The main structure of the

*Corresponding Authors: 365836145@qq.com and 278393325@qq.com

DOI 10.1201/9781003349563-66

lock (including the navigation section) is J30 + 507.9 ~ J31 + 298.9, the upstream approach channel is J31 + 298.9 ~ J32 + 100, and the downstream approach channel is J29 + 530 ~ J30 + 507.9. The total length of the section is 2.57 km, and the maximum excavation depth of the foundation pit is 28.4 m, which belongs to the narrow and long deep foundation pit. A monitoring section is arranged every 100 m along the direction of large mileage, and each section is divided into left and right amplitudes. Three displacement measuring points are arranged at the upper, middle, and lower positions of the slope, respectively. The layout of monitoring points is shown in Figure 1.

The Shushan junction is divided into two parts by the section J30 + 000. Mileage ranged from J29 + 530 to J30 + 000 is weak expansive soil and mileage ranged from J30 + 000 to J32 + 100 is medium expansive soil. The typical geological strata distribution is shown in Table 1.

In this project, the silty loam is expansive soil, known as "engineering cancer." Based on the test, it has weak or medium expansion. Once the expansive soil softens and expands upon encountering water, the corresponding shear strength decreases sharply. The shrinkage deformation is serious after dehydration, which will cause soil cracking. Expansive soil under natural conditions, due to the alternation of wetting and drying, further leads to deep soil exposure and accelerates soil damage, which is very unfavorable to the stability of the slope and foundation pit. The harm of expansive soil is long-term, repetitive, and latent. The damage of expansive soil to foundation pit often starts from small rainwater channels and develops into a regional slope collapse or slope instability by cyclic wetting-drying alternation. When such soil is encountered in engineering, it is usually necessary to carry out water isolation and reinforcement protection. Otherwise, even the foundation pit slope with small slope angle is extremely prone to landslide. The commonly used reinforcement measures include soil layer replacement, admixture modification, and water isolation.

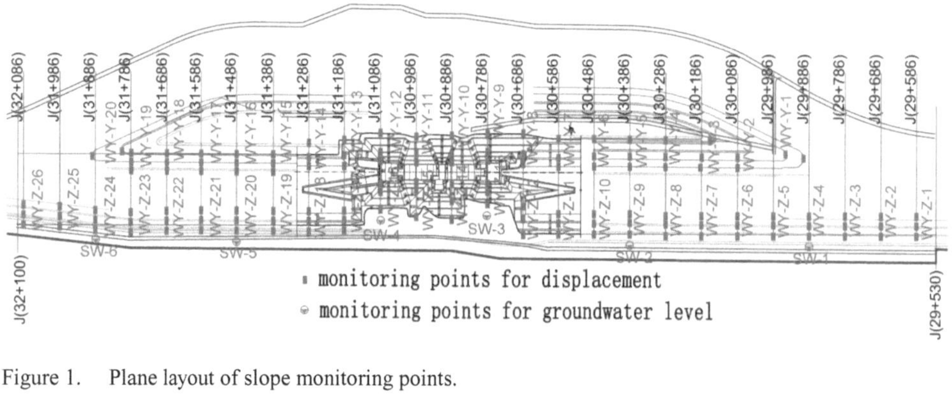

Figure 1. Plane layout of slope monitoring points.

The bedrock in this project is mainly fine sandstone. The local section has weak expansion. This kind of bedrock is soft and it will collapse encountering water. But dry cracking occurs after a water loss. When the water content of rock changes, the rock strength changes sharply. When the rock stratum is located at the bottom of the foundation pit, the settlement of the permanent structure is large without water insulation protection measures, the interaction from rainwater is prone to lead to uneven settlement. When the rock stratum is located on the foundation pit slope, due to the invasion of water, the shear strength of the rock decreases sharply which leads to the landslide risk.

The annual rainfall of the project site is large, and the distribution is uneven every quarter. The rainfall is mainly concentrated in the flood Month which is from April to August of each year. The annual average precipitation of the region is 969.5 mm belonging to the humid area. According to historical data statistics, the annual rainfall varies greatly, for example, the annual maximum rainfall is 1541.9 mm; minimum annual precipitation is 692.4 mm. The annual average precipitation is large and concentrated from April to August which will bring great challenges to the stability of the foundation pit. On the one hand, it is the continuous effect of physical forces such as erosion and infiltration of rainwater on foundation pit slope. On the other hand, the effective stress and

Table 1. Distribution of typical section strata.

Name of soil layer	Layer thickness(m)	Soil
Artificial fill (Q_4^{ml})	0.50~0.60	Medium and heavy silty loam, gray yellow, brown yellow, plastic, slightly wet
① Medium and heavy silty loam (Q_4^{al})	0.40~9.20	Brown yellow, gray yellow, soft ~ plastic, slightly wet ~ wet
② Silty heavy silty loam (Q_4^{al})	0.60~4.10	Gray, Soft Plastic, Saturated, Soft Plastic ~ Plastic Heavy Powdery loam
⑤ Heavy silty loam and silty clay (Q_3^{al})	0.50~11.50	Brown yellow, gray yellow, brown yellow clip gray white, yellow, hard plastic ~ hard plastic, local plastic, wet, widely distributed
⑤$_1$ Light and medium silty loam, sandy loam (Q_3^{al})	0.10~0.50	Brown yellow, brown yellow, gray white, slightly dense ~ medium dense/plastic ~ hard plastic, wet ~ very wet, local fine sand
⑤$_2$ Medium-fine sand (Q_3^{al})	1.00~4.70	Brown yellow, gray, brown red, medium dense ~ dense, saturated, locally sandwiched light, medium silty loam, and fine gravel
⑤$_5$ Fine sand in medium and heavy silty loam (Q_3^{al})	0.05~0.50	Brown yellow, gray yellow, brown yellow, plastic/ medium dense, wet, fine sand
⑨$_1$ Fully – strongly weathered powder and fine sandstone (K_2qz)	0.00~9.10	Fully weathered sandy clay and sandy soil, strongly weathered in short columns
⑨$_3$ Medium Weathering ~ Fresh Powder and Fine Sandstone (K_2qz)	0.00~24.90	Columnar and long columnar, locally intercalated with mudstone

shear strength of slope soil may be reduced due to the infiltration of rainwater. The internal and external factors accelerate the landslide failure of foundation pit slope.

3 ANALYSIS OF RESPONSE MEASURES

To cope with the expansive soil and disintegration rock foundation pit with confined water under the narrow and long deep foundation pit, the multilevel slope construction measure for the foundation pit is adopted. The slope ratio above 7.2 m elevation is 1: 3 and the slope ratio from – 8.8 m elevation to 7.2 m elevation is 1: 1.5. The slope height of each layer is 6 m and the slope platform which is 2 m wide is set between layers. Multistage slope caving, on the one hand, can reduce the economic cost of the slope caving construction, on the other hand multilevel slope caving is convenient for the setting of the drainage ditch at different positions forming the three-dimensional drainage network. In addition, compared to the single-stage slope with the same slope rate, the multistage slope can unload gravity of soil as much as possible and enhance the slope's self-stability.

Drainage ditches are set on the inside of the platform and the drainage ditches at all levels of the platform are connected vertically. A catchment is set at the bottom of the foundation pit and water in the catchment is pumped from time to time. The damage to expansive soil and disintegration of rock from rainwater is obvious, remarkable, and long-term. It is necessary to organize drainage facilities in advance to ensure draining rainwater timely and reduce water infiltrating soil.

When the rock and soil body was exposed after cutting concrete C20 with a thickness of 0.05 m was sprayed for temporary protection. At the same time, Φ80PVC drainage pipe was inserted at the seepage point on the slope and geotextile was wrapped inside the pipe for filtration. Then steel mesh with Φ10 steel were lain down and concrete C20 with 0.15 m thick was casted for slope protection. The surface soil on the top of the slope was replaced with modified soil which was fully mixed with

4 % cement and the thickness of the modified soil was 1.5 m to prevent surface rainfall to infiltrate the expansive soil layer. The process would cause damage to the soil and affect the slope stability.

Anchors were set up by drilling in the middle layer slope, the strongly weathered layer, and the moderately weathered layer of the underlying bedrock. GFRP bars with a diameter of Φ22 anchors had an anchorage depth of 5 m and a spacing of about 2.5 m. Prefabricated ecological slope protection or turf slope protection was adopted on the top of slope and they were 0.3 m thick.

4 ANALYSIS OF MONITORING RESULTS

The daily monitoring of the foundation pit is carried out according to the layout of monitoring points shown in Figure 1. The numbering rule of measuring points is type – direction – row number – vertical slope position. For example, WY-Y-4 – S represents the displacement measuring point above the slope in the fourth row on the right amplitude position.

4.1 *Analysis of settlement results*

To analyze easily, the vertical downward displacement of the measuring point is "+", and the opposite is "–". The horizontal displacement and vertical displacement share one measuring point and a total of 135 measuring points were set up. Due to a large number of measuring points, the top 10 measuring points with the largest cumulative value of horizontal displacement and vertical displacement are selected, and the two are combined to form 14 typical measuring points and analyzed. It can be seen from Figure 2 and Table 2 that in the early stage of foundation pit excavation, due to the release of stress, the settlement rate of the measuring point is large. During the construction of permanent structure of ship lock, the settlement rate of the measuring point tends to converge. The monitoring period of the vertical displacement of the whole foundation pit slope lasted for more than three years, and the maximum cumulative vertical displacement was 35.18 mm, which was 41.5% of the warning value of the cumulative displacement. The corresponding measuring point was WY-Y-4 – S and the maximum change rate was 1.31 mm/d, far less than the warning value of the design rate of 5 mm/d.

4.2 *Analysis of horizontal displacement results*

Since there are many measuring points, the 14 typical measuring points in the analysis of vertical displacement are also selected for analysis. It can be seen from Figure 3 and Table 2 that in the early stage of foundation pit excavation, the horizontal stress begins to release. The excavation has a great influence on the horizontal displacement. The horizontal displacement rate of the measuring point

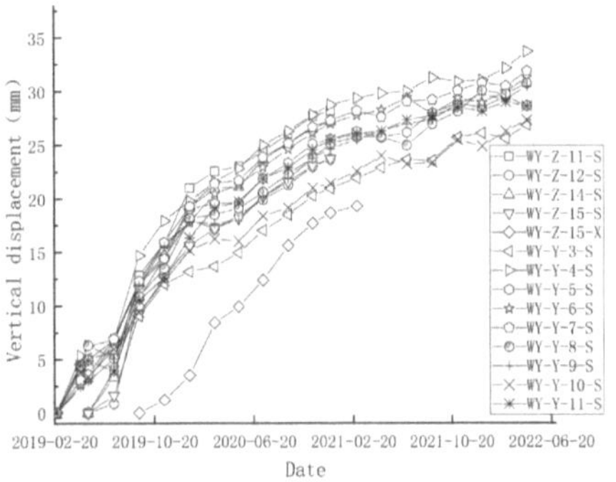

Figure 2. The whole process curve of vertical displacement of typical measuring points.

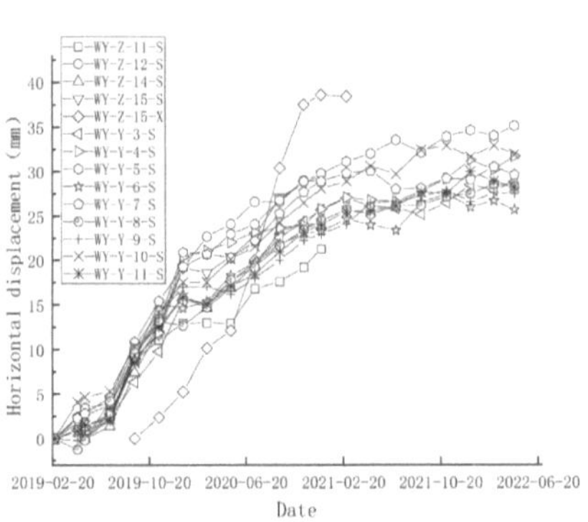

Figure 3. The whole process curve of horizontal displacement of typical measuring points.

is large, and the curve is steep. During the construction of permanent structure of ship lock, the displacement rate of the measuring point tends to converge, which is completely consistent with the changing trend of vertical displacement. The whole monitoring period of horizontal displacement of foundation pit slope lasted more than 3 years. After long-term construction, the maximum cumulative horizontal displacement reached 40.8 mm, which was only 40.8 % of the cumulative displacement warning value. The corresponding measuring point was WY-Z-15-X and its maximum change rate was 2.5 mm / d, which was far less than the design rate warning value of 10 mm / d. During monitoring no slope landslide, ground collapse, slope and surrounding ground cracking have occurred.

Table 2. Summary analysis of typical displacement measuring points.

Measuring point	Maximum cumulative vertical displacement (mm)	Maximum cumulative horizontal displacement (mm)	Cumulative displacement warning value (mm)	Maximum vertical displacement rate (mm/d)	Maximum horizontal displacement rate (mm/d)	Warning value of vertical displacement rate (mm/d)	Warning value of horizontal displacement rate (mm/d)
WY-Z-11-S	28.56	22.1	50.3	−0.92	1.2	5	10
WY-Z-12-S	23.89	29.6	57.5	−0.94	−1.0	5	10
WY-Z-14-S	27.41	25.2	124.6	0.96	1.3	5	10
WY-Z-15-S	23.88	30.2	100.0	−1.18	1.0	5	10
WY-Z-15-X	21.76	40.8	100.0	0.83	2.5	5	10
WY-Y-3-S	27.30	29.4	84.3	0.68	1.2	5	10
WY-Y-4-S	35.18	32.7	84.8	0.88	−1.1	5	10
WY-Y-5-S	29.93	35.7	85.3	1.27	1.2	5	10
WY-Y-6-S	31.73	28.2	86.8	0.94	0.9	5	10
WY-Y-7-S	32.66	31.6	86.6	1.31	1.8	5	10
WY-Y-8-S	31.16	29.5	84.4	1.22	−1.3	5	10
WY-Y-9-S	31.02	29.4	112.2	0.97	−1.6	5	10
WY-Y-10-S	27.80	33.4	109.9	−0.88	0.9	5	10
WY-Y-11-S	29.97	30.0	113.7	1.12	1.3	5	10

4.3 *Prediction of settlement results*

To further illustrate the reinforcement effect, the hyperbolic model is used to predict the settlement data from surviving measuring point and the final settlement is obtained. The model is as follows:

$$S_t = S_0 + \frac{t - t_0}{a + b(S_t - S_0)} \tag{1}$$

$$S_\infty = S_0 + \frac{1}{b} \tag{2}$$

$$\frac{Vt}{VS} = a + b \cdot Vt \tag{3}$$

In the formula, S_0 is the settlement under full load corresponding to the selected starting time t_0, t is the time , a and b are constants to be determined, S_t is the settlement under full load corresponding to time t, and S_∞ is the final settlement.

According to the prediction results shown in Table 3, it can be seen from the table that the model can well fit the settlement curve of the surviving measuring points. The fitting coefficient is between 0.842 and 0.982. The maximum final settlement is 64.64 mm and the corresponding measuring point is WY-Y-9-S, which does not exceed the warning value. The prediction results show that the final settlement will not exceed the warning value even if the monitoring continues. The reinforcement measures adopted can effectively prevent the settlement deformation of the foundation pit.

Table 3. Predicted settlement results of typical surviving measuring points.

Measuring point	*a*	*b*	Fitting coefficient	Current settlement (mm)	Ultimate settlement (mm)	Warning value (mm)
WY-Y-3-S	16.434	0.024	0.969	26.85	42.32	84.3
WY-Y-4-S	10.545	0.021	0.971	33.69	48.33	84.8
WY-Y-5-S	12.371	0.023	0.982	28.64	44.40	85.3
WY-Y-6-S	15.502	0.017	0.889	31.28	58.90	86.8
WY-Y-7-S	14.051	0.018	0.936	31.86	55.72	86.6
WY-Y-8-S	14.047	0.021	0.938	30.75	48.27	84.4
WY-Y-9-S	18.235	0.015	0.855	30.45	64.64	112.2
WY-Y-10-S	16.464	0.023	0.977	27.33	43.78	109.9
WY-Y-11-S	15.739	0.019	0.842	28.61	53.19	113.7

5 SUMMARY AND RECOMMENDATIONS

During the monitoring, there is no slope landslide, ground collapse, slope, and surrounding ground cracking happening in the foundation pit. At the same time, the cumulative value and rate of horizontal displacement and vertical displacement of the foundation pit are far less than the cumulative warning value and rate of warning value. It shows that in view of the geological distribution, climate, and hydrological conditions of the foundation pit of the ship lock in the Jianghuai communication section, relevant protective reinforcement, drainage, and water sealing measures have remarkable effects on improving the stability of the foundation pit slope.

Whether horizontal displacement or vertical displacement, the maximum cumulative displacement values appear in the middle expansive soil section. It is indicated that although the foundation pit has taken reinforcement and waterproof measures, the influence of middle expansive soil on the vertical and horizontal displacement of foundation pit slope is still greater than that of weak expansive soil.

The settlement prediction results show that the hyperbolic model can well fit the settlement curve of the surviving measuring points and the final settlement will not exceed the warning value even if the monitoring continues. The reinforcement measures adopted can effectively prevent the settlement deformation of the foundation pit.

While some measures have proved effective, it is more important to ensure the construction quality of the relevant measures.

ACKNOWLEDGMENTS

This study is sponsored by Guangzhou Pearl River Science and Technology Star Special Fund (Grant No. 201806010162), and South China Marine Science and Engineering Guangdong Laboratory, and (Zhuhai) Estuary coast and reef engineering innovation team construction project (Grant No. 311020009).

REFERENCES

Chen Jian. (2015). Stress and deformation analysis of narrow and long deep foundation pit retaining structure in expansive soil [J] . *Journal of Underground Space and Engineering.* 11 (2): 440–445.

Kan Yanju. (2017). Protection scheme of collapse rock in Jianghuai communication section of jianghuai diversion project [J]. *Jianghuai Water Conservancy Science and Technology.* (1): 17–18.

Liang Houbing. (2016). Analysis and treatment of main geological impacts of Yangtze-Huaihe river diversion project [J]. *Jianghuai Water Conservancy Science and Technology*. (4): 17–19.
Liu Xingyuan, Yong Lei, Jingwen Kang. (2015). *Slope Engineering-design. Monitoring. Identification and Reinforcement* [M], Beijing, China building industry press.
Tang Guangxian. (2016). Deformation control of deep foundation pit construction in expansive soil area [J]. *Foundation Foundation*. 38 (12): 1647–1649.
Zi Hongli. (2017). Experimental study on expansive soil (disintegration rock) of Yangtze-Huaihe river diversion project [J]. *Jianghuai Water Conservancy Science and Technology*. (5): 40–41.

Advances in Civil Engineering and Environmental Engineering – bin Yahya & Bin Daud (Eds)
 ISBN: 978-1-032-39397-1

Numerical simulation study of dam break flood in dayang reservoir

Lin-feng Luo*
Zhejiang Water Conservancy Science and Technology Promotion Service Center, Hangzhou, China

Jian-wu Zhu*
Suichang County Water Conservancy Bureau, Suichang, China

Wei-jun Pan*
Zhejiang Water Conservancy Science and Technology Promotion Service Center, Hangzhou, China

ABSTRACT: Most of the small reservoirs in Zhejiang Province were built before the 1970s. They have reached or exceeded their service life and entered a period of high risk of disease. The risk of dam failure always exists. Combined with the proposal of the water conservancy department, forecasting, rehearsal, early warning, and pre-planning should be taken as the key to the safety of reservoirs during flood season. This paper uses numerical methods and two-dimensional hydrodynamic models to carry out simulation research on the dam failure process of Dayang reservoirs, simulate the development process of dam failure of small reservoirs, and draw a map of the impact range of dam failure. This paper provides technical support for formulating highly targeted safety emergency plans for small reservoir dam failures.

1 INTRODUCTION

Most of the small reservoirs in my country were built before the 1970s, and have reached or exceeded the service life (Li et al. 2020), and entered a period of high risk of disease. Taking Zhejiang Province as an example, there are 4277 reservoirs in Zhejiang, including 4084 small reservoirs. Although the safety of the reservoir has been significantly improved after years of risk elimination and reinforcement, the establishment of standardized management, and the strengthening of reservoir operation, management and protection, there are still 144 third-class dams, 511 second-class dams, and many small reservoir facilities in the province. Aging, lack of necessary safety monitoring facilities, there are weak links in the operation and management of the reservoir. These problems not only threaten the safety of life and property of the people, but also affect the coordination, stability and healthy development of the economy and society (Li 2018). To this end, Zhejiang Province proposes that forecasting, rehearsal, early warning, and pre-planning should be regarded as the key points of reservoir safety during flood season, improve reservoir monitoring and early warning facilities, strengthen rainwater forecast and forecast, release early warning information in time, carry out reservoir flood control scheduling, and simulate rehearsal. Continuously optimize and improve emergency management plans. In this paper, by means of digital means and two-dimensional hydrodynamic model, the dam failure process is generalized as the development process of weir hydraulic structures and breach (Tian et al. 2016), and the dam failure model based on the energy equation method is used to carry out the simulation calculation of the dam failure process of Dayang reservoirs. Simulate the development process of dam failure of small reservoirs, draw a map of the impact scope of dam failure (Chen et al. 2011), and strive for more emergency

*Corresponding Authors: 425700459@qq.com, 179399941@qq.com and 425700458@qq.com

 DOI 10.1201/9781003349563-67

response time when the reservoir is in danger, to minimize the harm caused by dam failure of small reservoirs.

2 OVERVIEW OF DAYANG RESERVOIR

The dam site of Dayang Reservoir is located in Xiangli Village, Keyan Street, Keqiao District, Shaoxing City. 1.6 kilometers away from the dam site is Xiangli Village and Keyan Scenic Area. The downstream villages are densely populated with many residents. The Dayang Reservoir was built in 1957 and rebuilt in 2003. The rainwater collection area is 2.50 km^2, the total storage capacity is 120,000 m^3, the dam height is 12 meters, the dam length is about 68 m, the dam crest elevation is 57.50 m, and the dam crest width is 4.40m. For earth dams, the slope ratio of the water-facing slope of the dam is 1:2, and the slope ratio of the back-water slope is 1:2. The discharge structure (spillway) is located on the left bank of the dam, with a width of about 6.0 m and an elevation of 56.60 m at the entrance. The diameter of the discharge culvert is 0.3 m. The surrounding mountains of the reservoir are continuous and gentle, the basin is surrounded by mountains on three sides, and the vegetation of the mountains is good. The slopes of the slopes on both sides are generally 30°C to 50°C, and the river valleys are mostly “U” shaped.

3 NUMERICAL MODEL BUILDING

3.1 *Overview of model scope and study area*

The selection of the scope of the two-dimensional hydrodynamic model (Long et al. 2008) of the Dayang Reservoir considers the reservoir itself and the submerged area downstream. After repeated trial and evaluation, the final model scope is about 2.4 km long from north to south, 3 km wide from east to west, and covers an area of 3.58 km^2. The collected terrain data includes high-precision contour lines and elevation points. In the project, according to the water level-storage-area curve, a funnel-shaped terrain is constructed as the underwater terrain of the reservoir.

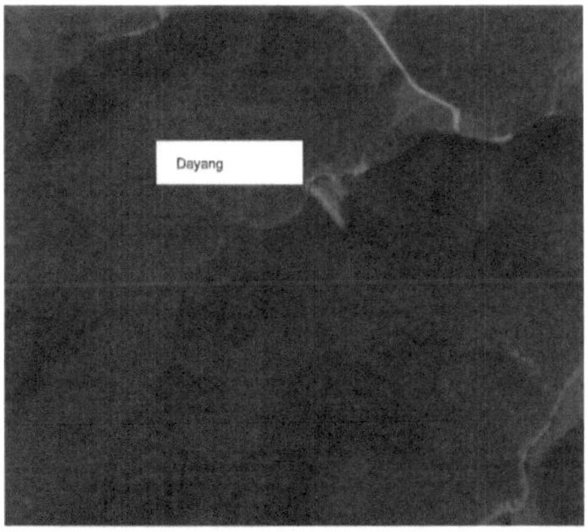

Figure 1. Geographical map of Ocean Reservoir.

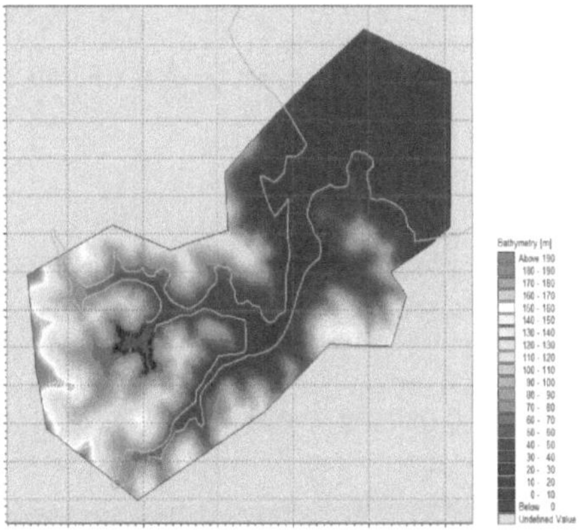

Figure 2. Dayang Reservoir Model Terrain Range.

3.2 *Water depth and computational grid*

When calculating the mathematical model, the water depth is determined based on the underwater topographic data and related report data. The computational mesh is composed of triangular

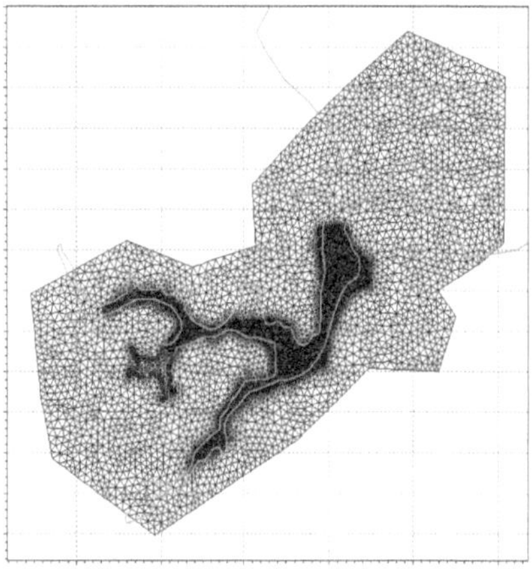

Figure 3. Model calculation area and calculation grid under current conditions.

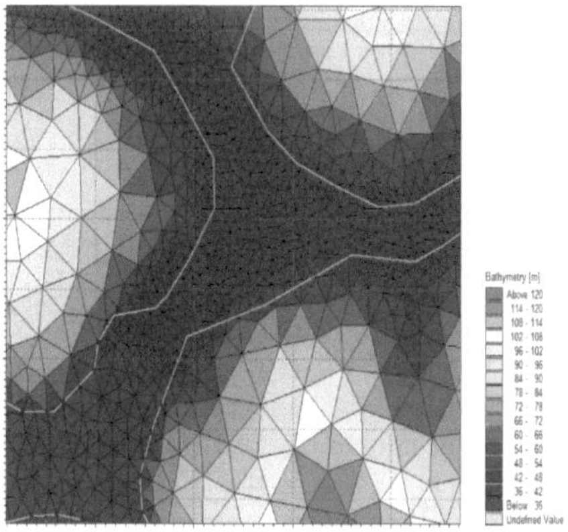

Figure 4. Terrain and its calculation grids near the breach of Dayang Reservoir.

elements, which can better fit the shoreline. According to the needs, the side length of the grid unit is 10~15m, and the grid of the breach and the residential submerged area is densified, and the side length of the grid unit in the densified area is 5~10m. The number of grid nodes of the mathematical model established in the area is 8858, and the number of generated grid cells is 17452. The calculation grid and underwater terrain are shown in Figure 3 and Figure 4.

3.3 *Determine boundary conditions*

The upstream boundary conditions of the two-dimensional hydrodynamic model of the Dayang Reservoir are set in the reservoir area. For the flood dam scheme, the design flood hydrograph that occurs once in 100 years is taken as the inflow; the downstream boundary is set at the downstream channel in the northeast corner of the model, and the setting is constant. The water level is 5m, and the initial water level is set by checking the water level.

3.4 *Calculated parameters*

According to the basic principle of MIKE 21 hydrodynamic model, the following main parameters need to be set in the model.

Roughness coefficient: The selection of the Manning value in this model refers to the general values of river dry flats and commonly used rivers, which are 0.033~0.05 (s/m^1/3) for flatlands and 0.02-0.03 (s/m^1/3) for rivers.

Eddy viscosity coefficient: estimated by Smagorinsky formula, the corresponding Smagorinsky coefficient is $0.28m^2/s$.

Time step: Dynamically adjust the model calculation time step according to the model grid size and water depth conditions, so that the CFL number is less than 0.8, which meets the requirements of model stability, and the calculation time step is between 0.01s and 1s.

Dry and wet boundary: The grid freezing method is used to process the dry and wet process of the beach in the calculation area. When the water depth of a certain point is less than 0.005m, the grid point is set as a dry point, and the beach is dry out, and does not participate in the hydrodynamic calculation; The grid point only participates in the calculation of the water flow continuity equation. When the water depth is greater than 0.1m, the grid point participates in the calculation of the continuity equation and the momentum equation.

4 CALCULATION AND ANALYSIS RESULTS OF DAM BURST INUNDATION

4.1 *Calculation scheme*

The calculation scheme of the dam failure model of the Dayang Reservoir includes: full dam collapse, half collapse of the dam body, and quarter collapse of the dam body.

4.2 *Analysis of calculation result of dam body's total collapse*

According to the simulation results, after the Dayang Reservoir completely collapsed, a dam-breaking flow of 3 to 5 m/s was formed near the breach, and the dam-breaking flood evolved along the valley. Within 2 minutes after the dam burst, the flood water first flowed to the northeast, and then turned to the due east. The water level was between 25 and 42m, and it gradually decreased from upstream to downstream. The dam-breaking flood reaches the downstream residential gathering area after 8 minutes, and enters the downstream river channel after 20 minutes; the maximum value of the submerged water depth during the flood propagation process is calculated, and the maximum submerged water depth map is obtained, as shown in Figure 7 and Figure 8 (Enlarged map of downstream residential area). It can be seen from the figure that the maximum submerged water depth near the breach is between 2 and 3 m; while in the downstream residential area, due to the influence of the topography of the bell mouth, the overflow section is large, and the maximum submerged water depth is between 0.5 and 1 m.

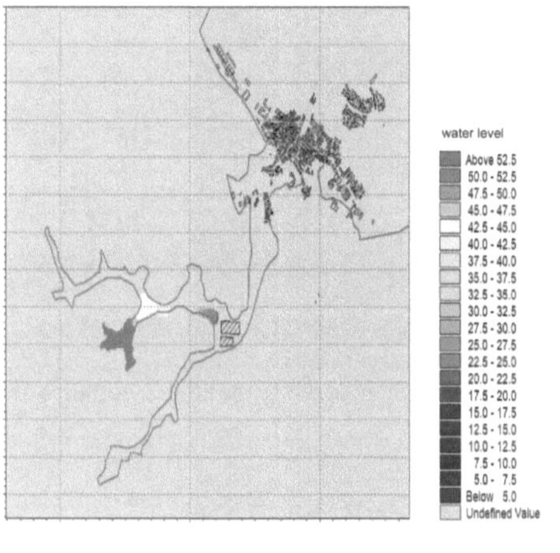

Figure 5. Submerged range after 2 minutes of full collapse.

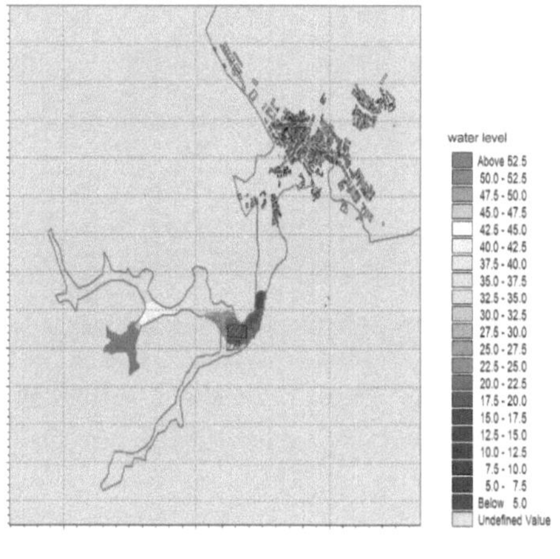

Figure 6. Submerged range after 4 minutes of full collapse.

4.3 *Analysis of calculation result of dam body's half collapse*

According to the simulation results, after the half-collapse of the Dayang Reservoir, a dam-breaking flow of 2-5 m/s is formed near the burst, and the dam-breaking flood evolves along the valley. Within 2 minutes after the dam burst, the dam burst flood first flowed to the northeast, and then turned to the due east. The dam-breaking flood reached the downstream residential area 10 minutes later, and entered the downstream river channel 22 minutes later. The maximum value of the submerged water depth during the flood propagation process is calculated, and the maximum submerged water depth map is obtained, as shown in Figure 11 (global inundation map) and Figure 12 (enlarged map of downstream residential areas). It can be seen from the figure that the maximum submerged water depth near the breach is between 2 and 3 m; while in the downstream residential area, due to the influence of the topography of the bell mouth, the overflow section is large, and the maximum submerged water depth is between 0.5 and 1 m.

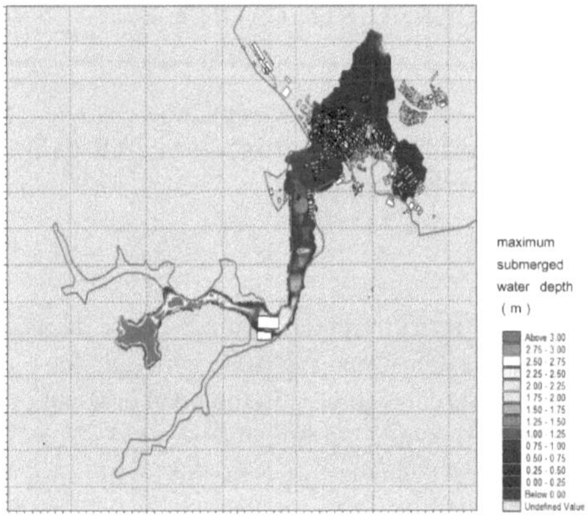

Figure 7. Full collapse maximum submerged water depth map.

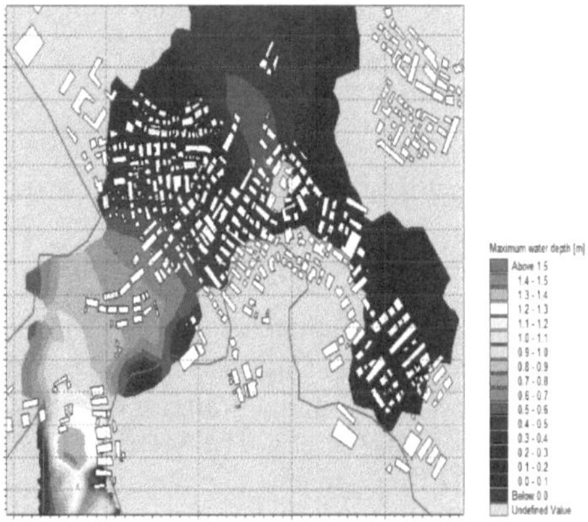

Figure 8. Full collapse maximum submerged water depth map.

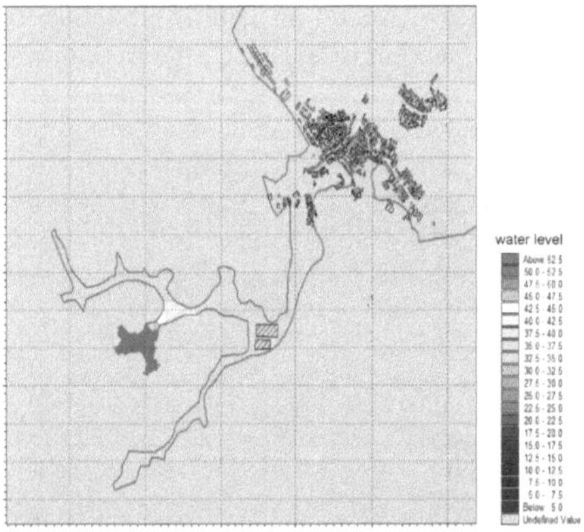

Figure 9. Submerged range after 2 minutes of half-collapse.

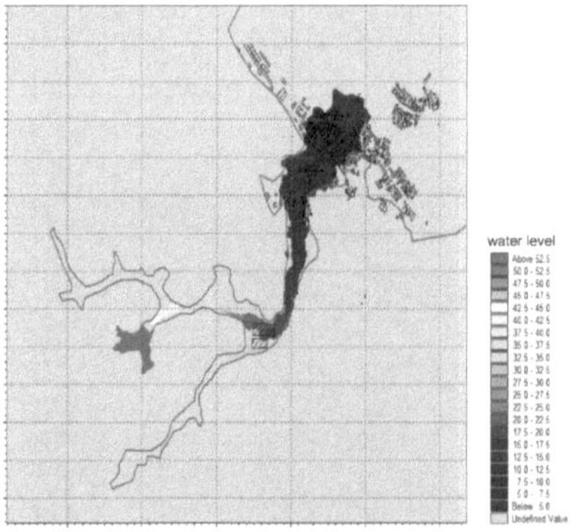

Figure 10. Submerged range after 22 minutes of half-collapse.

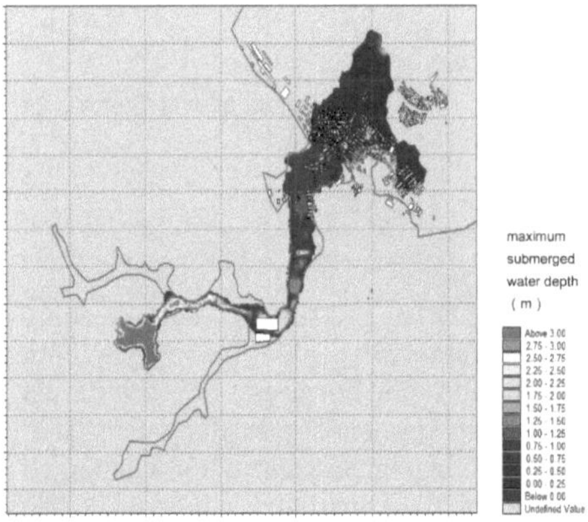

Figure 11. Half-collapse maximum submerged water depth map.

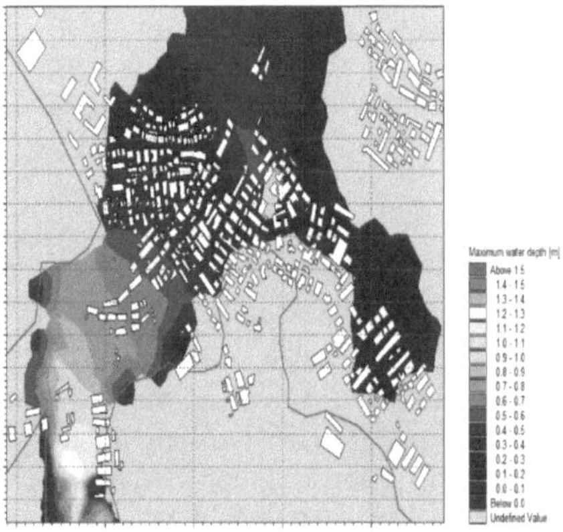

Figure 12. Half-collapse maximum submerged water depth map.

4.4 *Analysis of calculation result of the dam body's quarter collapse*

According to the simulation results, after 1/4 of the dam breach in Dayang Reservoir, a dam-breaking flow of 2-3 m/s is formed near the breach, and the dam-breaking flood evolves along the river valley. Within two minutes after the dam burst, the dam burst flood first flowed to the northeast, and then turned due eastward. The dam-breaking flood reached the downstream residential area 10 minutes later, and entered the downstream river channel 22 minutes later. The maximum value of the submerged water depth during the flood propagation process is calculated to obtain the maximum submerged water depth map, as shown in Figure 15 (global inundation map) and Figure 16 (enlarged map of downstream residential areas). It can be seen from the figure that the maximum submerged water depth near the breach is between 2 and 3 m; while in the downstream residential area, due to the influence of the topography of the bell mouth, the overflow section is large, and the maximum submerged water depth is between 0.5 and 1 m.

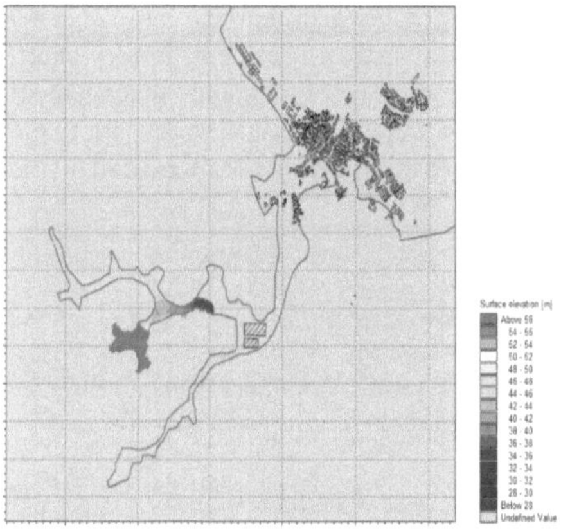

Figure 13. Submerged range after 2 minutes of quarter collapse.

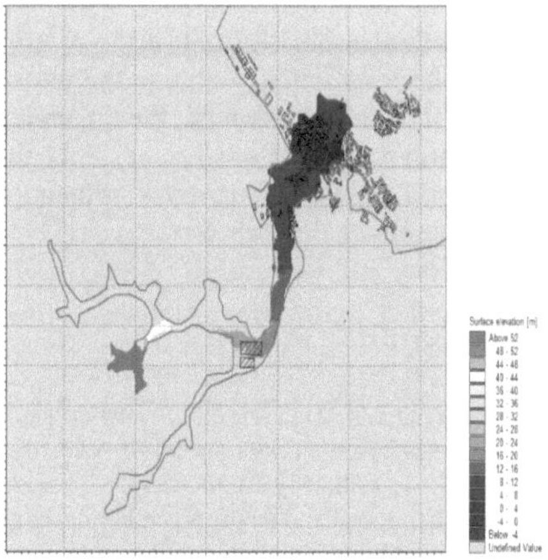

Figure 14. Submerged range after 22 minutes of quarter collapse.

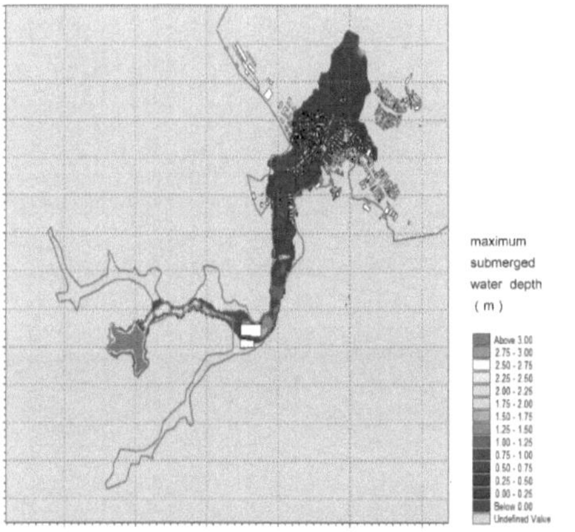

Figure 15. Quarter collapse maximum submerged water depth map.

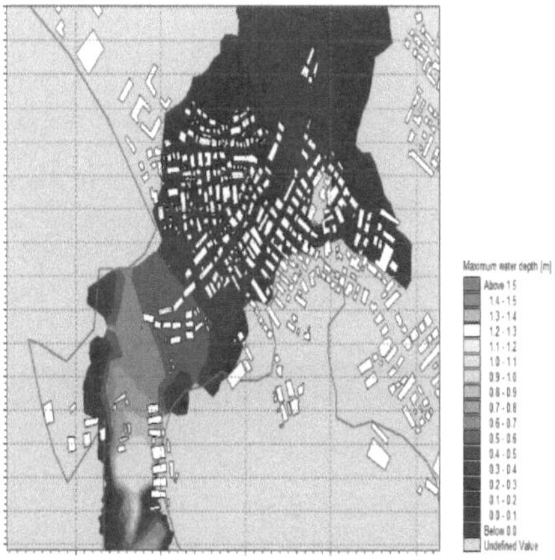

Figure 16. Quarter collapse maximum submerged water depth map.

4.5 *Summary*

According to the calculation results of the dam failure model of this Dayang reservoir, the dam body fully collapses, the dam body collapses in half, and the dam body collapses in a quarter, resulting in different dam breaking flow rates and submerged ranges. The maximum is 3-5 m/s. The dam-breaking flood reaches the downstream residential area after 8 minutes. Whether the dam body fully collapses, the dam body collapses in half, or the dam body collapses in a quarter, it will be left within 10 minutes for the downstream residential area, and the maximum submerged water depth has reached 1 m. It can bring huge losses to the safety of life and property of the downstream people.

5 CONCLUSION

Through the numerical simulation study of the dam failure and flood of the Dayang Reservoir, the huge and catastrophic impact of the reservoir dam failure on the downstream can be rehearsed. In the absence of timely warning, the retreat time left for the downstream residents is very short. The dam inundation risk map, the targeted development of the reservoir emergency management plan, inform the public in advance, and enhance the public's awareness of disaster prevention and mitigation. The scope of dam risk influence can be minimized to minimize the harm caused by the dam failure of small reservoirs.

REFERENCES

Jiang Li, Ying Liu, Ma Jun, etc. A new strategy for Xinjiang small reservoir earth-rock dams to deal with flood risks based on the concept of "overrun without collapse" [J]. *Water Conservancy and Hydropower Technology*, 2020, 51(12):78–85.

Juan-juan Li. Social and environmental impact assessment of small and medium-sized reservoirs dam failure [D]. *Zhengzhou University*, 2018.

Wenfei Long, Xinhua Zhang, Jinchi Huang. Research and application of flood prediction method for reservoir dam break [J]. *Journal of Sichuan University (Engineering Science Edition)*, 2008, 40(1): 21–26.

Yonggang Chen, Zhongmin Liang, Qingxuan Huang. Research on the preparation of flood risk map for reservoir dam failure [J]. *Hydropower*, 2011, 37(9):18–20.

Yu Tian, Yiming Zhang, Ji'an Li. Numerical simulation of reservoir dam failure flood and downstream risk assessment [J]. *Water Conservancy and Hydropower Technology*, 2016, 47(9): 130–133.

Advances in Civil Engineering and Environmental Engineering – bin Yahya & Bin Daud (Eds)
© 2023 The Author(s), ISBN: 978-1-032-39397-1

A study on the calculation method of building heat loads

Yajing Wang*, Zhimei Wen, Jiapu Yuan & Zhuangzhuang Qu
College of Energy and Construction Engineering, Shandong Huayu University of Technology, Dezhou City, Shandong Province, China

ABSTRACT: Heat load is the most fundamental basis for heating system design. To better calculate heat loads, different heat load calculation methods are studied here. With the central air conditioning industry college of Shandong Huayu Institute of technology as a study case, this paper adopts the unit area index method, the traditional heat load calculation method and the software simulation method to calculate the building heat load from the actual situation of the building. Finally, the different results are compared and analyzed. It is concluded that though the unit area index method is simple and convenient, its result is not as good as those of the other two methods; and the traditional heat load calculation method and software simulation method achieve similar performance. The research finding has practical significance for heat load calculation.

1 INTRODUCTION

At present, there are much research on heat load forecasting. Matteo caldera (Matteo, 2008) studied the influence of various parameters on energy consumption by using the sensitivity method through the relationship between building floor area ratio, window wall ratio, light transmittance, and heating energy consumption, and carried out simulation calculation by using a simplified model. The relative error is about 8%. Li (Li 2014) studied the indoor environmental heat load and the influencing factors of heating engineering and analyzed the variation characteristics of heat load. Zhang (Zhang 2017) deeply analyzed the influencing factors of heat load, which are mainly divided into three aspects: indoor conditions of buildings, outdoor meteorological factors, and variables. Based on determining the amount of heat supply, the heat load forecast is deeply analyzed to provide an effective reference for the actual heating. Zhao (2017) did some research on the existing heat load forecasting methods and predicted the heat load by using a genetic algorithm and BP neural network. The above research has greatly improved the prediction accuracy of heat load research, but the method is complex and lacks practicality, so it is rarely used in actual engineering design.

This paper studies the heat load calculation methods commonly used in the current engineering design, takes the central air conditioning industry college of Shandong Huayu Institute of technology as a typical case to calculate the building heat load and compares the results to draw a certain conclusion. This paper promotes the theoretical and practical research of heat load calculation, which is conducive to providing customers with high-quality heating services, achieving on-demand heating, heating energy saving, and creating a more suitable living environment.

2 METHOD

2.1 *The unit area index method*

The unit area index method is a static estimation method. This method is generally to estimate the load of every single building, and the calculation formula of the heat load is as follows:

$$Qj = qkF$$

*Corresponding Author: wyj@huayu.edu.cn

DOI 10.1201/9781003349563-68

where Q_j is the total load of the district building; F represents the area covered by different buildings; q_k represents the heat load index per unit building area

2.2 *Traditional heat load calculation method*

For the building, the heat load in winter can be divided into two parts: the heat consumption of the enclosure structure and the heat consumption of the cold air infiltrated into the room through the door and window. The calculation formulas are as follows

$$Qjh = AjhKjh\,(t_{rh} - twh)a$$

where Q_{jh} is the total load of the district building; A_{jh} represents the surface area of enclosure; K_{jh} represents heat transfer coefficient; t_{Rh} represents indoor calculated temperature; t_{wh} represents outdoor calculated temperature; a represents correction factor.

$$Qih = 0.278L\rho aohcph\,(t_{Rh} - t_{wh})$$

where Q_{ih} is heat consumption of cold air; ρ_{aoh} represents air density at outdoor calculated temperature; C_{ph} represents specific heat of air at constant pressure.

2.3 *Software simulation method*

The calculation software used in this paper is Tianzheng HVAC.

3 CASE CALCULATION

3.1 *Project introduction*

The building has a total length of 92.8 m and a width of 8.8 m from north to south. It is divided into two floors. The first floor has a clear height of 5.1 m and the second floor has a clear height of 3.9 m. There are 11 rooms on the first floor and 13 rooms on the second floor.

3.2 *Results*

(1) The calculation results using the unit area index method are shown in the following table.

Table 1. The results of the first floor.

Number	Heat Load (W)	Number	Heat Load (W)
1001	7464	2001	2800
1002*	2488	2002	2800
1006*	5038	2003	2800
1009	2388	2004	2800
1010	5026	2005	2800
1011	7389	2006	5600
1012	4995	2007	2800
1013	4976	2008	2800
1014	6608	2009	2800
1015	12800	2010	2800
1016	6608	2011	2100
		2012	2100
		2013	11358.9

The total heat load is 112138.9W.

(2) The calculation results using the traditional heat load calculation method are shown in the following table.

Table 2. The calculation results.

Number	Heat Load (W)	Number	Heat Load (W)
1001	4108.8	2001	2492.9
1002*	1369.7	2002	2492.9
1006*	2773.5	2003	2492.9
1009	1314.7	2004	2492.9
1010	2740.3	2005	2492.9
1011	4036	2006	4985.8
1012	2614.7	2007	2492.9
1013	2603.1	2008	2492.9
1014	6298.8	2009	2492.9
1015	14696.9	2010	2492.9
1016	6298.8	2011	1866.3
		2012	1866.3
		2013	20814.5

The total heat load is100824.3 W.

(3) The calculation results of the software simulation method are shown in the following table.

Table 3. The Calculation results.

Number	Heat Load (W)	Number	Heat Load (W)
1001	4108.8	2001	2175.3
1002*	1369.7	2002	2251.8
1006*	2773.5	2003	2251.8
1009	1314.7	2004	2251.8
1010	2740.3	2005	2251.8
1011	4036.3	2006	4503.6
1012	2751.3	2007	2175.3
1013	2739.4	2008	2175.3
1014	6510.7	2009	2175.3
1015	15241.7	2010	2175.3
1016	6510.7	2011	1686.6
		2012	1686.6
		2013	18814.1

The total heat load is 96671.7 W.

4 DISCUSSION

It can be seen from these tables that the results calculated by the unit area index method are quite different from other calculation methods: when the room area is small, the heat load calculated by the unit area index method is larger, while when the room area is large, the heat load calculated is smaller. The reason for this result is that in the unit area index method, the selection of unit area index is artificial, which is easy to cause the index too small or too large. The results obtained by using the traditional heat load calculation method and the software simulation method have little difference, indicating that the software simulation method has high accuracy, but the reason for the difference may be the difference in the selection of some coefficients and the enclosure structure.

The total load obtained by different methods is shown in Figure 1.

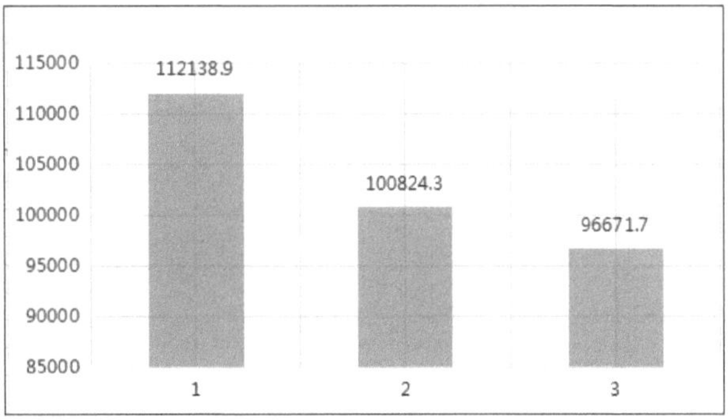

Figure 1. Comparison of total heat load.

It can be seen from Figure 1 that the total heat load of the building obtained by the unit area index method is the largest, about 112138.9 W; The result obtained by the traditional method is 100066.94 W; The result obtained by the software simulation method is 96672.05 W.

According to the calculation and analysis, the result of the unit area index method is 12.06% higher than that of the traditional heat load calculation method and 16% higher than that of the software simulation method; The traditional heat load calculation method is very close to the data obtained by the software simulation method, with a difference of only 3.51%.

ACKNOWLEDGEMENTS

This research is supported by Shandong Huayu University of Technology 2020 Annual Science and Technology Plan Project: Research achievement of "Experimental Study on Classroom Thermal Environment and Air Quality" (2020KJ22).

REFERENCES

Li ST. Research on heat load forecasting of district heating systems [D]. *Chang'an University*, 2014.

Matteo Caldera, Stefano Paolo Corgnati, Marco Filippi. Energy demand for space heating through a statistical approach:application to residential buildings [J]. *Energy & Buildings*, 2008, 40(10).

Zhang Qi. Heat load forcast method research of the central heating system [D]. *Changchun University of Technology*, 2017.

Zhao L. The research of dynamic balance adjustment in district heating system based on heating load forecasting. *Chang'an University*, 2017.

Advances in Civil Engineering and Environmental Engineering – bin Yahya & Bin Daud (Eds)

Research on the plan for disaster prevention and rescue of three-pipe parallel ultra-long high-altitude highway tunnel

Xuan Zhang, Xuefeng Li* & Xialin Liu
Research Institute of Highway, Ministry of Transport, Beijing, China
CCCC Second Highway Survey, Design and Research Institute Co., Ltd, Wuhan, Hubei, China

ABSTRACT: The Tianshan Shengli Tunnel is currently the longest ultra-long high-altitude highway tunnel under construction in the world. In order to explore the disaster prevention and rescue theory of the three-pipe parallel ultra-long high-altitude highway tunnel, this paper relies on the Tianshan Shengli Tunnel, on the basis of the division of disaster prevention zones, and according to the principles of evacuation and rescue command and mechanical and electrical system activation, the disaster prevention and rescue process adapted to the Tianshan Shengli Tunnel is obtained, and finally the specific disaster prevention and rescue plan is finally obtained. The results show that: ① The evacuation and rescue command principle of "combining self-rescue and rescue, waiting for rescue in service tunnel, and separating the evacuation of people and vehicles" is formulated for Tianshan Shengli Tunnel; ② The disaster prevention and rescue process is proposed for the three-pipe parallel structure of the Tianshan Shengli Tunnel; ③ Combined with the linkage control idea of electromechanical measures such as fans and protective doors in the large and small unit section of the smoke prevention zones and the disaster prevention zones, the disaster prevention and rescue plan of the Tianshan Shengli Tunnel is formulated.

1 INTRODUCTION

The highway tunnel space is small, the rescue and evacuation route is single, and the fire will cause serious consequences (Gao 2022; Zhou 2006). To reduce the loss of personnel and property caused by fire, it is necessary to establish a complete and effective disaster prevention and rescue plan for tunnels and fire characteristics to ensure the orderly evacuation of personnel and effective fire control (Ha 2017; Li 2018).

Ma Weibin et al. formulated disaster prevention and evacuation rescue precautions during the operation period of the railway long tunnel and proposed two types of evacuation and rescue schemes of "single-hole double-line tunnel+ flat guide" and "double-hole single-line tunnel" (Ma 2020). Zhao Qiang et al.took the extra-long shield tunnel of Beijing east Liuhuan road as an example and proposed a special tunnel evacuation and rescue plan through the study of the special evacuation rescue plan and the verification of the numerical simulation scheme (Zhao 2021). Sun Ce et al. took the Jihe expressway tunnel as an example to carry out comprehensive research on the emergency rescue of highway tunnel fire accidents and proposed a fire emergency rescue system for super-large cross-section double-deck highway tunnels (Sun 2021). Zhou Qing et al. summarized the basic principles of tunnel disaster prevention and rescue, established a set of tunnel disaster prevention and rescue management system, and analyzed the factors affecting fire evacuation and rescue and the process of tunnel disaster prevention and rescue system (Zhou 2009).

*Corresponding Author: xf.li@rioh.cn

DOI 10.1201/9781003349563-69

In summary, the current research on tunnel disaster prevention and rescue mainly focuses on railway tunnels and general highway tunnels, and there are few studies on three-pipe parallel high-altitude highway tunnels. Disaster prevention and rescue research is mainly based on theory, and no complete disaster prevention and rescue plan has been formed for different fire locations. This paper discusses the evacuation and rescue plan for a three-pipe parallel high-altitude highway tunnel for Tianshan Shengli Tunnel Project, to provide a reference for the evacuation and rescue process and plan setting of similar highway tunnels in the future.

2 PROJECT OVERVIEW

Tianshan Shengli Tunnel is located on the southwest side of Urumqi, Xinjiang, China, with a length of over 22km and an average altitude of about 2850m, which is an ultra-long high-altitude highway tunnel. The tunnel is composed of left and right lines plus intermiddle service tunnel, the distance between the left and right lines of the tunnel body is not less than 55 m, three turn lanes are set up, and the intersection of the main tunnel connected to the turn lanes are widened from two lanes to 4 lanes, and the length of the widening section is about 1700 m. To make the turning of the vehicle smooth, the angle between each turn lane and the main tunnel is 60°C. The schematic diagram of the turn lanes of the Tianshan Shengli Tunnel is shown in Figure 1, and the schematic diagram of the cross-sectional design is shown in Figure 2.

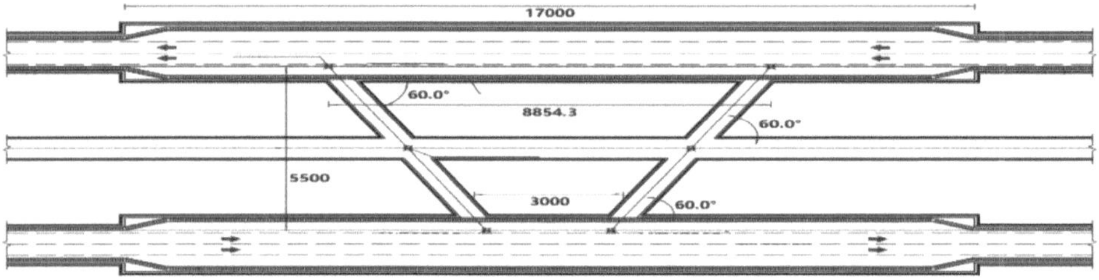

Figure 1. Schematic diagram of the Tianshan Shengli Tunnel turn lanes.

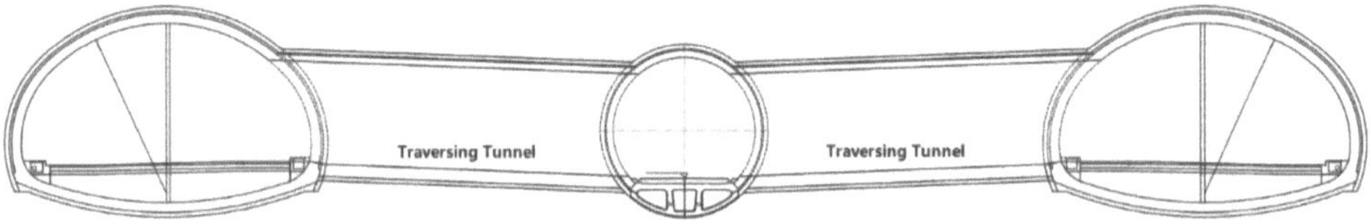

Figure 2. Cross section diagram of the Tianshan Shengli Tunnel.

3 DISASTER PREVENTION AND EVACUATION RESCUE COMMAND PRINCIPLES

The evacuation and rescue command of the Tianshan Shengli Tunnel should follow the principle of "equal emphasis on evacuation and rescue, separate evacuation of people and vehicles, wait for rescue in the service tunnel, evacuate vehicles through transverse channels and turn lanes efficiently, and rescue forces quickly rescue in the service tunnel". The vehicles downstream the fire quickly drive away from the tunnel. The vehicles that cannot be evacuated from the turn lanes upstream of the fire stop on the right side. The personnel in the vehicle get off the vehicle and enter the service tunnel near the vehicle's transverse channels to wait for rescue. The rest of the upstream vehicles through the front of the nearest vehicle transverse channels. Such as non-fire tunnel, the non-fire tunnel becomes a two-way drive, and according to the smoke control zones and the relative position of the fire, choose whether to return to the fire tunnel by turn lanes. According to the principle: on the premise of ensuring the shortest driving in the non-fire tunnel, crossing at least one smoke prevention zone, and ensuring that the fire smoke is not affected after returning to the fire tunnel. The schematic diagram of the vehicle transverse channels and the turn lanes is shown in Figure 3.

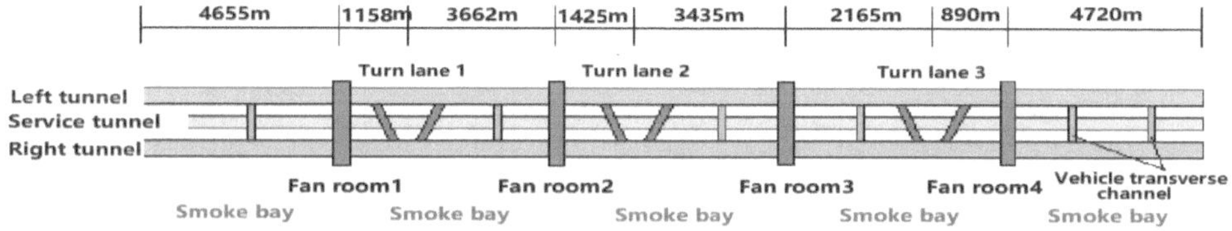

Figure 3. Schematic diagram of the vehicle transverse channels and the turn lanes.

The rescue process of Tianshan Shengli Tunnel can be organized and implemented, as shown in Figure 4.

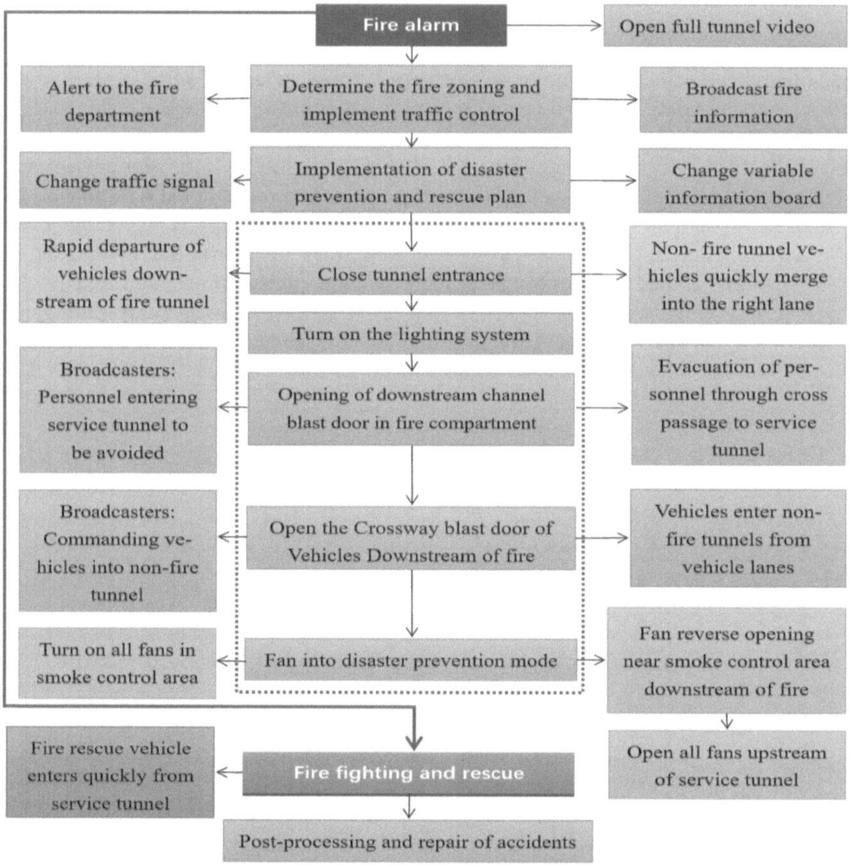

Figure 4. Organization chart of rescue workflow in Tianshan Shengli Tunnel.

When the fire occurs between the 2# turn lane and the 3# shaft, the smoke is discharged from the 2# smoking shaft. The downstream vehicle quickly drives away from the tunnel, and the upstream vehicles enter the non-fire tunnel through the vehicle transverse channels and the turn lanes as shown in Figure 5. The original vehicles in the non-fire tunnel merge to the right. After entering the fire tunnel, the vehicle will drive in the reverse direction from the left line, drive to the 1# turn lane, then return to the fire tunnel through the 1# turn lane, and finally leave the tunnel by the exit fire tunnel. At this time, because the smoke has been discharged through the 2 # smoking shaft when the vehicles return to the fire tunnel again, the fire tunnel has no smoke. Rescue vehicles enter through the service tunnel, and then enter the fire tunnel by the nearest vehicle cross channel or steering channel downstream of the fire to extinguish the fire.

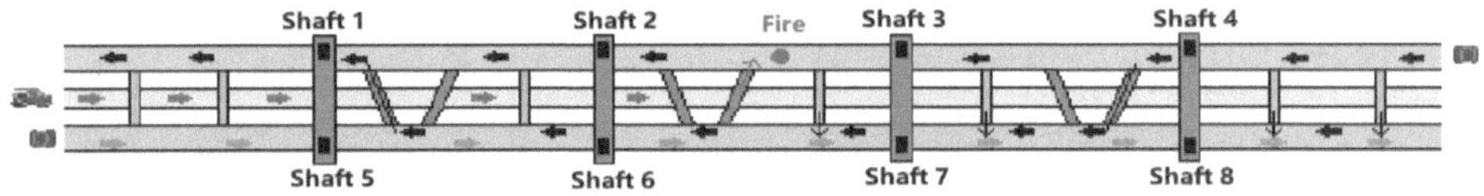

Figure 5. Vehicle evacuation diagram.

4 DISASTER PREVENTION AND RESCUE PLAN

4.1 *Division and labeling of disaster prevention zones*

Tianshan Shengli Tunnel is divided into 61 disaster prevention zones, including 31 disaster prevention zones in the left tunnel and 30 disaster prevention zones in the right tunnel. The tunnel smoke control zone left-1 and right-4, main tunnel, vehicle transverse channels, and pedestrian transverse channels are marked in Figure 6. Among them, "Z1-Z31" indicate the 31 disaster prevention zones in the left tunnel, "Y1-Y30" indicate the 30 disaster prevention zones in the right tunnel; "S_{Z1-1}, S_{Z1-2}" indicates 1# and 2# subzones in the disaster prevention zone No.1 of the left tunnel divided by the transverse channels, which is used to indicate the specific location of the fire; "R_{Z1-1}, R_{Z1-2}" indicates the 1# and 2# pedestrian transverse channels in the disaster prevention zone No.1 of the left tunnel; "C_{Z2-1}" indicates the 1# vehicle transverse channel in the disaster prevention zone No.2 of the left tunnel. The disaster prevention division of the service tunnel is divided according to the left tunnel, "F1-F31" indicate 31 disaster prevention zones for the service tunnel.

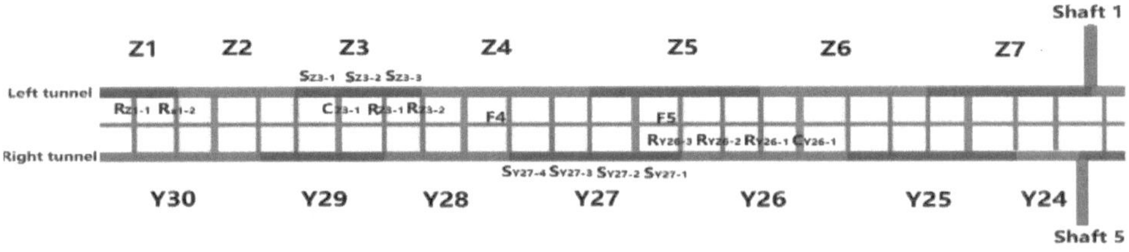

Figure 6. Schematic diagram of Smoking control zones left-1and right-4, tunnel, and transverse channels.

4.2 *Disaster prevention plan cases*

When fires occur in the left tunnel, plans are formulated for the 31 disaster prevention zones in the left tunnel, and the opening of the doors of the vehicle and pedestrian transverse channels, information boards outside the tunnel, signal lights inside the tunnel, vehicle indicators, the broadcast inside and outside the tunnel, and the opening of the axial fan and jet fan is explained. Taking the Z1 zone of the left tunnel as an example, the tunnel disaster prevention plan is shown in Table 1.

Among them, the scheme of opening the fan and the protective door respectively take the smoke prevention zones and the disaster prevention zones as the linkage control zones. The specific implementation principles are as follows: ① The fans in smoke prevention zones where the fire

Table 1. Left tunnel Z1 division Disaster prevention and rescue plan.

Disaster prevention zone	Z1	
Accident location	S_{Z1-1}	S_{Z1-2}
Protective door of the pedestrian transverse channel opens	R_{Z1-1}, R_{Z1-2}	R_{Z1-2}
Protective door of vehicle transverse channel opens	C_{Z2-1}-C_{Z31-1}, C_{Y2-1}-C_{Y30-1}	
Information board outside the tunnel	Prohibition of access to tunnel	
Signal light outside the tunnel	Red	
Lane indicator	Z1-Z31 are shown in red and Y1-Y30 are shown in green	
Broadcast in the tunnel	Z1, S_{Z2-1}: Personnel were evacuated by the nearest pedestrian crossing; S_{Z2-2}-Z31: A fire occurs and the nearest crosswalk ahead drove into the right tunnel; Y1-Y30: A fire occurs and quickly drove away from the tunnel	S_{Z1-2}, S_{Z2-1}: Personnel were evacuated by the nearest pedestrian crossing; S_{Z2-2}-Z31: A fire occurs and the nearest crosswalk ahead drove into the tunnel on the; S_{Z1-1}, Y1-Y30: A fire occurs and quickly drove away from the tunnel
Broadcast out the tunnel	Prohibition of access to tunnel	
Position of opening fan	The axial fans of the No. 1 supply shaft are opened 2, and the rest are opened according to the operating conditions; Z1-Z7 jet fans are fully opened (in→out), and the rest of zones are opened according to operating conditions; The F1-F31 jet fans are fully opened	

is located are all opened; ② The fans in the smoke prevention zones downstream of the fire are opened in reverse, and the fans in the rest of the zones are opened according to the operation mode; ③ The fans in the upstream service tunnel of the fire are fully opened, and the fans in the rest of the zones are closed; ④ The fans in the upstream transverse channels of the disaster prevention zones are fully opened, the fans in the upstream vehicle transverse channels are fully opened, and the fans in the crossing are closed.

5 CONCLUSION

Relying on the Tianshan Shengli Tunnel, through the study of the disaster prevention and rescue plan of the three-pipe parallel ultra-long high-altitude highway tunnel, the following main conclusions have been obtained:

(1) The evacuation and rescue command principle of "combining self-rescue and rescue, waiting for rescue in service tunnel, and separating the evacuation of people and vehicles" is formulated for Tianshan Shengli Tunnel.
(2) The disaster prevention and rescue process for Tianshan Shengli Tunnel is formulated, and the implementation plan is put forward, including fire alarm, traffic control, a closing tunnel

entrance, opening lighting, evacuating people for shelter, evacuating vehicles to change lanes, switching mode of disaster prevention fan, fire rescue, etc.

(3) Combined with the linkage control idea of electromechanical measures such as fans and protective doors in the large and small unit sections of the smoke prevention zones and the disaster prevention zones, the disaster prevention and rescue plan of the Tianshan Shengli Tunnel is formulated.

SUBJECT NUMBER

(1) Science and Technology Major Project of Xinjiang Uygur Autonomous Region 2020A03003-4,
(2) Major Project of China Communications Construction KJFZ-2017-073.

REFERENCES

Ce Sun (2021). Fire characteristics and evacuation rescue of double-deck shield tunnel with super-large cross-section: a case study on Shenzhen Bao'an international airport expressway tunnel [J]. *Tunnel Construction*, 41(8):1297.

Qi Li (2018). Research on reliability of personnel safe evacuation under high-speed railway super-long tunnel in fire [D]. *Southwest Jiaotong University*.

Qiang Zhao, Song Guo, Qian Bi (2021). Key technology of highway tunnel evacuation and rescue: A case study of extra-long shield tunnel of Beijing East Liuhuan road [J]. *Tunnel Construction*, 41(S2):429–437.

Qing Zhou, Xiangchuan Zhou (2009). Research on fire prevention and rescue for yanmenjie tunnel [J]. *Fire Science and Technology*, 28(08):599.

Weibin Ma, Xun Luo, Zhwei Wang (2020). Principles and key technologies of disaster prevention, evacuation and rescue for long and large railway tunnels [J]. *China Railway*, 164–172.

Xiaoqing Zhou, Xiangxiang Zhao, Yunfei Ding, Zhimin Zheng (2006). Numerical study on the influence of tunnel section shape on critical wind speed and smoke distribution during fire [J]. *Journal of Guangzhou University* (Natural Science Edition), 5(2):54–58.

Yuanyuan Ha (2017). Research and design on cross provincial highway tunnel management operation mode and rescue plan [D]. *Chang'an University*.

Yunji Gao, Zhisheng Li, Yueyang Luo, Hanwen Guo (2022). Experimental study on smoke characteristics of tunnel fire under effects of longitudinal ventilation and vertical shaft natural smoke exhaust [J]. *Fire Science and Technology*, 41(02):185–191.

Advances in Civil Engineering and Environmental Engineering – bin Yahya & Bin Daud (Eds)
© 2023 The Author(s), ISBN: 978-1-032-39397-1

Study on water quantity verification method for continuous water intake evaluation of construction projects

Caiming Chen*
Zhejiang Institute of Hydraulics and Estuary, Hangzhou, China

Guowei Zhou
Haiyan Water Conservancy Bureau of Zhejiang Province, Haiyan, China

Qifeng Li & Helong Wang
Zhejiang Institute of Hydraulics and Estuary, Hangzhou, China

ABSTRACT: Continuous water intake assessment is an important part of the water intake permit. Given the lack of relevant standards or technical specifications in the current continuous water intake assessment report, this paper proposes a method to calculate the water intake based on the production capacity multiplied by the water intake unit consumption (quota), which solves the problems of various methods for calculating the approved water intake and great human impact, It can be used as the basis for continuous water intake assessment and water volume verification of the construction project.

1 INTRODUCTION

According to the regulations on the administration of water intake license and water resource fee collection (Order No. 460 of the State Council) and the measures for the administration of water intake license (Order No. 34 of the Ministry of water resources), if the water intake license needs to be renewed after the expiration of the validity period, the water intake unit or individual shall apply to the original examination and approval authority 45 days before the expiration of the validity period, and the application for continued water intake shall be submitted when the application for continued water intake is submitted The original water intake application approval document and water intake permit. The water intake examination and approval authority shall make a comprehensive assessment of the originally approved water intake, the actual water intake, the water saving level, and the return water quality, as well as the average water consumption level of the industry in which the water intake unit or individual works, and the supply and demand of local water resources, and decide whether to approve the extension before the expiration of the water intake license. If the extension is approved, a new water intake license shall be issued; If the extension is not approved, the reasons shall be stated in writing. It can be seen that the continuous water intake assessment is an important work to implement Order No. 460 of the State Council and Order No. 34 of the Ministry of water resources, and it is also an important starting point to strictly control the total amount and efficiency of water use.

Different from the industrial standards for water resources demonstration of construction projects (guidelines for water resources demonstration of construction projects gb/t 35580-2017), there are no relevant standards or technical specifications for the continuous water intake assessment report at present, and the methods adopted by river basin institutions and local water administrative

*Corresponding Author: ccm1105@qq.com

DOI 10.1201/9781003349563-70

departments are different according to the actual situation in various regions. For example, the Yangtze River Water Resources Commission has formulated the outline for the preparation of continuous water intake application The Fujian Provincial Department of water resources formulated the outline of the continuous water intake assessment report, the Qinghai Provincial Department of water resources formulated the continuous water intake assessment form, and the Jiaxing Water Resources Bureau formulated the guidelines for the preparation of the Jiaxing continuous water intake assessment report (Trial), etc.

Given the problem that there are no relevant standards or technical specifications in the current continuous water intake evaluation report(Li 2017; Wang 2015), combined with the continuous evaluation of key water intake users in Haiyan County, this paper proposes a method to calculate the water intake based on the production capacity multiplied by the water intake unit consumption (quota), to solve the problems of various methods for calculating the approved allowable water volume and great human impact.

2 PROPOSAL OF VERIFICATION METHOD FOR CONTINUOUS WATER INTAKE PERMIT

According to the guidelines for water resources demonstration of construction projects (gb/t 35580-2017), the main points of water resources demonstration of construction projects are to answer how much water the construction project needs, whether there is so much water in the water source, whether the water intake has an impact on the third party, and whether the water withdrawal has an impact on the third party. The amount of water to be used in the construction project is calculated according to the water equipment and water use process in the preliminary work results such as the project feasibility study report, which will deviate from the actual production water. The core of continuous water intake assessment is to reasonably verify the allowable water intake, which is closely related to the actual water equipment, water use technology, and water-saving management level of water users.

The water consumption elasticity of the construction project proposed in the water resources demonstration stage is large, which is greatly affected by the technology and concept of the staff, and it is difficult to use a unified formula to determine the water consumption; The continuous water intake assessment is generally a summary assessment after the operation of water users for 5 years. Generally speaking, the production of water users is stable, the water intake is stable, and the unit consumption of water intake also tends to be stable. Therefore, it is possible to set certain rules to determine the allowable water volume of continuous water intake, which not only avoids human influence but also ensures the water fairness of different water users (Li 2017; Yuan 2015).

The water quota of Zhejiang Province (2019), implemented in June 2020, covers 59 industries, 932 products, and 2875 water quota values, including agriculture, industry, urban life, and service industry. The general value and advanced value are formulated for the first time, providing a basis for the continuous assessment of water intake.

3 CALCULATION PROCESS OF CONTINUING THE VERIFICATION METHOD OF WATER INTAKE PERMISSION

3.1 *Continue the verification principle of water intake assessment*

(1) The assessment is only for self-provided water users. For some water users who have had tap water for production in previous years, it is distributed according to the proportion of self-provided water and tap water.
(2) Respect history, control reasonably, keep the water intake within the original allowable water intake, and reduce it appropriately.

(3) The water intake shall be calculated by multiplying the production capacity by the unit consumption of water intake and shall be appropriately expanded.
(4) If the quota corresponding to the original approved water volume is better than the advanced value of the existing quota and better than the actual water intake unit consumption of water users, the original approved water volume shall be maintained in principle.
(5) For water users whose actual production capacity is much greater than the approved design capacity for three consecutive years, the verified water intake is the average of the actual water intake in recent three years, and the unit consumption of water intake must meet the quota, otherwise it is reduced to within the specified quota; If the unit consumption of water intake is better than the advanced value of quota, the approved water volume shall be appropriately expanded.

3.2 *Detailed rules for verification of continuous water intake assessment*

1. The water intake shall be calculated by multiplying the production capacity by the unit consumption of water intake, and shall be appropriately expanded:

$$Q = Z_0 W_{Zd} \lambda \beta$$

Where: Q-water intake scale of water provided by water users; Z_0-Capacity (approved design capacity, actual capacity); W_{Zd}-Unit consumption of water intake (actual water consumption per unit product, advanced value of quota, average value of advanced value of quota and general value, general value of quota), m3; λ-Expansion coefficient (1.3, 1.2, 1.1, 1.0); β-Proportion of self-provided water in total water consumption.

The specific rules are as follows:

(1) Calculation capacity determination rules:

① If the originally approved design capacity is not fully produced, and the reason is that the production line is not fully completed or affected by market supply and demand, the calculated capacity is the design capacity;

② If the originally approved design capacity is not fully produced and the reason is that the water users compress the production capacity, the calculated capacity is the actual capacity;

③ If the originally approved design capacity is fully produced and the actual production capacity is greater than the design capacity, and the capacity increase is approved by the relevant departments, the calculated capacity is the cumulative approved design capacity;

④ If the originally approved design capacity is fully produced and the actual production capacity is greater than the design capacity, and the capacity increase has not been approved by the relevant departments, the calculated capacity is the original design capacity.

(2) Rules for determining unit consumption of water intake:

① The actual water consumption per unit product of the water intake user is less than the advanced value of the quota, the actual water consumption per unit product of the water intake is taken, and the expansion coefficient is not greater than 1.3, and cannot be greater than the general value of the quota;

② The actual water consumption per unit product of water users is greater than the advanced value of the quota and less than the average value of the advanced value of the quota and the general value. The unit consumption of water intake is the advanced value of the quota, and the expansion coefficient is not greater than 1.2, and cannot be greater than the general value of the quota;

③ The actual water consumption per unit product of water users is greater than the average value of the advanced value and general value of the quota and less than the general value of the quota. The average value of the advanced value and general value of the quota is taken for the unit consumption of water intake. The expansion coefficient is not greater than 1.1 and cannot be greater than the general value of the quota;

④ The actual water consumption per unit product of the water intake user is greater than the general value of the quota, the unit consumption of water intake takes the general value of the quota, and the expansion coefficient is 1.0;

⑤ The products of water users are not within the quota range, the unit consumption of water intake is the actual minimum value in recent 3 years, and the expansion coefficient is not greater than 1.1.

2. If the quota corresponding to the original approved water volume is better than the advanced value of the existing quota and better than the actual unit consumption of water intake by water users, the originally approved quota shall be maintained in principle.
3. For water users whose actual production capacity is much greater than the approved design capacity for three consecutive years, the verified water intake is the average of the actual water intake in recent three years, and the unit consumption of water intake must meet the quota, otherwise it is reduced to within the specified quota; If the unit consumption of water intake is better than the advanced value of quota, the approved water volume shall be appropriately expanded, and the expansion coefficient shall not be greater than 1.3.

4 CASE ANALYSIS

Haiyan County, Jiaxing City, Zhejiang Province is located in the Hangjiahu Plain, where the terrain is flat and the rivers and lakes are densely distributed, belonging to a typical plain river network area. As of November 2019, Haiyan County has a total of 40 self-provided water intake enterprises. Taking Haiyan County as a pilot county for the comprehensive reform of water resources regulation in Zhejiang Province as an opportunity, based on the recent need to carry out continuous water intake enterprises, after combing, a total of 17 enterprises were selected to carry out continuous water intake assessment. According to the analysis, the total water consumption of Haiyan County in the previous year did not exceed the total water consumption control index issued by Jiaxing City. All self-provided water intake households installed metering and online monitoring facilities. The selected 17 enterprises had no adverse events of water intake in the past five years, and the water intake and use of water intake had not changed. The return water was discharged into the municipal sewage pipe network and uniformly discharged to the sewage treatment plant, and the water resource fee was paid in time, Therefore, these 17 enterprises can adopt a simple procedure to calculate the amount of water allowed for continuous water intake (Chen 2020).

1. First, conduct on-site research, collect the existing approved licensed water volume and the corresponding approved design capacity of 17 enterprises, the actual total amount of water intake (self-provided water, tap water, other water, etc.), and the actual capacity in the past three years, and collect the corresponding water consumption of each product if there are multiple products. See the Table 1.
2. based on a complete collection of water consumption and capacity of the enterprise, calculate the originally approved water consumption per unit product and the actual water consumption per unit product in recent 3 years.
3. Compare the average value of the originally approved water consumption per unit product, the actual water consumption per unit product in recent 3 years, the advanced value of Zhejiang provincial water consumption (intake) quota (2019), the general value of Zhejiang provincial water consumption (intake) quota (2019), and the advanced value of Zhejiang Provincial Water Consumption (intake) quota (2019) with the general value:
   Case 1: the unit consumption of the product corresponding to the original approved water volume is better than the advanced value of the existing quota and better than the actual unit consumption of water intake of the enterprise. In principle, the originally approved quota is maintained, that is, the original approved water volume is maintained.

Table 1. Sample table of enterprise master data collection.

Enterprise name	Water intake permit (ten thousand m^3)	Actual water intake (ten thousand m^3)		
		2016	2017	2018
KingDom	120	71.74	77.68	54.36
Enterprise name	Approved design capacity (t)	Actual capacity (t)		
		2016	2017	2018
KingDom	7960	6962	6552	6786
Enterprise name	Product name	Water consumption per unit product (m^3/t)		
		Average value in recent 3 years	Advanced value of quota	General value of quota
KingDom	Flax yarn	101	100	150

Case 2: if the actual water intake unit consumption of the enterprise is better than the advanced value of the quota and better than the originally approved unit consumption, the approved water volume shall be appropriately expanded, and the expansion coefficient shall not be greater than 1.3; Except for a small number of enterprises whose water use technology is particularly advanced, the vast majority of the situation is that the enterprises are overproduced. For enterprises whose actual production capacity is far greater than the approved design capacity for three consecutive years, the verified water intake is the average of the actual water intake in recent three years, and the unit consumption of water intake must comply with the provisions of the general value of the quota, otherwise, it is reduced to within the general value of the quota.

Case 3: ① the actual unit product water consumption of the enterprise is greater than the advanced value of the quota and less than the average value of the advanced value of the quota and the general value. The unit water consumption takes the advanced value of the quota, and the expansion coefficient is not greater than 1.2, and cannot be greater than the general value of the quota;

② The average value of the actual unit product water consumption of the enterprise is greater than the advanced value and the general value of the quota is less than the general value of the quota. The average value of the advanced value and general value of the quota is taken for the unit water consumption. The expansion coefficient is not greater than 1.1 and cannot be greater than the general value of the quota;

③ The actual water consumption per unit product of the enterprise is greater than the general value of the quota, the unit water consumption takes the general value of the quota, and the expansion coefficient is 1.0;

Case 4: the enterprise's products are not within the 2019 quota, the unit consumption of water intake is the actual minimum value in the recent 3 years, and the expansion coefficient is not greater than 1.1.

After calculation, in addition to 6 enterprises maintaining the original approved water volume, the remaining 11 enterprises have reduced the water intake to varying degrees, with a total of 1.795 million m^3, accounting for 15% of the original allowable water intake.

5 CONCLUSION AND SUGGESTIONS

(1) At present, there are no relevant standards, specifications, procedures, guidelines, and guidelines to determine the allowable amount of continuous water intake, so it is very necessary to explore a simple process to carry out continuous water intake assessment and formulate the principles and methods for the verification of continuous water intake assessment, which is also an important starting point for the implementation of the strictest assessment of water resources management system. The method proposed in this paper can be used as the method for the verification of continuous water intake.

(2) The premise of adopting the calculation method studied in this paper is to master the approved water volume and the corresponding approved design capacity of the enterprise, the actual total water intake and actual capacity in recent three years, and the corresponding water consumption of each product; From the perspective of acquisition channels, the approved licensed water volume and the actual total water intake in recent 3 years can be obtained through the Water Conservancy Bureau, water affairs group or a third-party heating enterprise; Enterprises can provide the redesigned capacity and the actual capacity in recent 3 years by themselves; More complex is that enterprises producing multiple products cannot obtain the corresponding water consumption of products due to incomplete metering devices. It is recommended that water users with annual water consumption of more than 50000 m^3 improve their internal measurement facilities and carry out water balance tests every 5 years.
(3) In this study, the expansion coefficient of the interval method is taken as 1.0, 1.1, 1.2, and 1.3, which can meet the calculation requirements. However, through the analysis of the water quota of Zhejiang Province (2019), it is found that the interval between the general value and advanced value of water quota of some products is too large, for example, the general value of quartz sand has reached 44 times of the advanced value. In this case, the expansion coefficient needs to be adjusted, otherwise, the calculation result is unreasonable. It is also suggested that when revising the quota, an in-depth investigation should be carried out on the water quota with a large range between the general value and the advanced value to ensure its rationality.

ACKNOWLEDGMENTS

This research is funded by Zhejiang water conservancy science and technology plan project (RC2027, RB1911).

REFERENCES

Chen Caiming. (2020), Zhejiang institute of hydraulics and estuary. *Evaluation Report on Continuous Water Intake of Key Water users in Haiyan County* [R].

Li Haichen, Qin Tao, Qiu Ying, et al. (2017). Analysis of China's water intake license extension management [J] *China Population, Resources and Environment*, 27 (25): 80–82.

Li Jinjing, Zhang Lifeng, Zhao Yiwen, et al. (2017). Discussion on issues related to water intake permit management [J] *Water Resources Management*, 19: 42–45.

Wang Zhiqiang, Liu Changshun, Dai Xiangqian. (2015). Ideas and key points for strengthening the continuation management of water intake permit [J] *Research on Water Conservancy Development*, 10: 42–44.

Yuan Liu. (2015). Discussion on issues related to continuous water intake assessment [J]. *China Water Transport*, 15 (6): 118–119.

© 2023 The Editor(s), ISBN: 978-1-032-39397-1

Author index